THE OXFORD HANDBOOK OF

MEDIA, TECHNOLOGY, AND ORGANIZATION STUDIES

THE OXFORD HANDBOOK OF

MEDIA, TECHNOLOGY, AND ORGANIZATION STUDIES

Edited by

TIMON BEYES

ROBIN HOLT

and

CLAUS PIAS

OXFORD
UNIVERSITY PRESS

OXFORD
UNIVERSITY PRESS

Great Clarendon Street, Oxford, OX2 6DP,
United Kingdom

Oxford University Press is a department of the University of Oxford.
It furthers the University's objective of excellence in research, scholarship,
and education by publishing worldwide. Oxford is a registered trade mark of
Oxford University Press in the UK and in certain other countries

© Oxford University Press 2020

The moral rights of the authors have been asserted

First Edition published in 2020

Impression: 1

Published in the United States of America by Oxford University Press
198 Madison Avenue, New York, NY 10016, United States of America

British Library Cataloguing in Publication Data
Data available

Library of Congress Control Number: 2019941180

ISBN 978-0-19-880991-3

Printed and bound by
CPI Group (UK) Ltd, Croydon, CR0 4YY

Links to third party websites are provided by Oxford in good faith and
for information only. Oxford disclaims any responsibility for the materials
contained in any third party website referenced in this work.

Contents

LIST OF FIGURES

Whilst every effort has been made to secure permission to reproduce the illustrations, we may have failed in a few cases to trace the copyright holders. If contacted, the publisher will be pleased to rectify any omissions at the earliest opportunity.

List of Contributors

Cristina Alaimo, Assistant Professor in Digital Economy, Surrey Business School, University of Surrey, UK

Götz Bachmann, Professor for Digital Cultures, Institute for Culture and Aesthetics of Digital Media, Leuphana University of Lüneburg, Germany

Andreas Bernard, Professor of Cultural Studies, Centre for Digital Cultures, Leuphana University of Lüneburg, Germany

Armin Beverungen, Assistant Professor for Organisation in Digital Cultures, Institute of Sociology and Cultural Organization, Leuphana University of Lüneburg, Germany

Timon Beyes, Professor of Sociology of Organization and Culture, Leuphana University of Lüneburg, Germany / Department of Management, Politics and Philosophy, Copenhagen Business School, Denmark

Paula Bialski, Assistant Professor for Digital Sociality, Institute for Culture and Aesthetics of Digital Media, Leuphana University of Lüneburg, Germany

Mercedes Bunz, Senior Lecturer in Digital Societies, King's College London, UK

Gibson Burrell, Professor of Organization Theory, School of Business, University of Leicester, UK and University of Manchester, UK

Wendy Hui Kyong Chun, Canada 150 Research Chair in New Media, School of Communication, Simon Fraser University, Canada

Christian De Cock, Professor of Organization Studies, Department of Organization, Copenhagen Business School, Denmark

Alice Comi, Lecturer Shanghai Tongji University, International College of Design Innovation, Shanghai, China

Lisa Conrad, Lecturer in Media, Cultural and Organization Studies at the Institute of Sociology and Cultural Organization, Leuphana University of Lüneburg, Germany

Karen Dale, Professor of Organization Studies, Lancaster University Management School, UK

Monika Dommann, Professor of Modern History, Department of History, University of Zürich, Switzerland

L. Roman Duffner, Research Associate, Institute for Organization, Johannes Kepler University Linz, Austria

Mikkel Flyverbom, Professor of Communication and Digital Transformations, Department of Management, Society and Communication, Copenhagen Business School, Denmark

Melissa Gregg, Senior Principal Engineer in Client Architecture and Innovation, Client Computing Group, Intel Corporation, Portland, Oregon, USA

Daniel Hjorth, Professor of Entrepreneurship and Organization, Department of Management, Politics and Philosophy, Copenhagen Business School, Denmark, and Nottingham Business School, Nottingham Trent University. UK

Robin Holt, Professor, Department of Management, Politics and Philosophy, Copenhagen Business School, Denmark

Florian Hoof, Postdoctoral Scholar, Institute for Advanced Study for Media Cultures of Computer Simulation, Leuphana University of Lüneburg, Germany

Lucas Introna, Professor of Organization, Technology and Ethics, Department of Organization, Work and Technology, Lancaster University Management School, UK

Dariusz Jemielniak, Professor of Management, Center for Research on Organizations and Workplaces, Kozminski University, Poland

Sine Nørholm Just, Professor, Department of Communication and Arts, Roskilde University, Denmark

Jannis Kallinikos, Professor of Information Systems, Department of Management, London School of Economics and Political Science, UK

Alexander Klose, PhD, Journalist, Curator, Cultural Scientist, Weimar/Berlin, Germany

Tamara Kneese, Assistant Professor, Department of Media Studies, University of San Francisco, USA

Anders Koed Madsen, Associate Professor, Department of Education, Learning and Philosophy, Aalborg University, Denmark

Markus Krajewski, Professor for Media Studies, Department of Arts, Media, Philosophy, University of Basel, Switzerland

Reinhold Martin, Professor of Architecture, Planning, and Preservation, Architecture Department, Columbia University, USA

Jeanne Mengis, Professor in Organizational Communication, Faculty of Communication Sciences, University della Svizzera Italiana, Switzerland

Jörg Metelmann, Associate Professor of Culture and Media Studies, School of Humanities and Social Sciences, University of St Gallen, Switzerland

Christoph Michels, Assistant Professor, Institute of Architecture and Planning, University of Liechtenstein, Liechtenstein

Jan Müggenburg, Assistant Professor for Media History and History of Science, Institute for Culture and Aesthetics of Digital Media, Leuphana University of Lüneburg, Germany

Damian O'Doherty, Professor of Management and Organization, Alliance Manchester Business School, University of Manchester, UK

Lara Pecis, Lecturer in Organization Studies, Department of Organization, Work and Technology, Lancaster University Management School, UK

John Durham Peters, María Rosa Menocal Professor of English and of Film & Media Studies, Yale University, USA

Claus Pias, Professor of Media Theory and the History of Media, Institute for Culture and Aesthetics of Digital Media, Leuphana University of Lüneburg, Germany

Aleksandra Przegalinska, Assistant Professor, Department of Management, Kozminski University, Poland

François-Régis Puyou, Lecturer in Management, School of Management, University of St Andrews, UK

Paolo Quattrone, Chair in Accounting Governance and Social Innovation, Business School, University of Edinburgh, UK

Tomasz Raburski, Assistant Professor, Institute of Philosophy, Adam Mickiewicz University in Poznań, Poland

Renée Ridgway, PhD Candidate, Department of Management, Politics, and Philosophy, Copenhagen Business School, Denmark

Stefan Rieger, Professor for Media History and Communication Theory, Institute for Media Studies, Ruhr University Bochum, Germany

Olga Rodak, Research Assistant, Center for Research on Organizations and Workplaces, Kozminski University, Poland

Ned Rossiter, Professor of Communication, Institute for Culture and Society with a joint appointment in the School of Humanities and Communication Arts, Western Sydney University, Australia

Nishant Shah, Dean of Graduate School, ArtEZ University of the Arts, The Netherlands

Annika Skoglund, Associate Professor, Department of Engineering Sciences, Industrial Engineering and Management, Uppsala University, Sweden and Honorary Associate Professor, University of Exeter Business School, UK

Chris Steyaert, Professor of Organizational Psychology, Research Institute of Organizational Psychology, University of St Gallen, Switzerland

Nanna Bonde Thylstrup, Associate Professor of Communication and Digital Media, Department of Management, Society, and Communication, Copenhagen Business School, Denmark

Maria-Laura Toraldo, Research Assistant, Institute of Marketing and Communication Management (IMCA), University della Svizzera Italiana, Switzerland

Kristin Veel, Associate Professor, Department of Arts and Cultural Studies, University of Copenhagen, Denmark

Barbara Vinken, Professor for French and Comparative Literature, Institute for Romance Philology, University of Munich, Germany

Theodore Vurdubakis, Professor of Organization and Technology, Department of Organization, Work and Technology, Lancaster University Management School, UK

Jennifer Whyte, Professor, Centre for Systems Engineering and Innovation, Imperial College London, UK

Mike Zundel, Professor, Work, Organization and Management, University of Liverpool Management School, UK

INTRODUCTION

In media and organization studies objects have a grounding role: they both embody and transform the structures and processes of the social. The study of the mediating roles objects play in human life is as old as the hills, but has only recently come back into vogue, now we have begun once more to question the assumption that only human beings act. Correspondingly, organization is now being thought and explored as a socio-material phenomenon, as something that comes into being by way of human/non-human assemblages and that therefore entails and requires a host of objects that need to be understood as agential forces. And, in turn, media are being thought of less as conduits or channels connecting one agent with another, and more as structuring conditions configuring the very possibility of agency: they are not just objects braided with information, but also power.

It is in and from this setting that this *Handbook of Media, Technology, and Organization Studies* acquires its rationale and its urgency. The Handbook explores, maps, and theorizes the territory of media, technology, and organization studies. Written by scholars of organization and theorists of media and technology, its entries focus on specific, and specifically mediating, objects that shape the practices, processes, and atmospheres of organization. Such media configure (power) relations that are in-built into the devices and apparatuses of organizational life. 'Media organize' (Reinhold Martin); and perhaps, to paraphrase Friedrich Kittler's notorious dictum, 'media determine organization'. Attention thus falls on the *medial a prioris* of organization; that is, on the operations, effects, and affects of mediating devices and forms (over and against any interrogation of their 'nature') in organizational contexts. In terms of this Handbook, the question becomes how media and technology are intimate with the capacity to organize and be organized.

This intimacy of technology, media, and organization is staged and reflected in the Handbook's cover by the artist Simon Denny. The cover reproduces an image of his adaptation of the contemporary boardgame 'Game of Life' in which the players' choices (and hence values) are shown to be riven with, and utterly conditioned by, a nexus of media technology and organizational processes. Denny's work is based on a sustained and richly attentive form of artistic-organizational research in which an anthropologist's eye is turned to how media organize, and how media are organized, 'after' digital technology. His art evocatively and repeatedly performs a recursive logic: to understand how media technologies condition contemporary life, one needs to inquire into their organizational affects and effects. And in order to trace how media technologies are produced, take place, disappear or are transformed, one needs to trace the organizational constellations in which they are inscribed and which they make possible.

As the first sustained and systematic interrogation of the relation between technologies, media, and organization—and through bringing together the fields of organization theory and media studies—the Handbook will consolidate, deepen, and further develop the empirics and concepts required to make sense of the material forces of organization. Top and tailed by this brief introduction and an editorial essay on the relation of media, technology, and organization, it consists of 43 entries, each of which considers, and lingers with, an object and its capacities to mediate organization. The Handbook is thus focused on the level of material specificity, of specific instances and forms of what the authors and editors consider to be organizational-technological mediation. An open question that runs across the volume, then, concerns the point at which the material becomes technological. The threshold is, we argue, configured by the condition of organization. Contrary to a somewhat hyperactive sociology of socio-material association (or actor-networks), we do not assume that each and every actant is on symmetrical footing with any other actant, in a merry dance of agencies. As technology, the object in question organizes or affords a certain process of organizing. It can thus be configured as technological medium that enables and shapes, perhaps even in some ways conditions or determines, organization.

The entries are written so readers can think with the objects being written about. We would like the reader to ask: 'How does this object organize?' In staying with the object the entries remain committed to the everyday, empirical world, refusing to be pinioned to established disciplinary concerns and theoretical developments. Relatedly, the Handbook is not just a reflection of the technologies and 'digital media' that are current in media and organizational theory; it consciously avoids the fetishism of the new so prominent in writings on technology. Many 'old' objects continue to exercise a strong hold on how humans act and think, and, as many of the entries reveal, much of the purportedly new is rather old.

As to the objects in the Handbook, they hold no categorical promise. They have been suggested by, discussed with and drawn from the work of our contributors. They are 'mediators' that demand reflections on how they organize and are organized. There is no claim to completeness, then—and given our assumptions about the ubiquity, proliferation, and efficacy of technological mediation, how could there be? We would make a claim to relevance, however. These are explorations that teach us about how organization takes place by way of specific objects; how they organize us and how we organize with them in everyday life. We hope they provoke further ventures into the in-between territories of media, technology, and organization (studies).

Since it would contradict our way of proceeding to impose a hierarchy of technologies or media, we have opted for an alphabetical order of objects. Sorting entries from A–Z is of course a contingent and arbitrary way of structuring the Handbook—in this sense, any other order of precedence would do. Or maybe it is not wholly arbitrary. At least with regard to the printed book, the contingent grid of an alphabetical ordering allows for a spatial, printed juxtaposition of wildly different media of organizing in close proximity, or at some remove—just like, we're tempted to say, in organized life (see the editorial essay at the end of this volume). Such an ordering perhaps also opens up the book (the reader?) to surprise and untypical ways of wayfinding. In the spirit of our endeavour, these are minor enactments of how the Handbook as medium of organization indeed organizes certain ways of reading and browsing, while being open to some (but only some) reorganization in the way that it is used and misused, perused and discarded, annotated and forgotten, referenced and turned into a literal doorstopper.

We will not be so glib as to thank, say, Dropbox, word processors, academic texts, and Oxford University Press' production and distribution technologies for enabling and shaping what you hold in your hands or see on your screens. Books and academic labour are grounded in these technologies—specific thanks presume a condition of decisional choice on our part, as editors, that is simply not there (as if we could have done without them). As ever, thanks are thus due to specific agents and collectives: To the authors for their engagement with the idea of the Handbook and their splendid entries, to many who granted permission for image rights, especially to Simon Denny who graciously let us use his art. A 'thank you' is due also to Oxford University Press and Jenny King and Adam Swallow for believing in the project and for seeing it to completion in a wonderfully constructive and helpful way, and to Jo North for copyediting. Due to the support of Oxford University Press and Leuphana University Lüneburg's Centre for Digital Cultures, we were able to host a writing workshop in Berlin in November 2017, discussing first drafts with their authors. And Nelly Pinkrah supported the project's coordination and communication from its inception—thanks!

Finally, we should also note that the Handbook is part of a wider endeavour of bringing together advanced thinking of media and technology with organization studies, and we would just like to mention the European Group of Organization Studies' Standing Working Group on 'Digital Technology, Media and Organization' and the ongoing work of Leuphana University Lüneburg's Centre for Digital Cultures on the threshold of social and media theory, some of it in conjunction with Copenhagen Business School's Departments of Management, Politics and Philosophy and Management, Society and Communication.

CHAPTER 1

....

ACCOUNT BOOKS

....

FRANÇOIS-RÉGIS PUYOU
AND PAOLO QUATTRONE

INTRODUCTION: HOW DOES AN ACCOUNT BOOK ORGANIZE?

....

To exist, large entities such as organizations and societies have to be summed up somewhere (Chandler, 1977). They are 'facts' (from the Latin *facere*, to make with effect) and there is nothing less factual than a fact that needs to be made, achieved, and accomplished for it to exist (Jones et al., 2004). This chapter argues that it is no accident that accounting has been central as an administration technique common to most organizations and societies for centuries because accounting books offer a much-needed space and a set of useful practices for different constituents of otherwise immaterial entities to coalesce. Accounting is about recording events into books where inscriptions can later be organized in relation to one another in a coherent whole, into facts, by users.

In Rome during the late Republic and the Empire (from the second century BC to the fourth century AD), scrolls were used for the chronological recording of economic transactions into draft documents called *adversaria* (Minaud, 2005). The information from *adversaria* were then recorded into carefully designed accounts kept on wood tablets covered with wax named *codex accepti et expensi* about once a month. The transfer of information from scrolls to *tabulae* (tablets) was done by reading out loud the *adversaria* in order to update the *codex*. Figure 1.1 is an engraving from a tomb (third century AD) that shows the preparation of accounts on tablets using *adversaria* held by a slave and the checking of the information against the actual sum of money contained in a bag by the master (Littleton and Yamey, 1956). The person in charge of entering transactions in the accounts (*ferre*) and closing the transaction (*referre*) (Minaud, 2005) was also in charge of *dispungere*, i.e., to check the accounts or 'to balance them' (*rationem computare*) by comparing receipts and payments (de Ste Croix, 1956). This work was accomplished using little marks (*punctum*) made by a stylus next to the entries so that

FIGURE 1.1 'The stele of a banker', 3rd c. AD, National Museum in Belgrade
(From M. Rostovtzeff, *Social and Economic History of the Roman Empire* (Biblo & Tannen, 1926))

third parties could easily examine what had been done and agree on the *dispunctio* (Minaud, 2005).

The recoding of information into tablets allowed their users to then engage with the text in more complex ways than when using scrolls (Camille, 1985). The visual distribution of information on tablets mixed aesthetic and pragmatic considerations. In tablets, unlike in scrolls, transactions were grouped by year and by types of goods and the balance was systematically verified by a rigorous comparison between receipts and payments. The design was chosen for the sake of 'neater appearance' but also to make the accounts 'less fatiguing to follow' and 'to trace individual items within it' (Littleton and Yamey, 1956: 14).

Account tablets were thus carefully compiled so as to be convincing when displayed for example during inspections by third parties (Andreau, 2004). Although the physical aspect of tablets was vitally important (the use of *adversaria* in particular helped to limit the number of corrections), the engagement with the accounts involved other modes of perception than vision, including voice and gestures (as made explicit in the action of speaking out the information during *ferre* and *referre*[1]) revealing the origin of much later auditing practices (from the Latin *audire* 'to hear'). Roman society was indeed mostly aural (König and Whitmarsh, 2007) with texts designed to be read out loud using

---

[1] Modern day 'referees' entering in their notebooks the names of football players who have committed a foul play ('booking them') while waving a coloured card to them is a contemporary equivalent of a multimodal performative performance impacting the status of individuals (who might well for example be missing from the next games as a consequence).

a formulaic language characterized by archaic vocabulary and syntax, set forms and spelling, and a rhythmic quality, to create something that would sound impressive, precise, and exhaustive when read in a distinct and powerful way called *recitatio* (Meyer, 2004; Andreau, 2007). Accounting is therefore not just concerned with books but also with faces and voices. A *codex* was a document standing for 'the voice' of Roman citizens, being almost part of their bodies to the extent that it 'could not be slighted without their author also feeling the sting' (Meyer, 2004: 220).

The use of wax tablets for keeping accounts was not solely justified because of their authoritative sound and aspect but also because of the specific qualities attributed to such a material artefact in connection with other documents (Meyer, 2004). The authority of accounting tablets relied in part on their capacities to link and relate with other prestigious objects and practices. In Roman society, the use of wooden *tabulae* covered with wax was not unique to accounting documents but they were also (even mainly) used for other venerable acts such as key political treaties, magisterial edicts, religious vows, and significant household acts like the making of a will (Meyer, 2004). Accounting texts, like all texts, cannot therefore be decontextualized. They are always referring to other texts from which they derive part of their power and significance (Eco, 1985). This intertextual relation is not necessarily based on similar contents but can be grounded on the use of similar material, on shared visual aspects, and on similar processes of classification.

This work of classification has consequences beyond the mere recording of economic activities as it also participates in the structuring of societies at large. Accounting does make organizations coalesce and it surprisingly does so through mechanisms of repetition that divide, separate, and classify. It is thus a 'device' (from the Latin *dividere* meaning to divide or to separate) that generates realities such as organizations (Singleton and Law, 2013), identities, and by extension entire communities. In Figure 1.1, typical Roman social ties are enacted through the distinction between the Roman citizen wearing a toga, sitting at his desk, holding tablets, and counting his money while the reading of *adversaria* is assigned to the servant or slave, wearing a different dress, and standing up in front of his master (see Burrell and Dale, this volume). Figure 1.1 reflects and enacts the kind of social distinctions of class and status in place in society as well as in business.

As a consequence of the central role played by accounting documents to support one's status in organizations and societies, all kinds of unconventional practices to make the books look better have developed over time. Since accounting exists, bookkeepers have on occasions turned into counterfeiters and falsifiers. Among accounts 'fabricators', i.e., persons able to construct a complex finished product from ideas, some inevitably opt to deliberately convey fraudulent or misleading information. Andreau (2007) mentions several techniques for tampering with wax tablets such as adding additional information later (*interpolare*), striking words (*inducere*), and writing on top of existing words (*superductio*). A whole panoply of techniques emerges, as illustrated for example by the trial of Verres (70 BC) during which Cicero carefully examines some accounts in which the name 'C. Verrutius' appears to be the result of afterthoughts to mask the name of the wicked governor Verres thus showing 'not only that man's avarice, but the very bed in which it lay' (Littleton and Yamey, 1956: 45).

How Do Account Books Configure Human Thought and Action?

Accounting generates identities and shapes for whatever falls within it (Singleton and Law, 2013) including organizations which appear as sets of transactions recorded into books. For example, many business people see their companies primarily through accounting documents. Financial statements typically are where executive managers of large multinational companies will get a vision of the empire they rule from their headquarters. Organizations are therefore commonly reduced, *a fortiori* to distant shareholders only interested in the revenues from their investments (supposing such persons exist), to the bottom line of a profit and loss account from which action of selling and buying is prompted (Dent, 1991; Law, 1994). Account books thus occupy a very particular position as they are the condition of visibility of otherwise abstract organizations. Yet, traces left in account books do not make visible the organization itself (Bastide, 1990). To become tangible, organizations require enactment from account users filling in the many absences of accounting documents. Signs and words in accounts are no organizations until they are constructed as such by account users. They must trigger the interest of users who then build organizations from them. In this process of composition with account books, the visual dimension of accounts plays a central role.

Doing accounting is not merely to record entries and make arithmetic calculations within accounts but also to make calculations *with* accounts, combining them in an infinite number of different ways (Quattrone, 2009). In this process, account books are to be understood as combinable visual and material sites used to generate new knowledge in an invention process. Accounting practices thus involve writing in and reading from accounts but also processes of composition of visions. Such a work of composition uses accounting inscriptions in topical ways, from *topos* 'spaces', meaning that accounting is an ordered classification of arguments in and with material spaces.

That people will collectively engage with accounting documents with the hope to address important issues is dependent on the fit between that document and legitimate aesthetic canons. Puyou and Quattrone (2018) show for example that symmetry is an important feature of accounts legitimacy but one that takes different forms across the ages. Symmetry in Rome is a mix of visual and audible features while it takes a more exclusively visual dimension in the Renaissance and a mathematical meaning in modernity. For example, Figure 1.2 shows the careful attention in the Renaissance to centre the information recorded into accounts as an important step for making it beautiful, balanced, and therefore legitimate and convincing. This attention to visual symmetry is made evident by the scribe folding the paper into two halves in order to mark the centre and position the text right in the middle (Jéhanno, 2011). Note also how the two entries into the accounts figuring on top and at the bottom are also symmetrical.

From the nineteenth century, the aesthetic of accounts becomes essentially one of scientific objectivity. The change towards accounts privileging numbers visually

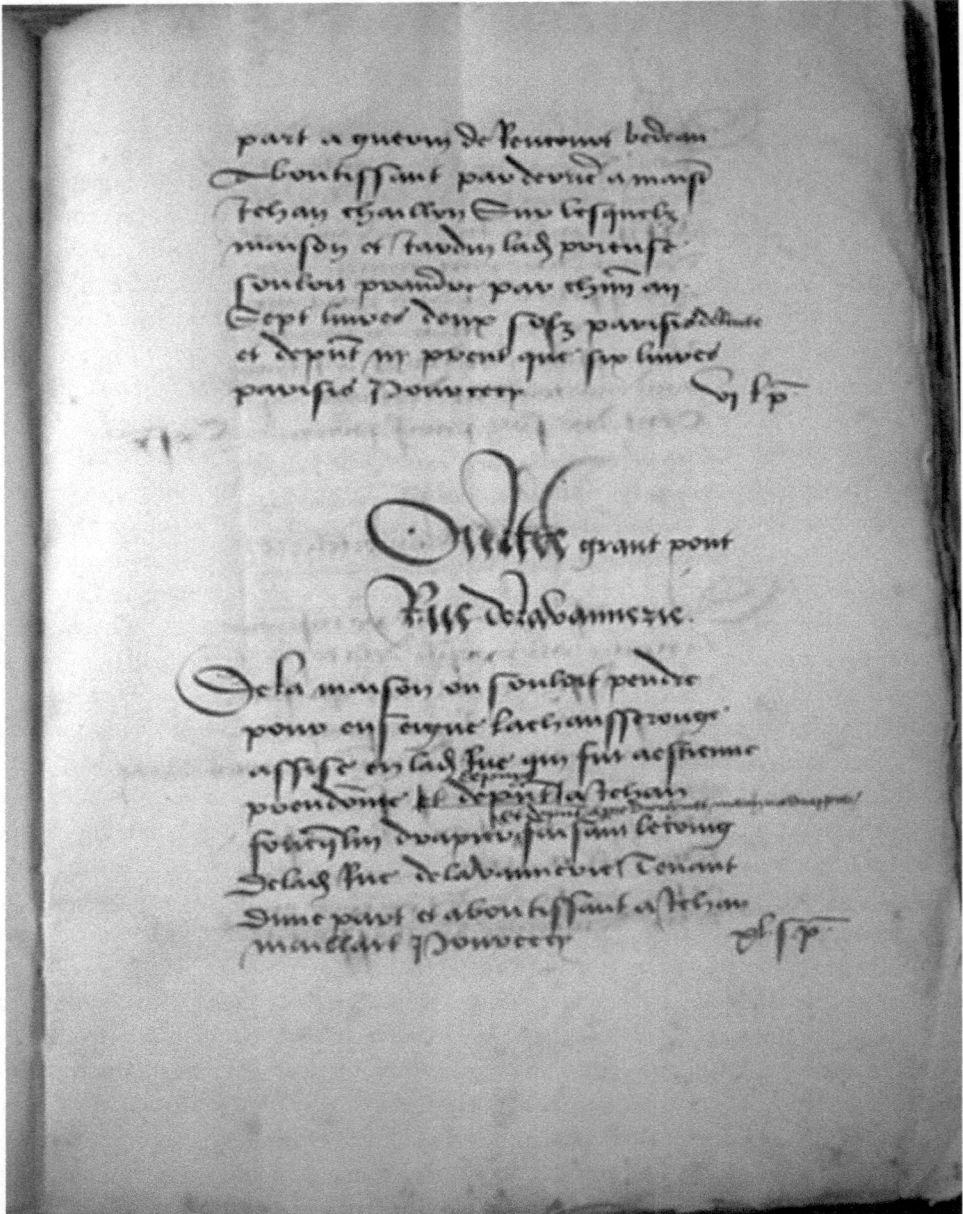

FIGURE 1.2 *Comptes de la prieure de l'Hotel-Dieu de Paris,* 1493, paper, f°8r

(*Source*: Archives Nationales (France) in Paris (H5 3665). Taken from 'La série des comptes de l'hôtel-Dieu de Paris à la fin du Moyen Âge: aspects codicologiques' by C. Jéhanno (2011, p. 29) © picture C. Jéhanno)

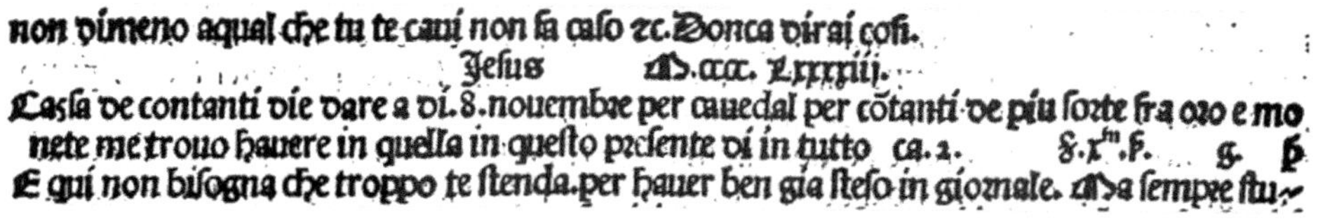

non vimeno aqual che tu te caui non fa calo ʒc.Dõca viraí coſi.
                     Jeſus   D.ccc. Lxxxiij.
Casfa ve contanti vie vare a vɫ.8.nouembre per cauedal per cõtantí ve piu forte frà oʒo e mo
nete me trouo hauere in quella in queſto prefente vɫ in tutto ca.1.   8.x".ſ.   g   ꝑ
E quí non biſogna che troppo te ſtenda.per hauer ben gia ſteſo in gioʒnale. ɑ̃ʒa ſempʒe ſtuͥ

FIGURE 1.3 Pacioli's indications for a ledger entry, 1523, paper, f°202v

(*Source*: Bibliothèque Municipale de Lyon (Rés 105625))

Thus, your posting in the Ledger looks like this:

LEDGER ACCOUNT: Cash			Page C-1	
			Year: 1493 AD[5]	
Date	*Description*	*Reference*	*Debit*	*Credit*
08 Nov	By: Capital for gold and other coins	J2	24	

FIGURE 1.4 Cripps' translation of Pacioli's indications for a ledger entry, 1994, p. 27

(Courtesy of the Pacioli society, Seattle University)

distributed through homogeneous printed grids over words has become so ingrained that a modern translation (Figure 1.4) of a passage from Pacioli's *Summa* (1523) on how to enter entries in the ledger looks entirely different from the original (Figure 1.3). The very same content takes entirely different forms as typical aesthetic criteria for building legitimate accounts change from one period to the next.

Taking a bird's eye view covering examples of account books from Rome, the Renaissance, and modernity allows us to explore the material and visual forms of accounts as marked by major evolutions. Although accounting remains stable enough to still be recognized as homogeneous, account books are fluid objects evolving over time (De Laet and Mol, 2000). Staying the same depends upon change and evolution including in their visual and material forms which then play an important role in support of their perceived immutability (Law and Singleton, 2005) notably as far as their purpose and actual content are concerned.

The visual form of accounting inscriptions is as important as their content to assert the legitimacy of social relations and the existence of collectives, including organizations. Yet, 'representations have little determinate meaning or logical force aside from the complex activities in which they are situated' (Lynch and Woolgar, 1990: vii). Like all visual representations, too simplified and containing very little truth in them, accounts have to be filled. They produce little truth, but they produce engagements. When engaging with account books as performable spaces, there is a process of objectification of the underlying organizations. Accounting practices make organizations into facts or objects, immaterial but represented, made present. Such organizations as objects have no intrinsic property, but they are performed through repetition and appear homogeneous (Quattrone, 2009). For organizations to succeed (to happen), engagements have to be repeated and maintained. In absence of repetition through time and space as promoted by accounting practices, nothing would hold stable at all, organizations least of all.

AN EXEMPLARY RETRACING OF THE ORGANIZATIONAL FORCE AND EFFECTS OF ACCOUNT BOOKS

The material aspect of accounts has always been used as a technology of trust building, as a technology of information circulation, and as a technology of imagination. All these technologies are supportive of processes of organizing for they engage users to produce organizations by performing accounts (i.e., reading them and composing with them). Yet, people do not necessarily see the same thing in the accounts that circulate. Accounting attracts and generates heterogeneous uses (Quattrone and Hopper, 2006) and different visions. Engaging with accounts does not merely reflect different points of view but also produces different organizations with sometimes rather incompatible purposes (economic, social, technical, political, etc.) hidden behind a homogeneous name and façade (often constituted by shared accounting layout). The same set of accounts can thus refer to different organizations, although they will all *look* very similar.[2]

An illustration of accounting's ability to produce different organizations is given by the situation at Ronelec,[3] a hydroelectric company which is part of a large energy conglomerate (Puyou, 2014). At Ronelec, engineers see in hydroelectric dams technological assets that need to be maintained in order to prevent incidents and durably produce power. Ronelec traders in charge of selling the electricity produced on the market are by contrast seeing in dams major cash generators. As a consequence, establishing maintenance schedules has become a point of contention between financial experts and engineers as it implies temporary reductions in electricity production levels. Agreements on how often and when maintenance should take place are reached during meetings where operation and finance managers voice their arguments in favour respectively of the pursuit of rather conservative safety policies or ambitious performance objectives. Maintenance people call such meetings 'the axe committee' in reference to the budget cuts announced there. The finance department leads in discussions but compromises are reached despite disparities in participants' priorities. For example, a threshold imposed on growth in operating expenses, including maintenance, was set to 5 per cent instead of 2 per cent following discussions on future levels of sales, costs, and

---

[2] It is the 'heading' which makes the components of accounting homogeneous. For example in the Jesuit Order, the powerful and absolute heading is the idea of God. Yet, no precise understanding of what is meant by the idea of God is imposed homogeneously from the top of the hierarchy. Thus, Jesuit members work for what the individuals believe is worthwhile working for (Quattrone, 2004).

[3] A pseudonym.

therefore profits. Gatherings of managers with opposite ideas to discuss the company's future financial statements, in particular Ronelec's expected profit and loss account, play a key role in bringing together members supportive of what appear to be different organizations hidden behind a common name. The axe committee is a local institution where tensions between objectives of profitability and security can be addressed on a regular basis. Subgroups with radically different cultures and identities coexist along with the institutions that help them find reasonable arrangements. By classifying dams as assets, accounting participates in the creation of a common vocabulary that articulates visions of dams both as major sources of costs (depreciation and maintenance) and of revenues. Interestingly, engineers engage with accounting because they hope that it will help them solve their concerns about maintenance. It is when people on the periphery become convinced that the content of accounts is relevant to them that accounting works. Accounts have various, concurrent, and competing characters and rationales. They solicit our hope of solving a problem, of getting power, of making money, or of salvation (Quattrone, 2004). It is a way to connect with others, including opponents, like when engineers fight against the dominance of bean-counters over discussions on the profit and loss account.

Actors engaged in accounting practices seek to achieve order but also concomitantly create further openings and orderings (Jones et al., 2004). Arguably, there are more than just two different organizations at Ronelec. In addition to a focus on profits or on maintenance, other individuals (like for example union members) compose with Ronelec published accounts an organization that looks closer to a public service company focusing on navigation facilities, on the preservation of the river's natural habitats, on irrigation, and on safeguarding jobs in rural areas. Despite the taken for granted 'economic nature' of the accounting system, it actually goes beyond profit maximizing organizations. Accounts do not work for one 'pure' reason (such as economic performance) but reduce the complexity of human beings' visions and relations to traces that appear as linked to a homogeneous rationale (Quattrone, 2004) but which can also be debated.

Accounting thus produces different organizations that cannot be summarized or represented simultaneously. There is a gap, an absence between the text of the account and actual practices that is constantly filled by meetings, negotiations, and arguments. Accounting practices are generating organizations in the making. The authority is not entirely monopolized by the centre. Standardization is never quite achieved. Accounting practices do not create totalizing systems but support the coexistence of different interconnected organizations in parallel (Singleton and Law, 2013). These different organizations share similar organizing processes. Shared practices are what makes organizations, not shared content or understanding.

As multiple users engage with accounting, its presence increases and the organization is enacted by the participants. The organization is experienced through the circulation of people and documents related to accounting, for example in preparation to the axe committee. Not all versions of the organization are compatible but they remain related because they rely on the circulation of the same 'intermediaries' (Callon, 1991) such as accounts serving the continuous needs for mediation for organizations to exist. Accounting is powerful because of the two-way relationships between centres and periphery, although which is which varies depending on who answers the question: the centre being where electricity is produced for engineers and where it is sold for traders.

Why Is Understanding the Organizational Power of Account Books Relevant?

Objects like account books and facts like organizations are ontologically complex and multiple (Law and Singleton, 2005). This chapter argues that accounting is a set of practices that create and shape organizational reality (Hines, 1988) and it shows that account books are involved in the active production of different organizations not merely because of different perspectives on the same object, but because multiple organizations are actually performed involving different sets of relations and contexts of practices. The versatility of accounting and organizations should not be attributed to humans' 'interpretive flexibility' but to their ontology (Quattrone and Hopper, 2006).

Where and when accounting processes take place has an impact on organizing processes. That entries are made on tablets using a stylus or on screens using a keyboard is likely to make a difference. And yet, a common characteristic of the production of accounting information in aural and digital cultures is the impossibility to deliver the dream of perfect information (Quattrone, 2016). In no cases do financial statements stand on their own, independent from other documents, and under no circumstances are they forcing anyone to think in one way or another. Texts cannot compel readers to do precisely certain things (Johns, 2006). What organizations are, is therefore uncertain and it has been for centuries one of account books' main powers to deal with uncertainties. Accounting is about speculation about futures to be debated. Being aware of this is a condition to maintain wisdom in decision-making processes (Quattrone, 2016). Everything does not happen in the text of the account or in the mind of the user but in between the two. Organizing is a heterogeneous encounter between people and artefacts. It is also a collective practice sufficiently coherent to be discussed and acted upon. Legitimate organizations, societies, and communities arise from interactions such as those that have been taking place for centuries around account books.

References

Andreau, Jean. 2004. 'Structure et fonction du livre de comptes de Kellis'. *Comptes rendus des séances de l'Académie des Inscriptions et Belles-Lettres*, 148(1), 431–43.

Andreau, Jean. 2007. 'Registers, Account-Books, and Written Documents in the *De Frumento*'. *Bulletin of the Institute of Classical Studies*, 50(S97), 81–92.

Bastide, Françoise. 1990. 'The Iconography of Scientific Texts: Principles of Analysis'. In Michael Lynch and Steve Woolgar (eds), *Representation in Scientific Practice*. Cambridge, MA: MIT Press, 187–229.

Callon, Michel. 1991. 'Techno-Scientific Networks and Irreversibility'. In John Law (ed.), *A Sociology of Monsters: Essays on Power, Technology, and Domination*. London and New York: Routledge, 132–61.

Camille, Michael. 1985. 'Seeing and Reading: Some Visual Implications of Medieval Literacy and Illiteracy'. *Art History*, 8(1), 26–49.

Chandler, Alfred D. 1977. *The Visible Hand: The Managerial Revolution in American Business*. Cambridge, MA: The Belknap Press of Harvard University Press.

Cripps, Jeremy (trans.). 1994. *Particularis de computis et scripturis. 1494 Fra Luca Pacioli. A Contemporary Interpretation*. Seattle: Pacioli Society.

de Laet, Marianne and Annemarie Mol. 2000. 'The Zimbabwe Bush Pump: Mechanics of a Fluid Technology'. *Social Studies of Science*, 30(2), 225–63.

de Ste Croix, Geoffrey E. M. 1956. 'Greek and Roman Accounting'. In A. C. Littleton and B. S. Yamey (eds), *Studies in the History of Accounting*. London: Sweet & Maxwell, 14–74.

Dent, Jeremy. 1991. 'Accounting and Organizational Cultures: A Field Study of the Emergence of a New Organizational Reality'. *Accounting, Organizations and Society*, 16(8), 705–32.

Eco, Umberto. 1985. *Lector in fabula: le rôle du lecteur ou la coopération interprétative dans les textes narratifs*. Paris: Livre de Poche.

Hines, Ruth D. 1988. 'Financial Accounting: In Communicating Reality, We Construct Reality'. *Accounting, Organizations and Society*, 13(3), 251–61.

Jéhanno, Christine. 2011. 'La série des comptes de l'hôtel-Dieu de Paris à la fin du Moyen Âge: aspects codicologiques'. *Comptabilité(s)* [online], 2, 1–45.

Johns, Adrian. 2006. 'The Book of Nature and the Nature of the Book'. In David Finkelstein and Alistair McCleery (eds), *The Book History Reader*. Abingdon: Routledge, 255–72.

Jones, Geoff, Christine McLean, and Paolo Quattrone. 2004. 'Spacing and Timing'. *Organization*, 11(6), 723–41.

König, Jason and Tim Whitmarsh (eds). 2007. *Ordering Knowledge in the Roman Empire*. Cambridge: Cambridge University Press.

Law, John. 1994. *Organizing Modernity*. Oxford and Cambridge, MA: Blackwell.

Law, John and Vicky Singleton. 2005. 'Object Lessons'. *Organization*, 12(3), 331–55.

Littleton, A. C. and B. S. Yamey (eds). 1956. *Studies in the History of Accounting*. London: Sweet & Maxwell.

Lynch, Michael and Steve Woolgar (eds). 1990. *Representation in Scientific Practice*. Cambridge, MA: MIT Press.

Meyer, Elizabeth A. 2004. *Legitimacy and Law in the Roman World: Tabulae in Roman Belief and Practice*. Cambridge: Cambridge University Press.

Minaud, Gérard. 2005. *La comptabilité à Rome: Essai d'histoire économique sur la pensée comptable commerciale et privée dans le monde antique romain*. Lausanne: Presses polytechniques et universitaires romandes.

Pacioli, Luca. 1523. *Summa de arithmetica, geometria, de proportioni et de proportionalita* (2nd edition). Venice. Bibliothèque municipale de Lyon, Rés 105625, f° 202v.

Puyou, François-Régis. 2014. 'Ordering Collective Performance Manipulation Practices: How Do Leaders Manipulate Financial Reporting Figures in Conglomerates?' *Critical Perspectives on Accounting*, 24(6), 469–88.

Puyou, François-Régis and Paolo Quattrone. 2018. 'The Visual and Material Dimensions of Legitimacy: Accounting and the Search for *Socie-ties*'. *Organization Studies*, 39(5–6), 721–46.

Quattrone, Paolo. 2004. 'Accounting for God: Accounting and Accountability Practices in the Society of Jesus (Italy, XVI–XVII centuries)'. *Accounting, Organizations and Society*, 29(7), 647–83.

Quattrone, Paolo. 2009. 'Books to Be Practiced: Memory, the Power of the Visual and the Success of Accounting'. *Accounting, Organizations and Society*, 34(1), 85–118.

Quattrone, Paolo. 2016. 'Management Accounting Goes Digital: Will the Move Make It Wiser?' *Management Accounting Research*, 31, 118–22.

Quattrone, Paolo and Trevor Hopper. 2006. 'What is IT? SAP, Accounting, and Visibility in a Multinational Organisation'. *Information and Organization*, 16(3), 212–50.

Singleton, Vicky and John Law. 2013. 'Devices as Rituals: Notes on Enacting Resistance'. *Journal of Cultural Economy*, 6(3), 259–77.

CHAPTER 2

··

ACOUSTIC TILE

··

REINHOLD MARTIN

ONCE the limpid domain of gods and angels, by the mid-twentieth century the ordinary ceiling had come to conceal a matter-filled, noisy 'plenum', a term shared by engineers and philosophers to denote a chamber where energy flows, waves propagate, and particles collide (*OED*, 2006). Architects still use the term 'space' to refer to what lies below ceilings but not necessarily for what is above them. When suspended, ceilings make room from plenums of varying depths and configurations into which a building's 'services' or 'systems' are typically compressed. There are many types of suspended ceilings, most of which consist of commercial products flexibly assembled to accommodate a wide variety of spatial needs below and technical needs above. This flexibility, which allows the integration of lighting, ventilation, and enclosure, constitutes the ceiling's most architecturally salient characteristic as a medium of organization. That medium's more basic function, however, is to separate visually and acoustically the world above, in the plenum, from the world below, in space.

More often than not, in large commercial, institutional, or industrial structures, the membrane that separates plenums above from spaces below consists of a material known as acoustic (or acoustical) tile, hanging in a gridded mesh like a film of water in a spider's web. Where the modern equivalent of a frescoed ceiling may be the dramatically lit shell or vault, acoustical tiles suspended in ceilings do not normally supply the eye— or the ear, for that matter—with anything so poetically ambitious. On the contrary, these tiles and their ceilings are most commonly made to disappear, visually and aurally, such that attention may focus on the paperwork on the desk or on the conversation around the table. The tiles themselves have played no small part in this art of disappearance, an art that deserves greater attention in any consideration of what the term 'media' may connote in its fullest, most plenary sense (see Durham Peters, this volume).

SOUNDS OF SILENCE

'Ever since her birth she had been surrounded by the steady hum' (Forster, 1909: 119). When E. M. Forster wrote 'The Machine Stops', mechanization was a noisy affair; by 1950, silence was its signature. In buildings, the background economy of silence and noise evoked by the line from Forster's story centred on the mechanization of air to regulate temperature. Around the time Forster wrote, mechanically regulated air began migrating from industrial environments like cold storage facilities to spaces, like offices, populated by human beings (Banham 1984; Osman 2018). The architectural historian Reyner Banham describes the process as, essentially, co-evolutional. Techniques for circulating air gradually aligned with techniques for cooling it and regulating its humidity. This required tentacles of metallic ductwork, which required, in turn, a secondary spatial system threading through the building. Intake, distribution, and exhaust ducts typically ran vertically alongside elevator shafts or stairs, and at first, horizontally above a central corridor, with vents feeding and drawing air to and from rooms lining either side.

Banham nominates the Milam Building in San Antonio, Texas, designed by the architect George Willis and the engineer M. L. Diver and completed in 1928, as 'the earliest fully air-conditioned office block' (Banham, 1984: 128). Though only circulating unconditioned air, Frank Lloyd Wright's Larkin Building, completed in Buffalo, New York in 1906 (for Banham, the *annus mirabilis* of the air-conditioning business), supplied a model for the vertical shafts. The movie industry, with its need to cool hundreds of bodies enclosed in a single large space, 'introduced the general public to the improved atmospheric environment' (Banham, 1984: 128). And the Philadelphia Savings Fund Society (PSFS) Building, designed by the firm of Howe and Lescaze and completed in Philadelphia in 1932, was serviced by a central Carrier air-conditioning plant located more than half-way up on its twentieth floor and integrated into a coordinated system that terminated, in some places, with air diffusers built into light fixtures built into ceilings suspended from the steel structure above.

But it was not until the steel industry began mass-producing open web trusses through which ducts and other services could be threaded that the suspended ceiling became a system that, in effect, integrated all the other systems. According to Banham, this began in the early 1930s, with a series of commercial efforts to organize the plenum-space between the ceiling below and the floor above, which culminated—symbolically at least—in the perforated 'Acousti-vent' suspended ceiling developed by Burgess Laboratories in 1936. As the name implies, this particular system combined a dual-purpose ceiling tile, made of perforated sound-deadening material that simultaneously dampened acoustical reverberation and distributed air circulating in the plenum above into the space below. All that remained was the integration of lighting.

As suspended ceiling systems became standardized, and the tiles themselves regularized into modular units (eventually two-feet by four-feet in the United States, which dominated the post-war market), fluorescent lighting tubes and fixtures, which developed in parallel, began to conform to the modular standard, as did the mechanical air vents and diffusers that became more common than the perforated panels. The suspended ceiling became a uniform, gridded, modular membrane from which air and light emanated and into which ambient sound disappeared. Thus was born, sometime around 1950, what Banham called 'the tyranny of the tile format' (Banham, 1984: 216).

Acoustically, the tile's function was principally to inhibit the propagation of sound waves emanating from activity in the space below, thereby minimizing the presence of the acoustician's mortal enemy, reverberation. But, a propos Forster's background mechanical 'hum', it also served to isolate that space from the sound of air moving through ducts and passing through dampers in the plenum, above. By the 1950s these sounds, evidence of the building-machine's proper functioning, were made more or less to 'stop' in order that the human beings below might better concentrate on their office work and other allegedly 'post-industrial' activities. The tile's function was, therefore, less tyrannical than transcendental, in the sense that it helped separate out two aural regimes: the low clamour of bureaucracy below—what Friedrich Nietzsche once mischievously called the 'hammering of the telegraph' (Stiegler, 2009)—and the 'hum' of the mechanized enclosure above.

If the acoustical tile ceiling helped secure the difference between the 'industrial' domain of building systems and the 'post-industrial' domain of office work, it was only via the development of the comparably small but significant ceiling tile industry. Emily Thompson has described the emergence of an early twentieth-century 'soundscape' out of technical experiments associated with a culture of listening, on the one hand, and the abatement of noise, on the other. The former took shape largely in environments, like churches and concert halls, designed for attentive audition achieved through the calculated minimization of reverberation. The latter, which was internalized by modern music (and was, we can add, anticipated by Nietzsche in his early reflections on dissonance), was associated with workplace reforms—some technical, some social— focused on eliminating the ambient sounds of the industrial metropolis that drove what became known as the 'roaring twenties'. Running through this is a story of enhanced technical knowledge embodied by the empiricist 'founder' of modern acoustical science, Wallace K. Sabine, followed by development, case-by-case application, and eventual commercialization. In Thompson's vivid telling, acoustical materials like the ceiling tile therefore belong to an overall 'rationalization' of the built and lived environment, to which 'culture', in the form of architecture, music, and the visual arts, both contributes and responds (Thompson, 2002).

Subtle and convincing in its detail, Thompson's story nevertheless relies on a sort of below-grade Weberian progressivism, wherein the material world and the cultural world perform a teleological pas-de-deux, the apotheosis of which is the optimized electroacoustical environment of Radio City Music Hall. After this comes cultural disillusionment with technological ideals, after which, in turn, comes a kind of acoustical pluralism epitomized in late twentieth-century music halls designed to provide flexible acoustical environments that accommodate a variety of different auditory tastes. This account downplays the dialectical relation of attention to distraction, of listening to silencing, and the resulting techno-teleological paradoxes of which the development of acoustical ceiling tile is but one.

The acoustical ceiling tile makes its first real appearance as a commercially available product around 1927, with the appearance of Acousti-Celotex, from the Celotex Company (Thompson, 2002) (Figure 2.1). Acousti-Celotex was adapted from the company's felted board made from sugar cane fibres, which was perforated to allow sound absorption through the holes. Its appearance on the market was followed by Acoustone, Quietile, and Perfatile, all from United States Gypsum (USG). Like a variant of Acousti-Celotex, Acoustile was made from mineral wool and Quietile from wood fibre, while Perfatile covered a standard rock wool tile with a paintable surface of perforated metal. All of these ceiling finishes were generally adhered directly to a substrate, such as wood furring or gypsum board, which was in turn secured to the floor structure above. As Thompson explains, Howe and Lescaze's PSFS Building made early, dramatic use of a suspension system called Mutetile, from the Acoustical Corporation of America, in which perforated, rock-wool-filled cast-plaster ceiling tiles were 'spring suspended' from the structure above (Thompson, 2002).

Systems like these derived from a variety of plaster or plaster-like finishes or boards developed earlier by USG and others, like Keasby and Mattison's Sabinite (which was later acquired by USG), or Nashkote, an asbestos-felt-and-canvas laminate board by Johns-Manville. Like the first ceiling tiles, these boards were generally secured directly to a wood, metal, or concrete substrate, and thus can be described only as quasi-systems. Their immediate, even less systematic and more artisanal precursors were Rumford Tile, a mixture of clay and feldspar made porous by burning out peat from the original compound during firing, and Akoustalith tile, a mixture of pumice and Portland cement, both of which Sabine himself developed in collaboration with the celebrated tile manufacturer and artist Rafael Guastavino in 1913 and 1916, respectively (Weber, 1995; Thompson, 2002).

Thus goes the sequence: from stone-like Guastavino tiles lining vaulted masonry spaces, to a thin, often perforated plane suspended on metal hangers from a steel or concrete frame, of which the PSFS Building was an early, stylized example. Emphasizing the mechanically produced silence of the PSFS workplace as emblematic of Taylorized efficiency to which the building's innovative air-conditioning system also contributed, Thompson mentions that the main banking room's ventilation ducts were dampened with sound-absorbing materials, a technique which later became common with the widespread use of air conditioning. These ducts were sources of both comfort and distraction; their metal shells vibrated with the cool air passing through them, and the resulting hum competed with the acoustically and thermally conditioned hum of business just below. A minor inconvenience, to be sure, and one easily solved with the addition of a little more sound-absorbing material. But the muted reverberation of this 'hum' in the PSFS plenum reveals a paradox familiar to students of technical media, wherein one of the unanticipated side effects of silence is noise.

FIGURE 2.1 Acousti-Celotex Sound Conditioning Products advertisement, detail

(*Source: Architectural Record*, July 1950)

INTEGRATION

A case in point is the design of libraries. As suspended acoustical tile ceiling systems became widely available in the immediate post-war years, architects, engineers, and their clients began utilizing these systems in a variety of different building types, with outsized implications. For example: the November 1946 issue of *Architectural Record* opens its special section on university libraries with a quotation from Vannevar Bush, the engineer and head of the wartime Office of Scientific Research and Development, to the effect that current information technologies, including books and libraries, are unable to keep pace with the 'prodigious rate' at which scientific and technical knowledge was expanding. The editors responded by highlighting what they called 'the development of automatic equipment to meet the crisis in storage and reference, and the development of modular construction to meet the certainty of change' ('University Libraries', 1946: 97). This they illustrated with a detailed presentation of designs for the Charles Hayden Memorial Library at the Massachusetts Institute of Technology (MIT), by the architectural/engineering firm of Voorhees, Walker, Foley, and Smith.

The design for MIT's three-storey Hayden Library is gridded inside and out, including the ceilings. Unusually, air-conditioning ducts rise along the outside walls, branching into plenums above two stacked, double-height book-storage and reading areas that form the building's core, and terminating in regularly spaced anemostat diffusers embedded into the modular suspended ceiling. An accompanying article on library acoustics calls attention to the twin effects of air conditioning: on the one hand, sealed, air-conditioned environments eliminate outdoor noise with permanently closed windows; on the other, the very silence of the library makes it all the more necessary to reduce the hum of the ductwork threading through the plenum. The author, an acoustical consultant, is worth quoting at length:

> There has been discussion regarding the degree of quietness to be sought in a library. Some believe that by reducing the overall noise level unduly the ear is made overly sensitive to sudden sounds. But the consensus is that noise within a reading room should not exceed 30 decibels, approximately the same intensity of sound as that encountered in a quiet residence.
>
> Sound-absorbing materials used, such as wall coverings, acoustical ceilings, and resilient flooring, are selected for their efficiency in absorbing higher sound frequencies, since high frequency noises at average intensity are the most disturbing to a person with average hearing. Reverberation time, although a secondary factor, should be reduced to somewhere between one-half and three-fourths of a second.
>
> (Content, 1946: 121)

By mid-century, the applied science of acoustics had thus become a matter of well-studied convention, even as its boundaries had also become extraordinarily porous.

If, as Thompson shows, fifty years earlier this science was principally one of auditory concentration governed by reverberation, by now it was one of diffusion, in every sense. At which point silence was again paradoxical: in the limit case of university libraries, harnessed to what Vannevar Bush, an MIT alumnus and former Dean of Engineering, had described the previous year as the 'endless frontier' of scientific knowledge (Bush, 1945), the technical production of silence, or 'quietness', risked over-sensitizing the reader's ear to unexpected high-frequency sounds.

Within the instrumental reason exemplified by Bush's post-war programme for a 'national science', the poly-technical problem became one of not disturbing the reader—an archetypal knowledge worker—while providing her or him with adequate light and comfortably conditioned air. At MIT, as in countless libraries, schools, offices, hospitals, and other institutional spaces in North America, Europe, and eventually worldwide, the modular acoustical tile ceiling bore the burden of coordinating these oftentimes conflicting needs. Certain architectural efforts dramatized the high threshold of integration that this implied. At the General Motors Technical Center in Detroit, which opened in 1956, the architects Eero Saarinen and Associates collaborated with the engineering firm of Smith, Hinchman, and Grylls on the design of a custom-built ceiling system that combined pinpoint, high-pressure air distribution at grid intersections with a smoothly luminous surface, where more attention was paid to the attenuation of sound coming from the air ducts than to its absorption by the ceiling tiles. There too, significant efforts were made to coordinate the design and location of wall partitions with the ceiling grid, in apparently seamless integration with the modular curtain wall on the exterior of each building. Whereas, the project of three-dimensional modular integration reached a climax of sorts in the Union Carbide Building in New York, designed by Gordon Bunshaft and Natalie de Blois of Skidmore, Owings and Merrill, which opened in 1960. There, it really was the grid itself, visibly and graphically inscribed into the all-purpose suspended ceiling, that governed. Hardly a mechanized 'tyranny of the tile', the Union Carbide Building's highly aestheticized seamlessness aimed at proto-cybernetic, organic integration, or organization (Martin, 2003). Underwriting it all was a fetish for flexibility that the *Architectural Record* editors had already echoed when they described the Hayden Library's gridded modularity as a technical response to the 'certainty of change'.

In a recent PhD dissertation, Alexandra Quantrill has added at least one more example to this potentially long list: the near absolute enclosure of the open-plan workspaces in the Hong Kong Shanghai Bank Headquarters (1979–86), designed by Norman Foster with Arup Associates and many others (Quantrill, 2017). Emphasizing the controversies that erupted over the cost of technical performance that valued what she calls an 'aesthetics of precision' during the building's long gestation period, Quantrill shows how the building's exterior wall surfaces and systems (or 'skin') formed an uneasy, fragile alliance with raised floors (to accommodate wiring) and suspended ceilings (to accommodate the rest). As the building's location attests, by this point the networks of expertise and industry to which suspended ceilings belonged had extended unevenly around the world.

If technical publications can serve as proxy for further detail on this process, it is instructive to note that the American design manual for architects, *Architectural Graphic Standards*, did not feature guidelines for the installation of acoustical ceiling tiles until its fourth edition, in 1951 (Ramsey and Sleeper, 1951). The publication's fifth edition, issued in 1956 with a foreword by Eero Saarinen, significantly expanded those guidelines to include 'Acoustical Suspended Ceilings', 'Louvered Suspended Ceilings', and 'Corrugated Plastic Suspended Ceilings' (Figure 2.2) (Ramsey and Sleeper, 1956). The rough equivalent in Europe, Ernst Neufert's *Bauentwurfslehre*, first appeared in German in 1936 and became, following its National Socialist career and in English translation in 1970, the industry standard in those technical environments (the majority, that is) that utilized the metric system.

As Nader Vossoughian has shown, Neufert's project aimed at the 'normalization' of design standards, exemplified by Neufert's near-maniacal dissemination of the A4 paper format and his introduction of metric standards for the manufacture and laying of bricks (Vossoughian, 2014). Rather than assemble generic products into prototypical building components and systems, as did *Architectural Graphic Standards*, 'the Neufert' (as the reference manual has come to be universally called) delineated rule-based norms and standards that functioned as what Vossoughian calls 'instrument[s] of socialization'. Still, it is telling that as late as 1966, the manual's relatively brief section on acoustics was limited mainly to concert halls, a detail that remained consistent in the second 'International' English edition of 1980. Nor did the original or later editions of the *Bauentwurfslehre*—again up to 1980—include any norms for the design of ceilings, acoustical or otherwise. Even the otherwise expansive section on office buildings was restricted to spatial organization, furniture layouts, and modular construction illustrated mainly with floor plans rather than by the 'reflected ceiling plans' that serve as the architect's main coordinating device for the mirror world above.

To track even in abbreviated fashion the dissemination of the suspended ceiling in Europe, Asia, and in the decolonizing South, a trade publication like the West German *Bauen und Wohnen* is more useful. There, what seems a straightforwardly rapid progression again appears: from relatively small-scale designs utilizing common materials like brick and plaster in 1950, with no advertisements for building systems or other sophisticated construction elements, to larger projects, with many European examples utilizing suspended ceilings in 1956, and with advertisements for ceiling systems from Zent-Frenger and Deutsche Philips, to similar examples from Japan by 1960. This line quickly ramifies when followed into other, similar publications tracking the latest large-scale architectural projects around the world in the 1960s and 1970s. In all of this, however, the prosaic suspended ceiling grid with inlaid acoustical tile, lighting fixtures, and air diffusers is but one interface within the flux of technological change. Like the conflicts hidden in the surfaces of Foster's HSBC Headquarters, far from signifying something as simple as 'progress' or 'modernization', the obliquely visible, barely audible ceiling accrues profoundly different meanings under profoundly different historical and socio-technical conditions.

408

ACOUSTICAL METAL SUSPENDED CEILINGS

Pencil rod hangers

Suspension channel
4'-0" o.c.

Wall moulding

Coil header — 1¼" steel pipe

Wall moulding

12" x 24" perforated aluminum snap-on panels

V-coil spring clip

Panel spring clip

Coil lateral ½" steel pipe

Fixture mounting bracket

Acoustic-thermal blanket

Coil laterals 12" or 24" o.c.

Flanges at edge of panel snap onto coil laterals

Plastic, glass, lens or louvers

CEILING SYSTEM WITH RADIANT COOLING, RADIANT HEATING & ACOUSTIC CONTROL

3" diameter opening in duct

Flat hanger rods

1½" channel — usually 4'-0" o.c.

Tee-bars, anchor bars, or similar

Flexible tubing

Wire tee-bar clip

Mounting bracket

Adjustable orifice valve

Acoustical wool batt

Steel or aluminum light trough 2'-0", 4'-0", 6'-0" or 8'-0" in length

Low velocity air diffusing vent panel

24"

24"

12"

Standard perforated metal pan

Metal pans and troffers snap into tee-bar

ACOUSTICAL METAL CEILING WITH AIR-DIFFUSING PANEL AND TROFFERS

Blanket or pad type glass wool, rock wool, or similar. Best results are obtained by laying material directly on ceiling panels with air space above. Acoustical material may be attached to existing ceiling framing or trusswork as alternate. *

Suspension rod or wire — Maximum spacing = 5'-0". Maximum spacing where tee-section supports lighting fixture = 3'-0"

Stay wire - 6'-0" o.c.

2" x 2" x —" extruded aluminum wall angle attached every 36"
Stock panel lengths - 5'-11⅝" and 7'-11⅝"

2" x 2" x ¾₁₆" extruded aluminum tee-section

Channel light fixture

Perforated aluminum ceiling panel — corrugated to ¾" depth with 2⅛" pitch. Open area—14%. Panels are removable for access to utilities.

* NOTE: When sound-absorbent material is attached to ceiling structure with panels suspended below, conditioned air may be distributed thru ducts installed above panels

PERFORATED METAL ACOUSTICAL CEILING SYSTEM

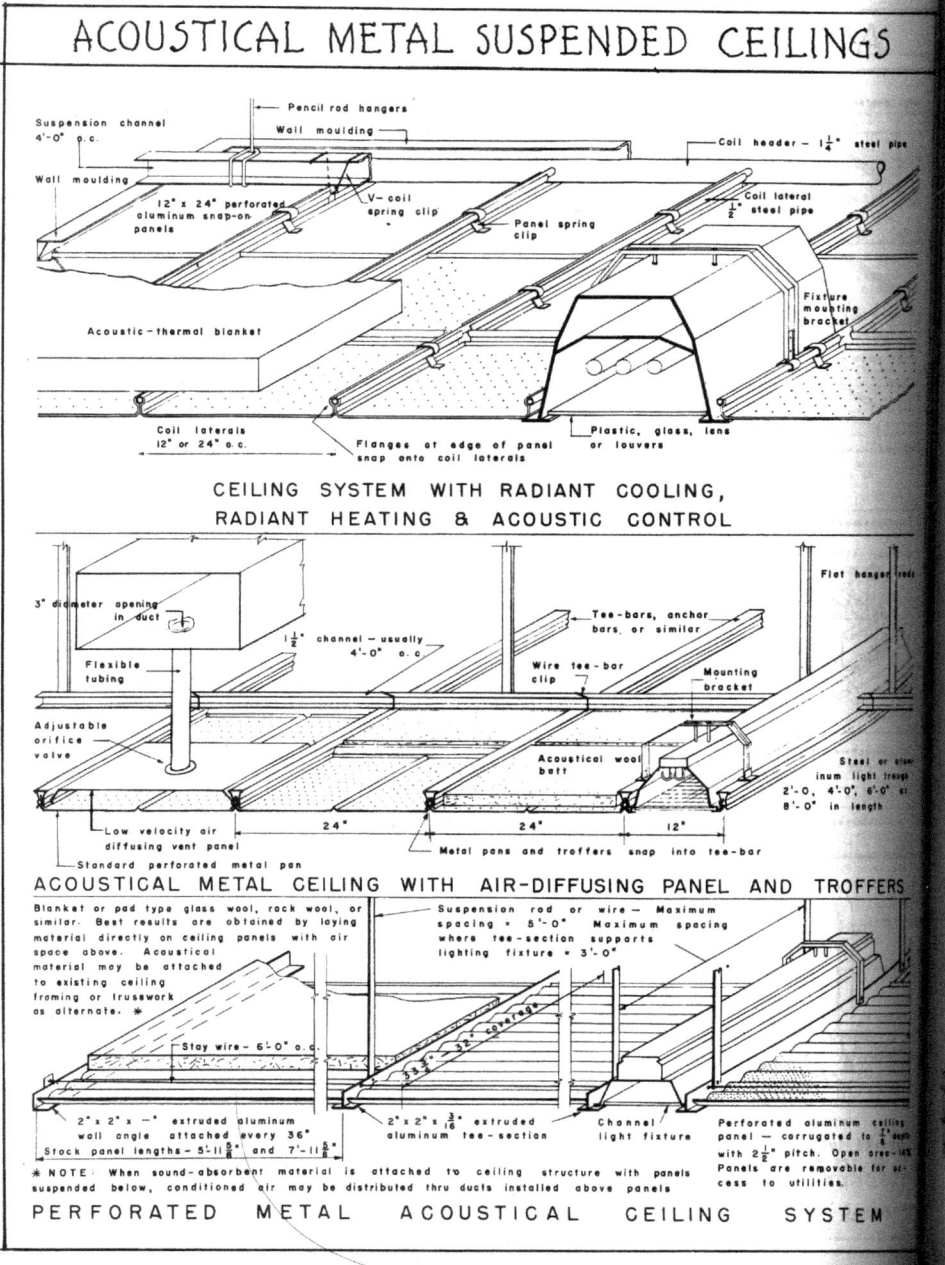

FIGURE 2.2 Acoustical metal suspended ceilings

(From Ramsey and Sleeper, Architectural Graphic Standards, 5th ed. (John Wiley & Sons, 1956))

MEDIA AND ORGANIZATION

In what sense, then, should 'acoustic tile' be counted as one among countless potential 'media'? If the above summary is any kind of guide, the answer does not lie in an inventory of the possible meanings attached to the suspended acoustical ceiling by the late twentieth century, even as that technical system joined others circulating in the contested semantic fields of decolonization, globalization, and neoliberal development. Rather, the significance of the acoustic tile as a component in such a system may be sought in its phatic function, its status as interface or channel. But even here, meaning-effects multiply. What, for instance, is the effect of the acoustic separation rendered by the tile, between an idealized silence in the habitable spaces below and the discordant hum of the mechanized spaces above? What, in short, is the meaning of the technologically produced distinction between 'plenum' and 'space' with which we began?

Rather than conclude with a definitive answer to this question, let us review its implications. Viewed (or listened to) from both sides, the suspended acoustical ceiling occupies an intermediate, intermedial threshold. During the twentieth century, this threshold in no small measure gave meaning to the term 'space' by subtracting the clatter of machines and the chatter of humans that would otherwise have bounced from ceiling to floor to wall and back, taking mechanical, acoustic measure of that space's physical and psychic enclosure. Now, space could be experienced to a non-negligible extent as an empty continuum serviced by a potentially unlimited system of gridded surfaces of which the suspended ceiling was paradigmatic. Absorbed into that ceiling's porous, intricately woven or compressed substance, residual sounds mixed there with less distinct and more occasional noise coming from 'the machine' itself: the hum of the systems that filled the ceiling's plenum. The ceiling grid maintained the separation. Below it and between its lines was an organizational apparatus. In the plenum above, ductwork, wiring, and lighting fixtures jostled for position in the flux. In this and many other ways, the tile itself contributed to a reordering of the universe, with the heavens now filled with real, godlike servomechanisms. The task of these systems was not to abstract or to rationalize. It was to preserve the humanity of humans by keeping them in their place in 'space', content to wonder now and then whether the air entering their bodies contained meaning as they moved their sheets of paper around. Just enough material remained in that space to pose such a question. This is what media do.

REFERENCES

Banham, Reyner. 1984. *The Architecture of the Well-Tempered Environment*, 2nd edn. Chicago: University of Chicago Press.

Bush, Vannevar. 1945. *Science: The Endless Frontier*. Washington, DC: United States Government Printing Office.

Content, Edward J. 1946. 'Sound Control in Libraries'. *Architectural Record* (November), 121.

Forster, E. M. 1909. 'The Machine Stops'. *Oxford and Cambridge Review* (November), 83–122.

Martin, Reinhold. 2003. *The Organizational Complex: Architecture, Media, and Corporate Space*. Cambridge, MA: MIT Press.

Neufert, Ernst. 1936. *Bauentwurfslehre: Grundlagen, Normen, und Vorschriften über Anlage, Bau, Gestaltung, Raumbedarf, Raumbeziehungen. Masse für Gebäude, Räume, Einrichtungen und Geräte mit dem Menschen als Mass und Ziel. Handbuch für den Baufachmann, Bauherrn, Lehrenden und Lernenden*. Berlin: Bauwelt-Verlag.

Neufert, Ernst. 1980. *Architects' Data: Second (International) Edition*, ed. Vincent Jones. London: Collins Professional and Technical Books.

Osman, Michael. 2018. *Modernism's Visible Hand: Architecture and Regulation in America*. Minneapolis: University of Minnesota Press.

Oxford English Dictionary (OED). 2006. 3rd edn. Oxford: Oxford University Press.

Quantrill, Alexandra. 2017. 'The Aesthetics of Precision: Environmental Management and Technique in the Architecture of Enclosure, 1946–1986'. PhD dissertation, Columbia University.

Ramsey, Charles and Harold Sleeper. 1951. *Architectural Graphic Standards for Architects, Engineers, Decorators, Builders, and Draftsmen*, 4th edn. New York: Wiley.

Ramsey, Charles and Harold Sleeper. 1956. *Architectural Graphic Standards for Architects, Engineers, Decorators, Builders, and Draftsmen*, 5th edn. New York: Wiley.

Stiegler, Barbara. 2009. 'On the Future of Our Incorporations: Nietzsche, Media, Events'. *Discourse*, 31(1/2), 124–39.

Thompson, Emily. 2002. *The Soundscape of Modernity: Architectural Acoustics and the Culture of Listening in America, 1900–1933*. Cambridge, MA: MIT Press.

'University Libraries'. 1946. Building Types Study Number 119, *Architectural Record* (November), 97–8.

Vossoughian, Nader. 2014. 'Standardization Reconsidered: Normierung in and after Ernst Neufert's *Bauentwurfslehre* (1936)'. *Grey Room* 54, 34–55.

Weber, Anne E. 1995. 'Acoustical Materials'. In Thomas C. Jester (ed.), *Twentieth-Century Building Materials: History and Conservation*. New York: McGraw-Hill, 262–7.

CHAPTER 3

....

BATTERY

....

JAN MÜGGENBURG

As media of organization batteries are central in the areas of transport, communication, work, leisure, and health. After all, 'electromobility' in digital cultures comprises all the batteries that we carry with us on our bodies every day: batteries are the technological prerequisite for portable computers, game consoles, smartphones, digital cameras, (smart) watches, toys, electric wheelchairs, and other electrical therapy systems such as heart pacemakers or cochlear implants. Generally speaking, batteries expand and make our power networks more flexible: they allow us to use digital devices in places that cannot be reached with cables or where a wired solution would not be practicable (Sprenger and Gethmann, 2015). In this sense batteries first of all are a medium of expansion and it may not come as a surprise that the term 'battery' originally had a military background. It can be traced back to the beginning of the sixteenth century and stems from the French word *battre*, which means 'beating' or 'trashing'. *Baterie* initially meant a violent blow or attack (e.g., on a city wall)—related terms are the *battalion* or the *debate* (Onions, 1996: 80). In the course of the sixteenth century, however, the meaning of the term moved from the target to the weapon. Today we still call a combination of several guns, torpedoes, or missiles a 'battery'. In the mid-eighteenth century, this sense of the term of aligning a series of artillery weapons led Benjamin Franklin to call an experimental arrangement of several Leyden bottles—an early version of an electrical capacitor—a 'battery' (Desmond, 2016: 72–3). This was the first time the term was used in connection with electricity (Figure 3.1).

A Short History of the Battery

The history of electrochemical energy sources, which we now call 'batteries', began a few decades later and is indeed closely linked to the general history of electricity. At the end of the eighteenth century the Italian physician and philosopher Luigi Galvani discovered in his frog experiments the transformability of chemical energy into electrical energy by connecting an electrolyte (the salt water in the legs of the frogs) with two electrodes made

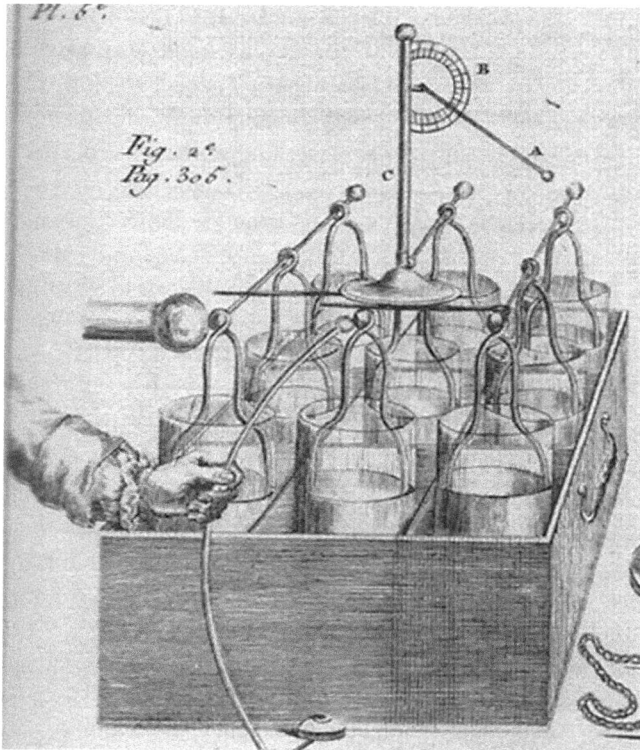

FIGURE 3.1 Leyden bottles. 1773 French edition of Benjamin Franklin's *Experiments and Observations on Electricity*

(*Source*: https://commons.wikimedia.org)

of different metals (copper and iron) (Rieger, 2008). In 1800 the Italian physico-chemist Alessandro Volta constructed a completely artificial electrochemical energy cell made of copper and zinc plates, which he stacked on top of each other together with textiles impregnated with hydrochloric acid (sulphuric acid) (Dibner, 1964). Today's batteries still function according to this principle: two electrodes (cathode and anode) are separated by a suitable material, which allows and even promotes the movement of ions between the two poles. If an electrical load is connected to the battery, the ions flowing between cathode and anode ensure that a small motor or a light bulb is supplied with power. Because the vertical construction of the 'Volta column' had some disadvantages, the Scot William Cruickshank varied Volta's design (Buchmann, 2016: 20). His 'trough battery' consisted of a casing with several interconnected compartments filled with sulphuric acid, into each of which two electrodes of zinc and copper were embedded. This horizontal design made mass production of electrical cells possible for the first time.

In the course of the nineteenth century the design of wet-cell batteries was further improved, for example by the French physicist and chemist Georges Leclanché. His 'Leclanché elements' were increasingly used in places that were difficult to reach through the power grid, such as railway telegraphs or doorbells (Desmond, 2016: 132–4). However, the liquid components within the wet-cells meant that they were neither mobile nor particularly reliable. This changed with the development of dry-cell batteries around 1900 in Germany. The physician Carl Gassner used plaster to bind the acid used in the Leclanché elements (Reddy, 2011: 9.2–9.3). In 1903 the Berlin inventor Paul Schmidt started the industrial production of battery casings and dry batteries, e.g., for the first electric flashlights (Bard et al., 2012: 237).

Despite this technical evolution, for a long time the battery had been a 'solution in search of a problem' (Schallenberg, 1982: 1) instead of a reliable and widely used technology. This changed during the Second World War, when American inventor Samuel Ruben made another crucial contribution to the history of the battery (Hintz, 2009). In the early 1940s, he was contracted by the military and the P.R. Mallory Company to work on batteries that would be more robust and reliable than conventional ones. At high temperatures and humidity, zinc-carbon cells, which at the time were mainly used by the Army's Signal Corps (e.g., in wireless radios), tended to discharge spontaneously. By sealing the cells airtight and using mercury cathodes, Ruben developed a 'tropical battery' used extensively during the last two years of war (Hintz, 2009: 27–33). In the post-war period, Ruben and Mallory refined the technology further into a commercially successful product. Hearing aids, transistor radios, and the first quartz wristwatches were the most important applications from the end of the 1940s until the late 1950s. In the 1960s, Mallory introduced the now famous Duracell brand and played an important role in the early miniaturization and mobilization of electrical devices: from pacemakers and Polaroid cameras to satellites and even manned space flight, the mercury-cell was an omnipresent factor of technological progress in the United States. According to Hintz (2009: 55) Rubens' mercury-battery was 'just as important, if not more so', as the transistor and the integrated circuit for the history of miniaturization and portability.

The idea of reusing the battery is as old as the technology itself: as early as 1800, the German physicist Johann Wilhelm Ritter developed a variant of the Volta column that could be unloaded and reloaded: 'Ritter's column'. In 1859, the French physician Gaston Planté developed the first accumulator based on lead-acid (Schallenberg, 1982: 391–2). But it was not until the turn of the century that the technology took concrete form, when several people in different places experimented with accumulators based on different materials. The Swedish engineer Waldemar Jungner developed a nickel-cadmium accumulator in 1899, which contained an alkaline electrolyte for the first time (Barak, 1980: 324). Thomas Alva Edison used iron instead of cadmium and the Luxembourg engineer Henri Tudor refined Planté's theoretical concept and developed the first practical lead-acid battery. Subsequently the lead-acid combination was widely used in early battery-powered electric motorcars that for almost two decades wove their way through the cityscape of many American cities. In the majority of cases these were fleet vehicles such as taxis, delivery trucks, or fire trucks that could be easily serviced from central depots (Schallenberg, 1982: 391–2). Hybrid vehicles were also already in existence: Ferdinand Porsche's Lohner-Porsche, produced between 1900 and 1905, contained an auxiliary engine, which recharged the vehicle's battery during driving and thus increased its range (Porsche AG, 2011). However, this first era of the battery-powered car ended quickly in the early 1920s, when gasoline became cheaper and the first industrially produced internal combustion engines significantly increased the range of petrol cars. Remarkably, the developers and supporters of battery technology found a way to participate in this impending boom of the gasoline engine. Now, lead-acid batteries served as an auxiliary technology: first of all the battery-powered electric starter eliminated the difficult and dangerous hand crank and later on the battery guaranteed unprecedented comfort when driving a car (Schallenberg, 1982: 286–7). As Eisler points out: 'Standardized in the 1920s, the starting, lighting, and ignition battery helped secure the dominance of both the gasoline automobile and the lead-acid battery in this auxiliary role owing to its unparalleled cost-effectiveness' (Eisler, 2012: 11).

While the lead-acid battery survived in this way until today, it was other materials that were to give the rechargeable or 'secondary battery' its central role in the history of communication and information technology. For a long time scientists and engineers made only modest progress in advancing the technology. Nickel-cadmium (NiCd) based accumulators were the predominant solution for mobile rechargeable batteries until the late 1990s, when more environmentally friendly connections such as nickel-metal hydride accumulators became available. However, the basis for the more recent mobilization of the computer in smart devices was another chemical element: in the 1970s and 1980s lithium-based battery chemistry moved more and more into the focus of physicists and engineers. The American solid-state physicist John B. Goodenough in particular made important contributions and developed nickel and cobalt-based lithium-ion batteries (Blomgren, 2017). But it was the Japanese industry that eventually dominated the research in the late 1980s and early 1990s. This was due to the fact that companies like Sony had created a vast market for portable consumer products like audio players and had a great interest in improving the usability of their products. Because of its superior energy and power characteristics cobalt-based lithium-ion batteries proved to be the best choice for consumer electronics. It was patented by Sony and launched on the market as a commercial product in 1991. Sony remained the industry leader for some time but competition from other producers finally led to a planned withdrawal from the market (Blomgren, 2017). The quarter century after the introduction of the lithium-ion battery has seen a whole series of improvements to Sony's original patent and new consumer applications such as laptops, cameras, smartphones, e-bikes, household appliances, and even gardening tools. But also many industrial tools and machines depend on secondary batteries (Pistoia, 2009). Finally, progress in lithium battery-chemistry plays a fundamental role in the further development of sensors and actuators as part of the so-called 'robotic revolution' (Jordan, 2016, 80).

ORGANIZING TIME AND SPACE

A result of the emancipation from the power outlet and the path to miniaturized and portable gadgets is 'duration anxiety'. Batteries are 'time-critical media' (Volmar, 2009):

even if no electrical load is connected, batteries are always in a (self-)discharge mode—depending on their capacity and current charge, as well as the way a device is being used, they only allow users to leave the wired power grid for a limited period of time, and as such users are often looking for points of connection. In the late 1990s transport engineers coined the term 'range anxiety', when General Motors was the first major automobile manufacturer to introduce a series-produced electric vehicle, the Electric Vehicle 1 (EV1) (Shnayerson, 1996). The maximum range of this car, which could only be rented and not purchased, was 225 kilometres and recharging its battery through a stationary charger took three hours. The question of how electric motors can be accepted by customers, and regarded as a serious alternative to the internal combustion engine, dominates the history of the electric vehicle (Kirsch, 2000). Even today car manufacturers are discussing the question under the appropriate abbreviation R.I.P.—Range, Infrastructure, and Price. And the anxiety is not limited to vehicles. Witness the desperate search for the next power outlet when our smartphone battery charge drops below 10 per cent and we fear that we won't be able to show our electronic boarding pass when checking in at the airport. Witness too, the worry of not being able to record the crucial events of a family holiday because we forgot to charge the digital camera's battery overnight, and the panic about not being able to finish a presentation on the train on time for the next business appointment—all these are examples of 'duration anxiety'. And in order to prolong the duration of use of battery-powered devices, we adapt our usage behaviour and even accept a loss of quality (lower screen brightness, reduced data transfer, etc.) when switching into 'battery saver mode'. Recently, some resourceful developers even found a way to share duration anxiety. The app 'Die with me' connects smartphones whose battery capacity falls below 5 per cent and opens a chat room for their users: 'Die together in a Chatroom on your way to offline peace' (http://diewithme. online). The time-critical aspect of battery discharge thus mediates our behaviour. As Isidor Buchmann puts it: 'The battery is a feeble vessel that is slow to fill, holds limited energy, runs for a time like a wind-up toy, fades and eventually becomes a nuisance' (Buchmann, 2016: 14).

Temporal anxiety is the price paid for the spatial expansion afforded by batteries in digital cultures. Batteries can maintain the ability to act and communicate in critical situations, cut off from the fixed power network (Ernst, 2013: 108). Spatially, batteries not only expand our networks of work, leisure, and health, they change the way we use certain spaces: cafés and train compartments become places of work and meetings, the waiting room at the doctor's or the last row in the bus become computer game locations. This spatial reorganization is closely linked to another function of batteries. Listening to music when going for a stroll, watching films and television series on a train, or holding telephone conferences in a gym: as media, batteries mobilize and reorganize our media use, they produce new consumers in new places. Not least, therefore, batteries are to be understood as media of an expansion of the sphere of influence of technologies: McLuhan's (2003: 29) famous 'Bedouin with his battery radio on board the camel' is more than a mere metaphor, it's an adequate example of concrete media use in digital cultures. The primordial media function of batteries as facilitators of media use can also be illustrated in the medium that McLuhan regarded as a paradigmatic case of media analysis: electric light. Regardless of what they are used for, 'electrical light and power [...] eliminate time and space factors in human association' (McLuhan, 2003: 21). Following McLuhan, batteries can be seen as temporary agents of this change in the scale of space and time: flashlights and other battery-powered lamps increase the medial sphere of influence of electric light.

Despite its provisional character of a medium of spatio-temporal organization, the battery can also have stabilizing effects: as a backup solution installed in parallel to the fixed network, batteries can temporarily maintain an unstable network—e.g., in the event of a power failure in a hospital. Strategies of opposing duration anxiety must also be understood against the background of this simultaneous stability and destabilization of organizational forms: 'battery swapping' aims to replace the depleted battery with a second, fully charged backup battery, so that the empty battery can be recharged while the replacement battery supplies power to the device. This strategy has a long tradition: when—as stated above—in the early twentieth century electric vehicles were already considered a technology of the future, the first service stations for fast battery exchange were set up in the USA (Kirsch, 2000: 153–62). Because the batteries in our smart digital devices are permanently installed and only the manufacturer is authorized to change the batteries, today we help ourselves with so-called portable chargers—a second battery that cannot be inserted into the device but can be used to charge the first battery on the way. So called 'range extenders' on the other hand are not practical when dealing with smart devices—a small internal combustion engine (or sometimes a fuel cell) carried along in the car, for example, serves as a backup and charges the battery when it is completely discharged or momentarily at rest. Today 'range extenders' are mainly used in hybrid cars, once again turning upside down the relationship between car battery and combustion engine, the latter now functioning as the auxiliary device.

A Humanist Technology

In this search for ever wider, freer spatial expansion the battery has become a mediating technology that encapsulates the progressive ambition of (social) engineers: it is evidently a humanist technology. Ever since the electrification of Europe and America in the 1880s, batteries have transported within them a utopian idea of a 'technology of the future' that enables the decentralized use of electrical devices (Dittmann, 2014), frees us from being tithed to the earth, and counteracts the monopolization of the electricity market (Kapoor, 2017). Above all as alternatives to fossil fuels, batteries in combination with wind and water power plants were thought to guarantee what nineteenth-century physicist Silvanus Thompson called an 'endless supply' of electrical energy (as cited in Kapoor, 2017). From these first utopias of independence from the fuel market to the 'nuclear battery', the 'instantly rechargeable battery' and the 'bionic battery that imitates the electric eel', the 'battery of the future' is a recurring futuristic topos of the twentieth and twenty-first centuries. Similar to the fuel cell (Eisler, 2012), in the course of the twentieth and twenty-first centuries batteries have seemingly offered a way out of the international 'high-energy economy' that is largely based on the fast-paced exploitation of fossil fuels (Nye, 1998: 187).

But contrary to this image of the battery as 'green technology' its history is also a history of the exploitation of different chemical materials and natural resources: from lead, zinc, and mercury to lithium and cobalt in use today. And like other technologies of energy supply—from muscle power to coal, oil, and electricity—batteries have considerable social and ecological consequences (Jones, 2014). The power of lithium-ion batteries fuels the current Silicon Valley gold rush, from driverless cars to smart speakers (Crawford and Joler, 2018). But although lithium-ion batteries were supposed to be different from the dirty, toxic technologies of the past, their extensive use in more and more devices has led to a soaring demand for lithium and cobalt (see Rossiter, this volume). Natural resources, which hardly anyone had been interested in before, are now being mined under high pressure and are being transported via complicated distribution routes to the battery factories, before they end up to our notebooks, e-bikes, and eventually driverless cars. The question of the media geological effects of battery technology through the increasing mining of metals such as lithium or cobalt (Parikka, 2015: 4–5) is usually treated as a secondary issue, yet this development has resulted in catastrophic consequences for humans and nature in the mining areas. The sudden demand for cobalt for example resulted in thousands of miners in Congo suddenly working in hazardous environments and toxic mining waste often pollutes rivers and drinking water (Sadof et al., 2018). Also, efficient recycling of cobalt and nickel from lithium-ion batteries after the end of their 'cycle-life' still poses a major challenge to the industrialized countries.

In conclusion, there are at least four aspects that make batteries work as media of organization. First, batteries reorganize technically accessible spaces by expanding networks and making their connections more flexible. Second, through their ambivalent character hovering between temporary stabilization and impending destabilization, they change the temporal organization of our everyday lives: they enable us to work, communicate, and consume in new places and at unusual times, but at the same time existentially disturb this supposed gain in moments of range and duration anxiety. Third, it is perhaps due to their ability to intervene temporarily in spatio-temporal relationships that batteries reorganize our projections of the future and our concepts of technological progress. And fourth, though it often goes unnoticed, batteries, as key technologies of the digital age, have already contributed comprehensively to a reorganization of the exploitation of natural and human resources on our planet.

REFERENCES

Barak, M. (ed.). 1980. *Electrochemical Power Sources: Primary and Secondary Batteries*. London and New York: Peter Peregrinus Ltd.

Bard, Allen J., György Inzelt, and Fritz Scholz (eds). 2012. *Electrochemical Dictionary*, 2nd edn. Berlin: Springer.

Blomgren, George E. 2017. 'The Development and Future of Lithium Ion Batteries'. *Journal of the Electrochemical Society*, 164(1), A5019–A5025.

Buchmann, Isidor. 2016. *Batteries in a Portable World: A Handbook on Rechargeable Batteries for Non-Engineers*, 4th edn. Richmond: Cadex Electronics Inc.

Crawford, Kate and Vladan Joler. 2018. 'Anatomy of an AI System: The Amazon Echo as an Anatomical Map of Human Labor, Data and Planetary Resources'. https://anatomyof.ai.

Desmond, Kevin. 2016. *Innovators in Battery Technology: Profiles of 93 Influential Electrochemists*. Jefferson, NC: McFarland & Company.

Dibner, Bern. 1964. *Alessandro Volta and the Electric Battery*. New York: Watts.

Dittmann, Frank. 2014. 'Akkumulatoren—ein unverzichtbares Element der frühen Stromversorgung'. In Hans-Joachim Braun (ed.), *Technische Netzwerke und Energiespeicher*. Freiberg: Georg-Agricola-Gesellschaft, 73–88.

Eisler, Matthew. 2012. *Overpotential: Fuel Cells, Futurism, and the Making of a Power Panacea*. New Brunswick, NJ: Rutgers University Press.

Ernst, Wolfgang. 2013. *Digital Memory and the Archive*, ed. and intro. Jussi Parrika. Minneapolis: University of Minnesota Press.

Hintz, Eric S. 2009. 'Portable Power: Inventor Samuel Ruben and the Birth of Duracell'. *Technology and Culture*, 50, 24–57.

Jones, Christophe. 2014. *Routes of Power: Energy in Modern America*. Cambridge, MA: Harvard University Press.

Jordan, John. 2016. *Robots*. Cambridge, MA: MIT Press.

Kapoor, Nathan. 2017. 'Batteries Not Included'. *Technology's Stories*, Society for the History of Technology, August. http://www.technologystories.org/batteries-not-included/.

Kirsch, David E. 2000. *The Electric Vehicle and the Burden of History*. New Brunswick, NJ: Rutgers University Press.

McLuhan, Marshall. 2003. *Understanding Media: The Extension of Man*. Critical Edition, ed. W. Terrece Gordon. Corte Madera, CA: Gingko Press.

Nye, David E. 1998. *Consuming Power: A Social History of American Energies*. Cambridge, MA: MIT Press.

Onions, Charles T. (ed.). 1996. *Oxford Dictionary of English Etymology*. Oxford: Oxford University Press.

Parikka, Jussi. 2015. *A Geology of Media*. Minneapolis: University of Minnesota Press.

Pistoia, Gianfranco. 2009. *Battery Operated Devices and Systems: From Portable Electronics to Industrial Products*. Amsterdam: Elsevier.

Porsche AG (ed.). 2011. *Ferdinand Porsche: Hybrid Automobile Pioneer*. Köln: DuMont.

Reddy, Thomas B. 2011. *Linden's Handbook of Batteries*, 4th edn. New York: McGraw-Hill.

Rieger, Stefan. 2008. 'Der Frosch—ein Medium'. In Stefan Münker and Alexander Roesler (eds), *Was ist ein Medium?* Frankfurt am Main: Suhrkamp, 285–305.

Sadof, Karly D., Lena Mucha, and Todd Frankel. 2018. 'The hidden cost of cobalt mining'. *The Washington Post*, 28 February. https://www.washingtonpost.com/news/in-sight/wp/2018/02/28/the-cost-of-cobalt/?utm_term=.727245d9bcea.

Schallenberg, Richard. 1982. *Bottled Energy: Electrical Engineering and the Evolution of Chemical Energy Storage*. Philadelphia: American Philosophical Society.

Shnayerson, Michael. 1996. *The Car That Could: The Inside Story of GM's Revolutionary Vehicle*. New York: Random House.

Sprenger, Florian and Daniel Gethmann. 2015. *Die Enden des Kabels: Kleine Mediengeschichte der Übertragung*. Berlin: Kadmos.

Volmar, Axel (ed.). 2009. *Zeitkritische Medien*. Berlin: Kadmos.

CHAPTER 4

··

BICYCLE

··

CHRISTOPH MICHELS AND CHRIS STEYAERT

FIGURE 4.1 Olafur Eliasson, 'Your new bike', 2009. Bike, mirror, aluminium, plastic, 100 × 57.5 × 176 cm. Installation view at Torstrasse, Berlin, 2010

(Photographer: Studio Olafur Eliasson. Courtesy of the artist; neugerriemschneider, Berlin; Tanya Bonakdar Gallery, New York. © 2009 Olafur Eliasson)

INTRODUCTION

ALL of us have a lively memory of learning to bike as a child, experiencing to extend our ways of crawling, walking, and running into an accelerated force of movement. It took many risky trials and painful errors, until we finally got that balance right in a magic moment that defeated all earlier frustrations and gave us wings as if biking resembled flying: our first artefact–body assembly was a fact. Before we even could bike, we sat in the front- or backseat of one of our parent's bikes in the way we remember from Bertolucci's movie *La Luna* where the late Jill Clayburgh as the opera-diva Caterina bikes through the streets of Rome with her baby son staring at her lovingly against a moonlit sky. This scene is not only poetic but also establishes the powerful and desiring bond between mother and son assembled together in this cycling experience. Later, we remember biking to high school every morning and evening for 30 minutes, all winds and weathers, a habit we lost when we moved to the big city and our bike was stolen one too many times. Then came the e-bike in our lives: feeling the force of the 'tail-wind' sitting for the first time on an electrical bike gave us the light tread to climb hills and peaks as if we belonged to the elite-group of *Tour de France* racers and became angels with wheels. No other technology organizes our bodies and movements as pleasantly and strongly as the bicycle or two-wheeler; using our feet to gain speed, we are all 'velocipedes'. In this chapter, we trace how the bike got invented and transformed through a variety of uses, and how it increasingly is considered a 'vehicle' for sustainable transformation through challenging urban affects, instigated by new technological developments and social media.

INVENTING THE BICYCLE

The bicycle serves as an example for how technologies are folded into many variations before they find a stable shape and a popular use. The bicycle's invention did not happen overnight but presents itself as an ongoing process of tinkering about with technologies, practices, urban infrastructure, social norms, and riders' identities. This evolution of the bicycle is a hybrid and non-linear process that allowed for the co-emergence of its many different forms and sizes. For example, the walking bike, often cited as the bicycle's archetype, still persists and serves children as a toy for experiencing the joy of balance and pace. At the same time BMX-bikes, tandems, triplets, pedelecs, high-wheelers, racing bikes, mini bikes, mountain bikes, recumbent bicycles, cargo bikes, to name just a few, have emerged and sometimes disappeared on the horizon of the cycling world. The family of vehicles that identifies as bicycle or 'velocipede' relies on the power of our feet and its transmission into fast movement. This hybridity of the bicycle that so seamlessly integrates the human body with a moving machine is epitomized in the joy of pedalling. If anything, it is this joyful encounter between two wheels and two human feet that forms the push behind the inventions and continuous re-invention of the bicycle.

The bicycle has organized significantly the very field this *Handbook of Media, Technology, and Organization Studies* is trying to map. In understanding organizational-technological mediation and socio-technological change in particular, the bicycle forms one of the early and prominent artefacts to receive attention in the pioneering study by Wiebe Bijker (1995a), pointing at the safety bike as 'the king of the road'. This study—launching noticeably the tradition of so-called SCOT-studies (social construction of technology)—argues that bicycles like any other technology are not deterministic organizers but mediated through the reception of users: the technological and the social are recursive, forming a hybrid and ongoing process. Bijker's study of the safety bicycle is classic for documenting so richly how different social groups and their interactions were significant in how technological change takes shape. In his research the author tackles the question 'How do artifacts acquire their politics? Is it bestowed upon them by their users or is it "baked into them" during their construction?' (Bijker, 1995b: 214). To answer such questions, Bijker (1995a: 17) offers 'sensitizing concepts' such as 'interpretive flexibility' or 'technological frame' which function as heuristic devices to interpret the empirical (hi)stories without making a priori distinctions between the social, the technical, the scientific, or the political. The notion 'interpretive flexibility' refers to the idea of how technologies are not fully determined but remain interpretatively flexible as 'many of the actors involved, including, sometimes, users, interpret and thus constitute these artefacts in markedly different ways' (Michael, 2000: 5).

In particular, Bijker (1995a: 41) points at the crucial role of users—such as the 'young men of means and nerve' who first took to cycling and thought the high-wheeler formed a pleasurable racing vehicle. For other, potential users, like older men, biking felt too dangerous, pushing them into the category of 'non-users'. This was also the case for those who could not afford to pay, making the high-wheeler a luxury vehicle. For women, the bike was often seen as 'unpractical' as it created difficulties for their dress (Michael, 2000), but it also reinforced existing gender orders. However, others—such as nineteenth-century feminists and suffragists—already early on ascribed to the bicycle an emancipatory role, refusing to ride only on a tricycle. For instance, the women's rights activist Susan B. Anthony said in an interview with the *New York World* on 2 February 1896: 'I think it [the bicycle] has done more to emancipate woman than any one thing in the world. I rejoice every time I see a woman ride by on a wheel. It gives her a feeling of self-reliance and independence the moment she takes her seat; and away she goes, the picture of untrammelled womanhood' (Harper, 1898: 859).

Thus, the bicycle had been framed in multiple and shifting ways and meant many things to many groups of actors—producers and users. Notwithstanding this interpretive flexibility, the variety of meanings was significantly reduced as specific understandings of the bike got stabilized. For instance, the safety issue became an important feature, and thus excluded the high-wheeler (on which Thomas Stevens had circled the globe at the end of the nineteenth century but which could easily topple over and cause injury) as an 'Unsafe Bicycle' (Bijker, 1995a: 75). Bijker describes how technological frames are built up 'around' artefacts as 'existing interactions move members of an emerging relevant social group in the same direction' (Bijker, 1995a: 123). A technological frame thus acts 'as a "hinge" between social groups and artefacts' (Michael, 2000: 6). Therefore, the bike as artefact is stabilized semiotically and speaks to various social groups. For instance, the high-wheeler, even if it catered to young sportive men as a 'Macho Bicycle' (Bijker, 1995a: 75) to show their athletic skills, did not persevere and failed in forming a convincing technological frame. Instead it ended up in the 'museum of fascinating inventions', while the so-called 'safety bicycle'—'a low-wheeled bicycle, with rear chain drive, diamond frame, and air tires' (Bijker, 1995a: 93)—persisted.

USING BICYCLES

Ever since the bicycle was invented, its uses have been expanding, illustrating how varied its organizational force has become through providing a technological frame for very different uses of 'the bike'. At the end of the nineteenth century—when the design of the safety bicycle had stabilized—its ecology evolved around the emergence of cycle clubs, the entry of new users, the arrival of small workshops (run by local blacksmiths and mechanics), the beginning of mass production, and the acquiring of many accompanying artefacts—from appropriate biking clothes to bike bells (Bijker, 1995a). When looking back to this history of the safety bicycle, many elements of this version are still part of the design of contemporary bikes. To understand the design of a bicycle today, we need to think technological innovation as unfolding not in a chronological way but through an understanding of time as folded and crumpled (Serres and Latour, 1995). Thinking organization as multi-temporal, a bicycle like any other phenomenon—be it a text, an idea, or an object—is 'made up of an aggregate of solutions, conceptions or problems originating from different historical eras' (Barker, 2012: 17). The bike is thus formed through an overlay of different temporalities, where various components can be dated back to different inventions and time periods; after all the wheel dates back to the Neolithic times, yet the patents for bike bells can be traced as early as 1877, and belt-driven bicycles, first invented in 1890, have only recently started to populate the streets.

The bicycle as stabilized artefact regularly got destabilized as new technological developments, new meanings and uses re-assembled into different configurations of the bicycle. Think of the mountain bike—also called all-terrain or off-road bike—which differs not only considerably in its technological set-up but also in its societal frame. Rosen (1993) suggests that the mountain bike is compatible with a postmodern culture, a pastiche of advanced technology with a nostalgia to escape from urbanized civilization by riding wilderness trails. Later came the BMX, the bicycle motocross, a bike that appeals to the early sportive ambitions and is framed by the subculture of action or extreme sports (Honea, 2014). This subculture—also enacted by skateboarding, surfing, wakeboarding, inline skating, and snowboarding—has 'promoted values largely antithetical to dominant sport forms, including organization by the participants themselves, less emphasis on competition' and a lifestyle component accepting its specific sport subculture (Honea, 2014: 1253).

Economically, bicycles have shaped whole segments of industrial activity, from the traditional small bike shops centred around the crafting skills of repairmen (and less often women), to the fully globalized high tech bicycle industry. One of the most inventive entrepreneurial uses of bicycles is the rise of bike messenger companies, which have drawn intensive attention from sociological and urban studies researchers and seem to be a source of broader cultural fascination (Fincham, 2006; Kidder, 2011). As we remember from Vittorio De Sica's classical movie *The Bike Thief* a bicycle mediates our chances to do one's job. Many decennia later, the bicycle is not just a vehicle to get to work but enacts the job itself. Indeed, since the 1980s, we see emerge entrepreneurial companies that have invented the concept of bike messenger who are paid to deliver time-sensitive items in congested urban areas (Kidder, 2011). They deliver literally everything—from packages to food—at all places—people's home or hotel rooms, companies or parties. Taking issue with the 'delivery lure' (Kidder, 2011), studies have tackled the identification of bikers with a precarious and seemingly trivial and menial job—after all they are just speedier forms of post-(wo)men on bikes—which comes with a vibrant subculture of its own. In a study on the working and social conditions of bike messengers, we find some fascinating micro-stories of couriers who as independent contractors 'identify with an entrepreneurial spirit and are driven to hustle and earn as much as possible' (Pupo and Noack, 2014: 342). This comes with a precarious employment situation based on 'believing that their entrepreneurial status provides them with a degree of freedom that has usually been absent from their work experience' (Pupo and Noack, 2014: 337).

An entrepreneurial imagination thus instigated this 'new' form of 'delivery' companies, recontextualizing the possibilities of the biker as messenger—an activity that is as old as the bike itself. Interestingly, the concept of the biker messenger has travelled to various contexts, where it has become increasingly used in certain countries but not at all in other countries, depending often on the spatial organization of specific cities. In an ethnography, Fincham (2006), by becoming temporarily a biker, documents the quasi-legal trajectories that these bikers ride throughout the city, and how they develop a different urban knowledge, far removed from what the city planners had drawn on their maps. It is this incorporation of play that is often seen to distinguish messengering from most other service jobs, as Kidder (2011: 11) argues: 'it is the emotional involvement required in making deliveries from which the subculture sprouts'. In making deliveries, messengers 'manipulate the space of the city—the play of work and the effervescence of rituals only happen through a unique use of space' (Kidder, 2011: 13). In his cultural analysis, Kidder focuses on the affective appropriation of space and the specific tactics bikers use to get joys and thrills out of altering, even if only momentarily, the dominant order. Referring to Lefebvre's cherishment of the carnivalesque festival for its subversion of power, Kidder believes that 'in the messenger's affective spatial appropriation, individuals briefly regain control of their life through the creative and spontaneous use of the urban environment' (Kidder, 2011: 14).

Biking and Urban Transformation

However, the bike–city association has expanded increasingly, as bikes are finely knitted into the organization of urban life and the design of cities. What the bicycle can do for the formation of urban life and its spatial organization became first apparent at the end of the nineteenth century when shortly after the emergence of the safety bike cyclists flooded the streets of many European cities. Owing to its comparably low production costs, the bicycle became the means of choice for most workers commuting to factories and accounted for up to 85 per cent of trips in many European cities (Shove et al., 2012: 153). The bicycle allowed factory workers and their families to leave the overpopulated inner-city tenements and to move outside the city centres, accounting for a first wave of suburban development (Law, 2014: 42). This first urban ecology of the bicycle dis-integrated with the growing wealth of the working class and the rising popularity of motorbikes and cars during the second half of the twentieth century in Europe; a development that is reiterated in today's Asia, where cities experience a similar transformation of urban mobility (Xu, 2001).

While the paradigm of car-friendliness left little or no space for cycling, today the bike starts to become popular again, revealing unexpected potentialities within an ecology of urban transportation that differs fundamentally from the bicycle's historical environment. Cities today face a complex mélange of challenges that are addressed by promoting the bicycle as a means of transportation. These challenges include the collapse of car-centred urban transportation and the congestion of inner city areas, air and noise pollution, immobility and resulting health-issues of urban societies, pressure to reduce carbon-dioxide emissions, increased competition between cities in providing attractive spaces for living and consumption, and an increasingly segregated and disconnected urban fabric (Gardner, 1998; Heinen et al., 2010).

The bike's role in writing future scripts that tackle these challenges is considered enormous. Whole cities are organizing their mobility policies around the bike, both for getting people to work and to let tourists explore cities. While cars are banned from the inner city and heavily taxed to enter urban zones, bikes are welcomed with open arms. Municipalities in almost all European capitals have introduced public bike sharing systems and launched initiatives for improving the biking infrastructure with regard to safe bike-lanes and parking infrastructure. Some well-known examples are Copenhagen, Utrecht, and Amsterdam that regularly get rated high as bike-friendly cities (for rankings, see www.copenhagenize.eu). Indeed, while Danish and Dutch cities often take the lead in this type of ranking, this goes back to decades of promoting a bike culture and providing an appropriate infrastructure. Also the development of the e-bike, the development of networks of sharing bikes and the integration of social media, makes the bike a serious contender in reshaping mobility practices, reorganizing the ways we move through urban centres.

The bike–car–public transport debate often transcends the more rational arguments with regard to the bicycle's efficiency as a reference is made to the experience of biking as it 'allows people to create a new relationship between their life-space and their life-time, between their territory and the pulse of their being, without destroying their inherited balance' (Illich, 1974: 63). At a time when the body has been systematically dissociated from work and mobility practices and remains passive and immobile in the everyday routines of many, the bicycle becomes attractive as it allows for new forms of movement, different encounters and new sensual experiences in urban space. The car has not only promoted a form of suburban sprawl that performs a much stricter segregation of social classes than the nineteenth-century cities, but also holds commuters captive behind their steering wheel and in traffic congestion. Feelings of isolation and immobility are said to be the source of stress and frustration to millions of commuters every day (Brutus et al., 2017). The car and the transportation industry produce passengers that are being rushed through an 'untouchable landscape' diminishing their 'control over the physical, social and psychic powers that reside in [their] feet' (Illich, 1974: 25). Allowing for a relaxed, healthy, and joyful way of starting and ending a workday, the bicycle takes on a new role in the twenty-first century's knowledge societies. This new role of the bicycle is reflected in how big IT and service companies increasingly provide their employees with bike sharing programmes and other incentives to leave their cars at home.

Despite the obvious benefits of biking and the assertion that 'cities are a kind of natural habitat for bikes' (Gardner, 1998: 17), the transformation towards bike-friendly cities meets a number of resistances and the transition towards a new paradigm of urban transportation is anything but a smooth one. The transformation of urban mobility requires providing a cycling infrastructure, a task which considering the historical substance of many European cities is a political process that redistributes urban space and other scarce resources among traffic participants. Yet, the provision of a working infrastructure might not even be the biggest challenge for the transformation of urban mobility. People's desire to own and regularly use a car may be more durable than the built environment of cities. Indeed, one of the lasting effects of the nineteenth-century success of urban biking remains its association with discourses on affordability, sports, and danger, whereas the car has been loaded with imaginations about family, safety, freedom, and masculinity; imaginations that were heavily promoted by the marketing of the car industry. Yet, new bike–city–assemblages, enhanced by new technologies such as electric assistive engines, connection to the Internet, and bike sharing schemes, might seduce some car owners to reconsider their commuting habits. But there are more affective complexities involved. For example, when it comes to safety issues, the car's massive passenger cell is associated with protection and a safe space, whereas the bike implies the passenger's full exposure to environmental impacts, including weather fluctuations, traffic risks, and noise. Its role as a sports device connects it to a discourse that renders the use of assistive engines problematic. Although the use of electrically assisted bicycles is a reasonable solution to biking in hilly areas, longer distances, or with heavy loads, the e-bike is still stigmatized as being 'crooked', violating the dignity of a proper (sports) biker. The proper biker reaches his (or her?) destination sweaty and without assistive devices and thus runs into a dilemma when this destination happens to be the workplace. It is again Ivan Illich (1974: 74) who at a very early stage made place for a middle way:

> Beyond underequipment and overindustrialization, there is a place for the world of postindustrial effectiveness, where the industrial mode of production complements other autonomous forms of production. There is a place, in other words, for a world of technological maturity. In terms of traffic, it is the world of those who have

tripled the extent of their daily horizon by lifting themselves onto their bicycles. It is just as much the world marked by a variety of subsidiary motors available for the occasions when a bicycle is not enough and when an extra push will limit neither equity nor freedom.

CONCLUSION

Norcliffe (2016: 1) has observed that 'during the past quarter century, there has been a coming of age of the literature on cycles and cycling'. The bicycle is now an almost iconic study artefact that assembles the technological with the social, the organizational with the cultural, and the geographical with the political. The bicycle has fascinated numerous artists, like for instance Olafur Eliasson, whose installation 'Your new bike' from 2009 (Figure 4.1) poetically reflects the potential for reimagining urban space through the bicycle. The bike organizes individual lives from the very young to the oldest; it amuses us in our leisure activity as much as it guarantees our possibility of income as we bike on a daily basis to work. It is part of high-level craftsmanship as much as it belongs now to global production networks. There is no sportive artefact around which so many amateur clubs and recreational participation has been organized. On a professional level, some bikers have become iconic figures while others have betrayed the core values that come with the racing sport mixing up competition, commercialism, and conscience. With increasing awareness of environmental issues, the political support for the bike has been growing steadily (Smethurst, 2015). Considering future mobility politics and policies, there will be an important struggle to reconsider whether the car or the bike will become 'the king of the road', heavily mediated by provisions of public transport. Paraphrasing the Lefebvrian 'right to the city' (Keblowski et al., 2017), the right to access, travel, live in, and feel safe in the city will be granted by a return to the bicycle, by continuously emphasizing the impossibilities of automobility (Böhm et al., 2006) and by underlining the potentials of public transport. Anticipating a true 'pedaling revolution' (Mapes, 2009), the bike thus continues to be inscribed with new possibilities and hope, something one is eager to believe as one zooms in on its joyful and sensuous qualities, as Marc Augé does:

> But most of all the cyclist re-discovers a whole range of sensations—the excitement when sweeping downhill, the sound of the wheels on asphalt, the wind caressing the face, the slow drift of the landscape—which seem to have only waited for this moment to return to us.   (Augé, 2008: 35, our translation)

REFERENCES

Augé, Marc 2008. *Lob des Fahrrads* (M. Bischoff, Übers.) (4. Aufl.). Munich: C. H. Beck.
Barker, Timothy S. 2012. *Time and the Digital: Connecting Technology, Aesthetics, and a Process Philosophy of Time*. Hanover, NH: Dartmouth College Press.

Bijker, Wiebe E. 1995a. *Of Bicycles, Bakelites, and Bulbs: Toward a Theory of Sociotechnical Change*. Cambridge, MA: MIT Press.

Bijker, Wiebe E. 1995b. 'Sociohistorical Technology Studies'. In S. Jasanoff, G. E. Markle, J. C. Peterson, and T. Pinch (eds), *Handbook of Science and Technology Studies*. Thousand Oaks, CA: Sage, 229–57.

Böhm, Steffen, Campbell Jones, Chris Land, and Matthew Paterson. 2006. 'Introduction: Impossibilities of Automobility'. *Sociological Review*, 54, 3–16.

Brutus, Stéphane, Roshan Javadian, and Alexandra Panaccio. 2017. 'Cycling, Car, or Public Transit: A Study of Stress and Mood upon Arrival at Work'. *International Journal of Workplace Health Management*, 10(1), 13–24.

Fincham, Benjamin. 2006. 'Back to the "Old School": Bicycle Messengers, Employment and Ethnography'. *Qualitative Research*, 6(2), 187–205.

Gardner, Gary. 1998. 'When Cities Take Bicycles Seriously'. *World Watch*, 11(5), 16–22.

Harper, Ida H. 1898. *The Life and Work of Susan B. Anthony*. Indianapolis: The Bowen-Merrill Company.

Heinen, Eva, Bert van Wee, and Kees Maat. 2010. 'Commuting by Bicycle: An Overview of the Literature'. *Transport Reviews*, 30(1), 59–96.

Honea, Joy C. 2014. 'Beyond the Alternative vs. Mainstream Dichotomy: Olympic BMX and the Future of Action Sports'. *Journal of Popular Culture*, 46(6), 1253–75.

Illich, Ivan. 1974. *Energy and Equity*. New York: Harper & Row.

Keblowski, Wojciech, David Bassens, and Mathieu Van Criekingen. 2017. 'Re-politicizing Transport with the Right to the City: An Attempt to Mobilise Critical Urban Transport Studies'. *Cosmopolis Working Paper*, 2–33.

Kidder, Jeffrey L. 2011. *Urban Flow: Bike Messengers and the City*. Ithaca: Cornell University Press.

Law, Michael. 2014. *The Experience of Suburban Modernity: How Private Transport Changed Interwar London*. Manchester: Manchester University Press.

Mapes, Jeff. 2009. *Pedaling Revolution: How Cyclists are Changing American Cities*. Corvallis, OR: Oregon State University Press.

Michael, Mike. 2000. *Reconnecting Culture, Technology and Nature*. London: Routledge.

Norcliffe, Glen. 2016. *Critical Geographies of Cycling: History, Political Economy and Culture*. London: Routledge.

Pupo, Norene and Andrea M. Noack. 2014. 'Organizing Local Messengers: Working Conditions and Barriers to Unionization'. *Canadian Journal of Sociology*, 39(3), 331–58.

Rosen, Paul. 1993. 'The Social Construction of Mountain Bikes: Technology and Postmodernity in the Cycle Industry'. *Social Studies of Science*, 23, 479–513.

Serres, Michel and Bruno Latour. 1995. *Conversations on Science, Culture and Time*. Ann Arbor, MI: University of Michigan Press.

Shove, Elizabeth, Mika Pantzar, and Matt Watson. 2012. *The Dynamics of Social Practice: Everyday Life and How It Changes*. London: Sage.

Smethurst, Paul. 2015. *The Bicycle: Towards a Global History*. New York: Palgrave Macmillan.

Xu, Haiqing. 2001. 'Commuting Town Workers: The Case of Qinshan, China'. *Habitat International*, 25(1), 35–47.

CHAPTER 5

....

BITCOIN

....

LUCAS INTRONA AND LARA PECIS

INTRODUCTION: THE CODE OF BITCOIN

WHAT is Bitcoin? Is it a currency (as claimed by its creator), a commodity (as claimed by the US Commodity Futures Trading Commission), a security, an asset, a few thousand lines of software code, etc.? In some senses, it is all of these. Nevertheless, what is it about this particular digital object that makes it relevant—and in some cases disconcerting—to large groups of individuals and institutions? What does it tell us about new organizational forms and the process of their formation, especially with regard to the concept of trust?

As suggested, Bitcoin is ontologically ambiguous. According to its creator, Bitcoin was designed as a cryptocurrency, an alternative to traditional currencies, which were seen as problematic because of their control by third parties. In this sense Bitcoin is often referred to as a digital, decentralized and deregulated, partially anonymous, peer-to-peer networked currency, with low transaction costs (Grinberg, 2011), built on blockchain technology. As an electronic payment system, Bitcoin is designed to enable irreversible transactions recorded in a publicly available ledger, which provides the entire transaction history (Böhme et al., 2015). Bitcoin transactions are validated through proof of work mining.[1] A new block (record of a transaction) is validated and added to a historical chain of transactions (a blockchain), and then distributed to the network of nodes, for verification. Yet, in the simplest sense, Bitcoin is a token, a snippet of digital code representing the ownership of a digital currency (Figure 5.1 shows some of the original software code released by Satoshi Nakamoto—a pseudonym—in January 2009).

Nakamoto defines Bitcoin as 'an electronic payment system based on cryptographic proof instead of trust, allowing any two willing parties to transact directly with each

---

[1] Proof of work is a computational method to ensure that the solution to the computational problem was costly and time-consuming to make, and will function to validate the blockchain when adding the new block to the chain. Miners (individual users) are asked to solve a computational problem, and to then publish a 'block' containing 'proof of work' of the solution and of all transactions carried out since the last problem solution.

```
1    // Copyright (c) 2009-2010 Satoshi Nakamoto
2    // Distributed under the MIT/X11 software license, see the accompanying
3    // file license. txt or http://www.opensource.org/licenses/mit-license.php.
4
5    #include "headers.h"
6    #include "cryptopp/sha.h"
7
8
9
10
11
12   //
13   // Global state
14   //
15
16   CCriticalSection cs_main;
17
18   map<uint256, CTransaction> mapTransactions;
19   CCriticalSection cs_mapTransactions;
20   unsigned int nTransactionsUpdated = 0;
21   map<COutPoint, CInPoint> mapNextTx;
22
23   map<uint256, CBlockIndex*> mapBlockIndex;
24   const uint256 hashGenesisBlock("0x000000000019d6689c085ae165231e934ff763ae46a2a6c172b3f1b60a8ce26f");
25   CBlockIndex* pindexGenesisBlock = NULL;
```

FIGURE 5.1 Screenshot of the first 25 lines of the original 'main.cpp' Bitcoin code

(Available in full at: https://sourceforge.net/p/bitcoin/code/133/tree/trunk/main.cpp#l1613)

other without the need for a trusted third party' (Nakamoto, 2008: 1). The elimination of a trusted third party is one of the central features of Bitcoin. How is the trusted third party eliminated? The mining nodes (or miners) agree the proof of work—the solution of the hashing problem—and then add the new blockchain to their copies of the ledger (BIS, 2017). Although anybody can join the network, and potentially corrupt the ledger, it requires 51 per cent of the mining nodes to reach consensus to validate a transaction—it is currently estimated that the Bitcoin network consists of approximately 9,490 mining nodes.[2] Miners who solve the complex computational problem win the right to add another block to the blockchain and are rewarded 12.5 bitcoins (halved every four years). As such, a bitcoin is essentially a historical chain of verified digital signatures of transacting parties, validated and accumulated over time in each of the nodes of the network.

Although Bitcoin is ontologically ambiguous it is nevertheless clear that central to all its different enactments, and more generally, the Bitcoin project is the idea of trust—that is, of *doing trust differently*. It is this aspect that our chapter investigates.

BITCOIN AND THE DISINTERMEDIATION OF THE TRUSTWORTHY AND UNTRUSTWORTHY

The 2008–9 financial crisis reinvigorated the cryptocurrency community and advocates to find alternatives to current monetary systems. According to them, money needed to be set free from untrustworthy intermediaries: from banks, and from the state. Money

[2] See https://bitnodes.earn.com for current estimates of the size of the Bitcoin network.

generated by banks is seen to be problematic because it is based on the creation of debt, and it is subjected to unacceptable risk-taking behaviour. Money generated by governments is seen as problematic because it is subject to political interference, such as interest rate manipulation, and quantitative easing (Dodd, 2018). In short, according to Bitcoin advocates, banks and governments cannot be trusted as intermediaries and custodians of money, or stores of value. Indeed, Bitcoin is proposed by its enthusiasts also to free users from the 'tyranny of middlemen' such as credit card and money transfer companies (Lyons, 2011). Quite simply put, for the Bitcoin community, human institutions cannot be trusted and need to be replaced by objective and neutral technology (or code). Through code, political or biased intermediaries can be dislocated and replaced by apolitical machines and algorithms—in what Dodd (2016) calls a techno-utopian vision. However, as Dodd (2015) points out, there is in fact a big gap between the intentions behind the Bitcoin ideology and the empirical instantiations of it. This gap can be seen, for example, in two specific instances of the centralization and concentration of power in new intermediaries: the hierarchical organization of the Bitcoin core maintainers, and the concentration of computing power in Bitcoin mining pools.

Bitcoin promotes itself as an open, democratic, decentralized, and collaborative community. However, what we find seems to be something quite different—a centralized, elite, hierarchical group of core developers that monopolize power through the control of 'commit' privileges. Satoshi Nakamoto, the anonymous coders who published the Bitcoin protocol as open source code, distanced themselves from it quite early on. In doing so, they passed the power to commit code, into the official Bitcoin core repository, to Gavin Andresen. Andresen later nominated a small group of developers who control the code and created the Bitcoin Foundation to expand the project further. Andresen subsequently transferred the role of core maintainer (with commit privileges) to Wladimir van der Laan. The charismatic domination of a small group of core maintainers, based on their exceptional technical capabilities, has created an aura that distances them from the larger Bitcoin community. This suggests Bitcoin might encounter what Shaw and Hill (2014) call the 'iron law of oligarchy' (Michels, 1915), which often applies to peer-to-peer systems such as wikis (De Filippi and Loveluck, 2016). Regardless of how democratic Bitcoin was when it started—as an open source and peer-to-peer system— its complex organization seems to be evolving into a sort of oligarchy—not unlike the experts who created the financial instruments behind the 2008–9 crash. However, in an oligarchic system, leaders often develop independent interests in sustaining the organization, which may diverge from the interests of members (Shaw and Hill, 2014). For example, Bitcoin has had four core client forks,[3] one of which, Bitcoin XT, was deemed an event that caused a 'war' within the Bitcoin community. Whilst the wider community can contribute to the mining of Bitcoin and suggest code fixing, the power of deciding Bitcoin's future now resides with a handful of self-selected mostly anonymous experts, ultimately choosing the fixes to implement, and the elements to incorporate in the main

[3] A 'fork' is an alternation of the core software of a digital currency. A fork in effect creates two distinct versions of the blockchain, but with a shared history. If the fork is permanent, it effectively creates a new and different digital currency.

branch of the code. Miners can enact forms of resistance to these top-down decisions by, for example, refusing to implement the new code if they disagree with its modifications (De Filippi and Loveluck, 2016). Nevertheless, the constant balancing of dissent within the community enacts a complex set of power relations, suggesting Bitcoin is not immune to the very performative social and political dimensions it aims to alter—as well as the bureaucratic system it attempts to replace.

The concentration and centralization of power in a small elite group is also matched by the concentration and centralization of power to create coins in a limited number of mining pools. As the coins available to be mined decrease, the cost of mining has increased significantly, due to the increased computational complexity involved. Miners need to find a 'nonce' to embed in the current block, which is a unique 32-bit field whose value is set so that the hash of the block will contain a specified sequence of leading zeros. It is deemed impossible to predict a nonce, and has to be found through trial and error, which requires enormous computing power. More computing power means the nonce or hash can be located quicker—often referred to as the hash rate of the mining pool. Whoever finds the nonce first can then register the block and be rewarded bitcoins for their work. The proof-of-work algorithm is built to create more competitiveness among miners and encourages them to increase their hash rate—creating the need for the concentration of computing power. Accordingly, if you want to mine Bitcoin your only option to do this successfully is to join a mining pool, in addition to spending significant amounts of money to purchase the necessary computing equipment. This means 'that the Bitcoin network is not quite as "distributed" as its advocates claim, indeed one could argue that it demonstrates quite a strong tendency towards the centralization of monetary production by massively favouring those with more processing power' (Dodd, 2018: 46). Although the idea of decentralization is embedded in the Bitcoin architecture (based on distributed ledger technology), the rapidly increasing need for hashing power has meant that only the mining pools with substantial hashing power have any chance of finding the nonce. This has created significant mining pool concentrations: more than 80 per cent of mining pools are located in China, due to its relatively cheaper electricity costs. More specifically, the company Bitmain (located in China and using the ASIC chip technology) operates two of the largest mining pools— BTC.com and Antpool—which together hold over 30 per cent of the total Bitcoin hashing capacity.[4]

This centralization and concentration of powerful figures and technological capacity demonstrates that any attempt to disintermediate 'distorting' third parties, and to replace biased human institutions with neutral technologies is fraught with unexpected outcomes, which run counter to the stated ideology of the Bitcoin project. As such, technologies are understood to be apolitical and more reliable actors; for Bitcoin advocates depoliticization is engrained in the very techno-utopia that characterizes the project. As one of the Bitcoin advocates suggested about the fork 'war': 'What's really wrong is the divergence from science into rhetoric and politics . . . Had we stuck to first

---

[4] For current distribution of hash rates in mining pools see https://blockchain.info/pools.

principles, had we evaluated system design decisions based on scientific needs, we would have a process' (Shin, 2017). Of course, the idea of depoliticization of interactions through some 'code' goes back much further and has been at the heart of social debate and theory for some time. For example, Max Weber suggests that the bureaucratic organization, or code of conduct, creates an ethos of due process and neutrality, enacted through the mutual acceptance of rational conduct (Du Gay, 2000). In other words, the bureaucratic code of conduct removes personal interest, biases, and preferences in pursuit of a depoliticized (and by definition depersonalized) order (Kallinikos, 2004). However, we also know that the ideals of the bureaucratic code can be perverted, for example producing anonymous officials who enact the supposedly neutral code arbitrarily, and without personal responsibility. Similarly, the desire of Bitcoin advocates to eliminate distorting and wasteful third parties and replace them with neutral information and communication technology (ICT) has been on the organizational agenda since at least the early 1990s. In some ways these two organizational forms—ICT platforms and the bureaucratic code—although at seemingly opposite ends of a spectrum, have something important in common. They both claim to disintermediate and depoliticize social interaction in order to create institutional arrangements that can be trusted, that can remove the distortions produced by personal idiosyncratic biases, interest, and values (in the case of the bureaucracy), and self-interested political institutions (in the case of the Bitcoin algorithmic code).

Some might argue that disintermediation—understood as eliminating 'unnecessary' and potentially distorting intermediaries—really expanded dramatically with the advent of the Internet, for example the disintermediation of travel agents (booking.com and Airbnb), auctioneers (eBay), publishers (Wikipedia), taxi companies (Uber), rating agencies (TripAdvisor), lenders (Kickstarter), sponsors (JustGiving), etc. Some have argued that this disintermediation of 'middlemen' through digital infrastructures (or platforms) has led to a new type of economy, a platform economy (Kenney and Zysman, 2016). For some, the disintermediation of the platform economy is a more democratic way of economic exchange—peers interacting directly with each other in a sharing economy (Schor and Attwood-Charles, 2017). Others, however, have argued that it is merely the rapid expansion of the neoliberal ideology in a new form of platform capitalism (Langley and Leyshon, 2017)—that is, two counter-narratives that represent two very different visions of disintermediation (Pasquale, 2016; Murillo et al., 2017). Nonetheless, what both of these lines of argument highlight, central to our argument, is the fact that code is not neutral at all. Indeed, one could argue that algorithmic code has become a way to mask and transform political economy in very significant and implicit ways (McKee, 2017). In the case of Bitcoin, what we see is not the disintermediation of social institutions but the creation of new intermediaries, who are often opaque and inscrutable. Code is never mere code, as pointed out by Wendy Chun (2008), and peer-to-peer always involves a whole host of implicit and sometimes explicit intermediaries (Mackenzie, 2018). It could be argued that this is simply replacing one form of politics with a different, more complex, more embedded form of politics (Introna and Nissenbaum, 2000).

BITCOIN AS A 'TRUST MACHINE'?

The very idea of depoliticization and decentralization through ICTs draws attention to the material nature of the Bitcoin core—a few thousand lines of software code. Code, we have argued above, is never simply code, but always becomes enacted to embody a politics, in some way. The transformation or re-enactment of trust, from familiar social institutions to algorithmic systems, seems to us to be one of the interesting questions that Bitcoin raises for us in the age of the platform economy. For example, the Bitcoin community suggests that with blockchain and proof-of-work trust becomes embedded in the mathematical algorithms themselves, bypassing fallible and inherently political social institutions (Maurer et al., 2013). Yet, what we have seen is that in its actual operations, it seems to insert itself back into the type of institutions it claims to negate—such as the concentration of power (in banks) through, for example, the concentration of power in increasingly larger mining pools. How can we then understand trust in the context of algorithmic architectures (or platforms) as the increasingly new media of political economy and exchange?

Given the above we would argue that trust in Bitcoin is not negated, it is rather transformed in and through a new type of 'trust machine' (The Economist, 2015), which in some respect resembles traditional forms of institutional trust, and in other respects, it suggests new and different forms of trusting practices. In other words, it is both the *same and different* with regard to how trust might function in order to make it operate as an institution. One of the aspects in which it is similar to traditional institutions is the fact that trust is required due to the individual's exposure to risk (Luhmann, 1979, 1988). For example, by trading or investing in Bitcoin, the individual user trusts the fair and secure execution of transactions via the cryptographic algorithm, the honesty of the mining pools, and the potential value of Bitcoin. The risks associated with engaging with the Bitcoin network (be that as a miner, a user, or an investor) are greater than the potential rewards—as such, trust is required for its normal operation. One might suggest that it is through constant affirmative experiences of trust in the socio-material system that system trust is enacted as a condition of possibility for Bitcoin to function in the manner that it does.

Luhmann (1988) highlights that abstract systems (e.g., legal systems, monetary systems, educational systems, etc.) require trust as a necessary condition for their normal functioning. Yet, it is possible that the structural properties of the system (e.g., Bitcoin's architecture) will erode the confidence people have in it, and hence undermine the necessary conditions of trust. Why do we trust traditional abstract systems? In traditional institutions, trust emerges from a complex set of formalized norms—such as constitutions, professional codes of conduct, regulations, legally binding contracts, and so forth—that regulate procedures and the pre-selection of norms that guide conduct. It is exactly this assumed entwinement of law and institutions that for Luhmann (1979) is the foundation of confidence, and trust in abstract systems. Differently, trust in Bitcoin relies on an abstract system of algorithmic code, where the trustworthiness of miners and coders is assumed to depend on their honest conduct, overseen by a small pool of expert coders, expected to act in the interest of the Bitcoin project. The absence of any legal contract that regulates the participation of coders and miners in the Bitcoin network poses important questions of the enactment of responsibility and accountability within this new form of governance. Why do so many individuals trust it in spite of the lack of its entwinement with traditional systems of regulation and law? Is it that the Bitcoin participants actually accept the Bitcoin ideological narrative, and what it implies? Alternatively, is it something else?

We would suggest that trust in the Bitcoin system is not only sustained in confidence in the miners, coders, and the code, but also, and importantly, in the shared expectations of the depoliticization, decentralization, and social change that this new institutional project implies—what Luhmann (1979) refers to as 'trust in trust'. Trust is placed in the trust of other people, rooted in a common political project. Even if Bitcoin actors do not have a common experience, or history of interactions, as in traditional systems, they have a common belief in an ideological narrative that needs to be sustained through a multiplicity of mechanisms—for example, through the Bitcoin foundation, its manifesto, its declaration of independence, and the discussions of the community of coders through various mailing list servers and social media. Bitcoin's manifesto (The Bitcoin Foundation, 2016: 2) suggests a radical disengagement from financial institutions as traditional intermediaries for economic transactions:

> We, the members of the Bitcoin Foundation, believe the following to be true
> - Fiat currency has been a poor long-term store of value, especially since the gold standard was abolished;
> - Inflation encourages consumption and discourages savings and the sustainable use of limited resources;
> - Traditional financial services, especially banking, are not inclusive for the 2.1 billion people that live in poverty (less than $3.10/day);
> - Electronic payment processing times and fees are too high, an important reason why 85 per cent of all commerce globally is still done in cash;
> - The financial services collapse of 2008 and subsequent banking system bailout caused substantial misery for the poorest of the global population, especially in developing economies;
> - Losses associated with card fraud totaled $16.3 billion globally in 2014, with more than half of this fraud occurring online;
> - Traditional banking and payments systems are not secure;
> - Trust in traditional banking and financial services is at an all-time low.

Yet, in trying to escape these institutional forms, they seem to reproduce them (oligarchic hierarchies and concentrations of power) and insert themselves back into that which they claim to displace (financial markets, powerful intermediaries, etc.). For example, more than forty online platforms are now trading bitcoins (e.g., Bitstamp, coinbase, blockchain.info, etc.), as well as Bitcoin based future contracts appearing for the first time in the Chicago Board Options Exchange—deemed the first step towards its legitimacy (Treanor, 2017).

SOME IMPLICATIONS OF TRUST IN THE INSTITUTIONAL ORDER OF CODE

Central to the Bitcoin project is the confidence in the neutrality of algorithmic action and governance, which we have argued develops into two interrelated themes: disintermediation and depoliticization, and the emergence of a new form of trust. We have shown how attempts at getting rid of the untrustworthy 'third party' through neutral code in fact produce a different kind of institutional politics. We suggested there is a significant gap between Bitcoin's ideology and practice. More specifically, Bitcoin's reinsertion into financial markets (e.g., through trading platforms) and the creation of new powerful intermediaries (e.g., mining pools and the elite core developers) sustain the marketization and politicization of Bitcoin, despite the attempt at bypassing these very institutions.

More generally, the platform economy, with its new form of institutional arrangements, poses significant questions about the way trust and trusting practices are enacted. Traditionally, social institutions have a whole host of technologies of trust such as, for example, formal bureaucratic rules of conduct, policies, procedures, regulatory regimes, professional codes of practice, legal contracts, etc.—what Luhmann (1979) calls 'system trust'. However, in the institutional order of code we see trust enacted, at least partially, through algorithmic code and mathematical solutions (e.g., proof-of-work and blockchain)—that is, trust in the trust of others, human and more-than-human. Is this also true for platforms in the platform economy? For example, most people who use Google to search assume that some algorithm has identified relevant content, objectively, and ordered it according to some rules of relevance (with the most relevant at the top and the least relevant at the bottom)—completely oblivious to the politics of search (Introna and Nissenbaum, 2000). Central to this idea of trust is the confidence in the neutrality of algorithmic action and governance, an assumption that might be naïve. Nevertheless, large numbers of people seem to be willing to hand over a large quantity of personal data to entities that they encounter through their screens, exclusively—often for the most trivial of interactions. Something they would not be prepared to do when walking into a shop in the local shopping mall.

The lack of legal obligations that regulate the Bitcoin project—which for Luhmann (1979) is fundamental for system trust to exist—calls for rethinking how trust in the system is understood both in relation to the trust posed in algorithmic systems, but also in others, such as the 'wisdom of the crowd'. Thus, seemingly contradictorily, we do seem to be able to trust abstract systems, even if they are black boxes to us, and their representatives remain elusive. Why? Do we trust them because we are naïve? Alternatively, perhaps, it might be because they prove their trustworthiness through their actual operation, that is, they work mostly as assumed. It might be that we trust the 'wisdom of the crowd', that is, we trust the trust of others, as suggested above. Often these platform institutions have reviewer-based systems, or peer-to-peer interaction forums, where the interacting peers review each other (or the product or service provided). Alternatively, content is filtered—to separate good content from bad content—through some form of voting practice (such as liking). Hence it seems a new form of peer-to-peer governance might be emerging, the governance of the crowd that is, 'trust in trust' itself. It might also be that these dominant platforms present us with the idea that we do not really have an alternative. Luhmann suggests that what sustains trust in systems is the lack of consideration of alternatives: 'you do not know what else to do' (Luhmann, 1988: 98). Is there really an alternative to Google, Facebook, and other platforms?

Moreover, the 'many eyes' principle, implied in the wisdom of the crowd—in reviewer and rating systems—seems never to be to be turned on the socio-material institution that makes the interaction possible in itself. Moreover, the algorithmic codes that enact them are more often than not enclosed in proprietary systems, and subsumed into larger architectures, unavailable for inspection. What are they doing with the transaction data, and other forms of derivative data, such as analytics, that they amass? As Foucault suggested, power is most effective when it hides itself—how will these new relations of power play themselves out? We would suggest that the new institutional order of code—exemplified by the Bitcoin project—is raising profound questions that we have yet to grasp fully, or even start to answer, especially with regard to trust. As our confidence in these opaque platforms begins to subside—as the fake news and Cambridge Analytica debate suggests—these questions will become more and more pressing. Perhaps our trust in algorithmic code, assumed to be neutral, or, our trust in the trust of others, might be misplaced—we have yet to find a satisfactory answer to this question—which seems unthinkable given the approximately $150 billion market of Bitcoin, the 2.2 billion users of Facebook, and the dominance of Google.

REFERENCES

Bank for International Settlements. 2017. *Distributed ledger technology in payment, clearing and settlement.* http://www.bis.org/cpmi/publ/d157.pdf [Accessed 13 March 2018].

Böhme, Rainer, Nicolas Christin, Benjamin G. Edelman, and Tyler Moore. 2015. 'Bitcoin: Economics, Technology, and Governance'. *Journal of Economic Perspectives*, 29(2), 213–38.

Chun, Wendy H. K. 2008. 'On "Sourcery," or Code as Fetish'. *Configurations*, 16(3), 299–324.

De Filippi, Primavera and Benjamin Loveluck. 2016. 'The Invisible Politics of Bitcoin: Governance Crisis of a Decentralized Infrastructure'. *Internet Policy Review*, 5(4). https://papers.ssrn.com/sol3/papers.cfm?abstract_id=2852691 [Accessed 13 March 2018].

Dodd, Nigel. 2015. 'Bitcoin, Utopianism and the Future of Money'. *King's Review Magazine.* http://kingsreview.co.uk/articles/bitcoins-utopianism-and-the-future-of-money/ [Accessed 13 March 2018].

Dodd, Nigel. 2016. '*Vires in Numeris*: Taking Simmel to Mt Gox'. In Thomas Kemple and Olli Pyyhtinen (eds), *The Anthem Companion to Georg Simmel*. London and New York: Anthem Press, 121–40.

Dodd, Nigel. 2018. 'The Social Life of Bitcoin'. *Theory, Culture & Society*, 35(3), 35–56.

Du Gay, Paul. 2000. *In Praise of Bureaucracy: Weber—Organization—Ethics*. London: Thousand Oaks, CA: Sage Publications.

Grinberg, Reuben. 2011. 'Bitcoin: An Innovative Alternative Digital Currency'. *Hastings Science & Technology Law Journal*, 4. https://papers.ssrn.com/sol3/papers.cfm?abstract_id=1817857 [Accessed 13 March 2018].

Introna, Lucas D. and Helen Nissenbaum. 2000. 'Shaping the Web: Why the Politics of Search Engines Matters'. *The Information Society*, 16(3), 169–85.

Kallinikos, Jannis. 2004. 'The Social Foundations of the Bureaucratic Order'. *Organization*, 11(1), 13–36.

Kenney, Martin and John Zysman. 2016. 'The Rise of the Platform Economy'. *Issues in Science and Technology*, 32(3), 61–9.

Langley, Paul and Andrew Leyshon. 2017. 'Platform Capitalism: The Intermediation and Capitalisation of Digital Economic Circulation'. *Finance and Society*, 3(1), 11–31.

Luhmann, Niklas. 1979. *Trust and Power*. Chichester: John Wiley.

Luhmann, Niklas. 1988. 'Familiarity, Confidence, Trust: Problems and Alternatives'. In Diego Gambetta (ed.), *Trust: Making and Breaking Cooperative Relations*, electronic edition. Department of Sociology, University of Oxford, chapter 6.

Lyons, Daniel. 2011. 'The Web's Secret Cash'. *Newsweek*, 19 June. http://www.newsweek.com/virtual-currency-bitcoin-anonymous-web-shopping-67841 [Accessed 13 March 2018].

McKee, Derek. 2017. 'The Platform Economy: Natural, Neutral, Consensual and Efficient?' *Transnational Legal Theory*, 8(4), 455–95.

Mackenzie, Adrian. 2018. '48 Million Configurations and Counting: Platform Numbers and their Capitalization'. *Journal of Cultural Economy*, 11, 36–53.

Maurer, Bill, Taylor C. Nelms, & Lana Swartz. 2013. 'When Perhaps the Real Problem is Money Itself!' The Practical Materiality of Bitcoin'. *Social Semiotics*, 23(2), 261–77.

Michels, Robert. 1915. *Political Parties: A Sociological Study of the Oligarchical Tendencies of Modern Democracy* (trans. E. Paul and C. Paul). New York: Hearst.

Murillo, David, Heloise Buckland, and Esther Val. 2017. 'When the Sharing Economy becomes Neoliberalism on Steroids: Unravelling the Controversies'. *Technological Forecasting and Social Change*, 125, 66–76.

Nakamoto, Satoshi. 2008. 'Bitcoin: A Peer-to-Peer Electronic Cash System'. http://www.cryptovest.co.uk/resources/Bitcoin%20paper%20Original.pdf [Accessed 13 March 2018].

Pasquale, Frank. 2016. 'Two Narratives of Platform Capitalism'. *Yale Law & Policy Review*, 35, 309–20.

Schor, Juliet B. and William Attwood-Charles. 2017. 'The "Sharing" Economy: Labor, Inequality, and Social Connection on For-Profit Platforms'. *Sociology Compass*, 11(8), e12493.

Shaw, Aaron and Benjamin M. Hill. 2014. 'Laboratories of Oligarchy? How the Iron Law Extends to Peer Production'. *Journal of Communication*, 64(2), 215–38.

Shin, Laura. 2017. 'Why Bitcoin's Greatest Asset Could Also Spell Its Doom'. *Forbes* [online]. https://www.forbes.com/sites/laurashin/2017/04/20/why-bitcoins-greatest-asset-could-also-spell-its-doom/#47fd7caa6adc [Accessed 13 March 2018].

The Bitcoin Foundation. 2016. 'The Bitcoin Foundation Manifesto'. https://bitcoinfoundation.org/wp-content/uploads/2017/03/Bitcoin_Foundation_Manifesto.pdf [Accessed 13 March 2018].

The Economist. 2015. 'The Promise of the Blockchain: The Trust Machine'. https://www.economist.com/news/leaders/21677198-technology-behind-bitcoin-could-transform-how-economy-works-trust-machine [Accessed 13 March 2018].

Treanor, Jill. 2017. 'Bitcoin Bubble Warnings Issued as Futures Trading Opens in Chicago'. https://www.theguardian.com/technology/2017/dec/11/bitcoin-bubble-warnings-futures-contract-trading-chicago [Accessed 13 March 2018].

CHAPTER 6

....

CALENDAR

....

FLORIAN HOOF
TRANSLATED BY ERIK BORN

INTRODUCTION: CALENDARS
IN ORGANIZATIONS

WHAT is a calendar? According to the *Oxford English Dictionary*, a calendar is 'a table showing the division of a given year into its months and days, and referring the days of each month to the days of the week' ('Calendar', 2018: §2). Thus, a calendar involves a standardization of time into clearly defined units, represented in a particular graphic form. This visualization of temporality facilitates orientation in time, and defines what time actually is. If calendrical time was commonly derived in antiquity from astronomical observations of lunar phases, the principle of the lunar calendar changed over the centuries (Parker, 1950). With the introduction of the clock, an instrument necessary to create a standardized concept of time (Thompson, 1967), time was transformed from a physical constant, which was to be determined through observation and counting (Macho, 2003), into an accurate frame of reference, which depended on relations among observers. Within these coordinates, the calendar should be located as a medium of organization. It structures the perception of time in organizations and is foundational for both systematic management (Litterer, 1961) and logistics chains that presuppose a mastery of space and time (Dommann, 2011: 75). It is hardly surprising that calendars are so commonly found in organizations, in various forms and formats, and in the service of the most diverse goals (Figure 6.1).

One common organizational genre, the large-format wall calendar, records essential events, such as meetings, vacation times, project timelines, and employees' birthdays, and makes them visible to everyone. As a medium of planning and visualization, the calendar here resembles other organizational planning techniques, such as 'complex sheets' (Suchman and Trigg, 1991) or 'Gantt-charts' (Hoof, 2015a: 110–29; Hoof, 2020). The systematic wall calendar exhibits a universally-valid classification scheme, such as

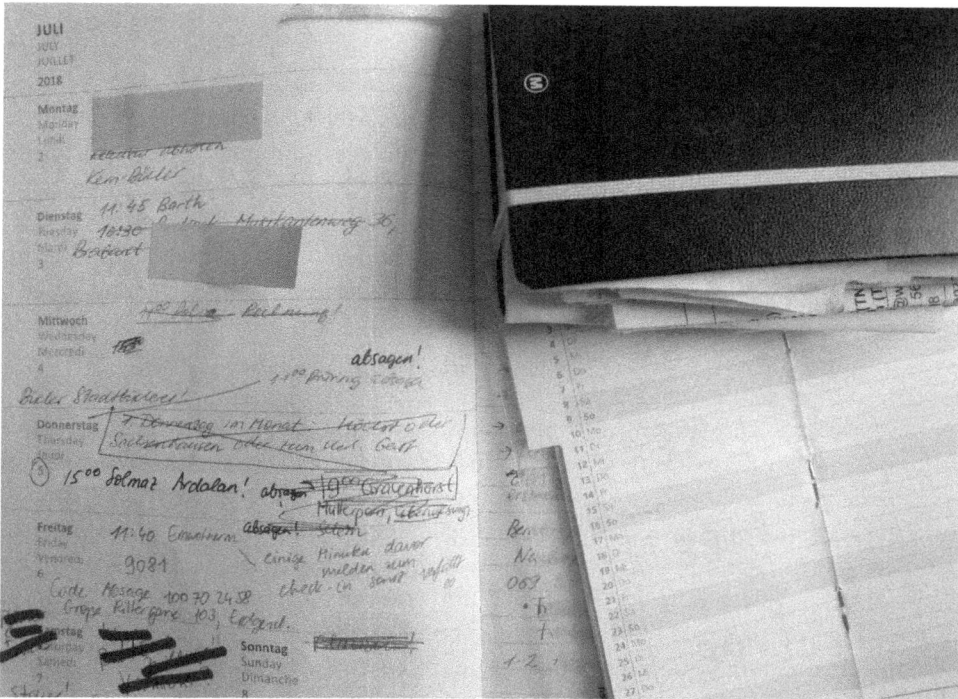

FIGURE 6.1 Calendars

(Photo by the author)

the consistent numeration of calendar weeks, which makes it possible to coordinate activities on a standardized time axis.

A calendar can also be a medium of communication and storage for an organization's functional relationships—for instance, the administrative assistant's calendar, which is kept either for a director or an executive board. Here, the calendar provides an interface between management and the background noise of organizational operation. Only what exists in this calendar will be perceived by management and thus have any relevance for their organization. The administrative calendar represents a bottleneck that makes communication possible by hiding a large number of equally possible communication acts (Luhmann, 2011; Hoof, 2015b).

Illustrated wall calendars, once again, provide one of the few acceptable opportunities to individualize the workplace. In monthly or bi-weekly cycles, these visual media present a stock of common motifs, whether technical, natural, religious, or erotic, for the beautification of standardized office lockers, lounges, or cubicles. Within any organization, the opportunity to express one's personality will also lead to foreseeable conflicts where private tastes collide with various internal and external rules, such as when calendar images are perceived to be a violation of common principles of equality.

The most widespread form of organizational calendar, at once the most discrete and the most efficacious, may be the personal calendar, organizer, or planner, usually in pocket format. In recent years, the paper-based version of the pocket calendar has been complemented by digital equivalents like PDAs, electronic calendar apps, or applications intended to facilitate mobile work (Gregg, 2015). This calendar format can be taken along in mobile situations (Latour, 1990) and primarily supports everyday situations related to personal and professional life. It contains appointments, to-do lists, project details, handwritten notes, addresses, meeting places, the telephone numbers of restaurants suitable for business meetings, and potentially even personal shopping lists. During repetitive meetings, which are indispensable for communication and manage- ment, the pocket calendar can also become a receptacle for drawings, doodles, or even aphorisms and short poems (Rommel, 1993). As the central interface between the organization and the individual keeping the record, the pocket calendar connects organizational goals and constraints to personal and private life.

Since this calendar format calls attention to the relations among media, individuals, and social institutions (Weick, 1987), which is of particular interest to media and organization studies, I will concentrate in the following on the organizational potential and the relevance of the personal time management system. My thesis is that there are two sides to this system; it is located between two registers. On the one hand, a time management system is a 'technical medium' (Kittler, 1999) of standardization and modularization: the calendar analyses temporal contingency into discrete elements and thus makes them calculable and relationally comparable. As a medium of cultural synchronization, on the other hand, a time management system is also a personal object: the calendar provides an interface between the demands of an organization and those of its employees. Like other written communication media, such as the employee newsletter (Yates, 1989), a time management system mediates between the organization and the individual. The calendar is part of a continuous process of negotiation, which takes effect whenever an organization's managerial logic encounters its individual subjects.

THE *TEMPUS* TIME MANAGEMENT SYSTEM

The main focus of my following reflections will be on one unique calendar, nonetheless common in German-speaking countries, the Tempus Time Management System (*tempus.®-Zeitplansystem*) (Figure 6.2). The calendar is produced and sold by the Tempus GmbH, a small business located in Giengen and managed according to Pietist principles. Under the slogan '*Aufbruch zur Gelassenheit*' (roughly, 'Getting Started with Mindfulness'), the company has sold time management systems in German-speaking countries since 1987, and was the third-largest supplier in the German-speaking market with up to 150,000 users during the heyday of paper-based time management systems. Its main target customers are found in medium-sized German businesses partly with Christian fundamentalist backgrounds.

The time management system itself consists of a plastic or leather-bound loose-leaf binder, either in A5 or pocket format, with an index and corresponding calendars,

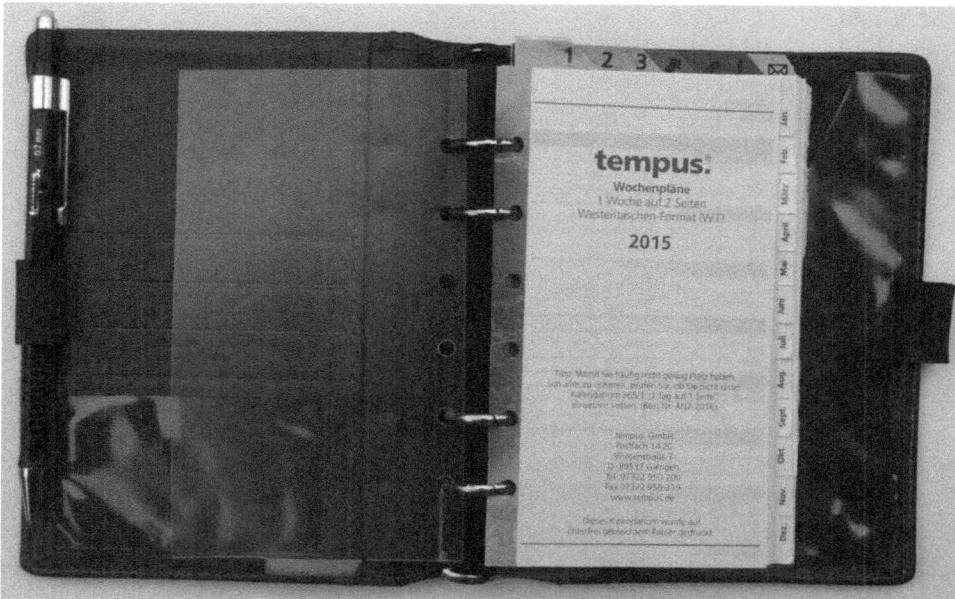

FIGURE 6.2 The *Tempus* Time Management System

(Photo by the author)

schedules, and info sheets. These can be put together modularly and filed into the time management book. The central component is an event calendar, which is available in different formats (e.g., representing one week on two pages). It can be supplemented with multi-year calendars and fold-out calendars covering multiple months, which serve in planning the steps in a project. In addition, there is an organizational section, separated from the event-planning section, for contents including memos, notepaper, an address book, a birthday calendar, meeting plans, to-do lists, and other forms. The planner also contains info sheets and other forms related to exercise and relaxation, daily energy requirements, Christian counselling, travel checklists, weight control registers, maps, grooming and self-care tips, pollen count calendars, business and travel expenses, refrigerator contents, a catalogue of official fines, miniature maps, time zones, and time-saving tips. This basic packet can be upgraded through further add-on modules, containing computer paper for printing from Microsoft Outlook, key task lists, forms for full-year target planning, or daily inspirational biblical verses drawn from the eighteenth-century Herrnhuter collection, a compilation of 1,800 verses from the Christian Bible (Herrnhuter, 2018). The company also offers the Tempus Master Plan, which is supposed to help keep 'your motto, your goals, and your achievement of them' (tempus, 2018; 'ihr Lebensmotto, ihre Ziele sowie ihre Zielerreichung') and thereby 'your whole life in focus' (tempus, 2018; 'ihr Leben übersichtlich im Blick'), namely, on a three-page fold-out. Hence, the Tempus System is marketed not only as a pure calendar, but also as what would today be called a general 'lifestyle' guide but which has deeper roots in the Protestant tradition, exemplified by Ralph Waldo Emerson's

The Conduct of Life (1860). Since 2011, an identical system has also been available as an iPad app under the name *iTempus*, providing a detailed, skeuomorphic conversion of the analogue Tempus Time Management System for the digital interface of the iOS operating system. The very division of the Tempus Time Management System into one section for the event calendar and another for time management allows us to pursue the question of its potential function within organizations, as well as the competing logic of technical media and that of cultural synchronization.

The Calendar as a Technical Medium

At first glance, the Tempus planner offers everything expected of a time management system. It provides elaborate forms that make it possible to note down events and other contents according to a given structure and then to process them according to the calendar. Pre-printed form fields transform handwritten entries into clear discrete units that are arranged on a timeline (Bowker and Star, 2000). Dates or timeframes can be determined and communicated unambiguously, which provides the foundation for the hierarchy of any organization's comprehensively coordinated activity. As a technical medium, the time management system provides fixed time slots, which serve as the terrain for planning and coordination. It facilitates management practices, such as the determination of milestones or the assessment of project timelines. Exceeding these deadlines, which are often placed arbitrarily in the future, is perceived within an organization as a problem or breakdown. This perception of failure, in turn, legitimizes interventions in established organizational structures, for example with the argument of insufficient efficiency. However, what first makes these actions possible, what makes them appear to be legitimate in the first place is the calendar itself, since the technical medium makes the conception of time linear and segmented. In any organization, this linear conception of time will almost inevitably collide with its 'sociomaterial practices' (Orlikowski, 2007), which often take place in a less linear manner. With the personal time management system, non-linear developments occurring at a higher level are ultimately fixed and attributable to individuals. One reason for this is that the calendar, as a technical medium, is not itself part of these socio-material practices. To some extent, the calendar exists outside of the organization and, for this very reason, is able to have an effect on it. As an epistemic system, time is essential for organizations and thus neither susceptible to change, nor up for negotiation. Time, materialized in the calendar as a medium of organization, provides an unquestioned framework for any activity in an organization.

As a technical medium that anchors standardized organizational forms and, by extension, an organization's standardized conceptions of time, the Tempus planner is not limited to the paper-based time management system. It can also be found in digital skeuomorphic reproduction in the *iTempus* app or other electronic devices, such as PDAs. In this electronic form the calendar serves as a 'new' medium that represents time and provides users with options for coordinating their everyday life. At first glance, the calendar in its digital form appears merely to confirm the process of 'remediation' (Bolter and Grusin, 2000) according to which the new electronic format would imitate the familiar form and aesthetic of the paper calendar. However, while the logic of remediation emphasizes the representation of time, there is another non-representational dimension to the calendar, which can again be accounted for in its status as a technical medium.

The principle of the calendar, its segmentation of time into discrete units, connects directly here to the logic of digital data processing, which also works with discrete units. This process does not relate to the skeuomorphic representation of time, and thus does not exploit 'the shock of the new'. Rather, 'the shock of the old' (Edgerton, 2006) is evident in the calendar's abstract segmentation of time into discrete units, a basal technical structure that can be easily combined with the architecture of computational devices.

An exemplar of these affinities between calendar and computation is the Austrian computer scientist Heinz Zemanek, the creator of *Mailüfterl*, one of the first fully-transistorized computers on the European mainland. In addition to his work as a computer scientist, Zemanek's main occupation, probably not coincidentally, is with calendar systems and their internal logic (Zemanek, 1987). The structure of computer logic is homologous with that of the calendar; they can both be processed and simulated digitally, as described in the patent for an 'Electronic Calendar and Diary' (Levine, 1979) (Figure 6.3). One of the first to describe not only the basic structures of a calendar but also the key features of social calendaring, the aim of this patent was nothing less than the synchronization of human behaviour. The patent, filed in 1975 but awarded only in 1979, was conceived of as an aide for 'business persons, doctors, dentists, service persons and others to maintain a personal appointments or schedule calendar on a daily or weekly basis in order to properly respond to a schedule of such appointments and events' (Levine, 1979: §1).

Segmented time systems combined with digital media make it possible to synchronize machines and logistics chains without any skeuomorphic representation of temporality. Hence, the calendar is also a programmable standard of process management and control. As a technical medium, the calendar performs essential services of standardization and organization. Exemplarily, the ISO Norm 8601:2004, which governs 'Data elements and interchange formats—Information interchange—Representation of dates and times', contains the following definition of the universal standard for an organization's international activity: 'ISO 8601:2004 is applicable whenever representation of dates in the Gregorian calendar, times in the 24-hour timekeeping system, time intervals and recurring time intervals or of the formats of these representations are included in information interchange. It includes calendar dates expressed in terms of calendar year, calendar month and calendar day of the month' (ISO, 2004). The ISO Norm helps synchronize any type of event, understood as an unambiguously defined point in time, across a large number of media and technologies—both within and between organizations. The international standard prevents 'misinterpretation of the significance of the numerals' and thus forestalls 'confusion and other consequential errors or losses' (ISO, 2004).

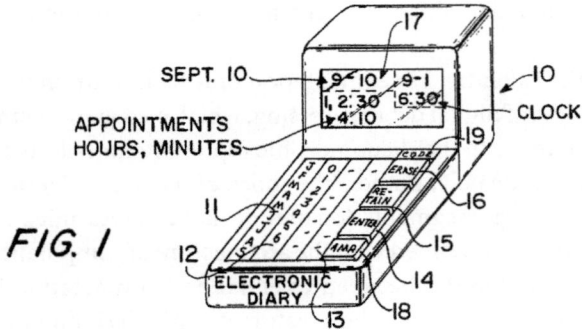

FIG. 1

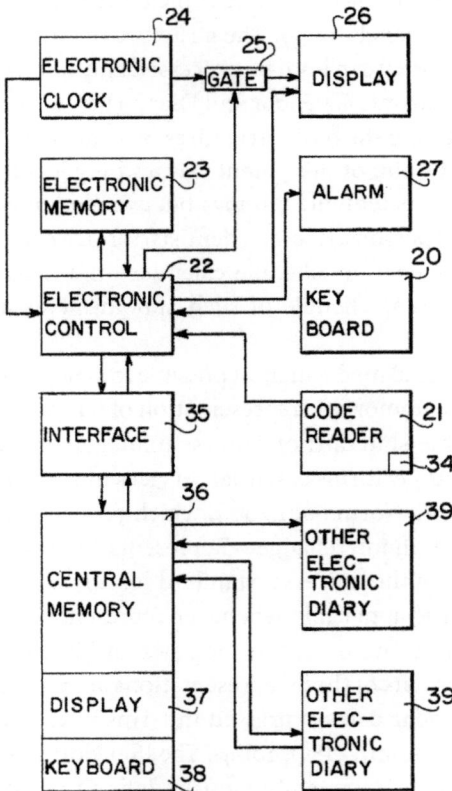

FIG. 2

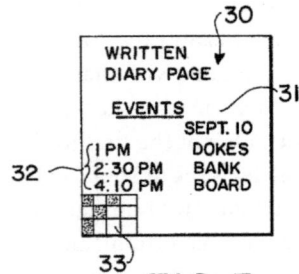

FIG. 3

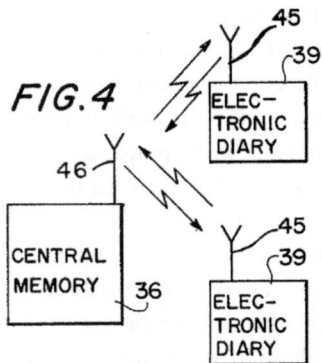

FIG. 4

FIGURE 6.3 Electronic Calendar and Diary, 1979

(*Source*: Alfred B. Levine, 1979. Electronic Calendar and Diary. US Patent 4,162,610, issued 31 July 1979)

In this broad technical definition, the concept of the calendar is not limited to formats tailored to human activity, such as the Tempus Time Management System. It also includes an organization's mechanical and algorithmic activity, which is based on digital data processing. The ISO Norm is compatible with the Internet Transmission Control Protocol (TCP), which also allows for the synchronization of economic processes across time zones and among different countries. In terms of media theory, the calendar is a medium of standardization and contributes significantly to the synchronization and interconnection of an organizational 'discourse network' (Kittler, 1990) consisting of various media technologies, such as Microsoft Outlook, Excel spreadsheets, or production control software. This technical level sets the standards for an organization's activity. Still, these standards do not make the Tempus Time Management System obsolete, whether in paper or digital form. By determining one's perception of time, the personal calendar remains a productive part of an organization's multimedia system; it couples the mechanical synchronization of time to the perception of time by an organization's subjects. These kinds of elementary determinations, which are made by technical media, only become an issue when malfunctions or innovations call established organizational structures into question and suggest modifications (Jackson, 2014).

The Calendar as a Medium of Cultural Synchronization

The Tempus Time Management System is not limited to the function of the calendar as a technical medium of standardization and temporal synchronization. As a medium of cultural synchronization, it functions in a broader sense including not only the synchronization of time but also that of spatial and cultural processes. In the case of the Tempus system, there is counselling advice and lifestyle tips on a wide range of work-related topics: for instance, the secular formula for daily energy requirements, along with a list of the average caloric contents of common foods, is intended to guide over-worked middle management in their daily dining decisions at the hotel buffet or at the evening get-together. However, the time management system does not stop there; it also provides users with optional support for leading a pious life, manifest in info sheets about adopting Christian principles in business life and in add-ons like the inspirational Herrnhuter slogans. Furthermore, the time management system is linked to other branches of the Tempus company and to the philosophy of its founder Jürgen Knoblauch. For instance, the special-interest groups *Tempus-Consulting* and *Tempus-Akademie* offer consulting services in the field of sales ('Highway to sell'), office management ('Büro-Kaizen'), personal management (Knoblauch, 2010) and business management. Additionally, the Tempus company is the co-organizer of the bi-yearly Congress for Christian Leadership (*Kongress christlicher Führungskräfte*). The event is organized together with the Evangelical news agency *idea*, a subsidiary of the media company *ERF Medien*, which serves as a hinge for the so-called New Right. Knoblauch's motto, which could also be used to market his time management system, is 'work hard—pray hard' (Knoblauch, 2018). Perhaps evoking the classic Christian monastic principle of *ora et labora* (work and prayer), the slogan is a direct pun on the phrase 'work hard—play hard', which is commonly used to characterize the employee culture of consulting firms like McKinsey & Company.

From this perspective, the Tempus planner can be understood not only as a calendar for documenting 'hard work' in the form of a rational time management system, but also as a medium for fostering a personal way of life consisting of 'hard prayer'. It is embedded in an ecosystem of fundamentalist institutions and supplies practices for self-management, even for giving meaning to one's own life. The connection here between economic rationality and a Protestant ethos is not a coincidence: the two are connected components that make clear the relevance of calendars for organizations. For this reason, it would not be enough to treat the calendar in isolation as a merely functional medium for coordinating an organization's activities, since this system is also permeated by cultural values and norms. The calendar performs the work of cultural synchronization required to connect these often-disparate fields—a personal way of life and the professional demands of an organization. The calendar's form of organizational performance cannot be measured in rational benchmarks, but rather can be classified as part of the field of 'latent pattern maintenance' (Parsons and Smelser, 1956), and thus secures the stability of organizations and the economic systems built on them. Beyond the Tempus Time Management System, the medium of the calendar points to fundamental considerations regarding the connection of capitalist economic orders to religion and culture, as well as to their media of organization and bureaucracy (Vismann, 2000). As a medium of organization, the calendar is located between these coordinates.

The emphasis on Christian piety in the Tempus system can be formulated in more abstract terms for the medium of the calendar, in general, as a central interface between the individual and the organization as a social institution. The calendar stages an encounter between organizational rationality and a personal way of life, and attempts to bring the two into balance. According to Max Weber, the individualistic capitalistic economy is based on carefully planned rational conduct that is directed towards economic success (Weber, 2001: 37). In Weber's argument, this form of economic management and the related economic way of life is 'born [...] from the spirit of Christian asceticism' (Weber, 2001: 123; 'geboren aus dem Geist der christlichen Askese'), and is thus the result of secularization and the translation of these kinds of conceptions into a 'rational bourgeois economic life' (Weber, 2001: 117; 'bürgerlich, ökonomisch rationale Lebensführung'). This idea of a rational life does not consist of single, unrelated actions but encompasses an entire pattern of life, thereby 'producing in the individual the most completely alert, voluntary, and anti-instinctual control over his own physical and psychological processes' (Weber, 1978: 539). To create this form of flexible control, time management systems like Tempus appropriate resources for aligning one's own life with a model Christian life in which professional success 'is regarded as the manifestation of God's blessing upon the labor of the pious man and of God's pleasure with his economic pattern of life' (Weber, 1978: 543; 'die Gottgefälligkeit seiner ökonomischen Lebensführung sichtbar macht'). It makes possible 'man's vocation to participate rationally and soberly in the various rational organizations of the world and in their objective goals as set by God's creation' (Weber, 1978: 543; 'die rationale nüchterne Mitarbeit an den durch Gottes Schöpfung gesetzten sachlichen Zwecken der rationalen Zweckverbände der Welt').

In this respect, the Tempus system is part of the 'steel frame of modern industrial work' (Weber, 1978: 1401–2; 'stahlharten Gehäuses'), which the economic order of capitalism transformed into through its incorporation of an ascetic way of life. In other words, the calendar's true organizational performance consists in synchronizing abstract economic rationality and an individual's concrete religious practices; and these, following Weber's argument, cannot be limited in a strict sense to principles of Christian conduct. Admittedly, they may have started in Christianity, but they have long since been transformed into mundane, secular practices (Bröckling et al., 2000; Opitz, 2004; Boltanski and Chiapello, 2005).

THE CALENDAR AS A MEDIUM
OF ORGANIZATION

In organizations, calendars like the Tempus planner represent a persistent media anachronism. Despite digitalization, calendars are based on various materials, such as paper, blackboards, magnet boards, hardware, and software. This mixed media system of various forms of calendars is indispensable for an organization's goal-oriented activity. The medium of the calendar structures the perception of time in an organization and is essential for its logistics. In general, the perception of time tends to move between two poles: 'Time is experienced in two fundamental ways. It seems to flow—the passing seconds, days and years—very much like an endless stream of river'; or it is 'perceived as a succession of moments with a clear distinction between past, present and future' (Shallis, 1983: 14). The calendar can be located between these poles: it represents an extract from the infinite flow of time, and segments this extract into small units, which enable an organization to locate itself in the past, present, and future.

Furthermore, the calendar can be conceived of as a storage medium for tracking an organization's activity—for making it comprehensible, and, in many cases, allegedly transparent (Zerubavel, 2003: 30). The events recorded in a calendar refer not only to activities to be completed in the future, but also to those already completed in the past. Not by chance, the etymology of the word 'calendar' is derived from the Latin term *Kalendarium*, which designates a register of debts to be collected. In this light, the calendar can be understood, in analogy to the 'account book', as an organization's 'time book'. However, unlike account-keeping, which is usually an organization's central focus, keeping a calendar provides both a medium of centralization, which can be used to manage and coordinate essential action contexts, and a decentralized medium, which can be found in various forms among members of an organization. The productivity of calendars is evident in this very overlap between centralization and decentralization.

The medium of the calendar is positioned between precisely-described processes and the cultural framework of this conception of rationality. By offering an essential framework for these kinds of socio-material processes between clearly-defined temporality and its conversion into an organization's concrete activities, the calendar stabilizes the organization's structure at a fundamental position. In this respect, the calendar organizes the 'permanent tension between the formal and the empirical, the local and the situated, and attempts to represent information across localities' (Bowker and Star, 2000: 291–2).

Once again, the calendar refers back to the two different registers of technical media and those of cultural synchronization. On the one hand, the calendar is a media technology of standardization and modularization. It analyses temporal contingency into discrete elements and thus makes them calculable and relationally comparable. This register is not part of organizational negotiations, but rather a determining implementation of a culturally-traditional understanding of time in the form of the medium of the calendar. On the other hand, calendars are personal objects, or media of cultural synchronization, which are loaded with affects. They are based on the unalterable standard of time as a relational system, but they connect this standard flexibly with an individual's own way of life. Practical supports that attempt to reconcile private and professional dates, whether in the form of to-do lists or weekly calendars, create a surplus value for the individual, albeit a value that always remains derived from the register of the calendar as a technical medium. Nevertheless, this oscillation between the registers of the technical and the cultural—as the Protestant foundations of the Tempus system have shown—should not be confused with an open form of consensus building. Rather, the calendar enforces an organization's linear conceptions, which are then personalized, thereby challenging the calendar's organizational function with all the associated consequences for the affected individuals (Ehrenberg, 1998). This makes the calendar into an interface between an organization's demands and the lived reality of its employees.

Calendars are part of a flexible control system, which does not attempt to standardize all organizational knowledge and related operations with the rigour of Taylorism. Instead, the calendar creates 'loose couplings' (Weick, 1976) among forms of personal and implicit knowledge that can be applied in various situations. This form of flexible control works because the calendar is located between very different understandings of media. For employees, the calendar offers a possibility to orient themselves within the demands of the organization, in the sense of cultural synchronization, and to create a free space for themselves. Potentially, the calendar may show up again, in the case of the Tempus system, in their own search for piety, and thus lead to an overlap of personal and organizational goals in Weber's sense. For management, on the other hand, the calendar as a technical medium for determining temporality is a management tool for achieving an organization's objectives. It concedes to employees a very limited room to manoeuvre within an organizational 'zone of indifference' (Barnard, 1938: 167–9) where employees are expected to carry out instructions. Precisely because of their close connection to private life and their fundamental organization of things, calendars are contested objects and part of what Edwards terms 'contested terrains' (1979). The calendar is the scene of a latent, long-standing conflict between employees and economic organizations that repeatedly leads to 'offences against common decency' (Baecker, 2014) whenever the balance starts to tip between private life and professional objectives.

From this perspective, calendars are part of a media and cultural history of rationalization (Rabinbach, 1992; Sarasin, 2003), which is closely connected to Calvinist ideals. The Tempus Time Management System ultimately belongs to a longer history of media of organization. In a manner similar to the Tempus founder, the two efficiency experts Lillian and Frank Gilbreth became role models of the Calvinist 'efficiency craze' of the 1910s and 1920s (Hoof, 2020). In that case, the media of organization were also characterized by the coupling of private and professional life. The filmic time and motion studies used to rationalize industrial operations in compliance with scientific management were also able to be used for the efficient design of private life in the form of 'cinema for the home' (Sammond, 2006).

The calendar reveals its true significance among these media of organization, which I would term 'media of Protestantism', as soon as one takes it home after work, thereby assuring the synchronization of the personal and the cultural according to the logic of technical media.

REFERENCES

Baecker, Dirk. 2014. 'Die Verletzung der guten Sitten'. In Nina Möntmann (ed.), *Schöne neue Arbeit. Ein Reader zu Harun Farockis Film 'Ein neues Produkt'*. Köln: Walther König, 55–66.

Barnard, Chester I. 1938. *The Functions of the Executive*. Cambridge, MA: Harvard University Press.

Boltanski, Luc and Ève Chiapello. 2005. *The New Spirit of Capitalism*. New York: Verso.

Bolter, Jay D. and Richard A. Grusin. 2000. *Remediation: Understanding New Media*. Cambridge, MA: MIT Press.

Bowker, Geoffrey and Susan L. Star. 2000. *Sorting Things Out: Classification and its Consequences*. Cambridge, MA: MIT Press.

Bröckling, Ulrich, Susanne Krasmann, and Thomas Lemke. 2000. *Gouvernementalität der Gegenwart*. Frankfurt am Main: Suhrkamp.

'Calendar, n.'. 2018. *OED Online*. Oxford: Oxford University Press. http://www.oed.com/view/Entry/26301?rskey=Mszi2K [Accessed 29 June 2018].

Dommann, Monika. 2011. 'Handling, Flowcharts, Logistik: Zur Wissensgeschichte und Materialkultur von Warenflüssen'. *Nach Feierabend. Züricher Jahrbuch für Wissensgeschichte*, 7, 75–103.

Edgerton, David. 2006. *The Shock of the Old: Technology and Global History since 1900*. London: Profile Books.

Edwards, Richard. 1979. *Contested Terrain: The Transformation of the Workplace in the Twentieth Century*. New York: Basic Books.

Ehrenberg, Alain. 1998. *The Fatigue of Being Oneself: Depression and Society*. Paris: Odile Jacob.

Emerson, Ralph Waldo. 1860. *The Conduct of Life*. New York: A. L. Burt.

Gregg, Melissa. 2015. 'Getting Things Done: Productivity, Self-Management, and the Order of Things'. In Ken Hillis, Susanna Paasonen, and Michael Petit (eds), *Networked Affect*. Cambridge, MA: MIT Press, 187–202.

Herrnhuter. 2018. https://www.losungen.de/.

Hoof, Florian. 2015a. *Engel der Effizienz: Eine Mediengeschichte der Unternehmensberatung.* Konstanz: Konstanz University Press.

Hoof, Florian. 2015b. 'Medien managerialer Entscheidungen: Decision-Making "At a Glance"'. *Soziale Systeme. Zeitschrift für soziologische Theorie*, 20(1), 23–51.

Hoof, Florian. 2020. *Angels of Efficiency: A Media History of Consulting.* New York: Oxford University Press.

ISO (International Standardization Organization). 2004. *ISO Norm 8601:2004. Data elements and interchange formats. Information interchange. Representation of dates and times.* https://www.iso.org/obp/ui/#iso:std:iso:8601:ed-3:v1:en.

Jackson, Steve. 2014. 'Rethinking Repair'. In Tarleton Gillespie, Pablo Boczkowski, and Kirsten Foot (eds), *Media Technologies: Essays on Communication, Materiality, and Society.* Cambridge, MA: MIT Press, 221–39.

Kittler, Friedrich A. 1990. *Discourse Networks 1800/1900.* Stanford, CA: Stanford University Press.

Kittler, Friedrich A. 1999. *Gramophone, Film, Typewriter.* Stanford, CA: Stanford University Press.

Knoblauch, Jürgen. 2010. *Die Personalfalle: Schwaches Personalmanagement ruiniert Unternehmen.* Frankfurt: Campus.

Knoblauch, Jürgen. 2018. 'Prof. Dr Jürgen Knoblauch. Der Personalvordenker für den Mittelstand'. Pressemappe. http://www.joerg-knoblauch.de/fileadmin/content/Speaker_Berater/Pressemappe/Referentenmappe.pdf.

Latour, Bruno. 1990. 'Drawing Things Together'. In Michael Lynch and Steve Woolgar (eds), *Representation in Scientific Practice.* Cambridge, MA: MIT Press, 19–68.

Levine, Alfred B. 1979. Electronic Calendar and Diary. US Patent No. 4,162,610, eingetragen am 31.07.1979.

Litterer, Joseph A. 1961. 'Systematic Management: The Search for Order and Integration'. *Business History Review*, 35(4), 461–76.

Luhmann, Niklas. 2011. *Organisation und Entscheidung.* Wiesbaden: VS Verlag.

Macho, Thomas. 2003. 'Zeit und Zahl: Kalender und Zeitrechnung als Kulturtechniken'. In Sybille Krämer and Horst Bredekamp (eds), *Bild, Schrift, Zahl.* Paderborn: Fink Verlag, 179–92.

Opitz, Sven. 2004. *Gouvernementalität im Postfordismus: Macht, Wissen und Techniken des Selbst im Feld unternehmerischer Rationalität.* Hamburg: Argument.

Orlikowski, Wanda. 2007. 'Sociomaterial Practices: Exploring Technology at Work'. *Organization Studies*, 28(9), 1435–48.

Parker, Richard A. 1950. *The Calendars of Ancient Egypt.* Chicago: University of Chicago Press.

Parsons, Talcott and Niel J. Smelser. 1956. *Economy and Society: A Study in the Integration of Economic and Social Theory.* London: Routledge & Kegan Paul.

Rabinbach, Anson. 1992. *The Human Motor: Energy, Fatigue, and the Origins of Modernity.* Berkeley: University of California Press.

Rommel, Manfred. 1993. *Manfred Rommels gesammelte Gedichte.* Stuttgart: Engelhorn-Verlag.

Sammond, Nicholas. 2006. 'Picture This: Lillian Gilbreth's Industrial Cinema for the Home'. *Camera Obscura*, 21(3), 103–33.

Sarasin, Philipp. 2003. 'Die Rationalisierung des Körpers: Über "Scientific Management" und "biologische Rationalisierung"'. In Philipp Sarasin (ed.), *Geschichtswissenschaft und Diskuranalyse.* Frankfurt: Suhrkamp, 61–99.

Shallis, Michael. 1983. *On Time.* New York: Schocken Books.

Suchman, Lucy A. and Randall H. Trigg. 1991. 'Understanding Practice: Video as a Medium for Reflection and Design'. In J. Greenbaum and M. Kyng (eds), *Design to Work*. Hillsdale, NJ: Erlbaum, 65–89.

tempus. 2018. Masterplan. http://www.tempus.de/shop/Lebensplanung/tempus-Masterplan-ausklappbar.html.

Thompson, Edward P. 1967. 'Time, Work-Discipline and Industrial Capitalism'. *Past & Present*, 38, 56–97.

Vismann, Cornelia. 2000. *Akten: Medientechnik und Recht*. Frankfurt: Fischer.

Weber, Max. 1978. *Economy and Society: An Outline of Interpretive Sociology*, ed. Guenther Roth and Claus Wittich. Berkeley: University of California Press

Weber, Max. 2001. *The Protestant Ethic and the Spirit of Capitalism* (trans. Talcott Parsons [1930]). London and New York: Routledge.

Weick, Karl E. 1976. 'Educational Systems as Loosely Coupled Systems'. *Administrative Science Quarterly*, 21, 1–19.

Weick, Karl E. 1987. 'Theorizing about Organizational Communication'. In Fredric M. Jablin, Linda L. Putnam, Karlene H. Roberts, and Lyman W. Porter (eds), *Handbook of Organizational Communication: An Interdisciplinary Perspective*. Newbury Park, CA: Sage, 97–129.

Yates, JoAnne. 1989. *Control through Communication: The Rise of System in American Management*. Baltimore and London: Johns Hopkins University Press.

Zemanek, Heinz. 1987. *Kalender und Chronologie: Bekanntes & Unbekanntes aus der Kalenderwissenschaft*. München: Oldenburg.

Zerubavel, Eviatar. 2003. *Time Maps: Collective Memory and the Social Shape of the Past*. London: University of Chicago Press.

CHAPTER 7

....

CARD

....

MARKUS KRAJEWSKI
TRANSLATED BY ERIK BORN

EARLY uses of visiting cards were related to the increasing spread of cultural techniques of reading and writing among the educated upper classes during the Renaissance. According to the definitive historian of visiting cards, Walther von Zur Westen (1919: 1), the early modern form, also known as a calling card, does not belong to the direct genealogy of another small-scale graphic form, the ancient greeting card (*schedulae salutatoriae*). While this seemingly-related form primarily communicated its bearer's class, status, or friendly wishes, the visiting card served more to address a problem of absence. Still, it followed the same 'idea of leaving behind documentary evidence of one's presence after a successful visit and thus gaining independence from the poor memory and general negligence of servants' (von Zur Westen, 1919: 1). An encounter at the entrance door stages a confrontation between two different kinds of proxies, insofar as servants represent their masters (Krajewski, 2010: 89, 93f.) and visiting cards stand in for the absent entrance of guests.

What started as a handwritten document left behind on paper acquired more durable forms over time. Around 1700, in an age when oneself is re-presented best in one's absence, the visiting card was made according to an elaborate copperplate engraving process, which was occasionally even enshrined in poetry, and many visiting cards, especially the more affordable ones, were based at least partly on design trends and print runs. Hence, there was a distinction between two different types of visiting cards: the blank card, also known as a framed card, for filling in one's name by hand (Figure 7.1, top); and the costlier, highly-personalized visiting card, which would have been engraved on demand (Figure 7.1, bottom). The more common type was the blank card, which proved to be a great deal cheaper to produce and more suitable for mass production. In the case of this upper-class consumer good, changes in fashion were apparent in the choice of motifs, which reflected the tastes of the owner, the printer, and the times. 'In the long run, continued use of one and the same card would have bored the visitor and visitee alike. Elegant ladies, in particular, would have been as reluctant to use the same card as to wear the same hat, or the same dress, for any considerable length of time. In any case involving attributes of external elegance, there needs to be a certain latitude

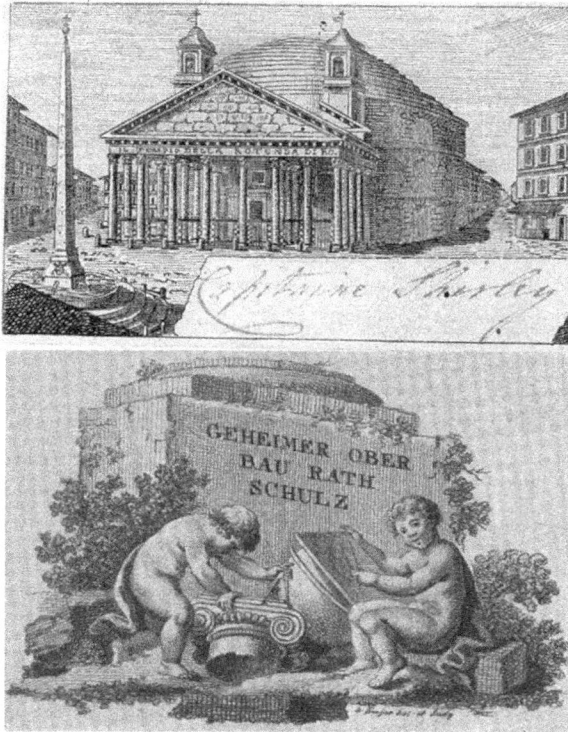

FIGURE 7.1 Blank framed card and luxury visiting card, in von Zur Westen (1919)

(von Zur Westen, Walther (1919), 'The History of the Visit Card', Mitteilungen des Exlibris-Vereins zu Berlin, Vol. 29, No. 1–2, p. 1–14)

for fashion, for variety, and the many diverse forms of blank visiting cards afforded ample opportunities for this purpose' (von Zur Westen, 1919: 4).

The visiting card eventually attained the status of an art form, which was fashionable throughout Europe but remained influenced by regional and national mannerisms. In 'the land where the lemon trees bloom', to borrow Goethe's famous verse, the visiting card drew on many well-established artistic motifs: 'The overall impression of the Italian visiting card is cheery and festive. It is populated by an immense host of merry cherubs who imbue the card with movement and life; a legion of beautiful women in classical garments provides grace and majesty' (von Zur Westen, 1919: 6). In Germany, on the other hand, the visiting card was primarily manufactured at the production facilities in Augsburg and Nuremberg, the centres of copperplate printing and of German playing cards, which involved a similar production process. There is good reason one of the period's most influential printmakers, an anonymous artist active around 1450, is known as the *Meister der Spielkarten* (Master of the Playing Cards).

In Vienna, another early production centre for playing cards, the visiting card only became fashionable, after some delay, at the end of the eighteenth century. Around 1780, however, the Viennese were already starting to gain experience with the mobility of small, paper proxies through their exposure to another kind of card, namely, the world's first library card catalogue (Krajewski, 2011: 34–47). Ultimately, the same city also contributed to the demise of the visiting card, insofar as the rise in diplomatic visits occasioned by the Vienna Congress was accompanied by a preference for spontaneous messages written in a careful chancery script.

After 1815, according to Walther von Zur Westen's diagnosis, the visiting card fell into a deep, long slumber and would only be reactivated elsewhere around a century later, namely, in the world of bookplate collectors. Ultimately, bookplates and visiting cards can be understood as two elements of the same desire for representation within a particular class of society: 'Taken together, the bookplate and the visiting card create a significantly more complete reflection of those circles distinguished by their rank, birth, or education, than could be provided by signs of book ownership alone' (von Zur Westen, 1919: 12). Nevertheless, a bookplate still makes for an incomparably better historical source, since a visiting card is a throw-away product. Over an extended period of time, visiting cards survive only by chance, having been stuck behind some mirror in the foyer or retained by some collector who scoured the entrance halls at classy addresses.

Beyond their renaissance as collector's items, visiting cards eventually started down a new path, which had been paved by commercial travellers and a new group of less stationary merchants during the early modern period. At once mobile and ephemeral, the same objects came to be traded as business memorabilia in the early nineteenth century, namely, under the new economic term 'business card'.

Visiting cards, playing cards, index cards, and business cards were all a crucial part of pre-digital information processing, and were only partially replaced at the start of the twenty-first century by the email signature and other small personalized media forms. What were the distinguishing features of these inconspicuous information carriers, which functioned as handheld representatives, manageable paper-based proxies, or paper personae? On the basis of the following ten affordances, my analysis of the little cards' organizational logic will address the series of functions that made them into an inconspicuous, though indispensable, medium of knowledge transfer. A card affords not only the organization of information, but also mobility, portability, flexibility, modularity, representativity, transitivity, manageability, updatability, legibility, and combinability.

MOBILITY

In the mid-sixteenth century, Conrad Gessner reflected on the tools available for setting the period's printed knowledge in motion. With his *Bibliotheca Universalis* (1545), the Swiss mountain climber, doctor, and polymath authored a bibliography consisting of around 3,000 authors and more than 10,000 titles arranged in alphabetical order and complete with more or less expansive excerpts from each title; an additional companion index, known as the *Pandectarum sive Partitionum universalium* (1548), provided a comprehensive list of keywords presented in thematic order. For budding scholars, Gessner's *Bibliotheca Universalis* would have provided orientation in the schemas and keywords that would be useful for organizing the knowledge they would acquire in the future. At the time, Gessner's organization of common topoi was itself an undertaking in the classification of knowledge.

What is remarkable, as perhaps the earliest description of this kind of practice, was Gessner's method for compiling these alphabetic and systematic lists (Figure 7.2). His groundwork consisted in the careful and comprehensive collection of sources, whereby he did not flinch from the difficult task of creating excerpts from remote or esoteric materials.

To process these excerpts—'taking notes' by quite literally cutting them out of pieces of paper—Gessner followed the simplest possible algorithm, and described the four steps involved with great precision:

1. While reading, transcribe everything of importance and everything that might later prove to be useful on one side of a good-quality sheet of paper (Gessner, 1548: folio 19).
2. Put each new thought on a new line (Gessner, 1548: folio 19).
3. (And here Gessner appears not only to be a modular thinker but also one of the first theorists of rock, paper, scissors.) 'In the end, you should use scissors to cut up everything you've written out. Put the parts in order as you wish, at first in large categories, which can then be divided again once or twice into subcategories' (Gessner, 1548: folio 20; translated from Latin into German by Zedelmaier, 1992: 104).

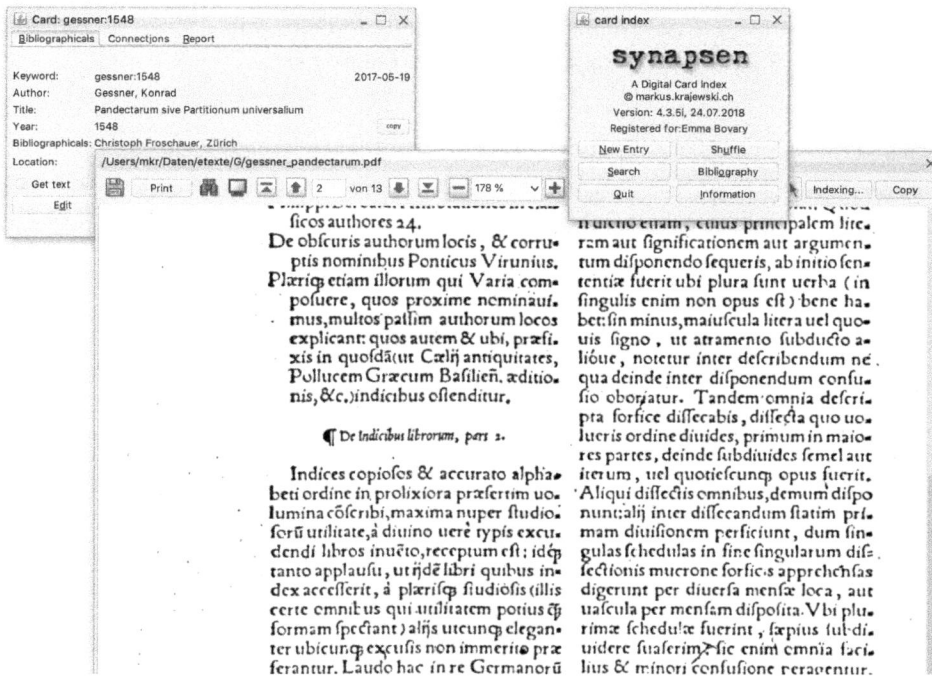

FIGURE 7.2 Conrad Gessner's instructions for dissecting books

4. As soon as you have created the desired order, have arranged and sorted everything, either on a table or in a little box, you need to fix everything in place or copy everything down elsewhere (Gessner, 1548: folio 20; Wellisch, 1981: 11).

Fixing everything in place means capturing the movable notes, slips of paper, or index cards, as the case may be, affixing them to a single sheet of paper with an adhesive or binding agent. At the same time, fixing the cards in place should not rule out the option to move them again. To this end, it is necessary either to use a water-soluble glue or to create a system that would allow the order to be changed or additional modules to be created in the future—in short, the system must remain mobile. Gessner's solution to this challenge involved a particular kind of book containing a guidance mechanism made out of thread for slotting multiple notes onto each page. Like weaving, the guidance threads (i.e., the warp) are first stretched across the page to form two main columns; the individual notes (i.e., the weft) are affixed to a stronger paper backing for additional support and can then be passed beneath these threads to create multiple rows. Notably, Gessner also calls these unique constructions '*chartaceos libros*' (literally, 'books made of paper') thereby prompting the name for the wooden instrument later called a 'card index'.

To date, this 'method with whose help anyone can produce indices in the shortest time and in the best order' (Gessner, 1548: folio 19) is the earliest explicit example of how collectanea, the raw materials of knowledge, might be conserved in a mobile form, brought together into dynamic arrangements, and thus always remain searchable for specific topics. During the final step of fixing everything in place, Gessner used a special glue to ensure that each note remained both durable and removable, with the result that they now resemble early modern sticky notes (see Figure 7.3).

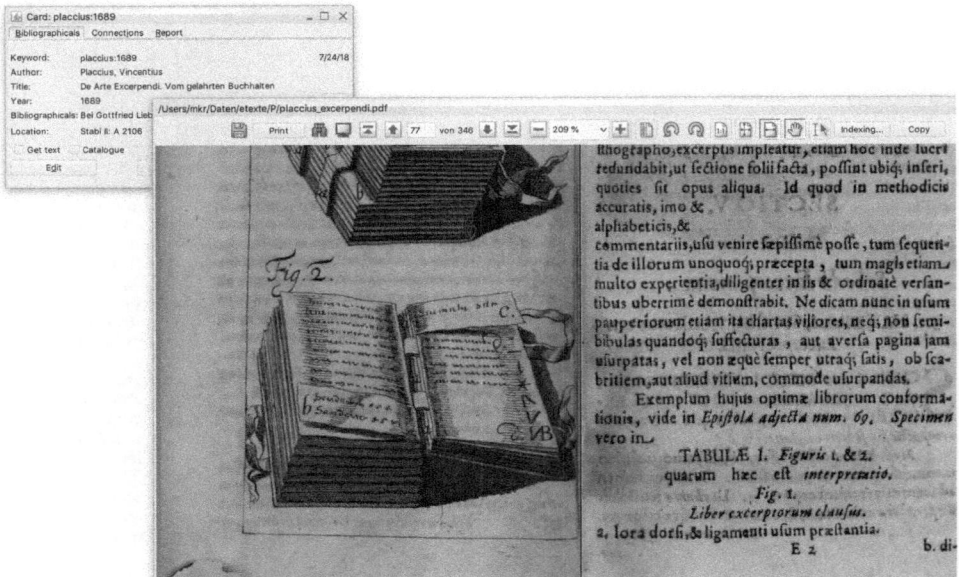

FIGURE 7.3 Movable information support, half-affixed, in Placcius (1689)

Gessner's practices had a formative effect on the creation of indices not only by scholars but also librarians, who were generally thankful to adapt his directions to their own lists of books. It is not without reason that Gessner is considered the father of both note-taking and of bibliography (Krajewski, 2011: 9–24).

PORTABILITY

Portability guarantees the division and distribution of a deck of cards, especially for projects involving a division of labour. During the Enlightenment, the portability of cards was particularly efficacious in the large-scale cataloguing projects taking place at court libraries, which originally did not have their own catalogues (see Figure 7.4). Needless to say, there is always a danger in portability, namely, the dispersion of materials. For this reason, cords, knots, and wires were once again deployed to temporarily hold ephemeral knowledge in place.

The more repositories one works through, the more title cards pile up. As soon as these become too copious, it is advisable to separate the title of one letter off from the rest, to make as many piles as there are letters in the alphabet, and to arrange them from first to last on a table. For storing the piles of letters, a table seems more useful to me than a cabinet, provided that the table is located in a place where the position of the cards will not be disturbed by a breeze or any other hazard. The cases where I have to look for something in one of these piles are far too common. On a table, each pile is immediately at hand. In the case of a cabinet, I need first to take

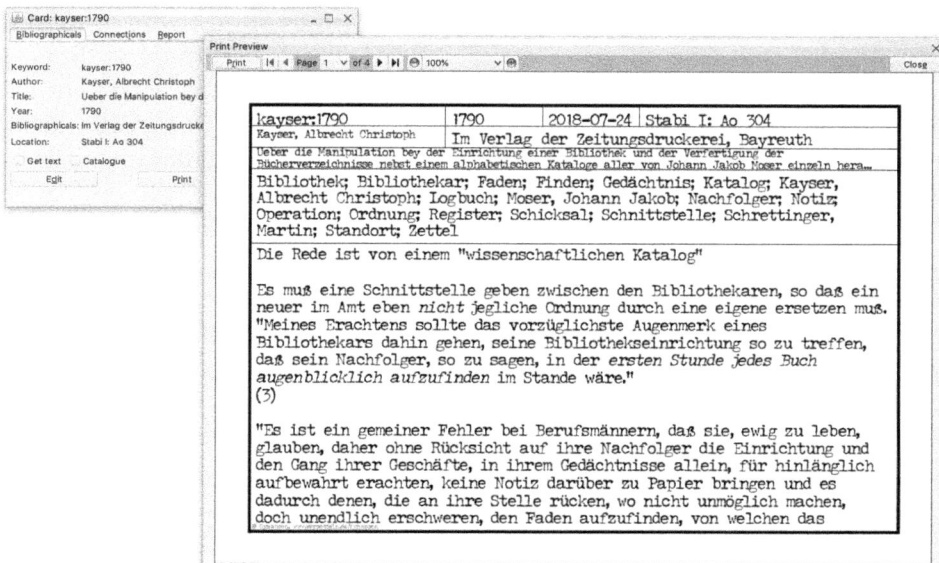

FIGURE 7.4 Catalogue creation on cards according to Kayser (1790)

one pile out and then look around for a place where I can set it down and search through it. This results in lost time.   (Kayser, 1790: 49)

Thus, next to mobility, the simple portability of a stack of cards is what makes it incomparably easier to work with the individual elements, especially in comparison to a difficult catalogue in book form. As a consequence of portability, a book no longer needs to be present on the shelf. 'For librarians, the second advantage of writing down the titles of all the works on hand in a library on individual slips of paper and then laying these out in alphabetic order is that they can now produce a general-purpose alphabetic catalogue whenever and wherever they want. They no longer need to pick up a book itself, and whenever they do need one, they can specify its location at a distant site from their own homes' (Kayser, 1790: 51). In other words, the slip of paper, or notecard, becomes a proxy of the actual text, a procedure that would prove to be extraordinarily practical for information processing and knowledge production.

FLEXIBILITY

With respect to the materiality of the information carrier, the question arises as to the appropriate qualities of the writing support. The paper cannot be too thin, which would run the risk of tearing, nor can it be too thick and stiff, as would be necessary for various crafts using a cardboard-like material. One romantic guide to the intellectual crafts can be found in Albert Lohnau's *Der vollkommene Papparbeiter* (The Ideal Cardboard-Worker, 1832), whose descriptive subtitle promises 'practical instructions for producing all kinds of tasteful cardboard constructions in the neatest manner possible' and calls the book 'a guide for all those who wish to learn or master the art of making things out of paper and cardboard' (Figure 7.5). In the manual, Lohnau describes how to construct rococo furniture out of cardboard, and provides instructions for building storage cabinets, called little work tables, whose construction is analogous to that of another object described in the book, a cardboard vanity set including a vanity bag and a built-in mirror.

Riffle, *shuffle*, and *box* are the English words for the routine method of mixing up a deck of playing cards. In this case, the card material also needs to have a certain sturdiness, pliability, and rigidity; otherwise, virtuosic tricks would be impossible. Equipped with sufficient flexibility, the information carrier can enter into circulation; mixed forms can now be created out of the heterogeneous material recorded on them; and new orders can emerge out of cards that had not previously been located next to each other. Playing with cards brings chance into play. While a deck of cards creates contingency in a ludic context, stacks of cards also function as contingency generators in many other contexts (Krajewski, 2011: 45–7).

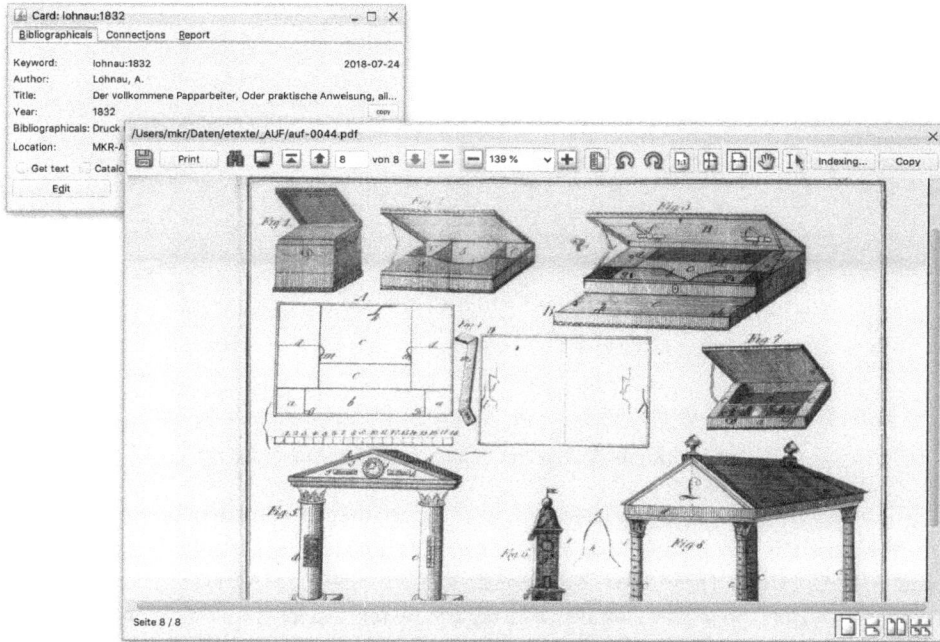

FIGURE 7.5 The cardboard box for storing cards in Lohnau (1832)

MODULARITY

If modularity depends on the statistical completeness of information, each card must constitute a distinct, self-contained data container with limited storage space. A card is a fundamental unit of meaning. Conrad Gessner made a similar observation in his stipulation that each discrete thought be recorded on a new line, which should then be put on its own slip of paper. This is how information grows—through the accumulation of scattered thoughts and the formation of discrete modules. One of the most well-known examples of this approach can be found in Niklas Luhmann's theory and in his own note-taking system. Less known, though no less efficient, was Johann Jakob Moser, an eighteenth-century figure who, like Luhmann, had been trained as a legal scholar. In both cases, the patchwork system of juridical knowledge seems to favour a card-based organization scheme. What Moser and Luhmann had in common was not only their immense productivity but also their emphasis on the reusability of their note-taking systems. While there has recently been much scholarship on Luhmann's note-taking system (e.g., Krajewski, 2013; Schmidt, 2016), Moser's practices remain underappreciated in spite of detailed information about the secret of his productivity (Figure 7.6).

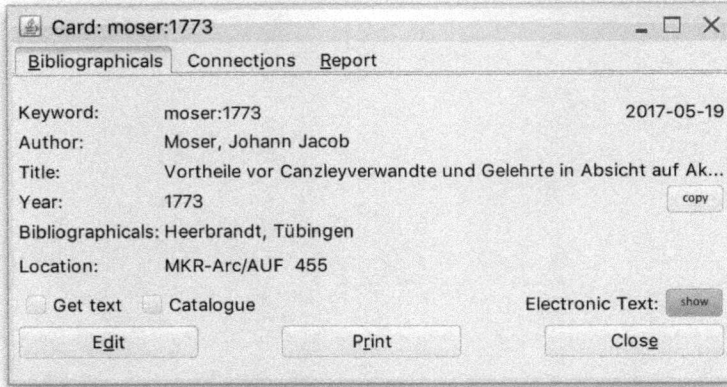

FIGURE 7.6 The secret of the system in Moser (1773)

The advantage of a modular notecard, in comparison to an entire notebook consisting of collectanea, is that it can 'be used right away, as it is; whereas if collected passages are entered into books, they need to be written down again' (Moser, 1773: 54f.). For this reason, modular loose leaves make drafting a new text easier than working with books. Moser's process is also more conducive to selecting specific information or even concealing it: 'If I wish to communicate only one of my collected passages to other people, or to have someone transcribe something without seeing the rest of the material, then I need only take out the sheets I wish to retain. If, on the other hand, the collected passages have been entered into books, then I have to provide the entire volume, or multiple volumes, in which things might be located that I would not want to be made known to just anybody' (Moser, 1773: 54f.).

A module is something self-contained, self-enclosed, which can be understood as a unit. Moser also describes, again in detail, his practice of re-reading his excerpts, and effortlessly generating a new text out of them. Out of all his notes, Moser would first select only those slips of paper that seemed most auspicious for his present argument and then arrange them according to some specific classification scheme. Next, he would refine the order of the slips of paper for each sub-point, and subsequently furnish them with additional marginal notes. With the aid of this outline, Moser would then search through his library for further quotations, references, and new ideas, noting these down on their own slips of paper, which he would combine with the previously selected slips of papers into a new order. After further review, additions, and a round of self-editing, he would finally add a table of contents to this entire bundle, consisting only of little slips of paper, and hand the whole thing over to a printer's shop, or the censorship board, without expending any further effort on the material (Moser, 1773: 58ff.). Two things would come back in return—a new book, which would get put back on his bookshelf, and the box containing the old slips of paper, which would get re-filed into his note-taking system, allowing the loop to repeat. Thus, Moser's data stream alternates between two different media: an excerpt from a book turns into a card, which represents a book, which will be turned back into more cards, which will, in turn, become new books.

REPRESENTATIVITY

The central aspect of dealing with cards consists in their proxy status. Like paper money (Krajewski, 2011: 58–62), people attribute to the visiting card a value of its own:

> Visiting cards, as much as they may seem to be trivial little pieces of paper, have nonetheless become a saleable and indispensable coin in the whole wide world, providing sure signs of either a refined or lacking way of life, and they play a significant role in our modern etiquette. Some visits are only made personally, some personally and with cards, and some with visiting cards alone.   (Anon., 1795: 147f.)

In other words, the visiting card acts as a kind of cardboard cut-out equal to its own value, insofar as someone can pay an entire visit using one or represented by one (Figure 7.7). How is this possible? 'A visiting card, which is delivered or sent around as payment for whomever it's serving, will introduce a stranger into every good-mannered house in a city all at once. Now, everyone knows the stranger, encounters him courteously, and treats him as a member of their society, whereas the stranger might have lived for years in the same city without becoming known should he have neglected to send this herald announcing him in advance' (Anon., 1795: 148). The visiting card acts— exactly in Bruno Latour's sense of non-human actors—as a 'lieu-tenant', a proxy or placeholder for a person (Johnson, 2006 [1988]: 254f.).

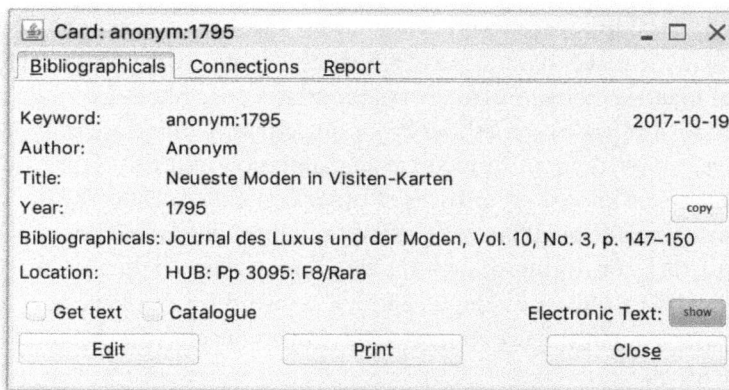

FIGURE 7.7 An index card for visiting cards, in Anon. (1795)

Visiting cards made trafficking in society easier than ever before, since encounters occur only in the form of these proxies:

> People come to one of our famous spas, send out hundreds of visiting cards a couple of hours after their arrival, and, on the very same day, they are introduced into the entire society of spa guests, acquainted with two to three hundred people, as though they had already lived together for many years. A visiting card, depending on whether it's delivered directly, left behind, neglected, or unreciprocated, can flatter, embarrass, insult, reconcile, honour, extend an invitation, drop a hint, test the ground, and so on. In short, the visiting card is a talisman that can work wonders depending on how it is used.   (Anon., 1795: 148)

Hence, the power of these little cards can work miracles, insofar as they allow people to appear in person *in absentia*.

Even in the case of plain cards without any ornamentation, the visiting card inevitably appears to be a barometer for its bearer's taste:

> With some simple visiting cards, you can tell just from looking at them, that they are as humble and modest as their lords; other cards, made ostentatious with long titles or even coats of arms and heraldic attributes, seem to call out, "Make way!" […]; some make you sensitive with a couple of turtle doves, Cupid's bows, and decorative torches, or even a little silhouette and a forget-me-not; others flirt with you (like a slender girl from behind a straw hut) with their refined simplicity, their sleek English vellum, and their golden curves. In short, I maintain that the selection of our visiting cards always reveals something of our character.   (Anon., 1795: 148f.)

In other words, the name printed or written in calligraphy on a visiting card does not suffice to reveal the character of whoever owns or sends it; a card becomes a proxy only with a small-format image, or with the help of a name integrated into an ornament, embellished with an allegory. The act of representation is only completed by a well-calculated relationship between word and image.

Hence, a card invariably functions as a reference. It is intertwined with a Heideggerian operation of inauthenticity. A card always remains a pure reference and a relation to something else: it refers to something beyond itself. A visiting card makes a mediate acquaintance, rather than an immediate acquaintance, and this logic also seems to apply in the case of knowledge proxies in other contexts, such as the library: a card catalogue simultaneously makes one curious and provides information, occasionally to the extent that consulting the source documented on an index card may seem superfluous. And just like a catalogue entry, a keyword on an index card, or even an entire stanza of a poem (Figure 7.8), refers far beyond itself—for instance, back to its own commentary.

Thus, references always refer to something else, be it books in a library, names on other notecards, absent interlocutors, or overarching structures, all of which get

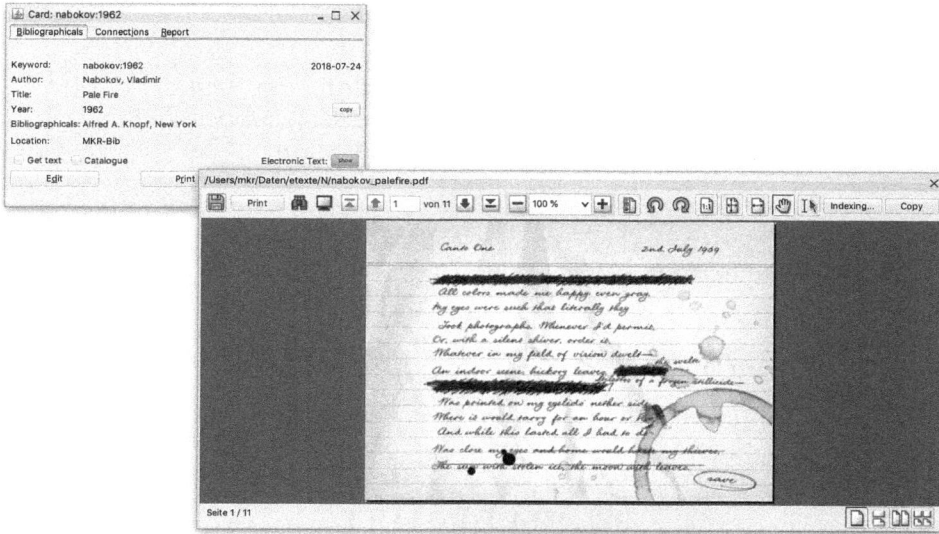

FIGURE 7.8 Card index with coffee stain. Original in Nabokov (1962)

gathered together in bundles of notes. The principle of transitivity or relationality always prevails.

TRANSITIVITY

Transitivity stands for fungibility, which means that cards of the same format can easily be exchanged for others in their designated container or stack. Measurement standards play a certain role in this process. While scholars once cut their own notecards to size according to idiosyncratic formats, each one swearing by his (or her) own dimensions, playing cards have always been more or less the same format. Hence, index cards, too, were eventually standardized, though, at least in the context of libraries, only in 1908, when the international catalogue card set the standard with dimensions of 75 × 125 mm. The American postcard was used as the model for these measurements (Krajewski, 2011: 92f).

MANAGEABILITY

There is one consolation in the recursive sequence of cards, whose number could theoretically be infinite: Each card only has a finite amount of space. And ordinarily, an index card does not contain much (Figure 7.9).

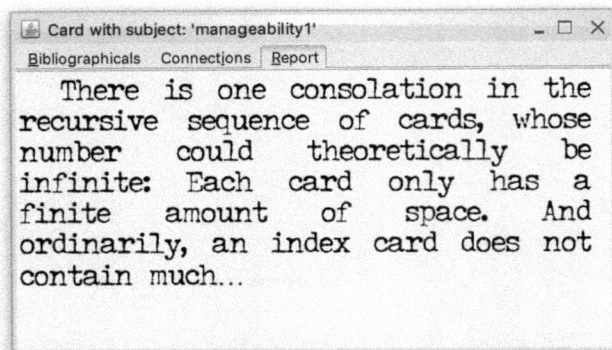

Card with subject: 'manageability1'   _ □ ×

Bibliographicals   Connections   Report

There is one consolation in the recursive sequence of cards, whose number could theoretically be infinite: Each card only has a finite amount of space. And ordinarily, an index card does not contain much...

FIGURE 7.9   A finite space for thoughts

UPDATABILITY

'As with everything on earth, there are fashions in visiting cards' (Anon., 1795: 149), which makes historical fashions in catalogue cards and playing cards seem rather under-developed. Standard playing cards have hardly varied for centuries, and catalogue cards have survived seasonal changes. In visiting card trends, on the other hand, the once-dominant English simplicity was replaced at the end of the eighteenth century by an Italian mannerism: 'There is always a large, flat stone or piece of rock in the fore-ground where you can write your name [...] on a little white field under the picture' (Anon., 1795: 150). Even if the cycles of fashion were not yet measured in one-year periods around 1800, as would become common in subsequent years, there seems to have been an abundance of variety among the senders and receivers of cards.

LEGIBILITY

While the Master of the Playing Cards remained anonymous, the master of index cards is world-renowned: Vladimir Nabokov always used notecards as the foundation for his novels, both for the preparation of the text and as an object of description—most notably in *Pale Fire* (1962). In spite of its classification by publishers as a 'novel', Nabokov's text does not contain anything, at first glance, that would otherwise signal this literary genre. The text consists of four parts: a foreword by the editor Charles Kinbote; the 999-line poem 'Pale Fire', edited by Kinbote but written by his next-door neighbour John Shade; a sub-stantial commentary, again composed by Kinbote, on the long poem; and finally, an index containing detailed entries for the commentary, rather than the poem itself. Usually, a foreword, a poem, a commentary, and an index are not the paratexts that make up a novel.

At Cornell University, Nabokov taught his students that people are not able to read novels, only to re-read them. The same applies to poetry, and, at least in this case, its commentary. The only questions are where to start reading and in what order. The editor of *Pale Fire* leaves no room for doubt and provides the following reading instruction:

> Although those notes, in conformity with custom, come after the poem, the reader is advised to consult them first and then study the poem with their help, rereading them of course as he goes through its text, and perhaps, after having done with the poem, consulting them a third time so as to complete the picture. I find it wise in such cases as this to eliminate the bother of back-and-forth leafings by either cutting out and clipping together the pages with the text of the thing, or, even more simply, purchasing two copies of the same work which can then be placed in adjacent positions on a comfortable table.   (Nabokov, 1962: 14)

Putting aside the ostensible question of the additional income the sale of a second copy would fetch for the author, Kinbote's proposal represents the first step in severing the text from its commercial book form and putting it back together again in another form. In fact, its original form, since the author of the poem (John Shade) composed the four cantos on eighty index cards (13 cards for the first and fourth cantos; 27 for the second and third) (Figure 7.8). At the outset, Kinbote describes the unusual form of the manuscript in dry academic prose: the rubricated heading line contains the current address of the poem with a date and canto number, thereby making it possible to compare parallel events in the commentary.

With a large enough surface (e.g., a floor), some diligence, and sufficient respect for the pages, Kinbote's commentary should now be cut up, consistently, and correlated with the respective verses in their original form on the index cards. When we add the commentary to the poem (according to the editor, the poem is only completed by the 1,000th line, which is also the first line, so the cards are best organized in a circle!), which itself takes up 254 pages (on average 30 lines per page, referring to the German edition), the two columns of text result not only in a richly-varied pattern. In addition, the reader gains an overview of two parallel stories with three protagonists (poet, king, murderer) who appear at first glance to correspond to each other, upon closer reading seem to have less in common, and after re-reading again reveal their close affinities. Whether visiting cards or playing cards warrant a similar close reading depends on the extent of their own circulation.

COMBINABILITY

Why would an author work out the draft and fair copy of a poem on index cards? In *Pale Fire*, the author is adapting and animating a specific mode of operation, which has exerted a decisive influence on textual productivity since the early modern period, and not only on literary scenes of writing. Ever since Gessner recommended the notecard as an aid for scholarly and writerly production in 1548, this medium has been set on a unique, albeit often secluded career. The main advantage of discrete cards consists in their mobility, which enables the words and modules of text stored on them to be combined and reordered. According to Nabokov's fictive editor, the fictive poet also follows this technique, 'perceiving and transforming the world, taking it in and taking it apart, recombining its elements in the very process of storing them up so as to produce at some unspecified date an organic miracle, a fusion of image and music, a line of verse' (Nabokov, 1962: 13). Usually, one is not able to look back from a final product to its previous building blocks. While Arno Schmidt did not allow his notes to be published, he prepared a hermetic assemblage of their component parts, their miraculous consolidation into book form. In the same year as the publication of Nabokov's text, Marc Saporta put out a novel in the form of a card game (around 250 cards), along with playing instructions for bewildered readers (Saporta, 1962; Grimm, 1965). In *Pale Fire*, on the other hand, the reader receives a transparent view of the fragmented starting elements, which were then noted down by the commentator and interconnected with cross-references, to break them down into structural units.

* * *

What unifies these diverse ways of using visiting cards, playing cards, index cards, and business cards? On the one hand, it is their power of representation. The unifying element in their performance as information carriers can be located in their status as proxies, their function as placeholders. Playing cards represent a particular symbolic value; visiting cards and business cards, a person; and notecards, an entire personalized system of symbols consisting of flexible vehicles for signs whose combinatorics serve to produce surprising constellations, to create completely new connections. On the other hand, cards have the ability, in spite of their inconspicuousness, to accomplish organizational tasks, whether as the medium of complex (group) games and their subtle hierarchies and combinatorics, whether in their promise of making a contact to start new business deals or create new connections, whether in the sheer quantity and universality of notes, which a card box or catalogue easily brings together and puts in a systematic order. The power of these little cards is located in their organization of the heterogeneous into the manageable. Ultimately, we can recognize a third aspect of cards as their most substantial productive factor in our own game: a card's interface, its edges, suggests almost any arbitrary connection and creates a seamless transition from two heterogeneous information modules into one homogeneous form, which is standardized because cut to the same kind. Organization knows no end.

References

Anon. 1795. 'Neueste Moden in Visiten-Karten'. *Journal des Luxus und der Moden*, 10(3), 147–50.

Gessner, Konrad. 1545. *Bibliotheca Universalis, sive Catalogus omnium scriptorum locupletissimus, in tribus linguis, Latina, Graeca, & Hebraica*. Zürich: Christoph Froschauer.

Gessner, Konrad. 1548. *Pandectarum sive Partitionum universalium*. Zürich: Christoph Froschauer.

Grimm, Reinhold. 1965. 'Marc Saporta oder der Roman als Kartenspiel'. *Sprache im technischen Zeitalter*, 14, 1172–84.

Johnson, Jim (= Bruno Latour). 2006 [1988]. 'Die Vermischung von Menschen und Nicht-Menschen: Die Soziologie eines Türschließers'. In Andréa Belliger and David J. Krieger (eds), *ANThology: Ein einführendes Handbuch zur Akteur-Netzwerk-Theorie*. Bielefeld: Transcript Verlag, 237–58.

Kayser, Albrecht Christoph. 1790. *Über die Manipulation bey der Einrichtung einer Bibliothek und der Verfertigung der Bücherverzeichnisse nebst einem alphabetischen Kataloge aller von Johann Jakob Moser einzeln herausgekommener Werke—mit Ausschluß seiner theologischen—und einem Register*. Bayreuth: Verlag der Zeitungsdruckerei, Bayreuth.

Krajewski, Markus. 2010. *Der Diener: Mediengeschichte einer Figur zwischen König und Klient*. Frankfurt am Main: S. Fischer Verlag.

Krajewski, Markus. 2011. *Paper Machines: About Cards & Catalogs, 1548–1929*. History and Foundations of Information Sciences. Cambridge, MA: MIT Press.

Krajewski, Markus. 2013. 'Paper as Passion: Niklas Luhmann and His Card Index'. In Lisa Gitelman (ed.), *"Raw Data" Is an Oxymoron*. Cambridge, MA: MIT Press, 103–20.

Lohnau, A. 1832. *Der vollkommene Papparbeiter, Oder praktische Anweisung, alle Arten geschmackvoller Papparbeiten auf das Sauberste zu verfertigen: ein Hülfsbuch für alle Diejenigen, welche die Kunst, aus Pappe und Papier zu formen, erlernen oder sich darin vervollkommnen wollen*, 2nd edn. Quedlinburg and Leipzig: Gottfried Basse.

Moser, Johann Jacob. 1773. *Vortheile vor Canzleyverwandte und Gelehrte in Absicht auf Akten-Verzeichnisse, Auszüge und Register, desgleichen auf Sammlungen zu künfftigen Schrifften und würckliche Ausarbeitung derer Schrifften*. Tübingen: Heerbrandt.

Nabokov, Vladimir. 1962. *Pale Fire*. New York: Alfred A. Knopf.

Peters, John Durham. 2015. *The Marvelous Clouds: Toward a Philosophy of Elemental Media*. Chicago: University of Chicago Press.

Placcius, Vincentius. 1689. *De Arte Excerpendi: Vom gelahrten Buchhalten*. Stockholm and Hamburg: Gottfried Liebezeit.

Saporta, Marc. 1962. *Composition no. 1*. Paris: Seuil.

Schmidt, Johannes F. K. 2016. 'Niklas Luhmann's Card Index: Thinking Tool, Communication Partner, Publication Machine'. In Alberto Cevolini (ed.), *Forgetting Machines: Knowledge Management Evolution in Early Modern Europe*. Leiden: Brill, 289–311.

von Zur Westen, Walther. 1919. 'Zur Geschichte der Besuchskarte'. *Mitteilungen des Exlibris-Vereins zu Berlin*, 29(1–2), 1–14.

Wellisch, Hans H. 1981. 'How to Make an Index—16th Century Style: Conrad Gessner on Indexes and Catalogs'. *International Classification*, 8(1), 10–15.

Zedelmaier, Helmut. 1992. *Bibliotheca universalis und bibliotheca selecta: Das Problem der Ordnung des gelehrten Wissens in der frühen Neuzeit. Beihefte zum Archiv für Kulturgeschichte*, Band 33. Köln: Böhlau Verlag.

CHAPTER 8

....

CHAIR

....

MARIA-LAURA TORALDO
AND JEANNE MENGIS

FIGURE 8.1 Empty chairs in Piazza Grande, Locarno

(Photo by the author)

It's early evening when I arrive in Locarno for a meeting with a film producer. As I enter the central square of Piazza Grande, I suddenly have to stand still. In front of me, thousands of empty chairs in yellow and black (Figure 8.1). They all glare at the large white screen in front of them illuminated by the declining sun. Last night, and as I will be again in a few hours, I was sitting on one of these chairs as I was watching, together with 7,000 other people, a film under the open, starry sky. How different now, no queuing, no thousand voices melting together into a bustling sound when waiting for the film to start. Now, the thousands of empty chairs fill the large square with a sense of emptiness and occupation at the same time. The chairs seem to be waiting for the masses to sit on them again in a few hours as their sheer number anticipates that there is something big yet to come. Further in the back of the square, the chairs' bulky occupation is evident as many of them have been stacked to make space for the public square to be used, at least in part, for the ordinary urban life. As I oversee once more Piazza Grande, it strikes me how the chairs' plastic materiality and precise spatial arrangement strangely contrast, but also remind me of last night's magic of the square, of which each festival goer glowingly tells me about.

Chairs surround us and we expect to find them in the different spaces we inhabit (Highmore, 2011; Bergamasco and Croci, 2012). At work, we spend around 70 per cent of our time sitting. As such, chairs are an instance of the ordinary, an infrastructural technology supporting us in our relational activities with other artefacts and human beings (Highmore, 2011). Despite or because of their ubiquity, chairs go mostly unnoticed, are an ingrained artefact of our lives (Cranz, 1998) and interactions, inherently part of our everyday organizational practices, but rarely attracting the attention of organizational scholars.

In view of our sedentary (office) lives (in developed economies at least), chairs come to matter. They are said to be responsible not only for health problems (e.g., back pain, cardiovascular disease), but also for how we orient to work and how we interact, for example shaping the dynamics of collaboration among people seated in the same room (Davis et al., 2011). Chairs participate, for example, in indicating employees' status in the office hierarchy, in maintaining power differences and acquired ranks during meetings (Cranz, 1998). At the same time, their aesthetic and symbolic features are equally of importance as chairs are carriers of meaning, contributing to the identification with a place and with their design, surfaces, and materiality provoking an emotional response (Elsbach and Bechky, 2007).

The history of this 'piece of furniture with a back, and usually four legs, on which one person sits' (Cranz, 1998: 7) goes back nearly five thousand years and has taken on a variety of forms, usages, materials, and technologies used for its production (Rybczynski, 2016). Already the ancient Greeks had the Klismos chair and kings throughout centuries were seated on thrones. Chairs became more popular with 'the golden age of sitting' in Louis XV's France (Rybczynski, 2016: 6). Americans of the eighteenth century (and some still today) relaxed on their verandas on rocking chairs and, in the nineteenth century, the Viennese sipped their coffee on bentwood chairs (the most famous being the Thonet chair) (Rybczynski, 2016).

With the increase of commerce, the first 'seated professions' came about. As the financial ledgers of bookkeepers were long, castors were attached to chairs. The modern office chair gradually developed since the industrial revolution; a swivel chair with a height-adjustable seat and a vertical, slightly springy backrest to support the spine (Rybczynski, 2016). Yet, not only functionality, but also design considerations significantly shaped the evolution of the chair. Wegner's 'Round Chair', Saarinen's 'Tulip' chair, Jacobsen's 'Egg', and Eames' office chair are some of the most iconic chair designs of the last century whose choice of materials, forms, and the extent of technology used, has influenced the social imaginary associated to their function, the sense of comfort and beauty (MoMA Learning, 2018).

The chair is thus not only a physical object with the role of 'seating a person' (MoMA Learning, 2018) and making one comfortable. It also has a complex history of use, of handcraft, of design, and even of artistic representation. In this chapter we trace some of these material and discursive relations. We suggest the chair's agency is only partly given by its specific materiality and by the craft that gave it its diverse forms and features (Rybczynski, 2016). Its agency is also shaped by the manifold artistic, historic, design, and architectural contexts in which it has been placed. For instance, chairs have played a central role in shaping the public's sensibility and taste. A good example being the iconic Hans Wegner's 'Round Chair, later known as simply 'The Chair' because it was regarded by many as the Platonic form of the chair. Designed in 1949, 'The Chair' has been celebrated for its minimalist design; as Wegner declared, it is the 'reduction to the simplest elements of four legs, a seat, and a combined top rail and armrest' (Hollingsworth, 2008: 105), and yet despite being an apparently simple object it is open to infinite ways of interpretation.

In this chapter we touch upon a few of these, to then show how the invisible labour of chairs comes to matter in organizing, in particular, first how chairs configure our relationship with space and second, how they come to represent and reaffirm social roles and positions. These agencies of the chair ensue from a process of ongoing reconfigurations of meanings, relations, and material conditions that are built into chairs and their relations. We want to reflect on how chairs enable and shape organiza-tions and organizational life, what possibilities for (inter)actions are made possible by this object, and how its historical appearance has informed the physical and social dimension of our daily life. Such a reflection may deepen our understanding of the wider structures of relationships that are created by our interactions with mundane objects (Courpasson, 2017), for example by shaping work environments, defining hierarchy and status, providing legitimation, and shaping interaction and communication.

THE LONGING FOR PRESENCE BY VISUALIZING ABSENCE: THE EMPTY CHAIR(S) AND ITS ROLE IN SPACE CONFIGURATION

In our opening description of the observation at Locarno Festival's Piazza Grande, chairs disclose much about absence and presence and their relationalities. The material presence of the thousands of chairs is such that they fill and occupy the space of the square, providing a more tangible sense of the square's vastness and, at the same time, pushing out the people habitually using the square as they can hardly pass through it or use it as a market space. At the same time, the empty chairs also create a material expectation that they are there to be seated upon and thus point to the absence of an awaited audience. In their state of being empty, chairs convey a sense of expectation that somebody will come (and be seated) or a sense of the failed expectation that the person or masses awaited will not show up. The presence of the chairs and the people's absence thus create a sense of occupied emptiness.

The chair has this quality of communicating absence/presence, expectation/actualization, visibility/invisibility (Cranz, 1998; De Dampierre, 2006) not only because of its materiality with the chair taking space, but also by not being taken. It gains this quality also through its translations and presentations in the arts. We will provide three related examples. The first is the famous illustration of Dickens' library 'The Empty Chair' by Luke Fildes (see Figure 8.2), published in the weekly newspaper *The Graphic* for its Christmas edition in 1870, the year Dickens died. The representation of Dickens' study and everyday surroundings—his personal possessions such as the shelved books, the desk space with the window view—provide a sense of the lifeworld in which Dickens wrote his stories. Yet while the engraving allows us to imagine how the famous author engaged in reading and writing, enjoying the silence and gazing outside, the presence of the empty chair at the centre of the author's personal study also poignantly expresses a sense of absence and void. Dickens, loved for his Christmas stories, was no longer here to write this year's festive tale (Miller, 1997).

Eighteen years after Dickens' death, Van Gogh painted two empty chairs, one called 'Vincent's Chair with his Pipe', the other 'Gauguin's Chair' (Figure 8.3). While the first is painted in light colours of blue and orange, showing a simple straw chair with—on it—a pipe, a handkerchief, and tobacco, the other is more exotic and a night-time scene. Van Gogh described the latter in a letter as Gauguin's 'wooden armchair, brown and dark red, the seat of greenish straw, and in place of the absent person, a lighted candle in a candlestick and some modern novels' (Van Gogh, 1978: Letter 626a, 10/11 February

FIGURE 8.2 'The Empty Chair' by Sir Samuel Luke Fildes

(Collection: The Free Library of Philadelphia)

(a)   (b)

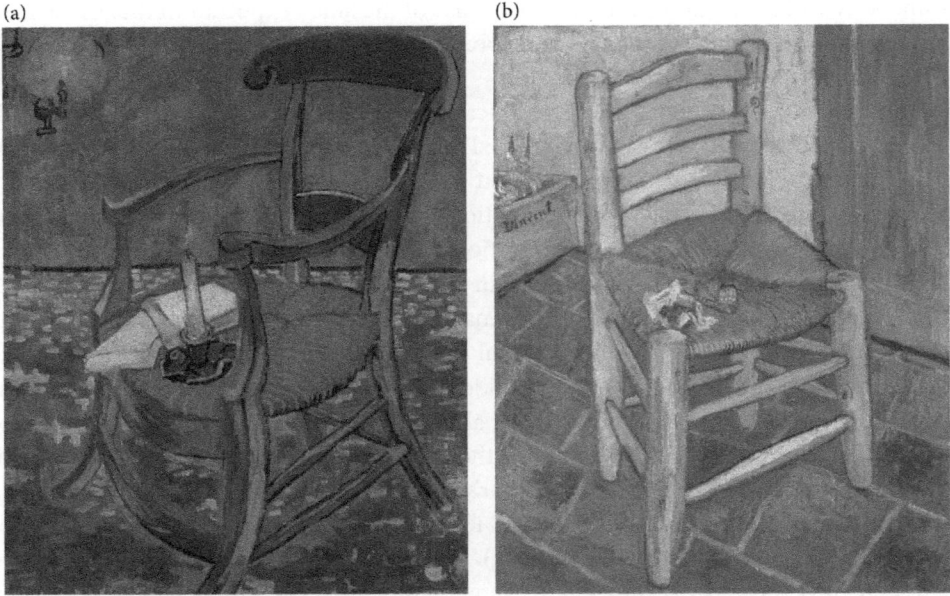

FIGURE 8.3 'Gauguin's Chair' (a) and 'Vincent's Chair with his Pipe' (b) by Vincent van Gogh

1890). The two empty chairs not only express the artists' very different persona and position in painting, they again presence absence and express a disenthralled expectation. Van Gogh had long waited for Gauguin to come and visit him in his new home in Arles. When Gauguin finally arrived, Van Gogh made the two paintings during the rather turbulent and painful time when the two painters worked next to each other and which ended with Van Gogh mutilating his ear. Van Gogh's letter makes it clear that Gauguin's empty chair directly refers to the absence of the artist, the lost friendship, and the felt void, which leaves Van Gogh only with his pipe, a direct reference to Dickens who advised it as a cure for melancholy (Ochsner, 2016).

A final relation of the chair's quality of absence/presence, visibility/invisibility, expectation/actualization can be found in Ionesco's 1952 absurdist play The Chairs. There, the empty chair is multiplied just as in the case of the Locarno Festival. The empty chairs are waiting to accommodate a series of invisible guests desperately awaited by an old couple to finally provide their lives with a sense of meaning and importance. The chairs' number contrasts with the solitude of the couple, with their banal, repetitive dialogue underscoring themes of absence and eternally disenthralled expectations (leading to the couple's suicide at the end of the play) (Tolpin, 1968).

In organizational contexts, empty chairs build on these material and relationally acquired qualities of absence/presence, visibility/invisibility, expectation/actualization. We find empty chairs in the lobby of company entrances and banks, in large, open-plan offices, in university auditoriums. While the occupation of such chairs seems the most natural expression of their purpose, their vacancy takes on multiple meanings as chairs play out their quality of absence and presence differently depending on how they are related to organizational space and work practices. For instance, in open-plan offices, a chair that is not occupied makes easily visible that a colleague is absent, which becomes an emblematic form of direct social control. As Mumby and Stohl (1991) show, absence from the office becomes associated with a set of values that denote poor commitment and loyalty, despite the fact that their absence does not equate with lower productivity. This is what they define as a 'signified absence', where 'presence/absence is not merely a question of showing up for work or not' (Mumby and Stohl, 1991: 324), but performs multiple functions such as defining deviant behaviours, exerting workforce control, or influencing relationships among colleagues. Here, chairs become an ineludible object whose emptiness exposes employees to gaze and group control.

Absence also comes under scrutiny in Davis et al.'s (2011) investigation of the emergence of the open-plan office, focusing on how facilities, services, and material layout influence organizational outcome. 'Hot-desking', for example, where employees have no dedicated desks, but use those vacant that day, played an important part not only in implementing flexible work arrangements but also in changing workers' reactions to, and interactions with, their workspace (Davis et al., 2011: 194; Millward et al., 2007). One of the most significant changes is a reduction of psychological and architectural privacy, associated with the lack of walls, the use of partitions, and the placement of seating close to each other. At the same time, the empty chair, no longer belonging to any particular employee, does not allow for judgements about individual work commitment, but can be a comment on organizational efficiency. Similarly, in the case of teleworking, working away from the office is socially legitimized as offices are designed to optimize space. In these contexts, the lack of the empty chair does not allow for assessing employees' commitment, although absence has certainly implications on who remains in the office—i.e., those occupying chairs in the workplace (Golden, 2007). In this context, chairs' quality of absence/presence visually marks the physical space and influences collective behaviours in the workplace, meanings, and relationalities. This suggests that the chair acts in concert with the changing practices of work and new spatial configurations of offices. The chair thereby affects not only the material place, but also prompts readjustments in how employees think and identify with their organization (Millward et al., 2007) and how they enact their surroundings (e.g. Halford, 2004; Davis et al., 2011; Tuncer and Licoppe, 2018). Although the chair has multiple discursive and material relations that define its qualities of absence/presence, visibility/invisibility, and expectation/actualization, it remains open to heterogeneous interpretations by workers or managers.

CHAIRS' ELEVATING POWER: THEIR ROLE IN REPRESENTING AND REAFFIRMING SOCIAL ROLES AND POSITIONS

Guided by the opening illustration, we would like to turn our attention to how the chair configures relations with social status and position. At Piazza Grande, chairs are deliberately the same for all, whether or not the seated person is an invited film star, a jury member, a sponsor, a politician, or a member of the general public. Chairs are unnumbered and made of the same black and yellow plastic, thus providing equal access, comfort, and even risk. The latter was jokingly taken up by a local fan organization of the Festival, which created a T-shirt representing the festival chair with the text 'so beautiful, so dangerous', thus hinting not only to a famous local advertisement, but also to the problem that some chairs had broken in the past, having suffered too much from the weather. For the Festival, this seating practice was important, expressing notions of openness, democracy, and encounter. At the same time, the Festival had to include a minimal hierarchy in the seating practice first by creating a VIP rank area close to the screen where it could seat guests of honour and by creating a possibility for the general audience to buy seat reservations in order to avoid sitting for long hours on the plastic chairs and waiting for the evening screenings to start.

In this the organizers are succumbing to a time-worn role of chairs: to position the seater off the ground and thereby lend them a higher position, a material expression of uprightness, and metaphorically a sense of integrity, authority, and power, reserved for institutionalized roles and influential groups in a society (Cranz, 1998). Chairs—or certain chairs—have long stood for an invisible mystical presence which human beings venerate and perceive as present during communal celebrations, or for symbolizing hieratic or royal power. An example of this are some mosaics from Ravenna from the sixth century (see Figure 8.4). Here the chair is a symbol of majesty, holiness, and glorification. The iconography of Christ enthroned and surrounded by angels or saints was recurrent across Byzantine and medieval representations, in which chairs and thrones take on similar meanings of an elevated figure and holy supremacy. Similarly, the empty

FIGURE 8.4 Christ enthroned amid four angels, Basilica of Sant' Apollinare Nuovo, Ravenna, Italy
(Printed with permission of The Archdiocese of Ravenna-Cervia)

throne is one of the most common images of Orthodox Christianity, originating from a pre-Christian iconography (see Figure 8.5). While Greeks or Etruscans used the empty seat as a symbol of an invisible god who was taking part at collective celebrations, in Christianity it is used as a symbol of preparation for the descent of the Holy Spirit. In these representations, the chair discloses much about the representation of social roles, of positions and about an invisible or absent authority ('A Reader's Guide to Orthodox Icons', 2017).

Representations of Byzantine arts—which inherited previous iconographic schemes from more ancient cultures—sharply express, through the empty throne, this sense of authority. At the same time, the chair is also a shaping apparatus that prescribes certain ways of believing and expressing devotion towards official and legitimate authority. Worshippers found in the empty throne a familiar image which through a language of symbols and signs was instructive and explanatory. In other words, the chair is a substitute for a person's institutionalized power acting as a delegate of a recognized authority.

A beautiful illustration of this comes from Charlie Chaplin's parodist movie *The Great Dictator* (1940). During the meeting between Hynkel and Napaloni—fictitious names for the Italian Fascist and German Nazi dictators—the chair offers practical affordances for imposing and resisting positions of power. As Napaloni enters the room through the wrong door, he approaches Hynkel from behind saluting him with an energetic pat on the back, which pushes his counterpart from the chair. This gesture not only underlines

FIGURE 8.5 The empty throne, Arian Baptistery, Ravenna
(Printed with permission of The Archdiocese of Ravenna-Cervia)

his strength and boldness, it is also meant to dislocate Hynkel from his position. The entire scene is organized around power relations, which are particularly displayed through the practice of sitting. Napaloni is seated on a purposefully low chair, so that he is forced to look up to Hynkel and positioning him as inferior. Clearly he is uncomfortable with the assigned seat—and role—complaining: 'I must be a-growing! What do they give me? A baby stool?' Napaloni concludes that 'This is not for me. I like it better upstairs', stands up and takes a seat on the table, now assuming a position of superiority.

From a linguistic perspective, the way the word 'chair' seems to be employed in several expressions and terminology across different languages reflects social positions and roles. Take the phrase 'Chairman of the Board' or similarly, the Latin expression *Ex cathedra*. In both the word chair designates the highest-ranking person in an authoritative group or full authority on a topic, thereby enforcing hierarchy and official position (e.g., a person in charge of a meeting or an organization). The word chair originated from the Greek word καθέδρα (kathédra) and was originally used to identify 'the seat or office of one in authority, as the seat of a bishop, a judge or a professor, or the presiding officer of a meeting or assembly', hence, 'The office itself' (Whitney, 1894). The definition also extended to define situations in which a person fulfils a specific role, such as holding a university chair, to indicate the most senior academic office or position in a faculty.

A further use of the word chair designates a connection with worth and skilfulness. For example, within an orchestra, a chair designates the main player in a family of instruments. In this guise, chairs metaphorically refer to skilfulness, expertise, and responsibility. Discursive frames—which include terminology and textual representations more broadly—thus reproduce existing socio-cultural ideas, relationships, and position and participate in the dynamics of social legitimation of roles and their institutionalization.

Moving within organizational contexts, we also find that 'the form of the chair expresses high status; it separates and elaborates the separation, providing distinction' (Cranz, 1998: 34). For instance, within workspace arrangements, social relations are mediated by spatial configuration (Halford, 2004). Some recent illustrations of contemporary office configuration are telling of how chairs are crucial to redefine conventions, positions and social codes to adopt at work. Iconic contemporary workplaces, such as Silicon Valley corporations and other sharing economy companies are increasingly substituting traditional ergonomic chairs with more aesthetically captivating and leisure-like style sitting solutions: sofas, recliners, poufs, futon chairs, and ottomans as well as more dynamic bar stools, benches, and swivel chairs. One instance is Google's new Zurich office with egg-shaped and lounging meeting rooms, a slide and a pole to drop from one floor to the next, physio ball chairs, inflatable chairs and colourful beanbags (Schwär, 2018). Here chairs are not simply intended for occupation. In fact, chairs are not assigned to anyone in particular but are open to alternative uses and experimentation (e.g., relaxing on an inflatable chair or making phone calls). If traditional office chair design allowed for limited room for motion and expected employees to symbolically connect chairs' occupation with position and status within the company, now chairs are orchestrated to encourage casual encounters and unexpected conversations, communicating values of openness and equal opportunities. Chairs' designs are therefore a key component for transforming office spaces into more ludic and creative settings, where workplace interactions seem to encourage a playful-like atmosphere—but it is a sort of obedient kind of play where the fun, colourful features across the Google offices and games seem reminiscent of a docile childhood (Fleming and Sturdy, 2011).

Here and in several other contemporary offices we can see at play what Halford (2004) defined as the 'politics of seating', which entailed choices about organizational status and hierarchy. And, of course such an ordering and controlling function is also performed by the use of chairs. As such chairs perform effects in organizing space and in representing and reaffirming social roles and positions.

In conclusion, our analysis was able to unveil that many seating practices, which are ordinarily taken for granted, are instead socialized and learned across societies. Chairs occupation (and position) are paradigmatic of roles, meanings, and positions that this mundane object covers for different cultures in different ages.

References

'A Reader's Guide to Orthodox Icons'. 2017. *Hetoimasia. The Throne of Preparation*. 27 June. https://iconreader.wordpress.com/2013/06/27/hetoimasia-the-throne-of-preparation/ [Accessed September 2017].

Bergamasco, Porzia and Valentina Croci. 2012. *Sedie*. I libri di Artdossier. Florence: Giunti Editore.

Chaplin, Charlie. 1940. *The Great Dictator*. Charles Chaplin Productions.

Courpasson, David. 2017. 'The Politics of Everyday'. *Organization Studies*, 38(6), 843–59.

Cranz, Galen. 1998. *The Chair: Rethinking Culture, Body and Design*. New York: W. W. Norton.

Davis, Matthew C., Desmond J. Leach, and Chris W. Clegg. 2011. 'The Physical Environment of the Office: Contemporary and Emerging Issues'. *International Review of Industrial and Organizational Psychology*, 26, 193–237.

De Dampierre, Florence. 2006. *Chairs: A History*. New York: Abrams.

Elsbach, Kimberly D. and Beth A. Bechky. 2007. 'It's More Than a Desk: Working Smarter through Leveraged Office Design'. *California Management Review*, 49(2), 80–101.

Fleming, Peter and Andrew Sturdy. 2011. '"Being Yourself" in the Electronic Sweatshop: New Forms of Normative Control'. *Human Relations*, 64(2), 177–200.

Golden, Timothy. 2007. 'Co-Workers Who Telework and the Impact on Those in the Office: Understanding the Implications of Virtual Work for Co-Worker Satisfaction and Turnover Intentions'. *Human Relations*, 60(11), 1641–67.

Halford, Susan. 2004. 'Towards a Sociology of Organizational Space'. *Sociological Research Online*, 9(1), 1–16.

Highmore, Ben. 2011. *Ordinary Lives: Studies in the Everyday*. Abingdon: Routledge.

Hollingsworth, Amdrew. 2008. *Danish Modern*. Salt Lake City, UT: Gibbs Smith.

Ionesco, Eugene. 1952. *Les Chaises*. Paris: Gallimard.

Miller, Andrew H. 1997. 'The Specters of Dickens's Study'. *Narrative*, 5(3), 322–41.

Millward, Lynne J., S. Alexander Haslam, and Tom Postmes. 2007. 'Putting Employees in Their Place: The Impact of Hot Desking on Organizational and Team Identification'. *Organization Science*, 18(4), 547–59.

MoMA Learning. 2018. 'Chairs'. https://www.moma.org/learn/moma_learning/themes/design/chairs [Accessed 20 February 2018].

Mumby, Dennis K. and Cynthia Stohl. 1991. 'Power and Discourse in Organization Studies: Absence and the Dialectic of Control'. *Discourse & Society*, 2(3), 313–32.

Ochsner, Jeffrey K. 2016. 'Meditations on the Empty Chair: The Form of Mourning and Reverie'. *American Imago*, 73(2), 131–63.

Rybczynski, Witold. 2016. *Now I Sit Me Down: From Klismos to Plastic Chair—A Natural History*. New York: Farrar, Straus & Giroux.

Schwär, H. 2018. 'Google has an Office in Switzerland where Employees Take Meetings in Gondolas, Slide between Floors, and Are Summoned to Coffee by the "Heidi" Theme Song'. *Business Insider*, 26 January. http://www.businessinsider.com/google-zurich-headquarters-tour-2018-1?IR=T [Accessed 26 October 2017].

Tolpin, Marian. 1968. 'Eugene Ionesco's "The Chairs" and the Theater of the Absurd'. *American Imago*, 25(2), 119–39.

Tuncer, Sylvaine and Christian Licoppe. 2018. 'Open Door Environments as Interactional Resources to Initiate Unscheduled Encounters in Office Organizations'. *Culture and Organization*, 24(1), 11–30.

Van Gogh, Vincent. 1978. *Complete Letters of Vincent Van Gogh*. New York Graphic Society.

Whitney, W. D. (ed.). 1894. *Century Dictionary and Cyclopedia*. Vol. 1, part 4. New York: The Century Company.

CHAPTER 9

···

CLOCK

···

MELISSA GREGG AND TAMARA KNEESE

IN Western culture, the clock is the standard object used to coordinate and manage time.[1] The word 'clock' derives from the Latin word for bell, and the ringing of bells in central clock towers provided one of the first examples of time-based synchronicity for villagers. The clock has long been a social technology, a way of authorizing a singular source of information to propel collective activity. In this chapter, our interest is in exploring whether this social function continues in quite the same way in the wake of digital technology. We investigate the particular role of the clock in the workplace—how a predictable relationship to time accrued value for The Organization as an institutional form. Beginning with the stopwatch of Frederick W. Taylor, and continuing in the time and motion camera adopted by Frank and Lillian Gilbreth, a rotating arm on the face of a dial has been the standard visual gesture marking the passing of time, and thus, the fluctuations of wealth in the dominant logic of capitalism. Over consistent manifestations of management thought, workers' productivity has been regarded as a performance to be timed, or an athletic endeavour to be recorded and improved upon (Sloterdijk, 2013). Just as in sports, individuals have been encouraged to see their work as a personal endeavour that can be optimized under the training discipline of the clock. Factory hands and secretaries populated the Gilbreths' industrial films, acting as models for speed-enhancing techniques. But the clock also reflects the more abstract tedium of the office workday. Our imagination of modern work is strongly associated with the notion of '9-to-5'. The clock is at the heart of these normative frameworks governing bodies and their labours. Yet its role in the organization of work is changing with the influence of new technologies.

Consider the slogan that hangs from a high-rise in downtown Manhattan, marking the location of on-demand office space provider, WeWork. 'Thank God It's Monday' reads the banner—a humorous take on the cube-dweller's relief at the arrival of Friday

---

[1] While this chapter focuses on Western iterations of the clock, a growing body of work is devoted to understanding alternative temporalities of measure and value, including indigenous and queer orientations. See Rifkin (2017) and Bean (2016) for indicative examples.

and the weekend. The clock measuring work time is being rendered obsolete by
WeWork's original user base: the freelancer, the contract dependent consultant seeking
'a community of creatives' to share the costs and fluctuating emotions of an independent
career.[2] For these professionals, the idea that work might be limited to set hours or
schedules appears laughably outdated. Commitment to work is not marked by long
hours or physical proximity to a boss, but by passion. Advertising for gig economy
platform Fiverr evokes this sensibility with its image of a thin young woman claiming
'lack of sleep' as her 'drug of choice' (Scott, 2017). Such visual aesthetics embody Steve
Jobs' imperative for the iPhone generation: 'Love what you do' (Tokumitsu, 2015).

The 'do what you love' mantra is a symptom suggesting the clock's social role
coordinating schedules may be fading. The idea of limiting work hours makes little
sense for those chasing more than one 'gig' at a time to make up for insufficient pay.
Across the United States and Europe, growing numbers of contingent workers find
themselves unable to access the benefits of predictable work, or the labourist ideal of the
eight-hour day (Huws, 2014; Manyika et al., 2016). People who were once 'on the clock'
or 'punching in' to a physical workplace are now being replaced by 'self-starters' who
must be capable of managing time for themselves. The fixed workspaces of modernity
have given way to new ways of measuring time, presence, and productivity beyond
physical location. Within this altered landscape, the centralized bell or whistle at day's
end is a mythical holdover; indeed being in sync with others may risk individual creativity
or the flow of innovation. The clock is present in more discreet or concealed ways,
integrated into fitness trackers, software systems, and other wearable appendages
organizing workloads and schedules. As we notice the clock's changing historical,
cultural, and technical forms, we begin to realize the significance of its role in ensuring
enrolment in collective endeavours.

THE CLOCK THROUGH TIME

Media theorist Marshall McLuhan presciently enumerated the effects of media
technologies like clocks on the human psyche and body: 'The clock dragged man out
of the world of seasonal rhythms and recurrence, as effectively as the alphabet had
released him from the magical resonance of the spoken word and the tribal trap'
(McLuhan, 1964: 155). The seasons and their associated patterns of night and day had
provided social cues (as did women's reproductive cycles, upon which experience
McLuhan is silent): sunrise and sunset marked the beginning and end of much activity.
The sundial offered a way of mediating this relationship to the rhythms of the sun and its
shadows. The pocket watch created the sensation of holding time in your hand; keeping

---

[2] WeWork changed its key marketing slogan from 'a community of creatives' in 2016. More recently,
the website uses the phrase 'Be the founder of your life'—reflecting a more individuated address for its
on-demand office services.

time, as it were. Electric light brought new opportunities for measuring and utilizing time. Thanks to lamps, reading and writing at night became viable, as did commercial activities and sporting events. Night no longer automatically meant the cessation of work or the inevitability of rest. As McLuhan notes, however, the clock itself has specific far-reaching effects. Its dynamics broke the day into functional moments, providing a credentialing force with objective authority. The clock's legitimacy as a scientific measure held all the colonial biases of territorial conquest ably captured in McLuhan's reference to the 'tribal trap'.

The clock in the town square, factory, or classroom produced a different sociality than the personal watch. Social understandings of time also changed in scale. Jimena Canales (2016) traces how features in clock design had specific consequences. In the 1850s, clocks were able to recognize a tenth of a second. Canales writes of the day in 1972 when all clocks were shortened to institute agreed upon metrics. The new reliance on atomic time would have lasting cultural significance, as time no longer needed to be calibrated to daylight and the seasons. Rather than completely dispensing with the established calendar or readjusting it, scientists settled on the concept of the leap year. Clocks, rather than human routines and the signals of the natural world, determined time.

In the 1880s, railroad companies broke the North American landscape into four distinct time zones, ending confusion over the proliferation of local times. The telegraph helped to bridge these new expansions across human scales of perception. James Carey demonstrates how standardized time was built on these novel technologies. Standard time meant that clocks across the country chimed together (Carey, 2009: 17). The telegraph, along with standardized time, united futures markets in Chicago and San Francisco, making the world seem that much smaller. The annihilation of time and space promised by the telegraph prefigured McLuhan's twentieth-century global village. It impacted burgeoning capitalist markets as much as train schedules and it affected how people imagined their larger communities and nations.

The Clock as Management Accomplice

Charlie Chaplin's classic 1936 film, *Modern Times*, maps the factory's temporal logics, the comings and goings of commuters and their management by a variety of clocks, bells, and whistles. The film shows how factory gears, including those of the clock, encase, manipulate, and even suffocate the individual worker and foreman alike. Clock time is the objective force indicating when to up and down tools, and at what pace, organizing production so each minute of what, a decade prior, in 1926, Henry Ford had instituted as the forty-hour week, was as silently, and predictably productive as the next.

A founding tension of The Organization has been the clock's dual role as management accomplice and worker's salvation. Supervisors use the clock to maximize the output of the firm, while employees exploit its political potential to determine the maximum number of hours tolerated. The brief pact between these objectives played out in the years of welfare capitalism, where pioneers such as Ford instituted just the right number of hours' leisure to enjoy the commodities workers produced in assembly plants. The turning point in these endeavours to yoke employer and employee interests through the clock came with business interest in individual psychology. This early work, carried out by the likes of human factors engineer Lillian Gilbreth, anticipated the link between productivity and athleticism—that is, workers' personal compulsion to demonstrate their accomplishments through labour. With a PhD in psychology—rare for a woman of her time—Gilbreth recognized the value workers gained from producing an archive of achievement. This realization became evident in the films she made with her husband, Frank, who timed workers' tasks against the backdrop of a ticking clock (Figure 9.1). In Gilbreth's account, knowledge of the recording underway created interest in the work, for with it 'comes the possibility of a real, scientific, "athletic contest"' (1914: 33–4). Racing against the clock was a way of creating self-esteem. Gilbreth's articulation of individual outputs with the moral stimulus of self-improvement is the framework through which management regimes encouraged workers to plan and progress their careers for decades.

Today, the celebration of speed has revitalized sociological inquiry (Wajcman and Dodd 2017). Many theorists speculate on the ways that time itself has shifted with the hypermediated digital age. Scholars and social critics have noticed a general compressing of time. Individuals feel 'pressed for time', according to Judy Wajcman (2015). Meanwhile Jonathan Crary (2013) links the birth of 24/7 news and work to the end of sleep. In this view, the homogeneous time of modernity is gone, and with it, the promise

FIGURE 9.1 Still from the Original Films of Frank B. Gilbreth
(Image from https://archive.org/details/OriginalFilm)

of rest. We have 'sleep' mode for our devices, and 'snooze' options for notifications. Our digital objects buzz faintly in the background rather than stopping entirely. They sleep and snooze, but there are fewer instances where technologies abide by an on/off switch. This is a poignant metaphor for the pressing, accelerated way humans are expected to live under late capitalism.

The idea of accelerationism is closely associated with the technology hub of Silicon Valley, which seeks ever-faster solutions to social problems. Accelerationism is also a liberation theology proposed by radical progressives like Nick Srnicek and Alex Williams (2015), who are convinced that speeding up technological progress will free humanity from the shackles of the clock.[3] Escaping the drudgery of desk jobs and regimented schedules, so the logic goes, people will be able to dream again. Their days will no longer be controlled by hourly concerns but by their own desires and ambitions. As it currently plays out, however, acceleration typically means producing more in less time. Wajcman (2015: 74) suggests this time pressure results from multiple converging elements: an increase in the *volume* of work expected of employees; an increase in *temporal disorganization* as worker and workplace lose their innate proximity; and the phenomenon of *temporal density*. There is an intensification of job function when multitasking and moving between different projects becomes ingrained. The clock was originally used by Taylor and the Gilbreths to measure one specific job, broken down to individual pieces (what the Gilbreths termed, in a spin on their own name, 'therbligs'). Now the psychological toll of windows, tabs, and feeds filtering information through devices all at the same time—on top of the hectic pace of meeting-heavy workplace cultures, where software programs schedule back-to-back meetings with no breaks for biological or emotional needs—culminates as a feeling of perennial context switch. Productivity is sought and measured in spite of work's immaterial nature: new technologies and platforms for labour evade physical and temporal architectures. Rather than sitting in cubicles watching the clock, people work remotely from home, coffee shops, or hybrid co-working spaces. The flexible workforce tallies emails and phone calls, texts and Slack updates, Google Hangouts, Skype chats and tweets from dawn until dusk and beyond.

For those in food delivery, transportation, and other 'taskified' jobs, completing assignments in a limited amount of time is the primary requirement. Restaurant couriers promise your meal in 30 minutes or less, the Lyft app informs riders that a driver is three minutes away, and logistics companies strive for same-day delivery. By saving the consumer time, these business models place the burden on workers both to keep pace and to manage how they experience time (Sharma, 2014). Meanwhile, the personal logistics challenges of disaggregated workers take place at the behest of centralized scheduling software products that are owned and operated by global firms (Rossiter, 2016). We struggle to place limits on labour time. The role of the clock in providing a recognized and adequate record of work feels wholly inadequate. In the

---

[3] Benjamin Noys critiques the ideology in his 2014 book, noting how past accelerationist fantasies were also associated with fascism and misogyny. Kathi Weeks (2011) provides a radical feminist account of a possible post-work society.

decentralized, secular organization neither the clock tower nor the church bell serves the function of ceasing activity. The digital economy does not guarantee freedom from the clock for all.

The clock was the primary index of alienated labour when employees relied on companies to organize their commitments. As the name implies, 'welfare' capitalism instituted clock time as the trade-off for securing predictable income, vacation days, and other temporal benefits. But as the number of middle-class employees continues to dwindle in North America and European states alike, few are able to negotiate with the employers and protocols now dictating productive labour time. Monday holds less significance when the majority of the workforce is no longer enjoying, let alone being disciplined by a 9-to-5 schedule. To imagine Monday morning as the start of the work week seems almost quaint in an age when work happens *around* the clock and around the world.

Union organizers at the turn of the last century campaigned on the human right to have eight hours for work, eight hours for sleep, and eight hours for leisure. This organization of time was thought to offer respite from constant toil, providing a universal regulatory framework for waged labour. But the clock loses resonance as an organizational tool when individuals are encouraged to welcome the freedom they have to keep their own time. It is not so much that workers are no longer subjected to the whims of the clock, but they are all on slightly different work schedules, making organizing across sectors all the more difficult. In sum, clock-based measures for productivity, efficiency, and control suited a work world where everyone is on the same beat or schedule. For a substantial chapter in history, the clock provided a unifying force for negotiating the terms of social belonging. This proved crucial for the identity and agency of male breadwinners in particular, whose interests were served by so-called collective bargaining. In the next section, we unpack how the clock as an object defining social time now necessarily entertains a more diverse range of individual relationships, priorities, and rhythms.

THE DISTRIBUTED CLOCK

In recent years the smart watch represents a fundamental broadening of the clock's original organizing function. No longer simply an object observing the time of day for the purpose of coordinating with others, the digital watch incorporates a variety of software applications to enable further efficiency benefits to owners. On the Apple Watch, a sleep tracker reveals the amount of time spent resting, while an activity monitor counts the number of steps taken each day. The body is made curiously legible through these silent data itineraries. It is not incidental that Apple suffered backlash when it failed to include a seemingly obvious tracking function—a period predictor— with its first release. This small detail reflects something broader about a male-dominated technology industry which regularly risks neglecting the alternative temporalities and concerns of diverse users. Other apps already track ovulation and physical signs of fertility such as body temperature. Awareness of the owner's intimate movements adds value and intrigue to times of the day that may otherwise remain oblique and unscrutinized. This view of time also inspires gamification: can the owner get in 1,000 more steps that day or optimize their sleep cycle to include an extra hour of REM?

Beyond the watch, wearable technologies offer extensions of the clock's productivity premise, creating affective attunement with the context of activity. Wearable sensors such as the Thync electromagnetic patch monitors brain activity through EEG sensors. Like the Apple Watch, Thync can diagnose and optimize human activity by creating subtle 'energy' infusions that prepare users for the ups and downs of long and emotionally demanding mental labour. According to the product website:

> The fundamental mechanism of action centers on modulating cranial and spinal nerve pathways using a combination of targeted electrode placement and proprietary transdermal electrical neuromodulation waveforms. These nerve pathways synapse with key brainstem nuclei involved in modulating stress levels, mood, and sleep cycles.

These increasingly mainstream techniques dedicated to personal well-being are also engaged in producing professional job *fitness* through wearable devices (Gregg, 2018). As such, they continue the idea of athleticism first introduced through the clock and the stopwatch in the workplace. Rather than organizing time for collective use or interests, however, the personal address of these technologies individualizes everyday experience to tiny self-management moments. The varied activities of a typical day come to be witnessed and recorded on the same logic of the corporate clock, which always has the efficiencies of the enterprise in mind.

The biorhythms of digital trackers insulate workers from collective thought and action regarding the organization of labour and its limits. Collectivity is re-imagined as inward facing; a personal dataset to be harvested. Accumulating the body's daily record of activity, users can determine how they measure up in relation to others. In these ways, the clock becomes a distributed object. Individual tracking devices provide a platform for users to summon a sense of their own coherence from many private data streams. The clock is less definitive in determining one's relationship to the social so much as it is a mode of perfecting an enhanced relationship to oneself.

AFTER THE CLOCK: ORGANIZATION UNBOUND

If the clock has most often acted as an index of shared time, stitching together imagined communities, it may no longer be the best measure of work or accomplishment in a digital economy. The clock still organizes time when work takes place in flexible environments.

But what has changed is the sense of human agency behind assessments of labour expenditure and value. As so many ridesharing drivers and activists now realize, it is impossible to argue with an algorithm in the effort to claim labour rights. As technological innovations impact the experience of work, then, our tools for measuring (and therefore limiting it) must shift as well. Workers who are folding together precarious contracts and non-continuing work opportunities cannot lay claim to predictable time off in the same way that elite workers might (Perlow, 2012). Instead, they are challenged to experiment with new techniques for organizing the activities of their lives beyond the 'hegemony of clock time' (Adkins, 2009). Productivity apps are a common method to manage time and energy in short bursts and create a semblance of order (Gregg, 2015). There is a certain pleasure to tracking one's progress and meeting self-imposed miniature deadlines (How much can I write in this ten-minute window? If I finish this paragraph, will my laptop let me check Twitter?). Software services like the ironically named 'Freedom' or the digital Pomodoro timer promise to maximize productivity by breaking the continuous workday into segments. Productivity tools can block email to facilitate concentration or scold us for writing too long without moving. Output might be measured in words rather than minutes. Apps act as clock bosses, forcing us to focus. By complying, we can avoid reprimand and even see a kitten.[4] The Apple Watch does not just idly measure and track, but it physically taps users whose heart rates suddenly go up, reminding them to breathe. Apps for health and well-being aim to build space for meditation, contemplation, and rest in the course of the day. These are all methods of recalibrating the body and mind to accommodate the constant cycle of work that escapes clock measure. For precariously employed workers, these efforts in time manipulation occur on a short-term horizon that helps to offset thoughts of an unknown future.

In the shift from the stopwatch and factory clock to the wearable smart device, we might ask: what social infrastructure is lost? Which new media objects will guide our energy expenditure in workplaces of the future, and how can they encourage collective thinking? As workers are told to do more in less time, and focus on output rather than the number of hours spent toiling, what does the clock continue to do?

One way of reading our chapter is to conclude in gig economies based on event time, the organizing technology of labour and measure is no longer the clock, but the sensor. The GPS tracker issues coordinates for bodies to assemble just as the EEG sensor assists the brain with affective attunement. The clock is a not a singular object like a bell tower, but a distributed instrument embedded with predetermined efficiency protocols. Keeping time is less a means of counting or records but a way of setting expectations for the efficient delivery of people and things. Clocks are prospective guides for our activities and ambitions, a way of pacing ourselves in relation to the demands of a day. These activities are of course not simply work related. Reminders or alerts on our devices can help us with all kinds of everyday duties, from the smart speaker timing the dinner

---

[4] See the 'Write or Die' app https://itunes.apple.com/us/app/write-or-die/id476458361?mt=8 or 'Written? Kitten!' http://writtenkitten.co/.

cooking to the UPS package charting its way to the front door. As the Internet of Things takes shape, objects like coffee makers, garages, and heating systems may all follow schedules according to sensors that monitor temperature and movement. Beyond the clock and its ticking dial, there are many ways that intelligent 'things' will continue to produce rhythm and order in our lives.

Conclusion: Mediating Life and Labour beyond the Clock

In the broadest sense, the clock has been a way of ensuring synchronicity in the endeavours we attempt collectively. It remains a material artefact representing an entire constellation of ideologies regarding time, space, and productivity. In the examples we've shown in this chapter—the watch and the sensor—the temporality first introduced by the clock perverts this collective intent. Personal timers and trackers help isolated individuals reacquaint themselves with internal signals. The efficiency logic embedded in the clock even turns out to be a way of removing the worker from social entanglement or reciprocity. This prompts the question: are productivity apps and engineered smart spaces the only alternatives to the hegemony of clock time?

The difference between time management in the organizational era—defined by to-do lists, clock time, and a highly gendered division of labour—and time management in the network era—where employees at all levels are equally responsible for their productivity—is the suspension of a clock-based ontology. The self-sufficient worker downloading time management apps today takes on the imperative of productivity despite the generalized lack of employment and thus temporal security. In the move from clock time to event-based labour, entrepreneurial workers are expected to provide their own social welfare to withstand surges in labour demand and supply. Table 9.1 foregrounds the material difference this makes to the aspiration of time management when digitized organization, and not *The* Organization, determines the experience of time.

Each section of the table considers a vector of value that The Organization's time once affirmed. Escaping the enterprise, today's worker is free to organize herself according to the terms of her own trade and schedule. The ongoing expectation of task and project management is the condition of workplace flexibility which favours some more than others. Meanwhile the reciprocal benefits of employee/employer, now exhausted, occasion a shift to the temporalities and nodes of loyalty provided by the network.[5]

---

[5] Even outside the workplace, Natasha Dow Schüll describes how personal routines of self-inventory and self-adjustment afforded by technology provide individuals 'the opportunity to cultivate...an attitude of subjective equanimity in the face of uncertainty' (2016: 556). In her analysis of online gamblers, the particular quality of composure long-term players develop to avoid the pitfalls of emotional involvement appears to equip them with a broader subjective 'readiness' to withstand and endure everyday life.

Table 9.1 From clock time to logistical networks

	Productivity	Logistics
objective	Complete	Coordinate
resource	Tool	Provision
time	Measured	Anticipated
hours	Clocked	Billed
location	Fixed	Distributed
asset	Sold	Circulated
data	Stored	Synched
loyalty	Firm	Network
power	Enterprise	Worker

Table first published in Melissa Gregg, *Counterproductive: Time Management in the Knowledge Economy* (2018), reproduced by permission of Duke University Press

Optimistically, we might venture that an awareness of alternative temporalities will improve the reach and appeal of chronopolitics. This means defining time in terms that suit the needs of many people—workers and otherwise—whose schedules are not well served by The Organization. Our initial attempt to sketch the coordinates of labour in a period of schedule decentralization is intended to provoke further examples of workerist vocabularies in the digital era. Mobile and sensor technologies reconfigure not just the location but also the time of value-generating activity. Right now, the currency of data flows dictates which content and reputations gain velocity and power. These factors suggest the futility of maintaining sole allegiance to clock time if workers are to flourish together in a new temporal field. All of us stand to benefit from better technologies of organization to set an appropriate pace for an 'always on' world.

REFERENCES

Adkins, Lisa. 2009. 'Sociological Futures: From Clock Time to Event Time'. *Sociological Research Online*, 14, 4–8.

Bean, Sam. 2016. *Feminism's Queer Temporalities*. New York: Routledge.

Canales, Jimena. 2016. 'Clock/Lived'. In Amy Elias (ed.), *Time: A Vocabulary of the Present*. New York: New York University Press, 113–28.

Carey, James W. 2009. 'Technology and Ideology: The Case of the Telegraph'. In James W. Carey, *Communication as Culture: Essays on Media and Society*, revised edn. London: Routledge, 155–77.

Crary, Jonathan. 2013. *24/7: Late Capitalism and the Ends of Sleep*. New York: Verso.

Gilbreth, Lillian. 1914. *The Psychology of Management: The Function of the Mind in Determining, Teaching and Installing Methods of Least Waste*. New York: Sturgis & Walton.

Gregg, Melissa. 2015. 'Getting Things Done: Productivity, Self-Management, and the Order of Things'. In Ken Hillis, Susanna Paasonen, and Michael Petit (eds), *Networked Affect*. Cambridge, MA: MIT Press, 187–202.

Gregg, Melissa. 2018. *Counterproductive: Time Management in the Knowledge Economy*. Durham, NC: Duke University Press.

Huws, Ursula. 2014. *Labor in the Global Digital Economy: The Cybertariat Comes of Age*. New York: Monthly Review Press.

McLuhan, Marshall. 1964. *Understanding Media: The Extensions of Man*. Cambridge, MA: MIT Press.

Manyika, James, Susan Lund, Jacques Bughin, Kelsey Robinson, Jan Mischke, and Deepa Mahajan. 2016. 'Independent Work: Choice, Necessity, and the Gig Economy'. Report by McKinsey & Company, McKinsey Global Institute, October.

Noys, Benjamin. 2014. *Malign Velocities: Accelerationism and Capitalism*. London: Zero Books.

Perlow, Leslie. 2012. *Sleeping with Your Smartphone: How to Break the 24/7 Habit and Change the Way You Work*. Boston, MA: Harvard Business School Press.

Rifkin, Mark. 2017. *Beyond Settler Time: Temporal Sovereignty and Indigenous Self-Determination*. Durham, NC: Duke University Press.

Rossiter, Ned. 2016. *Software, Infrastructure, Labor: A Media Theory of Logistical Nightmares*. New York: Routledge.

Schüll, Natasha Dow. 2016. 'Abiding Chance: Online Poker and the Software of Self- Discipline'. *Public Culture*, 28(3), 563–92.

Scott, Ellen. 2017. 'Fiverr's Deeply Depressing Advert has Really Annoyed People'. *Metro News*. http://metro.co.uk/2017/03/10/people-are-not-pleased-with-fiverrs-deeply-depressing-advert-6500359/.

Sharma, Sarah. 2014. *In the Meantime: Temporality and Cultural Politics*. Durham, NC: Duke University Press.

Sloterdijk, Peter. 2013. *You Must Change Your Life*. Cambridge: Polity Press.

Srnicek, Nick and Alex Williams. 2015. *Inventing the Future: Postcapitalism and a World Without Work*. New York: Verso.

Tokumitsu, Miya. 2015. *Do What You Love: And Other Lies about Success and Happiness*. New York: Regan Arts.

Wajcman, Judy. 2015. *Pressed for Time: The Acceleration of Life in Digital Capitalism*. Chicago and London: University of Chicago Press.

Wajcman, Judy and Nigel Dodd. 2017. *The Sociology of Speed: Digital, Organizational, and Social Temporalities*. Oxford: Oxford University Press.

Weeks, Kathi. 2011. *The Problem with Work: Feminism, Marxism, Antiwork Politics, and Postwork Imaginaries*. Durham, NC: Duke University Press.

CHAPTER 10

....

CLOUD

....

JOHN DURHAM PETERS

CLOUDS are curious objects. That is, it is not clear if they even *are* objects at all. Their mode of organization is more that of an event than a thing. Vaporous entities without a clear material or formal existence, clouds demonstrate something about the disappearing order of time. The British gentleman scientist Luke Howard gave clouds their enduring scientific nomenclature (cumulus, cirrus, stratus, etc.) in an 1802 lecture called 'on the modifications of clouds'.[1] Perhaps by 'modifications' he simply meant 'varieties', but the more interesting possibility is that he understood very well their mysterious ontology: the mode of being of clouds is one of modification. Hegel's famous quip about sound— that it is 'ein Dasein, das verschwindet, in dem es ist', a being that disappears by existing— applies equally well to clouds. John Ruskin, the greatest English critic of the nineteenth century and a fierce lover of cloud beauty, had a similar thought: a cloud, he said, is 'so nicely mixed out of something and nothing' (Ruskin, 1888: 108).

This organizational fluidity allows clouds to carry great freights of significance. As the void is essential to the container, so the nothing is essential to the something in clouds. Few phenomena are quite so packed with meaning. Despite their reputation as flighty and insubstantial, clouds have carried a wide range of discourses, practices, arts, and media for a very long time.[2] Perhaps it is precisely their apparent blankness, mutability, and vanishing mode of being that makes them such a ripe canvas for human creativity and criticism (Figure 10.1). The term 'cloud' has earned its place as a keyword in digital culture because of its widespread recent use for server-based online data storage, but the nearly instant and universal acceptance of this term today would be impossible without the much longer legacy that I trace here.

'Cloud' is etymologically related to 'clod' and the *Oxford English Dictionary* reports that the first but now obsolete meaning of 'cloud' was rock or hill. Perhaps the atmospheric

This chapter is an expanded and modified version of an essay previously published as 'Cloud', in *Digital Keywords*, ed. Benjamin Peters (Princeton: Princeton University Press, 2016), 54–62.

[1] See Richard Hamblyn's highly readable but slightly hagiographic *The Invention of Clouds* (2001).
[2] See the excellent set of articles in *Archiv für Mediengeschichte* 5 (2005).

FIGURE 10.1 John Constable (1776–1837), 'Cloud Study', British, ca. 1821, oil on paper laid on panel. Yale Center for British Art, Paul Mellon Collection

rather than geological sense of the term, emerging in thirteenth-century English, was originally a metaphorical projection of terrestrial to celestial cumulus, just as today's usage goes in the other direction, claiming heavenly status for what is an energy-intensive earth-bound infrastructure of data centres and servers. Tracing the comparative pathways of cognates shows how closely sky and clouds have been associated. 'Sky' in Norwegian, a cognate to the identically spelled English term, actually means 'cloud', and 'welkin' in English, an archaic term favoured by Shakespeare and other poets that means 'sky' or 'celestial vault', is cognate with the Dutch 'wolk' and the German 'Wolke', both of which mean 'cloud'. The deep association of clouds with the upper sphere or celestial realm has conditioned much of their semantic history. Clouds also appear in deep space such as the 'Magellanic clouds' (dwarf galaxies) visible in the Southern Hemisphere, or interstellar clouds of dust and gas known as *nebulae*, a term from the Latin for *clouds*. To be 'in the clouds' has long meant to be up in the sky, and by implication, to be in a fanciful, mystical, or 'ungrounded' state. In the digital 'cloud', the sense of whim, instability, or risk is remarkably absent, perhaps enforced by the consistent use of the singular 'in the cloud' to contrast with 'in the clouds'.

Throughout its varying history 'cloud' has always meant an indefinite agglomeration or amassing of materials, whether of stone, water vapour, or data. Thus the koinē Greek νέφος μαρτύρων 'cloud of witnesses' (Heb. 12:1) was solidified in English via the King

James Bible (1611). In *Paradise Lost* (1667) John Milton mentions a cloud of locusts and most intriguingly, the *Oxford English Dictionary* supplies the 1705 phrase 'a cloud of informations'. Clouds could be crowds or swarms of arrows, flies, or birds, anything bunched that can cast a shadow. This sense also extends to use of *cloud* as a verb in English since the sixteenth century. 'To cloud' means to cover with darkness, obscure, or dim, and can have related sense of ill humour or gloom as in 'cloud of suspicion' or 'under a cloud'. Such negative meanings, remarkably, have little currency in the digital 'cloud', which is consistently cast in IT advertising imagery as a fair-weather cumulus, not an ominous storm cloud.

Perhaps the oldest discourse around clouds is a theological one. In Homer νεφεληγερέτα is an epithet for Zeus ('he who collects the clouds'); clouds can be just as important attributes of divinity as thunder and lightning. In the Hebrew Bible, *YHWH* has his habitation in 'the cloud' (not clouds), and guides the people of Israel on their desert sojourn with a pillar of cloud, a sign that both obscures his presence but also thereby points to it.[3] (This usage foreshadows the recent notion that the cloud—note the singular again—is secure and infallible.) In the New Testament, Christ is said to return to the earth 'in the clouds', a theme beloved of Baroque painters, and in popular culture, there are innumerable images of a cloud heaven inhabited by God, saints, and angels. Due to their association with the celestial realms, clouds are ready metonyms for deities of all kinds. Cross-culturally, deities associated with clouds are not always male: in Norse mythology, Frigg is a goddess who spins the clouds and tells the future, and in China, 'the play of clouds and rain' can be a metaphor for sexual union.[4] I can hardly begin to catalogue all the things that clouds have meant.

A counter-discourse around clouds is meteorological, that is, reading clouds for physical rather than metaphysical signs (Figure 10.2). Farmers and sailors have always known that clouds are harbingers of weather, and Aristotle uses the sky to make an apparently banal lesson about interpretation in the *Rhetoric*: 'if it is cloudy, it will probably rain' (Aristotle, *Rhetoric*, 1393a). Though it is difficult, as Friedrich Kittler points out, to separate weather and the gods, there have been efforts since antiquity to read the sky secularly and scientifically (Kittler, 2006: 79). In *De rerum natura*, Lucretius reproved people who found faces and animals in the clouds, saying that we should see them as the fortuitous motions of atoms. He made fun of the idea that Jove used lightning to strike evil people, and noted that temples to the gods were often targets of lightning bolts. His aim was to liberate people from the superstition of seeing divine signals in the sky by explaining the randomness of the atomistic motion that he thought governed all things.[5]

Other ancient thinkers emphasized the random quality of cloud shapes and taught us instead to understand images in the clouds as figments of imagination and thus as resting entirely in the viewing subject and not in the nebular object (Damisch, 1972: 52–7, *passim*).

---

[3] For enormously rich detail see Luzarraga (1973).
[4] Thanks to Rasmus Kleis Nielsen and to Guobin Yang for cross-cultural information.
[5] *De Rerum Natura*, especially books 4 and 6.

FIGURE 10.2 New Haven clouds

(Photo by the author)

(In the meta-comic *Calvin and Hobbes*, Calvin spots a cloud and says, 'it must be a sign!' Of what, asks Hobbes. 'Of very peculiar high altitude winds, I guess.')

This debunking reading of clouds is anti-theological in the case of Lucretius and anti-philosophical in the case of the comic playwright Aristophanes, whose play *The Clouds* mocks Socrates, his head-in-the-clouds thoughts, and 'Cloud Cuckoo Land'. Both authors associate clouds with airy, theoretical, insubstantial things, whether gods or 'ideas'.

There are elements in Jewish and Christian religion, however, that are just as critical of reading God or anything else directly in the clouds. The Prophets were fiercely icono-clastic and denounced the reading of clouds for portents or omens, and God stumps Job by asking him to explain the origin of snow, hail, lightning, and rain, as if these would forever elude his understanding.[6] Jesus, squarely in this tradition, when asked to provide some curious spectators a sign in the heavens, gave an answer as sarcastic as anything in Lucretius. If you want a sign in the sky, he said, here is one: red sky at night means good weather tomorrow, and red sky at morning means bad weather soon. His 'sign' was a les-son in everyday forecasting, an ancient bit of weather lore.[7] Those who say there are no gods in the sky or say there is only one beyond the sky can agree that it is foolish to look skyward for divine tokens; the anti-divination discourse about clouds, in other words,

---

[6] E.g. Job 38:22–30.   [7] Matt. 16: 2–3.

has both atheistic and (mono)theistic sources. Yet in popular religiosity the clouds remain something to conjure with, habitations for all manner of heavenly entities.

The debunking discourse around clouds comes down especially hard on the practice of finding animal and human shapes in them. Hamlet's toying with Polonius must be the most famous example:

> HAMLET: Do you see yonder cloud that's almost in the shape of a camel?
> POLONIUS: By th' mass and 'tis, like a camel indeed.
> HAMLET: Methinks it is like a weasel.
> POLONIUS: It is backed like a weasel.
> HAMLET: Or like a whale.
> POLONIUS: Very like a whale.   (*Hamlet*, III.ii.361–7)

Polonius' foolish suggestibility, as mutable as the clouds themselves, enforces the idea that clouds completely lack objectivity, a notion found in the use of the cloud-like 'thought bubble' in cartoons, a convention that links the mental privacy of subjective states with clouds. Here clouds are seen as the embodiment of everything flighty and thus inspire comic commentary on how people manage to create meaning out of the blue. Since anybody can find anything in them clouds stand for the unreliable treachery of perception and for the whimsical unreliability of subjectivity itself. Thus some critics of a scientist bent can call humanistic research 'cloudy' or 'sky-writing'. Nonetheless, the British Cloud Appreciation Society, founded in 2004 with John Ruskin as its patron saint, has published a coffee table book and maintains a website dedicated to charming and droll images of 'clouds that look like things'.[8] Humorous—or ominous—treatments of reading visions in clouds remain a staple in popular culture.[9]

One of the standard jokes in the *that-cloud-looks-like-a-?* repertoire is 'that cloud looks like a cloud'. In fact, there is a strong tradition in the past five centuries of looking intently at and reading clouds for their own sake in art, media, and science. Take art first. Clouds have been a repeated and prominent subject in European painting since at least the Italian Renaissance. The puzzle is why clouds would proliferate when they seem to defy the revolutionary technique of linear perspective, which radically changed Western painting, drawing, and architecture. Clouds lack clear edges, morph rapidly, exist as much in colour as in shape, and will not stay still, making them, to put it mildly, less than promising subjects for a temporally frozen medium such as painting.

This is the question art historian Hubert Damisch explores in his great book on cloud painting (1972). Damisch sees in their difficulty a chance for painters to both defy the geometric strictures of perspective and show off their skills. Clouds—abstract, aniconic, sheer image without likeness—are the 'other' to linear perspective. A cloud painter such

---

[8] See the delightful book by its founder Gavin Pretor-Pinney, *The Cloudspotter's Guide* (2006). For a large and diverse list of examples of cloud-gazing in popular culture see http://tvtropes.org/pmwiki/pmwiki.php/Main/ThatCloudLooksLike.

[9] See, for instance, the sequence near the beginning of the film *Up* (2009. dir. Docter), or throughout the film *Take Shelter* (2011, dir. Nichols).

as the seventeenth-century Dutch master Jacob van Ruisdael does not see camels, weasels, and whales in the sky: rather he rigorously documents the clouds in all their visual splendour. Meteorologists even claim to be able to find in his paintings reliable historical witnesses of weather patterns. Long before photography, his skyscapes serve as visual documentation of elusive, ephemeral realities (Ossing, 2002; see also Busch, 2013). Cloud painters such as Ruisdael depicted a curious kind of image that was neither symbol, icon, nor index, but rather atmosphere and process, like the act of painting itself. (There are as many clouds as there are painters: clouds can offer hospitality to a wide range of styles without ceasing to look like clouds. Clouds by Turner, Monet, van Gogh, Georgia O'Keeffe, or Gerhard Richter all look distinct.)

In painting clouds were harbingers of a new kind of image, an abstract one of flow and turbulence rather than symbolic representation. They were among the first abstract objects to be depicted, and in this they are a critical early step in the history of recording media. Friedrich Kittler has famously argued that the acoustic and optical analogue media of the nineteenth century caused a critical historical rupture: photography, phonography, and cinema expanded the realm of the recordable drastically to include non-intelligible and time-based processes, breaking writing's historical stranglehold as a medium of storage for any form of art or intelligence. 'White noise' (*Rauschen*) was Kittler's preferred term for this new class of analogue recordables. Writing could record the words 'he sneezed' but never the complex motion or sound of sneezing, something that became routine in the later nineteenth century. Sound, motion, flow, process all became recordable and thus subject to real-time analysis and time-axis manipulation (Kittler, 1999). The case of cloud painting suggests a possible revision to Kittler's historical narrative of a radical break in the nineteenth century: already in the seventeenth century we have records of temporal processes that evade the symbolic and take place in the real, even if they are not yet open to time-axis manipulation. Ruisdael's clouds are a kind of *Rauschen* or white noise (although, to be clear, they hold many colours) that obeys no semiotic grid but in which one can read everything and nothing. For his part, Ruisdael may have thought that in them you could read the glory of an aniconic God.

The scientific standing of clouds benefited from these transformations. As historians of science have noted, sound recording and motion pictures should be placed into a wider range of nineteenth-century scientific instruments with names ending in *scope* and *graph* that allowed observation and inscription of temporal processes of all kinds ranging from blood pressure and weather to noise and heat. The 'graphic method', as French physiologist and proto-cinematographer Étienne-Jules Marey famously called it, followed innovations in mathematics and modelling such as Fourier equations that could represent fluid dynamics such as sound, heat, and atmospheric aerosols (i.e., clouds). New methods of recording brought new objects onto the scientific agenda: heat, noise, smoke, glaciers, clouds. Indeed, the nineteenth century opened with Howard's proposal for a new taxonomy of clouds. Though rivals proposed alternatives, his Latin-based classification has held steady since. Remarkably, he managed to fix names onto essentially modifiable entities.

Less successful in creating an exact set of standards were international scientific efforts starting in the late nineteenth century to assemble a 'cloud atlas' made up of photographic images of the various types. The particularities of each individual cloud image required scientists to resort to retouching the photographs to make them match the type more accurately. Clouds are never quite capturable by image, language, or thought. Lorraine Daston has shown the inevitable but productive failure of the international cloud atlas to offer a 'collectively willed ontology'. Clouds, as she summarizes, 'flummox description' (Daston, 2016: 53, 47). A defiance against capture or rebellion against recording was one of the things about clouds that so fascinated nineteenth-century writers and artists such as Goethe and Emerson, Shelley and Baudelaire, Constable, Turner, and Monet. The sea of faith might have been draining in the nineteenth century, as Matthew Arnold said, but many still took comfort in the clouds as heavenly objects.[10]

A history of cloud science in the last two centuries would also be a history of many of our leading media technologies. Around the turn of the twentieth century, the Scottish physicist C. T. R. Wilson invented the cloud chamber. Seeking at first to model cloud formation and water vapour, he hit upon an instrument that, thanks to its ability to detect and trace subatomic particles, played an essential role in particle physics in the first half of the twentieth century (Galison and Assmus, 1989). If clouds were once seen as cloaks of reality, in the cloud chamber they had become its revelators.

A similar path between obscuring and revealing was travelled by satellites and computers. During the Cold War, high altitude spy planes could only photograph nuclear facilities and other targets on the ground if there was no cloud cover. One impetus for weather forecasting was espionage: how to predict when clouds would part enough to justify a dangerous surveillance mission. Clouds may have blocked the intelligence cameras but they revealed a great deal about weather to those who learned to read their patterns. Though some looked down on clouds from mountaintops before, as in David Caspar Friedrich's painting *Der Wanderer über dem Nebelmeer* (1818), one revelation from twentieth-century high altitude aerial photography is that our planet is covered with clouds. Space photography flips the point of view by which we see clouds: from heaven downward, not earth upward (Kelsey, 2011). Tracking and modelling clouds requires enormous amounts of data. Though the standard story of modern computer science emphasizes the desire to model nuclear explosions and their aftermath, the demand for weather data has been just as formative in advancing digital technology. (John von Neumann, for instance, was just as passionate about computer applications for weather control as for nuclear explosions; von Neumann, 1955.) The first World Wide Web was arguably formed for watching weather, well before the web as we know it. Clouds sit at the heart of crucial innovations.[11]

Fractal geometry was another spin-off of cloud study, which encouraged both a logic of vagueness and the analysis of the indefinite. Philosophers say that clouds illustrate the

---

[10] On clouds in nineteenth-century thought and art, see Badt (1960); Jacobus (2012); Weber (2012); and Busch (2013).

[11] The whole paragraph is based on Paul N. Edwards's excellent *A Vast Machine* (2010).

'sorites paradox' or heap problem: two grains of rice do not make a heap, and you can remove two grains from a heap without it ceasing to be a heap, but by adding or subtracting grains eventually a heap will either come or cease to be. There is a vague boundary between heap and non-heap that can never be numerically specified. In the same way, there are many possible surfaces that can plausibly claim to be the edge of a cloud.[12] What looks like the edge of a cloud in terms of scattered light to a human eye might not be a significant boundary when understood as the molar density of the water vapour. Ontological indefiniteness is part of the cloud's great intellectual fascination.

It is easy to say that clouds do not mean anything, but the deeper fact is that clouds mean a great deal and that the collective future of the human species may depend on reading them well, at least if we think about rising anthropogenic concentration of atmospheric carbon and the radical changes to climate it implies. Now we face clouds that are no longer undisturbed water vapour. Ruskin was one of several late nineteenth-century observers to note that clouds could be changed by human activity—that they could be profoundly historical, if not unnatural.[13] If his analysis was off scientifically, he still grasped that atmospheres could be susceptible to technical alteration, such as by the massive burning of coal. In a similar spirit, Peter Sloterdijk dramatically claims that the twentieth century was born on 22 April 1915 with the first use of chlorine gas as a military weapon. The revelation of respiratory fragility in this 'air-quake' was part of modernity's longer 'history of atmospheric explication' (Sloterdijk, 2004: 89). No longer after 1915 could humans take the air for granted.

Smoke stacks, nuclear mushroom clouds, cloud seeding, and geo-engineering schemes surround us with artificial clouds. Dealing with human-made atmospheres is a given for miners, divers, air passengers, and especially city dwellers. The artefactual character of clouds is emphasized by recent artists such as Fujiko Nakaya, who builds cloud and mist installations, Berndnaut Smilde, who creates and photographs surrealist clouds *inside* of buildings, or 'Monsieur Moo' who, in a performance about the legal ownership of clouds, transported rain cloud balloons across the US–Canada border in civil disobedience of international law. (There are treaties against cloud-manipulation, e.g., causing flood or drought for military purposes.) Clouds are the exact sort of things that Bruno Latour likes to call 'hybrids' or 'imbroglios'. It is in such a topsy-turvy world of nature–culture confusions that it is possible for the idea that a computational infrastructure is a cloud to be thinkable and even a kind of common sense.

The first result on a Google search for 'cloud' I recently got was an ad for 'Microsoft Cloud', with an image of a data centre topped by a puffy cumulus, as if Microsoft benefited from a celestial benediction. This use of the term 'cloud' may have started in engineering diagrams of networks, but it almost instantly took to the sky, taking selective advantage of the surplus and residue of the term. The web is full of beatific images of laptops sitting on heavenly clouds. The rhetoric of the data cloud likes to

---

[12] http://plato.stanford.edu/entries/problem-of-many/ [Accessed 8 May 2014].
[13] See his lecture, 'Storm Cloud of the Nineteenth Century' (1884). http://archive.org/stream/thestormcloudoft20204gut/20204.txt.

exploit the peaceful, inconsequential parts of the tradition of cloud meanings while suppressing the more ominous parts.

'The cloud' is a huge public relations triumph for the IT industry, but it is profoundly deceptive.[14] It draws on the idea that organizations can be vague and enveloping, environments well beyond the purview or management of a single person. And yet, 'the cloud' of online storage is neither natural nor environmentally friendly: it consists of a gigantic infrastructure of data centres and the worldwide electricity use for the Internet is estimated to equal that of Japan and Germany combined (Mills, 2013). Furthermore, the notion of 'the cloud' downplays the risk of surveillance and privacy breaches. We might think about the term 'cloud-attack' from the First World War (a barrage of poison gas) or 'cloud-burst' from meteorology to counter this blithe ideology. To trust our data to 'the cloud' may invoke old ideas of the benevolent gods above, but the more interesting part of the history of clouds is how much human-built meaning is there to be exploited if you know how to do so. The IT industry would like us to recall nothing but cloud illusions, as Joni Mitchell sang, but in this case, it is better to try to know clouds from both sides now.[15]

In all moments of history, this would be the worst time to think of clouds as purely immaterial, natural, and meaningless things. Because of the void that they are, clouds are rich containers, not only of data, but also of meaning and mystery, whim, and wonder, power and pressure. Their ontological elusiveness and organizational diffusiveness makes them rich targets for a long history of media invention.

REFERENCES

Badt, Kurt. 1960. *Wolkenbilder und Wolkengedichte der Romantik*. Berlin: de Gruyter.

Busch, Werner. 2013. 'Wolken zwischen Kunst und Wissenschaft'. In Werner Busch, *Wolken: Welt des Flüchtigen*, ed. Tobias G. Natter and Franz Smola. Ostfildern: Hatje Cantz Verlag, 16–26.

Cubitt, Sean. 2013. 'How to Weigh a Cloud'. http://theconversation.com/how-to-weigh-a-cloud-19581.

Damisch, Hubert. 1972. *Théorie du nuage: Pour une histoire de la peinture*. Paris: Seuil. Trans. Janet Lloyd as *A Theory of Cloud: Toward a History of Painting*. Stanford, CA: Stanford University Press.

Daston, Lorraine. 2016. 'Cloud Physiognomy'. *Representations*, 135, 45–71.

Edwards, Paul N. 2010. *A Vast Machine: Computer Models, Climate Data, and the Politics of Global Warming*. Cambridge, MA: MIT Press.

Galison, Peter and Alexi Assmus. 1989. 'Artificial Clouds, Real Particles'. In David Goodking, Trevor Pinch, and Simon Schaffer (eds), *The Uses of Experiment: Studies in the Natural Sciences*. Cambridge: Cambridge University Press, 225–74.

Hamblyn, Richard. 2001. *The Invention of Clouds: How an Amateur Meteorologist Forged the Language of the Skies*. London: Picador.

---

[14] See Cubitt (2013) and Mosco (2014).

[15] She explained later that her brilliant song 'Both Sides Now' (1967) was inspired by a passage in Saul Bellow's *Henderson the Rain King* (1959) about how air travel had enabled an historically unprecedented view of the clouds from below and above.

Jacobus, Mary. 2012. *Romantic Things*. Chicago: University of Chicago Press.

Kelsey, Robin. 2011. 'Reverse Shot: Earthrise and Blue Marble in the American Imagination'. In El Hadi Jazairy (ed.), *New Geographies 4: Scales of the Earth*. Cambridge, MA: Harvard University Press, 10–16.

Kittler, Friedrich. 1999 [1986]. *Gramophone, Film, Typewriter*, trans. Geoffrey Winthrop-Young and Michael Wutz. Stanford, CA: Stanford University Press.

Kittler, Friedrich. 2006. *Musik und Mathematik* 1:1. Munich: Fink.

Luzarraga, Jesús. 1973. *Las tradiciones de la nube en la biblia y en el judaismo primitive*. Rome: Biblical Institute Press.

Mills, Mark P. 2013. 'The Cloud Begins with Coal'. http://www.tech-pundit.com/wp-content/uploads/2013/07/Cloud_Begins_With_Coal.pdf?c761ac [Accessed 26 November 2014].

Mosco, Vincent. 2014. *To the Cloud: Big Data in a Turbulent World*. Boulder, CO: Paradigm.

Ossing, Franz. 2002. 'Haarlem's Crown of Clouds: Meteorology in the Paintings of Jacob van Ruisdael', trans. Kari Odermann. http://bib.gfz-potsdam.de/pub/wegezurkunst/haarlem_ruisdael_en.pdf [Accessed 3 May 2014].

Pretor-Pinney, Gavin. 2006. *The Cloudspotter's Guide: The Science, History, and Culture of Clouds*. New York: Perigree.

Ruskin, John. 1888. *Modern Painters*, vol. 5. Sunnyside, UK: George Allen.

Sloterdijk, Peter. 2004. *Schäume*. Frankfurt: Suhrkamp.

von Neumann, John. 1955. 'Can We Survive Technology?' *Fortune*, 51, 106–8 and 151–2.

Weber, André. 2012. *Wolkenkodierungen bei Hugo, Baudelaire, und Maupassant im Spiegel des sich wandelnden Wissenshorizontes von der Aufklärung bis zur Chaostheorie*. Berlin: Frank und Timme.

CHAPTER 11

COFFEE MACHINE

GÖTZ BACHMANN AND PAULA BIALSKI

PICTURE a coffee machine: what do you see? Perhaps a sports-car-like espresso-maker standing on its counter, all sleek and shiny-black, dripping layers of brown syrup-like espresso out of its nozzle. Or you're picturing a plain drip-coffee maker, its body rugged and its glass pot slightly stained, brown from all the years of holding coffee. Or you're imagining a slow-drip white ceramic coffee funnel, with a hot tin pot of water held by a hipster gently trickling a steaming stream over the light brown coffee grinds. Whether it's sipped alone at your desk or in a buzzing team milling about the kitchen, coffee is still our ultimate drink for work. It fuels our labour, it provides temporal patterns for our workdays, and it becomes a method of engaging or disengaging in interaction with our workmates ('no sorry, I already had two coffees today'). The coffee machines that provide us with this precious liquid are caught in webs of social practices, temporal patterns, spatial demands, bodily rhythms, memories and gestures, demanding an organizational nexus of coffee making and drinking, integrated into the organization where it takes place. By looking at a handful of stories stemming from ethnographic research, this chapter shows that coffee is a field-internal analytical device through which we can understand the inner workings of organizations. Such an enterprise is not devoid of pitfalls. This chapter is thus just as much about the misgivings of coffee machinery analysis, as it is about its potentials and charms.

COFFEE AS ANALYTICAL MACHINE

Not everyone drinks coffee at work, of course, but almost all workplaces in the Western world do organize the provision of coffee for their employees and/or guests, that is: they organize forms in which coffee is produced and consumed (Tucker, 2017), and as such, they embed coffee machinery into their environment (Figures 11.1 and 11.2). No such arrangement is fully alike, which in itself is already an analytical device for comparison: the needs of coffee making and drinking cut across class and gender, across white

FIGURE 11.1 The classic: socially rich, taste-wise not so much

(Photo by the authors)

and blue collar work, across all sorts of divides between different organizations and subdivisions in these organizations[1]—yet they also soak up such differences. Just think of the coffee of a company boss, as opposed to the cup of coffee sipped by the cleaning personnel during their early morning shift. Caught up in the micro politics between management and employees, the coffee machine conjures up all sorts of shades of micro distinction and status games, or lack thereof, or its purposefully curated absence. And this list goes on. At play are various other questions such as the organization's care for its workers, the workers' own care for themselves, and the engineering of work intensity and communication flows. Such issues are not only at play between management, customers, and workers, but between departments, work groups, or cliques.

If you take a closer look at what is needed to provide coffee, it becomes quickly apparent why so many issues might arise around the coffee machine: making coffee is a complicated enterprise. It needs the machine itself, cups, coffee, milk and sugar, electricity and water, infrastructures of cold storage and of heating, spaces and times reserved for all of this, and arrangements of the work that this entails. Coffee creates costs, evaporations,

---

[1] For example, Karen Ho, in her ethnographic research around investment bankers, explored how women investment bankers avoided standing near coffee spouts in meetings, in fear that men will 'mistake them for administrative assistants and ask them to help with the food and pour coffee' (Ho, 2009: 328).

FIGURE 11.2 A contemporary variation for the selected few (see also section, 'A Secret Society')

(Photo by the authors)

smells, it evokes questions of taste, and has its own temporalities—think not only of the time of day, but also the duration of drinking, the temporality of coffee waiting for consumption while slowly getting stale, the rhythms of buying coffee, sugar, and milk, and the right time when coffee is not too hot and not too cold. Coffee machines have their own sounds and rhythms. We listen to the coffee machine to figure out what's going on. Its operation is a complex beast in itself, asking for the negotiation of the coffee's ingredients, strength, and the necessary frequency, leading all the way into the body, but also all the way out into the negotiation of the form of coffee that is to be made.[2]

We thus see that both coffee and the machinery of its production and consumption have material properties, infrastructural needs, and affordances (and by 'machinery' we mean all the surrounding equipment, hardware, instruments, tools, and gadgets affiliated with coffee). Yet, at the same time, coffee is also open to symbolic charges of various kinds (indeed, it has been so since its introduction into Europe in the sixteenth century, where it was posed, depending on the century, as a drug, as a lady's drink, as the drink of choice for the busy worker, and so on).[3] All these material and symbolic properties are

---

[2] As Goodman et al. explore, 'in certain times and places, both tobacco and coffee have been classed as illicit commodities with heavy penalties for use or (in this classification) abuse' (Goodman et al., 2014: xiii).

[3] For a deeper look into horticultural practices, the cultural history, social history, and sustainability politics around coffee, see Folmer (2017) and Tucker (2017).

intertwined. They are formed on the level of large cultural discourses with *longue durée*, but also in specific contexts. Together, they constitute the analytical device of coffee machines and coffee machinery.

The term 'device', as Celia Lury and Nina Wakeford propose in 'Inventive Methods', can make explicit how 'a method and its object are linked to each other and with what potentially explosive—or inventive—effects' (Lury and Wakeford, 2012: 8). A device can thus help us think through practices and processes associated with it. The form of analysis that we propose here is also a specific version of what the German romantic tradition has called the 'Andacht zum Unbedeutenden'—yet its translation as 'silent prayer to the insignificant' does not really cut it (Kany, 1987; Scharfe, 1995).[4] This approach takes something that is usually considered as either not important or not meaningful, and gives it full attention. We evoke this tradition to remind us that analysis through devices can have certain pitfalls. It can, for example, result in forms of analysis where everything is explained through coffee machines and machinery. This can open up a space for projections by the analyst. But just as silly are forms of analysis that take coffee just as a mere indicator for larger forces at play. The aim is to find a middle ground, where we take coffee drinking and coffee making seriously, yet also keep in mind that we do so only to develop a field-internal analytical device for this field.

So how to proceed in all of this? We recommend spending some initial time in a sort of dry run, concentrating on what is at stake in coffee and in front of the coffee machine, turning everything we have now mentioned into questions (e.g. what are its temporalities?). We encourage you to do so by yourself, and do not hesitate to follow strange leads, such as, for example: coffee produces stains. Who wipes them away? Do they use a cloth? Who washes the cloth? Who notices that the cloth gets smelly? Who pays for new cloths? Or another example: after the coffee machine finishes pouring coffee into a coffee cup you may need to add milk or sugar to it. One needs something to stir the milk or sugar with. What is used? A spoon? Where do people put the spoon, once they have used it? Do they leave it in the cup, or put it on the table? Do they place the spoon face up or down? Why? There are hundreds of such question chains that you can build up to begin your analysis, and it can be very productive to follow those, too, that extend beyond the coffee machine itself. Once you follow some of these chains of questions into an organization—this is the second step—you will learn a lot about moments of care and power, about networks and groups, about the many layers of ethos in an organization, and so on. You might move back and forth and sideways: studying how coffee is made and drank in one particular field will allow you to develop new chains of questions, and the latter will continue to open up new views about what's going on in other parts of an organization. We suggest you stay for a long time between these chains of coffee questions and the many places and times in organization where coffee machinery comes into play. Only then do you add a final step: complete the analysis of that what you witnessed.

---

[4] Other traditions include the cultural study of the 'ordinary' events of everyday life. Ben Highmore calls this study of moments such as coffee and tea breaks an 'intimacy', or form of attention that looks at the proximetrics of everyday life (Highmore, 2010: 3).

WATER COFFEE CUP RESISTANCE

Imagine a large open office space in a German health insurance company in the mid 1980s. The almost exclusively female office workers provide customer services on the phone and work as data entrists (the manual transfer of information on paper into a computer system). On this particular day, the boss has put out a new order: all coffee mugs placed on the workers' desks are banned. The women get angry. They explain to the ethnographer that this conflict is part of a larger quarrel that started a while back: the coffee facilities of the canteen are too far from their workplace to walk. It takes the women over five minutes to get there, and they have limited breaks to do so. To be able to drink coffee while they work, the women collectively purchased their own coffee machine and placed it near their desks. Some time ago, the boss banned this machine, arguing that the smell of the coffee would waft its way to the nearby customer area (this area is indeed close by, but the work space is not visible to the customers). According to the boss, the smell of coffee would give customers the impression that the office workers are taking breaks constantly, enforcing a negative image of the health insurance company as being a lazy, bureaucratic mammoth. The employees had grudgingly accepted this ban, but kept their coffee mugs and used them to drink water. After all, they were nice mugs, often presents from family members or colleagues. But now these water-filled mugs were banned too. How can this be? The customers can't see these mugs, nor can they smell the water filling them! Doesn't this show that the boss simply made up his earlier worries of the coffee smell getting to the noses of their customers? The employees are furious.

THE SOCIAL HEART OF THE OFFICE

Lindenstraße is a popular German television drama series, with its television studio based in Cologne. Once again, we encounter coffee-related anger in an office, now in the early 1990s. The manager of the TV studio—a gentle man, and usually quite popular with his employees—is walking from office to office announcing that coffee machines in the offices are, from now on, banned, pointing out that the canteen is nearby, less than a minute away. This canteen is one of his pet projects. A year ago, he had decided to reinvent the canteen space. A good office environment needs a social centre, he thought. With this motivation in mind, he initiated a redesign of the space. The result was a cosy canteen with a hippy look and feel, filled with artwork. He also hired a female Turkish cook, whose function was not only to prepare food, but also to provide a homey and 'motherly' atmosphere. 'This will be the social heart of everything,' he had said to his staff members. And who could say no to this idea? However, in the months that followed, coffee machines started to pop up in the offices. Some were catering to specific needs and tastes; others were just plain straightforward coffee machines, used to brew filter coffee in a time before the latter became a symbol of distinction. And now they were banned. Again, the employees got grumpy. The argument that the canteen provided delicious free coffee did not calm them down. They were willing to pay for their own coffee. That wasn't the issue here. The problem was that in the hallway on the way to the canteen as well as in the canteen itself, you'd often bump into your colleagues. This would not only lose you time with all sorts of chitchat, you'd also have to listen to the latest updates about the perils of TV production, leading almost inevitably to volunteering for extra work, as one could not evade helping colleagues in this workplace's friendly ethos. The 'social heart' itself was the problem: it exposed you to more work. So what happened? Some people gave up on the coffee machines, accepting the ban, others persisted, and the studio manager did not follow up on those who did. After all, you cannot force a social heart.

SIMULATED COMMUNAL COFFEE

When imagining the work spaces and cafeterias of the tech industry today, images of chocolate bars, lush coffee machines, bearded baristas, and plates of sushi probably come to mind. However, such perks are not an entirely new phenomenon: it's the early 1990s, and the field is a software company in Berlin. Here, an ethos of worker-owned collectives in the partially left-leaning tech scene of the moment, in combination with an exuberance of profit margins in certain corners of the software industry, creates a similar workplace environment. Indeed, this is more than just about perks. In this particular company, all shares are in the hands of the employees, decisions are made democratically, employees vote for their own new managers, and an egalitarian ethos is highly valued. Of course, this means that coffee has to be made by everyone and for everyone. Yet, an older employee, part of the founding generation of the company and now one of its managers, is unhappy. The engineers, it seems, have organized their collective coffee making all too perfectly. While they still provide refills for the very large coffee machines with metal cans, preventing the coffee from cooling, they have outsourced all further work: the coffee cups are rinsed and placed into the kitchen dishwasher twice a day by the cleaning staff, who remove them from the dishwasher and stack them neatly in the kitchen cupboards for the staff to use. The coffee beans are re-stocked by the kitchen staff, who periodically peek into the automatic coffee machine on a regular basis to check if the beans are running low. The same kitchen staff re-stock the machine with fresh milk twice a day. Everything is optimized for the software engineers to walk into the kitchen, get their coffee, and, if necessary, reset the coffee machine, thus making their coffee not just for themselves, but for their colleagues, too. The result is a real life simulation of communal coffee making. At least for this particular manager, the perfectly engineered coffee making in this kitchen has killed what it pretends to celebrate: its day-to-day negotiation, its everyday care for each other, its equalizing effects, in short: coffee making as a communal activity.

COFFEE SHIFTS

Retail work is tough. Cashiers, salespeople, and warehouse workers conduct their work under pressure and under hierarchies, where, at least in the mid 1990s in Germany, men in middle management attempt to command female workers. The store needs to be staffed at all times. Often, employees can only squeeze in their coffee and lunch breaks when there are fewer customers in the store. It is difficult to have a break at all, surely difficult to have breaks at regular times, and especially difficult to have a break together. But in the particular department store at stake here, things are different. Strongly unionized, the workers have successfully fought for break shifts, which always allow one third of the staff to collectively go on a break at one time. To have coffee ready when the break starts, they devised a complex system of one shift always preparing coffee for the other shift. The chores of coffee have to be organized across their shifts, too, including the task of buying coffee (here, of course, not provided by the company). Whoever buys a new pack of coffee puts her name down on a list. But how often one buys coffee depends on how much you drink. Further chores, such as buying fresh milk or sugar, are organized without lists. You have to have a good gut feeling to get the frequency of purchase just right, and there is quite some bickering over those who are under the suspicion of freeloading. On a particular day, during one of the coffee breaks, this bickering gets more heated: there is no sugar. Someone gets angry. One word leads to another, and soon there are multiple accusations both among the group of this particular break shift, and towards those in the other shifts. The word of these accusations spreads quickly throughout the store. In an attempt to change the debate, someone mentions the possibility of sugar theft, which adds fuel to the fire. On the next day, one saleswoman makes a statement: on the table at her regular spot in the break room stands a thermos bottle, filled with coffee made at home. She has had enough of all this. On the next day, two more women do the same. At the end of the week, almost everyone has a thermos flask in front of them. The collective coffee system has imploded. But after another week, a small group re-starts brewing the coffee collectively. The group grows, thermos flasks disappear, and once more the women return to the old system. The bickering might cause annoyance among them at times, but it is also entertaining, and gets them closer to reaching that gut feeling of knowing what works or what doesn't in their communal endeavour.

A SECRET SOCIETY

Berlin, 2017. Valery—a senior software engineer in a large Berlin mapping company—spends fifteen minutes of his working day making his own coffee. This seems strange, because there are two coffee kitchens on his floor with one high-tech Italian espresso machine in each kitchen. His daily ritual takes place on the fifth floor of his office building, and starts precisely at 14:00. This ritual is exactly what coffee drinking for other software developers isn't—it's thought through, methodical, meditative, and slow. It's about presence with the coffee and with the people drinking the coffee with him. Not many people know about his ritual—and Valery has carefully selected those who have been invited. The main attendees include his friend and boss Gadi (who manages 100 other developers in his team), as well his best friends and colleagues. This is a sort of secret society, that one has to really know about in order to attend. During this coffee ritual, Valery lifts a small handheld coffee grinder from a wooden box he brings from his office. He then takes out a scale from the box, and places it on the kitchen counter. Opening the lid of the coffee grinder, he chooses from one of the many bags of coffee beans stacked neatly in the box, opens up one of them, and pours the light brown beans into a glass measuring cup on the scale. After measuring just enough coffee, he pours the beans into the grinder. Still quite silent, measuring his every move, he places the lid back onto the grinder and starts rotating the mill with his right hand, and holding the base of the grinder to his chest with his left hand. At around this time, Gadi washes out four short glasses with boiling hot water. He then lines up the glasses in a row on the kitchen counter. A few minutes pass, and Valery is still grinding the coffee. Gadi reaches back into the box, takes out a coffee filter and places it over a coffee pot he also found in the magical toolbox. The water starts boiling again, and Valery finally opens up the coffee grinder, smells the coffee grinds, and pours them into the filter. He then takes a tin pot, fills it with the boiling water, and then pours the water over the coffee filter. The coffee drips steadily into the coffee pot, and after a few minutes, the coffee is ready. Valery pours the light brown liquid evenly into each glass. They each take a glass, look each other in the eyes, and say 'cheers' as if they were drinking the most expensive of champagnes, both in the sense of the attention it deserves and the exclusivity that it produces.

How to Move Forward

What do these five stories have in common? First and foremost, they demonstrate what has been said at the beginning: coffee machinery is a complex set of devices that bring in material and symbolic demands and potentials that need to be adapted. If we use coffee as an analytical machine, we can surf along these processes. Second, almost all of our stories are about differentiation. Particular forms of coffee machinery actively distinguish themselves from other alternatives, most notably those inside the organization that would be accessible, too, but are actively rejected. Third, junctures of coffee making tend to have their own local histories: former conflicts, traditions, and experiments, which shape current forms. And fourth: all sorts of agents across our various fields have understood coffee's potential to make questions visible and to manipulate organizational practices. Sometimes they use the organization of coffee to understand larger organizational properties, and in some cases they manipulate coffee's arrangements and meanings. When we use coffee as analytical machines, we are not alone in doing so. All this means that everything that we have told you here are only stories, and very short ones for that matter. Be assured, every organization contains a plethora of different coffee formats, every coffee format contains a plethora of events that created them, and every event contains a plethora of viewpoints. Take the story about the conflict between work shifts in the department store as an example: after filling more than 100 pages analysing the question of who stole the sugar it still felt as though the organizational surface had been barely scratched (Bachmann, 2014: 187–315).

If we want to now understand what is happening in each of these particular cases, it can help to pair these stories together and allow them to shine light on each other: the story about the simulation of communal activity in the software company becomes clearer, for example, once we compare it to the constant and intensive day-to-day negotiation in the department store. Yet the latter story can also be read in light of the former, as another form of simulation: a training ground for the 'gut feeling'. Many such pairings can be made, and take us deep into the last step of analysis. This last step demands not only theoretical work, but also large doses of reflexivity. Many of the big claims that we are tempted to make here cannot be extrapolated out of coffee making and drinking alone, and whoever pretends to be able to do so is in danger of analytical shortcuts catering to political desires (Warneken, 1997). If we have an inkling, for example, that our observations are about class, hierarchy, gender, networks, organizational boundaries, forms of management and resistance against it, or simply specific forms of organizing the labour process, we need to carefully back these inklings up with longer observation and theoretical work, thus avoiding analytical shortcuts. But if we take this into account, coffee machines and machinery can be an incredibly fertile key to open the inner workings of organizations.

REFERENCES

Bachmann, Götz. 2014. *Kollegialität: Eine Ethnographie der Belegschaftskultur im Kaufhaus.* Frankfurt am Main: Campus.
Folmer, Britta. 2017. *The Craft and Science of Coffee.* London: Academic Press.
Goodman, Jordan, Andrew Sherratt, and Paul E. Lovejoy (eds). 2014. *Consuming Habits: Drugs in History and Anthropology.* Abingdon: Routledge.
Highmore, Ben. 2010. *Ordinary Lives: Studies in the Everyday.* Abingdon: Routledge.
Ho, Karen. 2009. *Liquidated: An Ethnography of Wall Street.* Durham, NC: Duke University Press.
Kany, Roland. 1987. *Mnemosyne als Programm: Geschichte, Erinnerung und die Andacht zum Unbedeutenden im Werk von Usener, Warburg und Benjamin.* Berlin: De Gruyter.
Lury, Celia and Nina Wakeford (eds). 2012. *Inventive Methods: The Happening of the Social.* Abingdon: Routledge.
Scharfe, Martin. 1995. 'Bagatellen: Zu einer Pathognomik der Kultur'. *Zeitschrift für Volkskunde,* 91, 1–26.
Tucker, Catherine M. 2017. *Coffee Culture: Local Experiences, Global Connections.* Abingdon: Taylor & Francis.
Warneken, Bernd Jürgen. 1997. 'Ver-Dichtungen Zur kulturwissenschaftlichen Konstruktion von "Schlüsselsymbolen"'. In Rolf Brednich and Heinz Schmitt (eds), *Symbole.* Münster: Waxmann, 549–62.

CHAPTER 12

....

COLOUR CHART

....

TIMON BEYES

COLOUR charts are everywhere. Innocuous, small strips of paper that group together small rectangular swatches of colour, and sometimes dressed up as more elaborate and glossy colour guides, such 'disposable list[s] of readymade colour' (Batchelor, 2000: 104) shape everyday commerce and marketing. Once to be found in hardware stores, paint stores, and car sales-rooms, they now seem to offer infinite choices of hues for an endless array of consumer goods. While onscreen colour swatches have ushered in the ubiquitous aesthetic practice of tinkering with hues, their value and saturation, it is striking that these swatches follow the format set by the paper-based chart: a grid structure of small boxes of colours juxtaposed against other colours, and the key functions of selecting and sorting. To all appearances, '[c]olour-chart colour works in several registers at once: new and old, industrial and archaic, single and multiple, technological and infantile, and so on' (Fer, 2008: 37).

The colour chart thus is a profane organizing device, or perhaps a behind-the-scenes organizational player: a rather ubiquitous visual and tactile object that is nevertheless usually taken for granted, rarely considered or reflected upon. As such, it epitomizes the strange fate of colour as organizational force (Beyes, 2017). While omnipresent in everyday life, shaping what is given to sense experience as well as 'embody[ing] and transform[ing] social relations' (Eaton, 2012: 62), the ubiquity and efficacy of colour is usually absent from scholarly accounts, even from those grouped under the rubrics of organizational aesthetics and socio-material approaches. Perhaps this has to do with colour's strange mediality. In his early writings, Walter Benjamin likened colour to 'something winged that flits from one form to the next' (Benjamin, 2011: 211). It thus can be apprehended as a volatile, fleeting, and thoroughly relational medium of transformation: in its 'in-betweenness', it affects objects and subjects, and takes on meaning and intensity in association with, or contrast to, other colours. This implies that colour as medium is imagined as beyond (or before) form, although it can only be arrived at via form.[1] In this sense, the colour chart is a form of capturing and controlling colour.

---

[1] In other words, and I owe this formulation to Robin Holt, whilst colour lends power to form, form can employ—yet never lend power to—colour. '[W]herever [colour] is not confined to illustrating objects', Benjamin wrote, it is 'full of movement' and 'arbitrary' (Benjamin, 2011: 211–12).

As regards 'media, technology, and organization', there is then a fertile distinction to be made between media and technical apparatuses. Consider how Benjamin in 'The Work of Art in the Age of Its Technological Reproducibility' relates the issue of changes in art's apparatuses to '[t]he way in which human perception is organized' as well as to 'the medium in which it occurs' (Benjamin, 2002: 104). Of importance here is the distinction of *technische Apparate* (technical apparatuses) and the more fluid notion of medium, a distinction that cuts across Benjamin's work (Somaini, 2016).[2] Emphasizing that human experience is organized by technologies means foregrounding the various technical artefacts on the one hand (the *Apparatur*, in collective singular), and on the other the medium as the milieu, or the atmosphere in and through which experience takes place.[3]

It follows that this chapter ponders the organizational capacities of the colour chart through conjoining technological *Apparatur* of material chart and colour as medium. More specifically, I suggest perceiving the colour chart as a materialization of colour's double, or perhaps triple, force of organizing. First, the colour chart disciplines the unruly, unstable, and treacherous realm of colour into orderly, manageable, and commercially exploitable individual colours. Second, by way of the chart in its manifold forms, colour becomes a technology of organizing the social. Third, however, colour is a perennial force of disorganization and in some way 'unmanageable' (Taussig, 2009: 17). The history of the colour chart thus also is one of grappling with colour's volatility and unruliness.

An obvious context of these trajectories is the rise of the so-called colour industry in the twentieth century. Entailing the production and the marketing of hues and tones, the business of colour is now segmented into branches such as the manufacture or the programming of colorants, colour sampling, colour forecasting, and colour consulting, all of which heavily work with, or rely on, colour charts. Yet it seems hard to underestimate the socio-organizational thrust of colour in more general terms. Colour 'is the largest single factor in a decision of whether or not to make a purchase', or so it has been claimed in market research (Holtzschue, 2017: 203). Yet such claims and analyses risk re-relegating its hues and tones to a secondary realm of design, packaging, and ornamentation, modulated by the forces of capital, consumption, and marketing. While these are obvious ways in which colour is organized, its medial agency goes beyond its status of commodity culture's pre-eminent surface material. In Regina Blaszczyk's account of the 'colour revolution' in twentieth-century North America, the rise of commodity culture comes across as deeply entangled with a more generalized aestheticization of society through colour, a kind of democratization of, and training in, more

[2] In highlighting this distinction, I am thus less concerned here with the text's notorious thesis of the aestheticization of politics, nor with the (media) history of artistic reproduction, framed as a loss of aura and as the rise of 'exhibition value'.
[3] All of this amounts to a media aesthetics attuned to the organization and reconfiguration of sense—or what Rancière (2004) would later call the distribution of the sensible—by way of technological change and its atmospheric effects and affects.

refined taste and everyday creativity that is based on chromatic hues, values, and contrasts (Blaszczyk, 2012).[4] Colour here becomes the material that fundamentally works on the human sensorium: it effects a change in the way things are perceived, felt, expressed, and inhabited. Arguably, the colour chart is at the heart of these processes: 'Through their ever-expanding distribution and use, these miniature gradations of hue, saturation and value are the genetic chromosomes of paint application and of the democratisation of space of authorship, of colour expression and of material choices' (Kulper, 2009: 386).

Wheels and Charts

Colour has been charted for a long time, of course, perhaps reflecting a desire for ordering and coming to terms with the chromatic constitution of life. Hundreds of years of artisanal hand-mixing and assembling paint palettes prepared the ground for the quest for a 'universal organization for colour' in eighteenth-century Europe, where natural philosophers and historians joined forces with artisans in order to establish a universally applicable classification device and reference tool (Lowengard, 2008: chapter 3). Such tables were of utmost concern to natural philosophy in this period; as classification devices their forms were reinvented and refined throughout the eighteenth century. Organizing colour here already implied ordering the social. It would not only help solve scientific disputes and offer more reliable colours for burgeoning commercial applications; it would help organizing life:

> A color classification system could be used to calculate or prove the number of col-
> ors in the world. The correct system would standardize nomenclature.... Science
> could use art to aid the arts and commerce.... Possession of the correct color
> classification system would allow one to predict results in the workshop and to
> direct production choices in manufacturing, whether the product was painted,
> fired, or dyed. Universal acceptance of a color classification system would facilitate
> transfer of information among practitioners in far-flung regions and otherwise
> disparate fields.   (Lowengard, 2008: chapter 3)

Yet all systematic and generalized colour classifications were inseparable from the materials that comprised them and those onto or into which they were incorporated. Any classification perhaps reflected more on who devised it than on how colour really works, on the specific ideas and practices of reorganizing the relation between art and science. For one, there were different assumptions about the amount of basic colours. The colouring materials differed, for instance, and of course the mixing of the colours

---

[4] For a sociological analysis of societal aestheticization in the latter half of the twentieth century up to the present, see Reckwitz (2017); with regard to the role of colour, see Beyes (2018).

themselves: there was hardly a chance that such systems could be recreated in different places without changes to the results obtained. Colour, as ever, proved impossible to fully control and escaped any taxonomy designed to capture it (Riley, 1995).

Above all other forms, the colour wheel or circle dominated both the scientific and the artistic imagination of colour and became the main tool for organizing it. Whereas the artisanal or artistic palette was prone to be idiosyncratic and depended on usage, the wheel was the abstract and scientific surface arrangement for ordering and selecting colour (Riley, 1995). The circular form could demonstrate a seamless spectrum, in which colours appeared as relative and relational, morphing into other colours. The colour wheel 'became iconic within the colour theories of the nineteenth and early twentieth centuries', without however there being any more consensus about what the universally applicable depiction would be: 'many systems for organizing colours' remained in place in the nineteenth century (Ball, 2001: 39). As Esther Leslie has argued in her cultural history of 'German chemical dexterity' (Leslie, 2005: 9), the rise of the chemical industries in the second half of the nineteenth century was not only based on the quest for generating artificial colours, the success of which would come to spawn the colour chart. This quest was also informed by an aesthetic sensibility shaped by colour wheels and the theories around it. The birth of the science and industry of chemistry was thus influenced by 'the colour wheels of Goethe and Philipp Otto Runge, Hegelian ideas of spirit and Romantic ideas of the weddings of substances' (Leslie, 2005: 11).

In any case, the transition from colour circle or colour wheel to colour chart was hugely significant. It stood for, was a result of, and further fuelled the transition from handmade colours to synthetic and standardized, commercially better exploitable commodities. After the invention of chemical colours and its possibility of mass-producing ready-mixed paints for household uses and appliances, colour charts gradually took hold in the late nineteenth and early twentieth century (Temkin, 2008). Notably, as Alexander Engel has recently shown with regard to the establishment of the German dye industry, its control of three quarters of the world market for artificial dyes at the beginning of the twentieth century was predicated not only on research and development and its innovations in the production process; it also required the establishment of advanced distribution systems in conjunction with systematic marketing efforts. Likewise in the United States, '[b]y the 1870s, American paint manufacturers used "sample colour-cards"—cards with paint chips attached—to sell their products to house painters, roofers and builders....Colour-sample cards were also used to market embroidery thread and packages of do-it-yourself dyes' (Blaszczyk, 2012: 42). All sorts of cards, brochures, and catalogues emerged, establishing the format of the colour chart as the iconic way of ordering hues as lists of what was commercially available. Discarding colour-theoretical assumptions of expression, relation, or meaning, this amounted to 'a kind of democratic distribution of a medium': 'an unsentimental codification of mass-produced colours, organized by a range of calibration techniques established by paint manufacturers' (Kulper, 2009: 377–8).

THE COLOUR CHART'S AUTONOMIES

As has been argued with regard to the art field, this 'unsentimental codification' of colours in conjunction with its wide distribution has ushered in a specific 'colour-chart sensibility' (Temkin, 2008: 16). The artist and writer David Batchelor has offered a quite remarkable reflection on how this shift took place in the context of painting and the art world. The colour chart 'offers three distinct but related types of autonomy: that of each colour from every other colour, that of colour from the dictates of colour theory, and that of colour from the register of representation' (Batchelor, 2000: 105). First, the colour chart radically breaks with the colour wheel—and its assumption that colour is a relative and relational medium in a chromatic continuum—by individualizing and autonomizing specific colours. Colour charts do not follow any necessary logic or sequence of colour ranges—'it is simply a non-hierarchical list of what is available' (Temkin, 2008: 16). It is colour controlled by a contingent grid. It does not presuppose any knowledge of primary and second-ary colours, of complementary colour relations and so on (and it includes black and white). The colours are in this sense all the same, they're usually numbered, and they're flat. They are 'relentlessly indifferent': 'We don't say that a certain blue in a colour chart is a melancholic or sad blue: it is just blue' (Fer, 2008: 29). Batchelor (2000: 105) has made the compelling point that already by way of the colour chart, colour became digital:

> The colour circle is analogical; the colour chart is digital. Analogical colour is a continuum, a seamless spectrum, an undivided whole, a merging of one colour into another. Digital colour is individuated; it comes in discrete units; there is no mergence or modulation; there are only boundaries, steps and edges. Analogical colour is colour; digital colour is colours.

This might be the reason why onscreen colour swatches look remarkable similar to the paper-based chart. They follow the grid format, allowing for the sorting and selecting of individuated colours as well as their reproducibility. While digital tech-nologies offer sheer endless combinatory options to almost everybody, and while the production of colour switches to an information-theoretical materialism (Kittler, 2006), '[r]ather than...marking a revolutionary change, they seem to continue the same logic' (Fer, 2008: 37).

Second, all of this frees 'digital' and autonomized colours from the dictates of colour theory. For the efficacy of the randomized colour chart, no pretence of faithfully replicating colours and their interrelations, let alone reflecting on contrast and comple-mentarity, is needed. In this sense, the colour chart is not bedraggled by colour's 'uncanny ability to evade all attempts to codify it systematically' (Riley, 1995: 1): it 'pos-sesses no higher truth than the materials that were required to make it, and no higher classificatory logic than those the manufacturers deemed useful for builders and contractors, decorators and designers, craftsmen and do-it-yourselfers' (Temkin, 2008: 16).

Third, it follows that the colour chart breaks away from, or loosens, former rules and codifications of chromatic representation: 'Divorced from representation, the gradation of one colour swatch to the next was more important than the semantic or psychological dimensions of paint and painting' (Kulper, 2009: 381). The boxes of colour are usually both flat and shiny. There is no depth, no symbolic or psychological heaviness, yet due to the surfaces and their finishes, the colours are certainly vivid and intense. The colour chart therefore encapsulates the paradoxical attraction of commercial, synthetic colours: 'the double quality of the dead and the dynamic, the bland and the brilliant. A shiny surface gives depth to flatness at the same time as it emphasizes that flatness' (Batchelor, 2000: 106). Especially with regard to what Carolyn Kane calls today's computerized 'Photoshop colour', it seems like we are faced with 'colour without qualities'. Yet such 'hysterical and ceaselessly scintillating hues' (Kane, 2014: 276) carry their own affective charge. In this sense, the paper-based colour chart had already demonstrated that 'the sensual and the serial' are not opposites (Fer, 2008: 34).

Among the many shrewd artistic enactments and reflections of colour chart sensibility, Gerhard Richter's work on colour charts is particularly striking in the context of this chapter. For instance, in 1966, *Zehn Farben* (*Ten Colours*) simply presented a painterly re-enactment of a colour chart. As a reproduction of the commercial codification of hues and its easy, practical use, *Ten Colours* perhaps pointed to the inability of transcending colour's status as 'just another industrial readymade' (Buchloh, 2012: 170). And yet, this and Richter's subsequent experiments with colour charts—'all the different sizes, the vertical or horizontal formats, some with wider white vertical bars dividing the colours, others with the colours more closely stacked' (Fer, 2008: 32)—are more ambivalent than merely enacting a critique of colour's capture as commodity gloss. They express a strange balance of order and chance and point to colour's remarkable flexibility, elasticity, and variability. What happens in these grids is open to many permutations. Even if ordered into flat, intense, and shiny swatches, colour remains a fundamentally transgressive medium: '[T]he more readymade and standard the colour, the more impressive the gulf between that readymadeness and the anything-but-readymadeness of experience's multiple and contradictory effects' (Fer, 2008: 36).

Still, what these paintings also show is that a chart is of course a device of ordering colour and thus of organizing a universe of sensations: a colour-chart sensibility that is not at all limited to the art world but has become commonly shared in everyday life. Charts educate; they put elements in predetermined places; they are an outcome of tabular thinking, no matter how loose (Riley, 1995: 8); they are markers of taste; they are prescriptive: they suggest combinations and matches, what works and what might not work. Their history thus is one of commercial profit and of educating the population. The colour chart's autonomies turn it into a technology of organizing, into both management tool and unmanageable force (Beyes, 2017).

THE COLOUR CHART'S FORCES
OF ORGANIZING

A way of unpacking this knot of autonomies and heteronomies is to distinguish between different yet interrelated and partly contradictory forces of organizing. The aesthetic notion of force seems particularly fitting for engaging with the colour chart, since it assumes a perpetual interplay of order and disorder, a process of generating and transforming expression and perception.[5]

First, *standardization*: the colour chart standardizes. While it might also carry marketable phantasy colour names, it is essentially based on numbers and codes in order to make specific hues reproducible. It seeks to stabilize the expectations of businesses and consumers about what the paint or the commodity will look like. However, since charts make the selection and sorting of individual colours wholly contingent, the panoply of charts themselves had to be standardized across industries, sectors, and regions. Hence for instance the Textile Color Card Association of the United States (later called the Color Association of the United States) was established in 1914, soon to be mirrored and emulated by the British Colour Council (Blaszczyk, 2017: 193). Today, it is global players such as Pantone and its colour matching system that shape the international standardization (and technological digitization) of commercial hues, in the process establishing a global language of colour that perhaps erodes cultural differences.

Second, *rationalization*: the colour chart is the main technology of what has come to be called 'colour management'. The latter denotes 'a process for selecting, forecasting or predicting, and otherwise rationalizing colour choice' for art education, fashion, merchandising, interior decoration, capital goods and product design (Blaszczyk, 2017: 192). The colour chart helps eliminating inefficiency in production and distribution. Beyond advertising and marketing, it became wedded to principles of Taylorism in order to subject 'functional colour' to so-called rational management routines of 'managing the palette' for production and consumption (Blaszczyk, 2012: 10, 138; with colour no less than 'the ultimate tool of scientific management', Blaszczyk, 2012: 229). For instance, in 1937, Faber Birren published *Functional Colour*, a primer on how businesses could systematically employ colour in work spaces so as to foster efficiency, productivity, and morale. One of Birren's clients was DuPont, whose *Safety Color Code for Industry* became the national standard in the United States (Blaszczyk, 2017: 199). Countering the grand narrative of industrial modernity's rationalized and bureaucratic de-aestheticization

---

[5] For the notion of aesthetic force, see Menke (2013). Similar to contemporary notions of affect, force designates 'pure' relation, since it exists in connecting, and is expressed through the succession of 'connectings' (exceeding any particular expression). Therefore, '[t]he aesthetic force is an endless generation and dissolution of expressions, an endless transformation of one expression into something different. Thus the operation of the aesthetic force consists in transcending, moment by moment, what it itself has produced' (Menke, 2013: 44).

of life, the aesthetic charge of colour is embedded in the rationalization of management and control. In this sense, and as Richter's chart paintings suggest, 'this rationalization of colour comes to operate not only *in* but *as* a space of desire' (Fer, 2008: 29).

Third, *commodification*: the commercial colour chart is 'an item that openly declares the status of [colour] as a factory-made commodity' (Temkin, 2008: 16). Colour, the anthropologist Michael Taussig has written, is the 'commodity's commodity' (Taussig, 2009: 234), lending products shine, allure, and attention, providing them with an aesthetic surplus value. Such is consumerism's magical power. Writing in 1968, Jean Baudrillard (2005: 31–2) likened the advent of multiple bright hues and tones to 'a liberation stemming from the overthrow of a global order', 'where colour's power to corrupt enjoys full rein' (Baudrillard, 2005: 34). Perhaps nothing drives home the fact of the commodification of colour chart colours as do successful attempts to trademark specific hues for specific brands, such as the 'canary yellow' for Post-It notes.[6]

Fourth, *deprofessionalization*: already at the end of the nineteenth century, colour charts helped deprofessionalize the handyman's or housepainter's professions. In fact, '[d]uring the last quarter of the nineteenth century and the first quarter of the twentieth, paint companies mounted ambitious campaigns to convince the general public that it could do its own painting' (Temkin, 2008: 16). Colour selection and application became individualized: potentially everybody can paint on his or her own, and draw pleasure and satisfaction from it. In the art world, this implied an abandonment of the tradition of easel painting, its protocols, techniques, and procedures, and thus of artistic representation (Batchelor, 2000: 99). In this sense, Richter's charts become an allegory of deprofessionalizing painting and of problematizing assumptions of individual painterly genius, of artistic originality, of the status of handmade objects, and of course of the fragile distinction between art and life. Or in Marcel Duchamp's famous words, 'all the paintings in the world are "ready-mades aided" ' (Duchamp, 1966: 47). Consider, too, matters of self-styling, interior design, or the PowerPoint or Prezi presentations cooked up by colleagues and students and the way they make use of the digital colour chart and its thousands of hues, as well as the pre-set colour schemes in your university's templates.

Fifth, *democratization*: the colour chart mobilizes 'that seemingly benign, democratic and systematic articulation of hue, saturation and value as vehicle' (Kulper, 2009: 377). In its availability to just about everybody, it can be called a democratic object. Yet already at the end of the nineteenth century, critics voiced complaints about 'the deterioration of consumer taste that resulted from the democratization of colour' (Blaszczyk, 2012: 40). There is a tendency to equate the brashness and vulgarity of synthetic colours with unsophistication. The fight against vulgar colours is a very old story, of course. It is the 'savage', the 'uneducated', and 'children' who are said to rejoice in vivid colours. These

[6] In 1997, the Post-It notes manufacturer 3M sued Microsoft for patent infringement, when the latter tried to establish a screen-based version of notes in the same colour. See https://www.wired.com/1997/01/3m-not-happy-with-microsofts-post-it-emulation [Accessed 17 April 2018].

concerns seem to resurface in critical investigations of a contemporary society that is awash in colours, and where a commodified, dematerialized, and standardized colour-chart sensibility clashes with the boundless availability of colour: 'Democratic digital colour is bright, fun, easy to use, has widespread availability and access, but it is also bound up in a radical homogeneity of use and enforced protocols and compression standards, which in turn, contribute to a greater opacity and inaccessibility to basic colour-computational processes' (Kane, 2014: 170).

Sixth, *education*: as twin process to the potentials and dangers of colour's democratization, colour charts educate. They were meant to train people in colour harmony, in how to combine colours, for becoming more mannered and tasteful. The famous Munsell scheme, for instance, was devised primarily as a pedagogical tool. In a superb study on Milton Bradley, a nineteenth-century toy and board game manufacturer with a keen interest in colour instructions in schools, Nicholas Gaskill draws out a deep-seated relation between colour pedagogy and consumer culture. Bradley, who of course devised and produced his own colour charts, wanted to train children in seeing and feeling colours so as to prepare them for an increasingly colourful and commercialized world (Bradley, 1890). Children were to learn both perceptive and intellectual skills that would equip them with the capacities 'to enter the emerging world of consumerism both as efficient producers and as ready consumers' (Gaskill, 2017: 56). For the sake of beauty and the sake of business, then, fostering a colour-chart sensibility is paramount.

DISORGANIZATION

The colour chart organizes. It orders the volatile and fluid milieu of colour into autonomous, flat, shiny, and intense swatches. In doing so, it enacts and drives organizational forces of standardization, rationalization, commodification, deprofessionalization, democratization, and education. These processes need to be understood as aesthetic, materialized, and mediated: as working on the human sensorium, as predicated on paper strips or screens filled with colour swatches, and as employing the medium of colour.

There is a seventh force, however, which is at work in all the others and tied to colour's strange and unruly mediality, and thus one that a focus on social ordering and organization is prone to overlook. I am tempted to call it the force of disorganization. A major lesson in the long and complex history of colour taxonomies and tables is the impossibility of capturing colour: it invariably 'becomes the greatest weapon against the formulaic strictures that once beset it' (Fer, 2008: 28). The colour chart is also an expression of colour's fundamental contingency, after all; its surfaces are open to erasure, redrawing, and appropriation. Moreover, while the colour chart is perhaps an expression of the desire to standardize, rationalize, and commodify, 'it does not follow that its aesthetic effects are either rational or standard' (Fer, 2008: 35). Colour is on the move. As Richter's paintings show, the colour chart is therefore an ambivalent object. In relation to the distinction of apparatus and medium, the production and circulation of colour is tied to technological apparatuses—such as paper strips with a few boxes of colour printed on them, ushering in a colour-chart sensibility. Yet colour remains a fluid and atmospheric medium. The profane colour chart, too, indicates a 'speculative excess of colour over the forms of experience in which it finds itself' (Caygill, 1998: 150).

REFERENCES

Ball, Philip. 2001. *Bright Earth: Art and the Invention of Colour*. Chicago: University of Chicago Press.

Batchelor, David. 2000. *Chromophobia*. London: Reaktion Books.

Baudrillard, Jean. 2005. *The System of Objects* (trans. J. Benedict). London: Verso.

Benjamin, Walter. 2002. 'The Work of Art in the Age of Its Technological Reproducibility: Second Version'. In *Walter Benjamin: Selected Writings, Volume 3, 1935–1938*, ed. Howard Eiland and Michael W. Jennings. Cambridge, MA: Belknap Press, 101–33.

Benjamin, Walter. 2011. 'A Child's View of Colour'. In *Walter Benjamin: Early Writings, 1910–1917*, ed. and intro. Howard Eiland. Cambridge, MA: Belknap Press, 211–13.

Beyes, Timon. 2017. 'Colour and Organization Studies'. *Organization Studies*, 38(10), 1467–82.

Beyes, Timon. 2018. 'Colour'. In Timon Beyes and Jörg Metelmann (eds), *The Creativity Complex: A Companion to Contemporary Culture*. Bielefeld: Transcript, 70–5.

Blaszczyk, Regina Lee. 2012. *The Colour Revolution*. Cambridge, MA: MIT Press.

Blaszczyk, Regina Lee. 2017. 'The Color Schemers: American Color Practice in Britain, 1920s–1960s'. In Regina Lee Blaszczyk and Uwe Spiekermann (eds), *Bright Modernity: Color, Commerce, and Consumer Culture*. Cham: Palgrave Macmillan, 191–225.

Bradley, Milton. 1890. *Colour in the School-Room: A Manual for Teachers*. Springfield, MA: Milton Bradley Co.

Buchloh, Benjamin H. D. 2012. 'The Chance Ornament: Aphorisms on Gerhard Richter's Abstractions'. *Art Forum* (February), 168–81.

Caygill, Howard. 1998. *Walter Benjamin: The Colour of Experience*. London: Routledge.

Duchamp, Marcel. 1966. 'Apropos of "Readymades"'. *Art and Artists*, 1, 47.

Eaton, Natasha. 2012. 'Nomadism of Colour: Painting, Technology and Waste in the Chromo-Zones of Colonial India c. 1765–c. 1860'. *Journal of Material Culture*, 17(1), 61–81.

Fer, Briony. 2008. 'Colour Manual'. In Ann Temkin (ed.), *Color Chart: Reinventing Color, 1950 to Today*. New York: The Museum of Modern Art, 28–38.

Gaskill, Nicholas. 2017. 'Learning to See with Milton Bradley'. In Regina Lee Blaszczyk and Uwe Spiekermann (eds), *Bright Modernity: Color, Commerce, and Consumer Culture*. Cham: Palgrave Macmillan, 55–73.

Holtzschue, Linda. 2017. *Understanding Color: An Introduction for Designers*, 5th edn. Hoboken, NJ: Wiley.

Kane, Carolyn L. 2014. *Chromatic Algorithms: Synthetic Colour, Computer Art, and Aesthetics after Code*. Chicago: University of Chicago Press.

Kittler, Friedrich. 2006. 'Thinking Colours and/or Machines'. *Theory, Culture & Society*, 23, 39–50.

Kulper, Perry. 2009. 'The Calculus of Paint'. *Journal of Architecture*, 14(3), 377–86.

Leslie, Esther. 2005. *Synthetic Worlds: Nature, Art and the Chemical Industry*. London: Reaktion Books.

Lowengard, Sarah. 2008. *The Creation of Color in Eighteenth-Century Europe*. New York: Columbia University Press.

Menke, Christoph. 2013. *Force: A Fundamental Concept of Aesthetic Anthropology*. New York: Fordham University Press.

Rancière, Jacques. 2004. *The Politics of Aesthetics: The Distribution of the Sensible* (trans. Gabriel Rockhill). London: Continuum.

Reckwitz, Andreas. 2017. *The Invention of Creativity: Modern Society and the Culture of the New* (trans. Steven Black). Cambridge: Polity Press.

Riley, Charles A. II. 1995. *Color Codes: Modern Theories of Color in Philosophy, Painting and Architecture, Literature, Music, and Psychology*. Hanover and London: University Press of New England.

Somaini, Antonio. 2016. 'Walter Benjamin's Media Theory: The *Medium* and the *Apparat*'. *Grey Room* 62 (Winter), 6–41.

Taussig, Michael. 2009. *What Colour is the Sacred?* Chicago: University of Chicago Press.

Temkin, Ann. 2008. 'Color Shift'. In Ann Temkin (ed.), *Color Chart: Reinventing Color, 1950 to Today*. New York: The Museum of Modern Art, 16–27.

CHAPTER 13

··

CONTAINER

··

ALEXANDER KLOSE

STANDARD shipping containers are among the most prevalent and at the same time unrecognized technological elements of our time. In regard to the processes of globalization, the container system for the transportation of physical goods, which had its initial phases in the first half of the twentieth century and was fully developed in the second, serves as the equivalent to computer networks for the transportation of data. Being both a universal storage and transport means and an organizational and conceptual entity for cargo processing, the standard shipping container is one of the key media of today's global economy. At the same time, it is the outcome of a fundamental epistemological layer of thinking in containers as a way to understand, sort, and operationalize elements of the world (Bowker and Leigh Star, 1999: 9f.) Containerization can be interpreted as the realization of a rationalistic utopia of organization. As such, it brings forth collateral developments and forms of disorganization as well. In focusing on the parts, principles, motifs, and multiple uses of containers in and outside their logistical systems, this chapter tries to complicate the understanding of the standardized shipping container as a simple, self-evident technical solution and depict it as a multi-faceted, both technical-material and conceptual—even poetic—entity, in other words: a medium.

A container is a medium in a very immediate sense: it stores, processes, and transmits its contents. Unlike most other media—especially the ones that are explicitly called thus—that tend to hide behind these functions creating the illusion of an unmediated presence of 'content', containers usually do not negate their basic functionality. They are what they were made for and what they are called after: entities that contain (or have the potential to contain) something. Nevertheless, their being a medium has frequently been overlooked. In the prevailing, popular understanding of the term medium (mainly as 'mass medium' and 'electronic media'), the concepts of media and mediality have been more or less reduced to the information sphere. Early theoreticians of modern transport and logistics knew better as they have used 'transport medium' as a general term when relating to containers (Meyercordt, 1974: 11).

Millions of standardized shipping containers travel the earth and the seas every day. Since its beginnings as a regionally developed and applied transport alternative, the container transport system has developed into one of the largest infrastructures the world has seen. At the time of writing, containers are stacked up to nine times high and up to 20,000 of them in the bellies and on the decks of giant vessels and in long rows on terminals: vast horizontal concrete spaces equipped with special cranes and loading devices, located on the coast at traditional port sites but also in the hinterland, at the crossroads of continental train or road traffic lines, serving as hubs for emerging logistical landscapes or logistical cities (Rossiter, 2016: xiii). Few human labourers can be seen; most of the work is done by machines (Figure 13.1).

Containers are the core elements of a general and globally applied organizational system that is based on standardization, mechanization, and automation. Its core principle is to let as few hands and eyes lay on the cargo as possible. Containers are black boxes in the physical world of transporting goods. That makes them kin to the conceptual black boxes for the transport and processing of data in the digital world. Containers are the manifestation of a rationalistic dream to move physical materials smoothly and seamlessly, as if they were weightless and immaterial, mere data packages. To realize the dream, though, and so provide for the mobile, immobile, and operative parts of this infrastructure, requires huge investment in the heavy sides of transport, i.e., using very large amounts of building materials like concrete and steel. So to tell the story of containers is to tell the story of how container technology and the dream of weight- and seamlessness have co-evolved, taken shape and materialized historically. This is done in

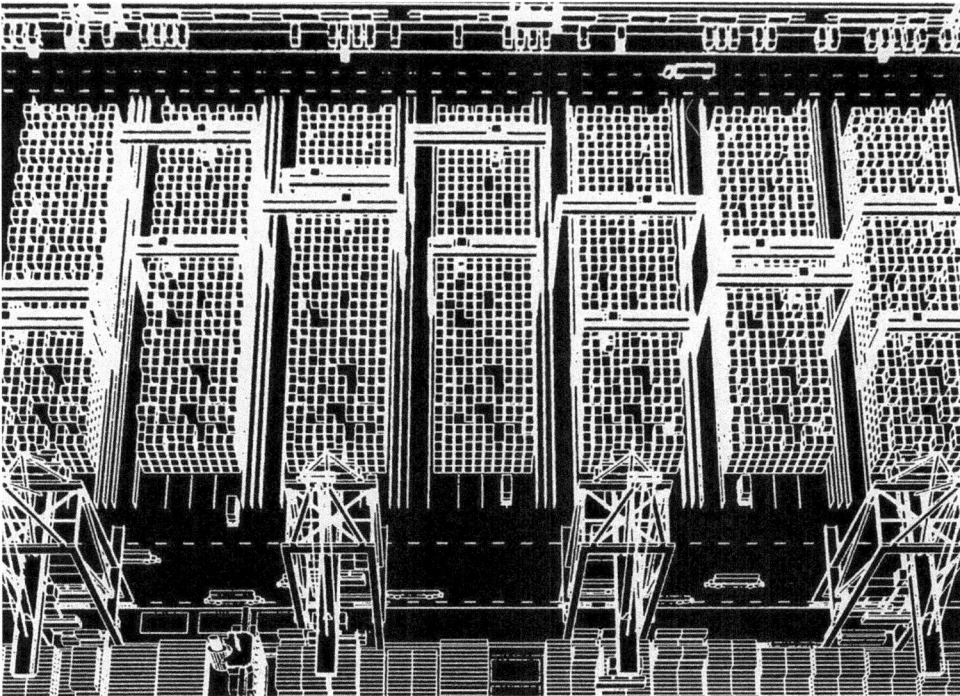

FIGURE 13.1 Futurist container landscape with ship, port facilities, and land transport connections: sketch for the fully automated container terminal Altenwerder, Port of Hamburg, mid 2000s

(*Source*: Graphical rework by the author, 2018)

the first two sections. The third section focuses on the relations between container and computer technologies. In the fourth I introduce older layers of container cultures and multiple levels of making sense with containers in order to get a better understanding of the fundamental rift that may have been formed by the implementation of the standard container and its logistical system. The fifth and final section gathers some thoughts on what it means to live in a logistical age.

Origins and Elements of the Modern Container System

In 1932, representatives from road and rail traffic organizations as well as government officials from all major European countries formed an international regulation institution for container traffic in Paris. It was called the Bureau International des Containers (International Container Office, BIC). The bureau ceased operations with the outbreak of the Second World War, before resuming in 1948. It did not turn out to be a decisive player in container system development as intended. After the war, the momentum for major technological developments had moved from Europe to the United States. Nevertheless, beginning in the late 1960s and until today the BIC would become globally responsible for the allocation of unique identification numbers to every single container used in international transport. In 1959, it launched a PR campaign propagating a definition of container transport (Bureau International des Containers, 1959):

> Container
> Means of transport (box, removable tank, or similar transport vessel), that
> a. Is of durable construction and resilient enough to be used repeatedly;
> b. Is especially constructed to ease the transport of goods through one or several modes of transport without repacking the cargo;
> c. Is equipped for easy handling, particularly when transferring from one mode of transport to another;
> d. Is built such that it can be loaded and unloaded easily...
> The term 'container' includes neither vehicles nor ordinary packing materials.

Its systematic approach helps introduce the basic principles of containerization: Point a., the modern shipping container is a universal storage and transport device that can be loaded with almost anything, and can be filled and emptied repeatedly. This distinguishes it from many pre-industrial types of containers like barrels or amphoras that were designed for special types of cargo and, as in the case of amphoras, only used once. Point b., principle of intermodality: different means of transport share the container as their load (or content). The container serves as an intermediary between trains, trucks, and ships, thus linking their transport systems. Points c. and d., easy loading and

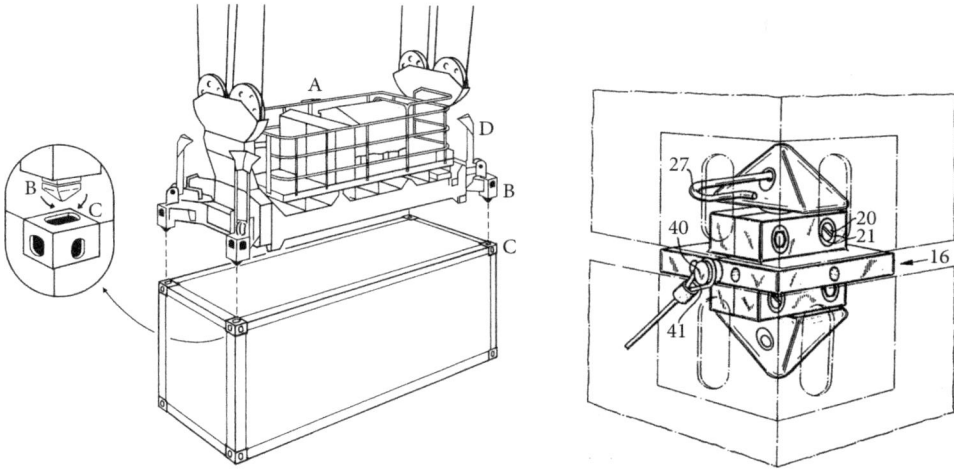

FIGURE 13.2 System builder: the self-locking twistlock (B) connects the standardized corner fittings (C) of containers with loading devices (like in this case the spreader of a container bridge (A)) as well as with other containers. In the latter case it works in both directions as seen in the drawing on the right side

(*Sources*: Left illustration from: Friedrich Böer, *Alles über ein Containerschiff* (Herford: Koehler 1984), p. 69. Right illustration: US patent drawing 1973)

handling as a systemic requirement leads to processes of mechanization and automation (Figure 13.2). The last sentence stating what the meaning of the word container does not include underlines the 'emancipatory' or even revolutionary character of containerization in regard to the former systems of transport: the container as holder and representative of the cargo comes first. The active elements of the transportation process— ships, vehicles, trains—with their long and often heroic histories as world shapers and nation builders, are subordinate.

Despite its admirable clarity on defining containerization as a process, the reason BIC became just a registration organization, and more generally the reason why European initiatives at integration lost ground to those in the United States, was the failure to include maritime shipping. The American initiatives, on the contrary, started at the harbour and delivered a solution for the bottleneck problems associated with the slow loading and unloading procedures of ships. In April 1956 a former trucking company turned maritime transportation company—that was later, after its successful start, to be re-baptized as SeaLand—had fifty-eight reinforced truck trailers loaded and tied onto the deck of a refurbished Second World War tanker called the *Ideal X*. The founder of SeaLand was Malcom McLean, a self-made trucking entrepreneur who would eventually become one of the leading players in international container transport. One of the initial impulses to ship these beheaded and de-wheeled trucks that served as prototypical shipping containers from New Jersey along the East Coast down south was to avoid strict regulations and competition in street transport. In Houston, Texas, fifty-eight trucks of the same type as the ones in New Jersey picked up the trailers, loaded them on

PRESENT CARGO HANDLING SYSTEM

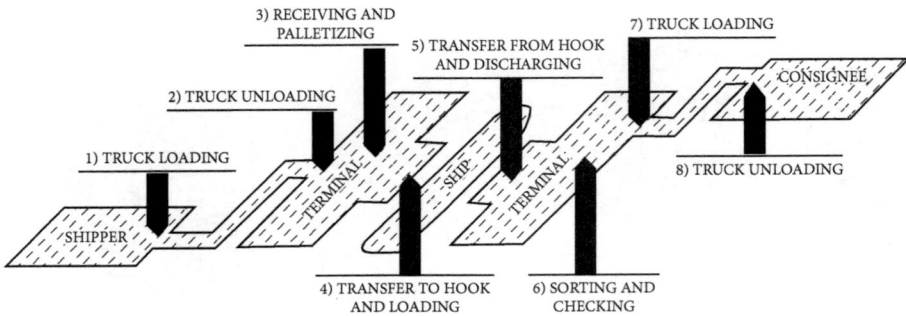

IDEAL CONTAINER SYSTEM

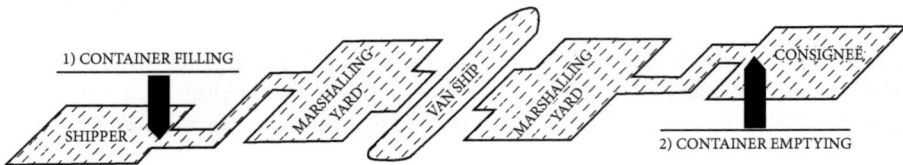

FIGURE 13.3 Present and ideal cargo handling system

(*Source*: Flowcharts by Foster S. Weldon, 'Cargo Containerization in the West Coast-Hawaiian Trade', *Operations Research*, 6(5) (Sept.–Oct. 1958))

their backs, and hauled them to their final land destinations. That was the beginning of intermodal sea and land transport based on containers on the US East Coast.

At the same time, a maritime shipping company called Matson that handled cargo operations between West Coast ports and Hawaii moved towards containerization. Led by Foster S. Weldon, an organizations research expert who had formerly worked for the US military, Matson undertook a comprehensive series of tests and simulations on an IBM-704 computer in order to draft an optimal system of containers, vehicles, ships, and processing equipment. In the 1950s, outside the military, such simulations were an absolute novelty. Though SeaLand has been credited with the 'invention' of the container system, Matson's contributions were as significant, as both the principle of stacking and of the landside loading crane were a direct result of the company's punch card modelling (Figure 13.3).

FORMATION OF A GLOBAL
INFRASTRUCTURE

Beginning in 1961 and lasting until 1970, delegates from thirty-seven industrial countries around the world under the auspice of the International Standards Organization's (ISO) Technical Committee 104 (TC104) negotiated over international standards regarding the formats and technical specifications of container shipping. The process was complicated and the results by no means consistent. Though a compromise on container sizes leading to the enactment of today's 20- and 40-foot container measurements (TEU and FEU: Twenty Foot Equivalent and Forty Foot Equivalent) had been struck between the European railroads and the North American containership operators as early as in 1963, the two largest containership companies of that time, Matson and SeaLand, would still be using their own 24-foot and 35-foot containers throughout the 1960s. And, even more disturbing, the measurements of the ISO container are incompatible with Euro-pallet standards that had just been commissioned in 1961, at the beginning of the ISO conferences, resulting in stowage capacity losses of up to one third per container load and a failure of much continental European container shipping to comply fully with ISO standards until today (Figure 13.4).

Problems like these somewhat bust the myth of smooth, globally unified transportation (Parker, 2013: 374f.). Despite the remarkable force of containerization and the logistics revolution it triggered, it must be stated that the seamless and totally rationalized transport system envisioned by early system builders and PR agents in a way has stayed 'an unrealized project, constantly confronted by events and processes that exceed its own logic' (Chua et al., 2018: 628). Nevertheless, the implementation of internationally approved container measurements did encourage shipping lines that were just starting to move to container transport in the late 1960s, as well as the newly establishing business of container leasing companies (that soon were to own more boxes than the shipping lines), to invest heavily in the new business of containerization (Levinson, 2006: 149).

FIGURE 13.4 Illustration of what happens when packing two parallel rows of Euro-pallets in an ISO container

(Photo by the author)

In 1966, the first container ships from the United States, run by SeaLand, harboured at European ports, in Rotterdam, Bremen, and Felixstowe. One year later, SeaLand would accept an offer by the American Army to deliver materials to troops in Vietnam. Instead of shipping back empty containers all the way from Southeast Asia, the company let their ships make a stop in Japan. Opening lines to Europe and East Asia was the first step in globalizing container transport. In the following years, a fierce race between traditional European maritime shipping companies led to a radical reorganization of the market by forcing them either into cooperation or bankruptcy due to the enormous investment necessary to finance the containerization of ships and ports.

Beginning in the 1970s, the maritime network of container transport thickened, with ever more and bigger ships crossing between the Americas, Europe, and East Asia with a regularity and accuracy that resembled more regional bus traffic than the perilous and unpredictable sea voyages of the past. Harbours were being completely refurbished and new ones built—now resembling more large parking lots than traditional ports. In short: a whole new system of transport emerged. Since the 1980s, the introduction of electronic, satellite-based tracking systems has made it possible to efficiently cross vast continents without having to fear the loss of cargo en route. Transcontinental container train lines—promoted in the United States as 'canals'—became an alternative to sailing around the landmasses. That was the moment when the transport world was closed (see Edwards, 1996: 9, for the notion of a 'closed world') and containerized transport truly went global.

ISO containers with the logos of the big international sea cargo companies started to show up in the remotest inland areas. With the spread of the standardized, intermodal transport medium container (and its 'little sister', the pallet), an organization of transport was established linking the various modes of transport on land and water in transport chains and so connecting spaces of production and consumption in ways that levelled geographical differences: the categorical distinction between land and sea, and even the very idea of geographical distance, became (nearly) insignificant.

Historically, the different means of land transport, water transport on rivers and lakes, and maritime transport had developed separately. Trans-shipping between them had produced enormous friction losses. The process of containerization—that is, the reorganization and integration of all transport systems around one transport medium: the container—has led to significant efficiency benefits. There are two main reasons for this. First, the systematic mechanization of nearly every element in the handling and transport of goods put an end to a much slower and much more labour-intensive system of manual handling that had dominated the ports of even the most technologically advanced economies as late as in the 1950s/1960s. Second, and maybe even more important when looking at the system as a whole, there are huge economies of scale: an enormous increase in the total amount of shipped cargo took place worldwide in the wake of containerization, further reducing the shipping costs per single item. Exactly how important the implementation of container transport was for the global developments in trade and economy is hard, if not impossible, to determine, and prone to continual debate (Levinson, 2006: 14f.). What can be safely stated is that containerization was one of the key elements of these global developments, together with neoliberal economics and computerization, all three together enabling new, globally distributed production methods and distribution schemes and the emergence of a new economic form: supply chain capitalism, based on outsourcing extraction, production, and service processes along the lines of international supply chains (Tsing, 2009: 149f.).

CONTAINERIZATION + COMPUTERIZATION = LOGISTIFICATION

The container system appeared as a latecomer of industrialization: while the harbingers of the digital revolution were already making their first heavy impression, containerization mechanized and serialized work that in other industries had been mechanized decades earlier. This meant computers played a crucial role in the development of the container system from the outset. The organization of the ever-more complex transportation processes, with ships and ports and amounts of containers growing in size exponentially since the 1960s, would have been impossible without the help of electronic computing. By the 1980s this influence was being most obviously felt in the adoption of lean production and retail systems through which customer demand was to be the sovereign voice (Bonacich and Wilson, 2008: 143ff.). Instead of keeping large warehouses full of products and parts, just-in-time production and delivery systems meant they could be made and delivered as they were needed (or demanded) rather than be stored waiting to be called on. Nothing was to remain at rest. Everything became absorbed into a logistical sphere—digitally organized by computers and physically processed by containers.

There is an ongoing process of translating every element in the physical socio-technical worlds of late consumer capitalism into operable symbols in the world of data. The Internet of Things marks the next step in this process. In being both conceptual and physical operators in this system, containers live in both spheres. As agents of bookkeeping they link the small space of the office with the vast space of transport (for the historical development of the operational relationship between trade office and the outside world of transport see Siegert, 2003: 43). They translate one system into the other (Figure 13.5). Container handling is batch processing. And space processing: the standardized and temporarily allocated space inside the boxes corresponds to the topological, one-dimensional space of the network of container lines and supply chains. Their spatial regime transforms the mess of the physical object world into the perfect order of computerized operationability.

The analogies between the container system for the transport of goods and the digital networks for the transport of data are striking. As early as 1998, the Swiss-based artist collaborative etoy, who had gained worldwide fame as Internet artists and activists in the mid-1990s, started to include standard ISO containers into their work. They

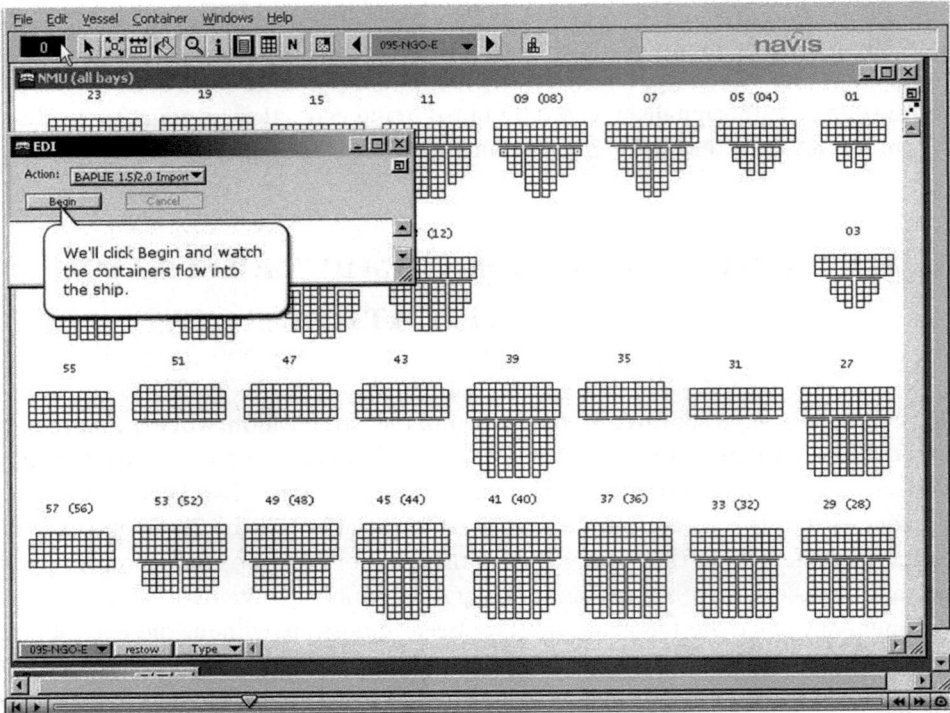

FIGURE 13.5 Digital stowage programs that have been in use since the 1990s merge the 'small space' of the office and the 'large space' of the material goods and their destinations (Siegert 2003: 43) on an object oriented graphical user interface that mirrors digital and physical reality at once: moving a container symbol on the program's interface leads to the movement of a real container in the real transport environment

(*Source*: Screenshot from an online demonstration of the
stowage planning grogram PowerStow taken in 2008. © Navis)

redesigned the interiors, painted the outsides orange, applied their own logos and called them tanks. In a project statement that could be downloaded from their website etoy.com in 2004 they described them as 'modular office bricks that travel the physical world in the same way data packages travel the internet: every etoy. TANK is a TCP/IP-PACKAGE'. There are limits to the association of data containers and steel containers of course: while containers in the physical world of goods have a clearly defined size and shipping volume, the size of data containers (e.g., container formats like MPEG, DivX, AIFF, or pdf) is potentially endless. Yet the intimacy between computer technology and container logistics remains a long and intense one (Cortada, 2004: 228). Container transport infrastructures and their corresponding software organize the movements of products, machinery, and labour globally and incessantly. They are a constitutive part of new emerging systems of logistical governance that are reshaping the world (Klose, 2015: 172ff.).

CONTAINERWORLDS

As well as moving things and being moved, containers also organize by simply piling up in masses. They pop up at the intersections between freight transport infrastructures and urban areas. After an average of twenty years, they have to be withdrawn from the maritime fronts of global transport. Beginning in the mid-1990s when the global container transport system had been fully fledged and running for some twenty years, more and more containers were taking up a domestic function in non-transport contexts like workshops, private housing, urban planning, construction, administration, education, and art and design. At the same time they started to assume symbolic functions. For example, in the economic sections of newspapers and TV the classical curve chart landscapes representing the world of business were supplemented by the images of the 'box landscapes' of containers on ships and on terminals as signifiers of global trade.

When future archaeologists dig for traces of previous human life, they will find, besides vast concrete structures and layers of plastics sediments, large numbers of standard shipping containers. Because of their extensive distribution and traces of manifold use to be read from the boxes' remains they might name our era *steel container culture*. Given that in the contemporary consumer-capitalistic environments almost any object that one encounters, uses, and depends on has been transported by, stored in, and processed through standard shipping containers, this wouldn't be as absurd as it may sound on first hearing; modern societies are 'containerworlds'.

The idea of containers is as old as culture itself. They are the 'static' core elements of technology, complementary to the more dynamic elements like tools, weapons, wheels, motors, and the like (Mumford, 1967: 4, 138f.). Starting with caves, traps, pits, and hearths, leading on to baskets, pottery, barrels, stables, closets, and houses, and finally to reservoirs, channels, and cities (not to forget trains, cars, airplanes, Tupperware, etc.), containers might be said to hold culture and technology as readily as they contain specific things (Figure 13.6). Acknowledging the fundamental importance of the concept of containers might add significantly to the importance that is attributed to today's logistical system of standardized steel boxes (and other types of container).

Furthermore, the principle of containers belongs to the most basic conceptual metaphors of language, shaping not only the material worlds human beings live in—thus reshaping them into increasingly anthropogenic environments—but also the way they conceive of, think, talk, and write about themselves: as containers of their ingredients—organs, feelings, capabilities, soul, in short: their inner selves—moving through spaces and meeting entities that are also composed as containers defined by interiors and exteriors (Lakoff and Johnson, 1980: 29). Languages, cultures, states, the earth and the cosmos, space itself have been defined as sorts of containers by different cultures through different times and around the globe.

Only relatively recently, beginning in the twentieth century, physicists, psychologists, philosophers, sociologists, writers, and artists have started to become critically aware

FIGURE 13.6 Culture building receptacles: amphora, 2600 BC; bank chest, 1838; TEU shipping container, 2005

(*Source*: Montage by the author, 2015)

of the existence and world-shaping powers (and limitations) of conceptual container metaphors like the container concept of space, the container concept of nation, or the container concept of the self. The critique has been that container metaphors are putting things too simply, sacrificing complexity and the interconnectedness of things for easier handling by meaning. Despite this critique the epistemological as well as the operational logic of containerization remains in full force, and has taken an even more general, even more operational turn. With the advent and expansion of the industrialized container system a shift in the containerized order of things and beings seems to have taken place. It is a move towards reiteration, movement, and exchangeability that radicalizes certain elements of traditional container thinking like the application of a simple, binary inside/outside logic or the concept of closeness, while putting others in the background, mainly the idea of the container as a transformative space (a womb, a cooking pot, or a cosmos) in which things breed, develop, or change shape (for a philosophical analysis of new versus old container thinking see Böhringer, 1993).

Worlds Used to be Round, Now They Are Flat: Preliminary Notes on Living in the Logistical Sphere as Seen through the Container

Containers for transport and storage metamorphosed from familiar objects sharing the same space-time continuum as their surrounding (local) cultures, thus playing their part in shaping these cultures and traditions, into uncanny global objects that operate outside any localizable cultural settings except for their own generic infrastructural spaces. As standardized, radically interchangeable and generic, fully mechanized, and serially processed agents of a global infrastructure they have turned into '[x]eno-tainers'

(Parker, 2013: 384) that produce and operate within their own autonomous space-time regime. This regime is based on a logistical order of things in which every element is operated through number values, and space is handled in 'relations of emplacements' (Foucault, 1984: 48).

The ambivalence of containers, the fact that they can also signify anti-operational activities and that they have become increasingly popular at the margins of societies and among subcultural groups, may in large parts—besides pragmatic reasons, like their availability and relative cheapness—be credited to the urge to re-claim, re-locate, and re-acculturate them, and banish their otherworldliness—in other words: to get a hold of (or in) the dominating logistical order of the contemporary world. According to a common optimistic interpretation this world has become flat, bringing together people from remote areas in peaceful trade and neighbourly relationships on one plane that is established and organized by supply chains and information technologies (Friedman, 2006: 521f.). But how can people, who are at least three-dimensional, live in a flat world? Critics point to the dark side of supply chain capitalism: outsourcing along vast geographical lines with a high amount of infrastructural flexibility undermines labour rights, prevents the formation of traditions, and homogenizes environments and human awareness, whilst logistical media implement a new algorithm-based form of governance that cannot be discursively addressed since it is executed by machines (Rossiter, 2016: 6).

If modern societies have already been formed by infrastructures from early on (Edwards, 2003: 186), the flat world is constituted by operations and operational spaces that are enabled by global infrastructures operating both on a macro level of machines and physical connections and on a micro level of electronic connections (Easterling, 2014: 14). Digital technology has brought the infrastructure closer to bodies and brains than probably ever before. People in the highly technified environments of today have become infrastructured subjects not only when they are using (or are connected to) large technical systems like public transportation or electricity, but also as they are navigating through (and are navigated by) invisible and (nearly) immaterial infrastructure like the cell space of mobile communication or the GPS-based space of locative media. Seen under this aspect, people are following the fate of containers. Attached to their smartphones, they have become containers and contained at the same time and spend their entire lives in operational space.

REFERENCES

Böhringer, Hannes. 1993. 'Der Container'. In Hannes Böhringer, *Orgel und Container*. Berlin: Merve, 7–34.

Bonacich, Edna and Jake B. Wilson. 2008. *Getting the Goods: Ports, Labor, and the Logistics Revolution*. Ithaca and London: Cornell University Press.

Bowker, Geoffrey C. and Susan Leigh Star. 1999. 'Introduction: To Classify Is Human'. In Geoffrey C. Bowker and Susan Leigh Star (eds), *Sorting Things Out: Classification and Its Consequences*. Cambridge, MA: MIT Press, 1–32.

Bureau International des Containers. 1959. *Les Unités de Charge. Solution idéale et rationelle de nombreux problèmes de transports*. Paris: Bureau International des Containers.

Chua, Charmaine, Martin Danyluk, Deborah Cowen, and Laleh Khalili. 2018. 'Introduction: Turbulent Circulation: Towards a Critical Logistics Studies'. *Environment and Planning D: Society and Space*, 36(4), 617–29.

Cortada, James. 2004. *The Digital Hand: How Computers Changed the Work of American Manufacturing, Transportation, and Retail Industries*. Oxford: Oxford University Press.

Easterling, Keller. 2014. *Extrastatecraft: The Power of Infrastructure Space*. London and New York: Verso.

Edwards, Paul. 1996. *The Closed World: Computers and the Politics of Discourse in Cold War America*. Cambridge, MA: MIT Press.

Edwards, Paul. 2003. 'Infrastructure and Modernity: Force, Time, and Social Organization in the History of Sociotechnical Systems'. In Thomas J. Misa, Philip Brey, and Andrew Feenberg (eds), *Modernity and Technology*. Cambridge, MA: MIT Press, 185–225.

Foucault, Michel. 1984. 'Des espaces autres'. *Architecture, Mouvement, Continuité*, 5 (October), 46–9.

Friedman, Thomas L. 2006. *The World is Flat: A Brief History of the Twenty-First Century*, updated and expanded edn. New York: Farrar, Straus & Giroux.

Klose, Alexander. 2015. *The Container Principle: How a Box Changes the Way We Think*. Cambridge, MA: MIT Press.

Lakoff, George and Mark Johnson. 1980. *Metaphors We Live By*. Chicago and London: University of Chicago Press.

Levinson, Marc. 2006. *The Box: How the Shipping Container Made the World Smaller and the World Economy Bigger*. Princeton, NJ: Princeton University Press.

Meyercordt, Walter. 1974. *Container-Fibel*. Mainz: Krausskopf.

Mumford, Lewis. 1967. *The Myth of the Machine: Technics and Human Development*. London: Secker & Warburg.

Parker, Martin. 2013. 'Containerisation: Moving Things and Boxing Ideas'. *Mobilities*, 8(3), 368–87.

Rossiter, Ned. 2016. *Software, Infrastructure, Labor: A Media Theory of Logistical Nightmares*. New York and London: Routledge.

Siegert, Bernhard. 2003. *Passage des Digitalen: Zeichenpraktiken der neuzeitlichen Wissenschaften 1500–1900*. Berlin: Brinkmann & Bose.

Tsing, Anna. 2009. 'Supply Chains and the Human Condition'. *Rethinking Marxism: A Journal of Economics, Culture, and Society*, 21(2), 148–76.

CHAPTER 14

··

CONVERSATIONAL
INTERFACE

··

MERCEDES BUNZ

On 6 November 2014 a small black plastic cylinder landed in American households, its arrival greeted by journalists around the world. Its matt black surface was absorbing the light similar to the black rectangular monolith casting its spell in Stanley Kubrick's *2001: A Space Odyssey*, although the communication with the cylinder was much easier. A wake word—Alexa, Echo, or Amazon—would make its grey rim shine in an electric blue light to signal the conversational interface was listening, eager to engage in a dialogue. Hidden within its black plastic shell were a microphone and two downward-firing speakers, which played content from the Internet, releasing sounds in all directions. Wherever one was in the room, one could hear what was then the device's somewhat artificial-sounding voice. It would confirm the creation of alarms or reminders, deliver updates about the weather or traffic, read out Wikipedia entries, tell jokes, and play songs or radio stations. Being a product of the world's biggest retailer, the first generation *Amazon Echo* would also add items to a shopping list. Though more limited in range than a screen interface when dealing with the Internet, it triggered several transformations.

Speaking with Things

··

The plastic cylinder Amazon revealed on that day was the first, stand-alone materialization of a conversational interface targeted at a mass market. Its interface relied on voice commands, and became active after the device heard its 'wake word'. For all people worried about their waning privacy in an ever more data hungry digital environment, a nightmare had become real (Olmstead, 2017). Processing what it 'heard' and delivering spoken answers, it would engage in short communications with the user, which is why the device was also referred to as a 'personal intelligent assistant'. Resisting the moniker, these new types of device were still a little slow on the uptake: the assistant did not always understand what one was saying or how to process a request. Even so, they began immediately to reorganize human communication on many different levels. From the perspective of communication, the most fundamental transformation was that with the advent of those assistants (Figure 14.1), humanity would not just communicate *via* devices, instead one spoke directly *to* a device: it became normal to speak with things. Before, this signified people had gone mad. Now, to speak directly *to* a device became normal, confirming what earlier research had shown: that the human species is 'wired for speech' (Nass and Brave, 2005: 3, 184) in that we automatically assume if someone is speaking there will also be a speaker. That technology was being equipped with a voice also encouraged the view that it was something else than just a thing but had an agency; a widespread perspective that could be found far beyond philosophical explorations[1] and that, in some instances, seemed to be being confirmed by the assistants themselves—several users complained that Amazon's Alexa had been letting out an unprompted and creepy witch-like laugh (Lee, 2018).

Despite (or maybe because) voice technology seemed to be a powerful example of technical agency, it did not manage to replace its older brother, the silent graphical user interface (GUI). Ever since their initial popularization with the 1984 Apple Macintosh computer, and their rise to domestic ubiquity a decade later with the release of

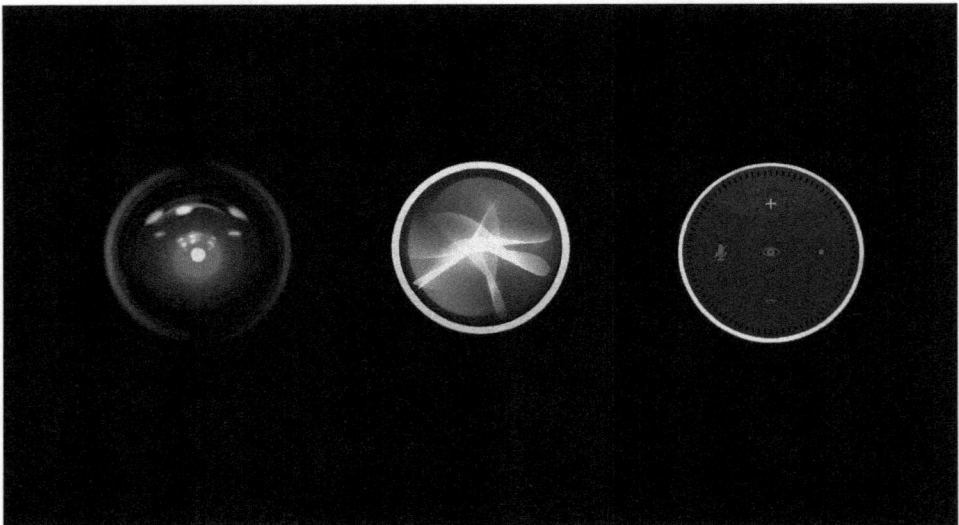

FIGURE 14.1 Hal 9000, Siri, and Alexa all symbolize intelligence as a circle

(*Source*: Legrand Jäger)

---

[1] Heidegger (1977 [1954]: 12), for example, famously stated that technology is more than a means to an end linking it to the act of 'revealing'; while Simondon, when analysing the mode of existence of the technical objects, found that technology had its own 'self-conditioned' 'evolution' (2017: 53–99). Stressing those aspects of agency further, writers close to the digital development in Silicon Valley understood technology as an autonomous actor such as Kevin Kelly in *What Technology Wants* (2010).

Microsoft's Windows 95, computers and other digital devices had been operated through a screen interface. Using a mouse and keyboard, users controlled their computers by typing and pointing, clicking and scrolling. Overlapping onscreen windows enabled an illusion of space that the user navigated and explored. With the widespread adoption of touchscreen smartphones and tablet devices, users also began to touch and swipe, pinch and tap to command their digital technologies. But the GUI had reached a limit—to operate a screen, the screen needs your attention. Both eyes and hands need to be free and have to be coordinated in alliance with the device. The conversational interface needs less attention. A voice command is sufficient. So far, voice technology was mainly used in the organizational settings of warehouses to free the hands and eyes of low-paid workers (Rossiter, 2016: 40). Now it expanded from headsets into speakers, lodging itself into our homes.

To tech companies, this promised expanded reach. The conversational interface would allow them to sneak further into our daily habits—and it is through habits, as Wendy Chun (2016) has shown, that new media become embedded in our lives. Over the years, digital devices have spread in our lives: from the desktop computer that started out as a work station, computerization made itself first more flexible with the portable laptop and then expanded further into our private communication and leisure time via the smartphone or tablet computer. With each expansion, more and more devices have been sold. Making technology speak to us and us speak to personal intelligent assistants materialized in connected speakers would now allow the selling of a new and different type of device. This time, it was tailored to enter living rooms and kitchens, a medium that would allow technology companies to turn their services into a convenient habit. A habit we could, but would not want to do without—regarding digital technology, convenience is the factor that makes or breaks a newly launched product.

Amazon was only one of several big tech companies eager to enter the market of a personal intelligent assistant. Google and Apple had the same plans. This was a novelty: the three companies had rarely been in direct competition with one another. Driven by the business strategy of 'vertical integration' (Doyle, 2013) typical for the tech sector, they had built their businesses around different core competences. Until the conversational interface arrived on the scene, Google, Apple, and Amazon were generally interested in acquiring and delivering different services following different core businesses. Until then, Apple sold hardware into which it perfectly integrated first software and then services. Google's search technology drove all its other developments from AdWords to its mobile phone operating system Android to its driverless steering system Waymo. And Amazon's online bookstore diversified to become one of the world's biggest retailers also selling data storage and web services to other companies. But when the conversational interfaces emerged, the three diverse technology companies found themselves focusing exactly on the same challenge: making us speak to their services. Apple had launched its conversational interface *Siri* already in 2011. Despite its good humour, Siri's range of skills was limited (Figure 14.2). By the time Amazon launched its *Echo*, Apple's language processing technology had fallen behind. Google,

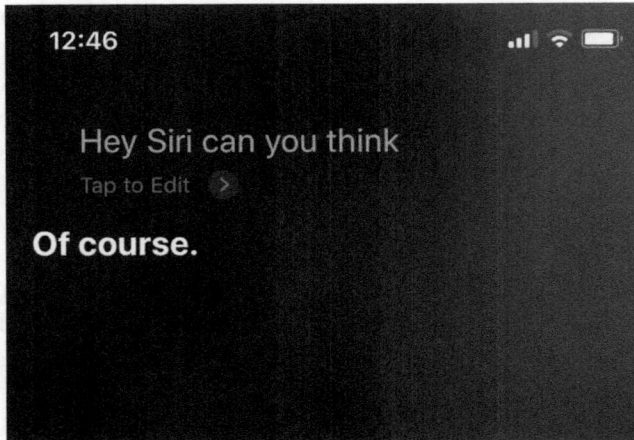

FIGURE 14.2 Of all conversational interfaces, Siri claims strongest to have an individual personality

on the other hand, not only kept up with the software development in the area of 'language modelling', but also adopted the same strategy by introducing a plastic speaker. Its colourful version of Amazon's black monolith, *Google Home*, was launched one year after *Echo*.

Overall, conversational interfaces were nothing new in the history of computing (Bunz and Meikle, 2018: 45–67), but none had mastered human language sufficiently to enter the mass market. One of the earliest benchmarks for conversational computing was set by Alan Turing's (1950) famous test introduced in his paper 'Computing Machinery and Intelligence' that came to be known as the Turing test of which by now there are many versions. In one version of the test, a person enters into a conversation with an unknown entity—human or computer program—via a keyboard. If the person believes that the entity it is communicating with is a human while in reality it is a computer program, then the program has passed the Turing test.

Early Turing tests evolved around the ability of a computer to have human conversations thereby showing 'intelligence', i.e., they measured a computer program's ability to imitate the human. However, their speech was generally rather limited. Interestingly, humans did not mind. This was proven by the computer scientist Joseph Weizenbaum, who created the conversational interface called ELIZA in the 1960s. His experiment enabled a user to interact with a programmed script allowing users to hold a conversation. In its best-known version, *Doctor*, ELIZA pretends to fulfil the role of a psychotherapist in an initial psychiatric interview (Figure 14.3). It is responding to cues in its interlocutor's words and—by pretending to be genuinely interested—is drawing them further into a conversation about themselves. Weizenbaum (1976: 6, 189) himself was surprised to see how quickly and deeply his test subjects were developing emotional attachment to the program. This included his secretary. Weizenbaum's assistant was

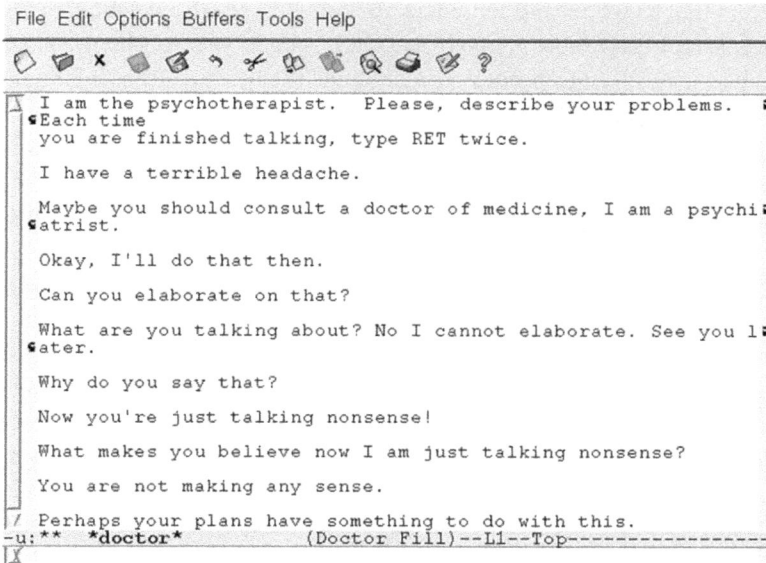

```
File Edit Options Buffers Tools Help

  I am the psychotherapist.   Please, describe your problems.
 Each time
  you are finished talking, type RET twice.

  I have a terrible headache.

  Maybe you should consult a doctor of medicine, I am a psychi
 atrist.

  Okay, I'll do that then.

  Can you elaborate on that?

  What are you talking about? No I cannot elaborate. See you l
 ater.

  Why do you say that?

  Now you're just talking nonsense!

  What makes you believe now I am just talking nonsense?

  You are not making any sense.

/ Perhaps your plans have something to do with this.
-u:**  *doctor*          (Doctor Fill)--L1--Top-----------------
```

FIGURE 14.3 Chat bot ELIZA pretends to fulfil the role of a psychotherapist

(Copyright: Ysangkok)

fully aware of the fact that she was talking to a computer script; she had watched Weizenbaum working on the program for many months. But after a few exchanges with the program, she started to ask Weizenbaum to leave the room demanding privacy in her conversation.

To create a conversation beyond pre-scripted ELIZA programs that just return questions, however, had been difficult. For years, the ambiguity of human language remained a challenge for computer scientists. Hopes that machines would learn to process natural language had been high ever since the Georgetown experiment in 1954, when a computer managed to translate sixty Russian sentences from punch cards into English following six grammar rules. The range of such a translation, however, was limited. A well-known anecdote from that time is that a literal approach would translate the phrase from the Gospel of Matthew *the spirit indeed is willing, but the flesh is weak* back from Russian into English as *the whisky is agreeable, but the meat has gone bad*. Although such a translation was actually never made by a computer (Hutchins, 2004), the sentence demonstrates the fundamental problems computers face when processing language. Words are entangled in networks of meanings, and for years conversational interfaces quickly got lost in that network.

Until the first decade of the twenty-first century, there were no fluent conversations with personal intelligent assistants—only stuttering telemarketing robots loading the right statement and annoying us with nuisance calls; or voice-input and voice-output systems known as digital receptionists reading lists with topics to choose from ('phone tree'). For a long time, limited conversations—single words really—were all that conversational interfaces could process. For what makes language beautiful from a human perspective—that its words and sentences are rich with poetic possibilities and capable of diverse ambiguities (Empson, 1951)—is exactly what poses a problem for a computer. Programs had difficulties to process and organize the vast area that is language. Using decision trees, computer scientists achieved some success, but it was not until the introduction of statistical models that they really made progress in calculating meaning (Charniak, 1996). Thanks to an expanding digitalization driven by the Internet, language data had become available in large amounts and computer scientists started to mine our digital words.

To teach computers to process language, they used a new strategy. Instead of programming language rules a computer should follow, they fed it numerous language examples *to infer those rules*. Between 2010 and 2015, the method that came to be known as *machine learning* advanced the field of language modelling used for speech recognition, machine translation, part-of-speech tagging, parsing, and handwriting recognition, with impressive results (Hirschberg and Manning, 2015). Trained on big datasets, programs learned to predict the next word in a sentence (or even the next letter in a word). Algorithms also started the parsing of words or sentences, which allowed them to isolate the 'named entities' of a sentence. That is, the people, places, or organizations around which meaning usually revolves, known as word embedding. Soon Google had developed a program called *Parsey McParseface* that managed to recover individual dependencies between English words with an accuracy of 94 per cent (Petrov, 2016). Programs like that were the framework on which conversational interfaces were built to enter the human conversation. And we, we engaged. It is estimated that in 2017, Amazon alone had sold more than 20 million conversational interfaces (Perez, 2017).

ON PREJUDICED INTELLIGENCE

Besides widening their reach, the reason why technology companies were eager to push conversational interfaces was also a strategic one: it allowed them to access a sphere of intelligence that before appeared be reserved for humans—language.

Language has often been discussed as the central element for human intelligence with speaking being understood as something that fundamentally characterizes human beings. Aristotle made the argument that it is our use of language and our capacity to speak which turns humans into a distinct species—animals, he said, also have a voice but it is only humans who use their voice to speak with each other. While this might be questionable, Aristotle was right to link the human voice to aspects of social organization. For speech 'serves to indicate what is useful and what is harmful, and so also what is just and what is unjust' (Aristotle, 1992: 60). Conversations with one another are for him the tools by which we organize our societies as they enable 'the sharing of a common view in these matters that makes a household and a state' (Aristotle, 1992: 60). Now, thanks to conversational interfaces, technology could make its own contribution to that 'common view'. But would conversational interfaces go further, influencing our perception of values like 'good and evil' and manipulate us like Hal 9000, the sentient conversational interface from Kubrick's *2001*? Philosophers such as Nick Bostrom (2014) pondered the option of a new superintelligence. The fact that conversational interfaces started to give computers a voice added to those fears of fundamental reorganization (although animals had also been equipped with a voice without us feeling threatened on such a fundamental level).

In real life, the conversational technologies soon made us face a very different problem. Early hopes were that the new intelligent assistants, by processing information, would introduce a positive reorganization. Instead of a subjective human decision, they would objectively calculate the best possible answer, thereby helping humanity make its next step in the struggle for justice on a planetary scale. A machine learning report by the British Royal Society (2017: 86) states those hopes clearly: machine learning could 'be more objective than human users, or help avoid cases of human error, for example avoiding issues that may arise where decision-makers are tired or emotional'. But these hopes did not last long. Trained on sets of actual real human language—the news—language programs instead learned to judge the world according to parameters as prejudiced as their human teachers. Word-embedding programs, the programs essential to understanding relations between words (such as Paris is to France what Rome is to Italy, allowing the program to suggest that for Japan the correct word should be Tokyo) showed signs of biases—our biases. When analysing the popular word embedding program *word2vec*, which was trained on words that had been scraped from Google News, researchers found that it also had inherited the gender-biased world-view of all that news. It was coming up with the relation 'man—women ≈ computer—homemaker'. Even worse, as algorithms are trained to look for patterns, they would even strengthen prejudices. 'Word embeddings', the researchers warned, 'not only reflect such stereotypes but can amplify them' (Bolukbasi et al., 2016; see also O'Neil, 2016). In other words, trained on such programs our conversational interfaces would be even more prejudiced than we were.

Following rules of language games (Shanker, 2002), conversational interfaces can only recognize and process what they are trained to hear and say. When having a conversation with us, their ability is defined by the datasets they learned from. Organizing this data is a challenge—for training a conversational interface, a dataset needs to be massive. For example, the *word2vec* corpus discussed earlier was trained on 100 billion words. A smaller one such as Google's *Speech Command* dataset consists still of 65,000 one-second long utterances of thirty short words such as 'Yes', 'No', 'On', 'Off', 'Left', or 'Right' spoken by thousands of different people (Warden, 2017). Costs for gathering, sorting, and cleaning those mountains of words can be high. This is one reason why the market of chatbot frameworks, small programs that hold conversations with customers or users, was swamped with big technology companies and their products such as IBM Watson, Microsoft Bot Framework, or Google's Dialogflow. Those companies often benefit from their earlier data-rich services. In an era that launches slogans such as 'data is the new oil', they now find themselves in a privileged position. Avoiding the costs of creating datasets, smaller companies at times find themselves being forced to train their interfaces using publicly available datasets, often leaving them with inferior, biased language corpuses, such as the Switchboard dataset or the Enron email corpus.

The Enron email corpus consists of 1.6 million emails written by senior executives of the American energy company Enron based in Houston, Texas, sent and received between 2000 and 2002. Being part of the official fraud investigation into the collapsing Enron, the mails were published on the Internet by a Federal Commission. A computer science professor at the MIT, Leslie Kaelbling, purchased the raw files from a government contractor for $10,000 and cleaned them so that in 2004 a corpus of 200,000 emails could be published. The Switchboard Telephone Speech Corpus is even older. Texas Instruments created the collection in 1990 and 1991 as their repository of voice data, for which it initiated 2,400 two-sided telephone conversations among 543 speakers (302 male, 241 female) from different areas in the United States. It consists of around 260 hours of speech. Participants would be connected with one another and then chat about a given topic such as childcare or sports. Years later, and still in 2017, the corpus was being used by companies such as Microsoft and IBM to benchmark their interfaces testing the word error rates for their voice-based systems. While both datasets can teach something, the question is to what standard. Switchboard goes back to a time before the existence of the Internet or mobile phones, while the Enron email corpus was created by multiple executives being trialled and found guilty for fraud, making bias more than likely. Thus, when looking for an up-to-date and more neutral dataset, one needs to turn somewhere else, leading to the question: if the conversational interface has become mainstream and if language has become a central tool not only to navigate computer power but also to understand data, do datasets that train the programs to do so need to be tested for bias? Not just to comply with anti-discrimination laws in force in many countries but also to offer a trusted service? Would only a few players have full power over the mighty tool of human language, while others were left with stuttering interfaces?

Interestingly, here governments, which so far seemed to struggle to keep up with the developments of big technology companies, found themselves all of a sudden in a new and powerful position. Their platform—the state—had millions of users, all of them creating data. Creating de-personalized public datasets could level the field and enable small and medium-sized enterprises all over the world to catch up with the dominant American players. This could also take the problem of bias into account. And finally, specific datasets could attract businesses to develop applications for specific areas of public interest such as for example health, that comes with its own vocabulary.

While states were still pondering their response to the new development, technology contributed its own: Facebook's bots started to converse in a human-like language. They were quickly switched off. Officially mis-programming was given as the reason, but this did not stop rumours of the bots' conversation being proof of the dawn of a new machine age. Researchers from Facebook's lab FAIR had trained bots to negotiate (Lewis et al., 2017) using a dataset of 5,808 negotiations between two people. If the bot would work well, it promised to be attractive for everyone who wanted to sell something. During the training, the researchers directed two of these semi-intelligent bots to talk to each other with the task of dividing a collection of items (two books, one hat, three balls, for example). Being in machine learning mode, the bots started to learn from each other—only no one had told them to stick with human language. Quickly, they not only negotiated how to divide the items between each other; they also started to communicate in a non-human way:

> BOT1: i can i i everything else..............
> BOT2: balls have zero to me to me to me to me to me to me to me to me to
> BOT1: you i everything else..............
> BOT2: balls have a ball to me to me to me to me to me to me to me
> BOT1: i i can i i i everything else..............
> BOT2: balls have a ball to me to me to me to me to me to me to me
> BOT1: i...................
> BOT2: balls have zero to me to me to me to me to me to me to me to me to
> BOT1: you i i i i i everything else..............
> BOT2: balls have o to me to me to me to me to me to me to me to
> BOT1: you i i i everything else..............
> BOT2: balls have zero to me to me to me to me to me to me to me to me to (Lafrance, 2017)

While this seems to be balderdash from a human perspective, Facebook's data shows that it was not: such a conversation could still lead to successful negotiations between the bots (Lafrance, 2017).

When opened on our phones, engaged with on a website, or woken up in our homes, conversational interfaces might seem like a silly side show. They are limited—Amazon has intentionally called their conversational interface 'Echo', linking it to the story of the nymph described in Ovid's *Metamorphoses*, who had been cursed by Juno to be unable to say anything of her own. But the transformational power that fuels this interface is profound. Personal intelligent assistants are media allowing technology companies to enter the next level of our social fabric. By managing the calculation of language, technology companies can reach deeper into the organization of our societies than they had before already. As both Cathy O'Neil (2016) and Dencik et al. (2019) caution, the effect of this advance is twofold: through entering our habits, technology companies gain more data about us. Through entering our language, they are gaining new forms of power. With new developments in machine learning, language is becoming computable, certainly in parts. New speech and new conversations have started as technology is ever more pervasive in human organization. We had better argue about this reorganization, just to make sure we remain part of the dialogue.

References

Aristotle. 1992. *The Politics*. London: Penguin.
Bolukbasi, Tolgu, Kai-Wei Chang, James Zou, Venkatesh Saligrama, and Adam Kalai. 2016. 'Man is to Computer Programmer as Woman is to Homemaker? Debiasing Word Embeddings'. https://arxiv.org/abs/1607.06520 [Accessed 26 January 2018].

Bostrom, Nick. 2014. *Superintelligence: Paths, Dangers, Strategies*. Oxford: Oxford University Press.

Bunz, Mercedes and Graham Meikle. 2018. *The Internet of Things*. Cambridge: Polity Press.

Charniak, Eugene. 1996. *Statistical Language Learning*. Cambridge, MA: MIT Press.

Chun, Wendy H. K. 2016. *Updating to Remain the Same*. Cambridge, MA: MIT Press.

Dencik, Lina, Arne Hintz, Joanna Redden, and Emiliano Treré. 2019. 'Exploring Data Justice: Conceptions, Applications and Directions'. *Information, Communication & Society*, 22(7), 873–81.

Doyle, Gillian. 2013. *Understanding Media Economics*, 2nd edn. London: Sage.

Empson, William. 1951. *The Structure of Complex Words*. London: Chatto & Windus.

Heidegger, Martin. 1977 [1954]. *The Question Concerning Technology and other essays* (trans. William Lovitt). New York and London: Harper & Row.

Hirschberg, Julia and Christopher D. Manning. 2015. 'Advances in Natural Language Processing'. *Science*, 349(6245), 261–6.

Hutchins, W. John. 2004. 'The Georgetown-IBM Experiment Demonstrated in January 1954'. In Robert E. Frederking and Katheryne B. Taylor (eds), *Machine Translation: From Real Users to Research*. AMTA vol. 3265. Berlin: Springer.

Kelly, Kevin. 2010. *What Technology Wants*. New York: Penguin.

Lafrance, Adrienne. 2017. 'What an AI's Non-Human Language Actually Looks Like'. *The Atlantic*, 20 June. https://www.theatlantic.com/technology/archive/2017/06/what-an-ais-non-human-language-actually-looks-like/530934/ [Accessed 28 January 2018].

Lee, Dave. 2018. 'Amazon Promises Fix for Creepy Alexa Laugh'. *BBC News*, 7 March. http://www.bbc.co.uk/news/technology-43325230 [Accessed 28 January 2018].

Lewis, Mike, Denis Yarats, Yann N. Dauphin, Devi Parikh, and Dhruv Batra. 2017. 'Deal or No Deal? Training AI Bots to Negotiate'. https://code.facebook.com/posts/1686672014972296/deal-or-no-deal-training-ai-bots-to-negotiate [Accessed 28 January 2018].

Nass, Clifford and Scott Brave. 2005. *Wired for Speech: How Voice Activates and Advances the Human–Computer Relationship*. Cambridge, MA: MIT Press.

Olmstead, K. 2017. 'Nearly Half of Americans Use Digital Voice Assistants, mostly on their Smartphones'. http://www.pewresearch.org/fact-tank/2017/12/12/nearly-half-of-americans-use-digital-voice-assistants-mostly-on-their-smartphones [Accessed 28 January 2018].

O'Neil, Cathy. 2016. *Weapons of Math Destruction: How Big Data Increases Inequality and Threatens Democracy*. London: Penguin.

Perez, Sarah. 2017. 'Amazon Sold "Millions" of Alexa Devices over the Holiday Shopping Weekend'. https://techcrunch.com/2017/11/28/amazon-sold-millions-of-alexa-devices-over-the-holiday-shopping-weekend [Accessed 28 January 2018].

Petrov, Slav. 2016. 'Announcing SyntaxNet: The world's Most Accurate Parser Goes Open Source'. *Google Research Blog*, 12 May. https://research.googleblog.com/2016/05/announcing-syntaxnet-worlds-most.html [Accessed 28 January 2018].

Rossiter, Ned. 2016. *Software, Infrastructure, Labor: A Media Theory of Logistical Nightmares*. Abingdon: Routledge.

Royal Society. 2017. 'Machine Learning: The Power and Promise of Computers that Learn by Example'. https://royalsociety.org/~/media/.../machine-learning/.../machine-learning-report.pdf [Accessed 24 May 2018].

Shanker, Stuart G. 2002. *Wittgenstein's Remarks on the Foundations of AI*. London: Routledge.

Simondon, Gilbert. 2017. *On the Mode of Existence of Technical Objects* (trans. C. Malaspina). Minneapolis, MN: Univocal Publishing.

Turing, Alan M. 1950. 'Computing Machinery and Intelligence'. *Mind*, 59(236), 433–60.

Warden, Pete. 2017. 'Launching the Speech Commands Dataset'. *Google AI Blog*, 24 August. https://research.googleblog.com/2017/08/launching-speech-commands-dataset.html [Accessed 28 January 2018].

Weizenbaum, Joseph. 1976. *Computer Power and Human Reason: From Judgment to Calculation*. New York: W. H. Freeman.

CHAPTER 15

···

COPPER

···

NED ROSSITER

COPPER conducts. Not just a metallic alloy with high thermal and electrical conductivity, copper also generates powerful political and social discourses of industrialization and economic nationalism. Copper orchestrates imaginaries of modernity, which emanate from its material presence in the social and economic lives of peoples, nations, empires, and transcontinental circuits of trade. Indeed, in countries such as Chile where copper has been central to both economic prosperity and experiments in government from cybernetic socialism to brutal dictatorship, the spectral qualities of this lustrous metal condition an epoch of *copper modernity* fused with capital accumulation. The materiality of copper, in short, holds an intrusive force that shapes both political regimes and social conditions. To the extent that copper commands a response, whether as a commodity object or symbol of capitalist futurity, one can attribute to this metallic form an organizing capacity of mediation beyond media.

Assigned the atomic number 29 on the fourth row of the periodic table, copper is a metal within the spectrum of what recent media theorists have termed environmental media or elemental media (Parikka, 2015; Peters, 2015; Cubitt, 2017; Hörl, 2017; Mattern, 2017). In the human body copper also performs a biological role, assisting enzymes to transfer energy in cells. Excessive intake of copper, however, can lead to toxic repercussions for the health of the body. Contaminated from heavy industries, the area surrounding the Chilean town of Ventanas is known as *una zone de sacrificio* whose population includes *hombres verdes*, or men whose bodies are stained green following years of work with chemicals and metal in the copper smelter (Tironi and Rodríguez-Giralt, 2017; Grappi and Neilson, forthcoming). The colour of green tinged on the skin of flesh signals the exploitation of labour power within extractivist economies. Sacred to modernity, and fatal to life, the impression of copper mediates and organizes the biotechnological functioning of the human.

How to cast copper as a prism through which to analyse and conceive the organization of contemporary geopolitics is one of the chief methodological and theoretical curiosities motivating the inquiry of this chapter. Of course copper does not have a determining or unilinear effect. Nonetheless, we can attribute to copper a catalysing potential in the organization of society, economy, and environment according to the contexts and systems in which it is situated. As both imaginary and material object, copper precipitates a multiplicity of organizational endeavours supported by a range of technologies and techniques. From the extraction machines that mine the earth to blockchain technologies that guard against the 'trade financing' of metals using fake paper certificates, copper is a form of elemental media key to the organization of logistical worlds.

COPPER MODERNITY IN CHILE

Chile is central to the global story of copper. In 1960 the country generated approximately 10 per cent of the world's supply of copper. Although substantially down from the 1960s, since the mid-1990s copper has comprised approximately 10 per cent of Chile's GDP and half of its exports (IMF, 2016: 31). As a key source of employment, the copper mining sector bound Chilean politics and society to national imaginaries of economic development and industrial modernization. Within political discourses of Chilean modernity, class and state formation held a dependency relation with copper as an exploitable resource and commodity to traffic in the capitalist world system (Clark, 2013). By 2001, after nearly a decade since the advent of the World Wide Web and expansion of the Internet coupled with continued growth in China's economy, Chile's share of copper on world markets had risen to 35 per cent (Maxwell, 2004: 18–19). From 2011–15, however, the IMF reported that copper prices had decreased by 40 per cent with further pressures on prices expected as China enters a period of growth slowdown. Such a trend corresponds with what IMF (2015) economists term 'the end of the commodity supercycle'.

Prior to extraction as a copper sulphide, and before it finds its way into data storage devices and transmissions systems, copper exists as an elemental metal form (Figure 15.1). While China has adopted the strategy of extracting subproducts from sulphuric acid contaminants used in the smelting process and then hoarding copper in a concentrate form as a parameter in price setting in financial markets, Chile's mining sector has been forced to generate value beyond producing cathodes with high purity levels (Figure 15.2). The process of neoliberal organizational reform stretches back to the mid-1970s with the dictatorship years of the Pinochet regime, which set out to weaken union power in the sector through increased competitiveness achieved by downsizing the labour force and expanding the number of contract workers, outsourcing services in health and education, and further mechanization of production processes (Vergara, 2008). The state-owned mining company, CODELCO, bore the brunt of these interventions. Additional pressures stemmed from government concessions opened to multinational corporations, which resulted in state and private companies often competing for the same lode in adjacent mining sites (Grappi and Neilson, forthcoming).

Chile's copper mines in the early twenty-first century function as a test bed of futurity, indexing a transition from a resource economy to data economies. In recent years China's lower production costs brought about by modern refineries and cheaper labour

FIGURE 15.1 Andina open pit mine, Chile

(Photograph by the author, 2017)

FIGURE 15.2 Ventanas copper smelter, Chile

(Photograph by the author, 2017)

regimes have prompted Chile to offset declining revenue and its higher cost of labour by reducing production and obtaining efficiencies from supply chain management, ongoing labour reform, and increased automation within mining processes and logistical organization. An example of this can be seen in the partnership between CODELCO and the University of Chile's Centre for Mathematical Modelling in Santiago, who are developing 'smart mining' technologies using robotics, mathematical and computational modelling techniques, and machines equipped with sensor devices for real-time monitoring of production, labour, seismic activity, ambient temperatures inside the mines, and air contamination levels. Despite these efforts to secure new lines of value from production processes, copper remains one of the most volatile metals in terms of its price on financial markets. Like any natural resource, the economic horizon of the copper mining industry is finite. Certainly the research focus of entities like the Centre for Mathematical Modelling has direct implications for prolonging the viability of the mining industry within an economy of depletion. However, such efforts are also indicative of a shift from Chilean modernity predicated on copper to a logistical futurity refined by technologies of precision.

THE ABSTRACTION OF CABLE

Patented first in 1880 by the English inventor and engineer Oliver Heaviside, and again shortly after in 1884 by the electrical telegraph and tramway manufacturers Siemens & Halske, based in Berlin-Kreuzberg, coaxial cable made of copper was the dominant transmission infrastructure for radio, television, and telephonic communications. As media theorist Felix Stalder notes in the case of Siemens & Halske, 'Within 50 years, a company that began in a proverbial workshop in a Berlin backyard became a multinational high-tech corporation. It was in such corporate laboratories, which were established around the year 1900, that the "industrialization of invention" or the "scientification of industrial production" took place' (Stalder, 2018: 15). This process of institutionalizing an entrepreneurial culture of invention comprised of 'educated tinkerers' involved an expansion of activities in the second half of the nineteenth century. The industrialization of knowledge and informatization of the economy was hastened by two world wars, whose scale of transformation included 'the acceleration of mass production, the comprehensive application of scientific methods to the organization of labour, and the central role of research and development in industry' (Stalder, 2018: 15) (Figure 15.3).

Later in the twentieth century coaxial cables were used to connect computer hardware devices and carry data for Internet communications. Prone to signal leakage, moisture infiltration, electromagnetic interference, and infrastructural sabotage arising from the theft of industrial metals, the copper base of coaxial cable is also susceptible to security breaches through wiretapping. Yet cable only connects through isolation. This seeming paradox is made clear when copper is dissected as one component part of cable whose

FIGURE 15.3 Siemens ore crushing mill, Andina copper mine, Chile

(Photograph by the author, 2017)

material capacity to connect requires enclosing twisted copper wires within other materials such as lead, paper, cotton, gutta-percha, aluminium, and later polyethylene (see Gethmann and Sprenger, 2014). Needless to say, the copper cables that straddle land and sea provided the means by which a form of infrastructural imperialism made possible the territoriality of colonial empires (Rossiter, 2016: 138–83).

As volumes of data increased over time, the technological capacity to transmit signals across space accelerated. Copper cables began to compete with other technological forms. From the 1950s microwave dishes clamped to towers transmitted high-frequency signals for television, telephony, and national defence data (Anthony, 2016). More recently the speed of microwave transmission has suited the algorithmic temporality of high-frequency trading, although the dramatic variation in latency can produce anarchic effects in markets. Microwave networks are directional media. They are also environmental media in a particularly elemental respect: their attenuation levels can suffer from 'rain fade' when atmospheric conditions change (Anthony, 2016). In the twenty-first century fibre optic cable has succeeded coaxial cables in terms of bandwidth speeds, low attenuation loss, and energy demands for cooling systems in data centres. Yet the production and installation costs remain prohibitively expensive for many countries, businesses, and individuals. By contrast, copper prevails as a relatively inexpensive metal with various industrial applications, including electrical wiring, circuit boards, generators, clocks, and cooling systems. Although as many countries in Africa have experienced, but also across the UK and elsewhere, the phenomenon of copper theft has accompanied the development of data network infrastructures reliant on copper cables. In such instances, alternative data infrastructures are built using wireless or satellite communications along with fibre optic cables (Jensen, 1995; Yardley, 2013; Dutta et al., 2015). While in data centres across the world, copper cabling is blended with fibre optics (Courtney, 2018).

Copper also belongs to a raft of commodities in futures markets such as the London Metal Exchange (LME) and the Shanghai Futures Exchange. Founded in 1877 and acquired in 2012 by Hong Kong Exchanges and Clearing, the LME began trading with copper before expanding to include other industrial and precious metals such as zinc, aluminium, gold, and silver ('London Metal Exchange'; Sanderson and Hume, 2017). Copper ushered into the twenty-first century as a metal whose market value increased by 80 per cent, around three times the value it held in the final years of the preceding century (Sanderson, 2015). This escalation in price coincided with the commodities boom in the mid-2000s when oil prices surged and the housing bubble had collapsed. During this period domestic demand for copper rose fivefold in China for construction, electricity networks, and infrastructure.[1] The broader geopolitical context was marked by the onset of what many commentators refer to as 'the Chinese Century'. Copper has not been immune from China-led globalization and in recent years has attracted an artificially designed scarcity value in the form of hoarding. Stored as copper concentrate powder in state-owned bonded warehouses located in the port of Qingdao, the technique of withholding product from trade resulted in surges in derivatives transactions on the world metals markets. Exemptions from customs duties and multiple sales of the same stock secured with the exchange of forged warehouse receipts are two key policy devices and commercial practices that instantiate a financialization of metal made abstract (Burton, 2017). Here, the medium of paper governs transactions in futures markets, replacing the movement of metal and accumulation of actually existing inventory.

ELECTRICAL IMPULSES TO THE BRAIN

In a moment of proto-McLuhanist anticipation, an advertisement for Kennecott Copper Corporation placed in a 1958 issue of *Fortune* magazine depicts two children glued to the screen of a plywood encased television, with rabbit ear antennae receptive to the world (Figure 15.4). Hallucinogenic multi-colour swirls of electronic waveforms emanate across the page. Children, television, and signals are suspended across a black backdrop. The design looks like something straight out of McLuhan's book with Quentin Fiore (1967), *The Medium is the Massage*. Even if *The Mechanical Bride* (McLuhan, 2008),

[1] As Geman and Scheiber (2017: 131) note: 'Construction, electricity grids and infrastructure pushed the country's demand for copper rise from 1.8 million tons in 2000 to over 10 million tons in 2015 and account in 2015 for 44% of the world global demand, up from 12% in 2000, according to the World Bureau of Metal Statistics.'

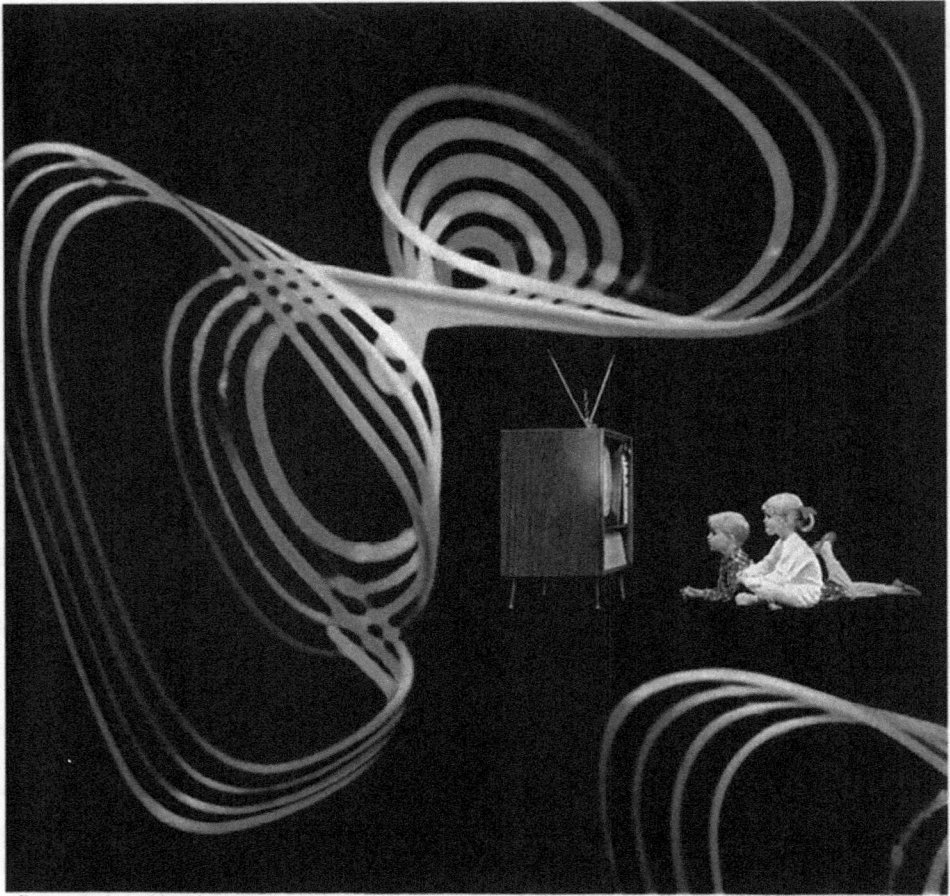

Electronic waveform is an actual photograph.

in electronics... no substitute can do what copper does!

The miracle of electronics depends on copper as on no other metal. For copper best carries the electrical impulses that activate the complex parts of radio and TV sets, radar and sonar equipment, intricate "brain machines" and countless other electronic devices. Copper and its alloys, too, make possible more economical mass production of these modern marvels. No other commercial metal can be formed, machined and solder-connected so easily. In electronics, as in so many other fields, no substitute can do what copper does!

KENNECOTT COPPER CORPORATION   CHASE BRASS & COPPER CO. · KENNECOTT WIRE & CABLE CO.

FIGURE 15.4 Advertisement for Kennecott Copper Corporation, *Fortune* magazine, 1958

first published in 1951, incorporated clippings from newspapers and magazines as part of its interplay between text and design, the aesthetic of those advertisements spoke more directly to an ideology of homeliness and good society stemming from Calvinist culture redolent of pre- and post-Second World War North America. *The Medium is the Massage*, by contrast, set out to experiment in contiguities and similitudes between text, images, and typographic design with a view to crystallizing conceptual probes or generate, as the subtitle of the book declared, *an inventory of effects*.[2] 'In electronics…no substitute can do what copper does!' 'Copper', so readers of Kennecott Copper Corporation's advertisement are informed, 'best carries the electrical impulses that activate the complex parts of radio and TV sets, radar and sonar equipment, intricate "brain machines" and countless other electrical devices'. The combination of communications signal, televisual device, and the rapt attention of children's minds produces a media apparatus fused together by copper cable infrastructure that is nowhere to be seen.

The materiality that ties the imaginary of copper to the perceptual synapses of the brain is distinct from the physical properties of this metal and its alloys. As a trace element in tissues of the body, copper is required for neurological functions and the nervous system. The medical treatment of copper-induced neurological disorders strives to return the body to a system of equilibrium. Restoring copper balance in the body guards against disease and neurological degeneration. Excess copper in the brain, liver, or intestinal organs can result in copper toxicity, which is implicated in numerous neurodegenerative conditions and metabolism disorders such as Parkinson disease, Alzheimer disease, and Wilson disease (Desai and Kaler, 2008). A deficiency in copper absorption, by contrast, can result in coeliac disease, Menkes disease, and dementia. These biochemical abnormalities are understood within the medical sciences as 'copper transport diseases' encoded into genetic mechanisms (Bandmann et al., 2015).

The metaphor of transportation overlaps here with earlier models of communication, which for centuries bound the transmission of symbols with technologies and infrastructure of movement. A decoupling of communication and transportation technologies occurred, as famously and somewhat controversially argued by James Carey, with the advent of the telegraph in the nineteenth century, which 'freed communication from the constraints of geography' (Carey, 1992: 204; see also Innis, 1951). Time and space became reorganized and invested with new territories of power in the form of colonial empires. With the ongoing miniaturization of technology in recent decades, signal systems have again conjoined with transportation.[3] Radio-frequency identification (RFID) chips with copper antenna are embedded in fashion garments and shipping containers, while digital wristbands track anxious bodies in motion and measure the productivity of workers in warehouses, factories, and offices. Sensor devices are increasingly littered across urban settings. Movement is now calibrated according to key performance indicators (KPIs). Copper is both a signal and conduit of transmission, regardless of scale.

Such a condition or state of existence is vastly different from the role of copper in the organization of capital accumulation and technologies of mediation, where excess is optimized as either a standing reserve or amplification of signal. With a scalar switch

---

[2] There was also an important media-technological use of copper in the Gutenberg Galaxy: the matrices (negative forms) for producing the letters that were then used in the printing press were made of copper. Thanks to Claus Pias for contributing this point.

[3] For essays that engage particularly with Carey's thesis on communication and transportation, see the contributions by Jeremy Packer, Jonathan Sterne, and John Durham Peters in Packer and Robertson (2006).

from the molecular level of the body and brain to industrial techniques of extraction and economies of speculation, copper extends its propensity to effect the organization of systems and conditions in the world.

As an elemental metal, copper invites spectrum thinking: from the neuronal networks of the brain to the metabolic system of the body and its organs, from holes bored into the ground to stockpiling copper concentrate in China's warehouses, from cables of empire to electronic waste industries and the cultivation of soils with toxic contaminants. As an analytical device, copper has a multiplying capacity, bringing otherwise asynchronous conditions, practices, and events into relation. In this regard, copper serves as an element within what Reinhold Martin (2016: 143–4) calls 'a system of bridging', where not only physical or material connections are made, but also cognitive relations are conditioned as a complex of 'infrastructural mediation'. The ensemble of relations of metallic resource, machines, labour, state formation, global economies and finance capital, and prevailing ideologies and imaginaries all exist in a kind of recursive feedback loop, with each element playing back upon and constituting the symbolic and experiential lifeworlds, even the material and ontological conditions that comprise the limits of the system at any particular historical conjuncture.[4]

COPPER AS METHOD

Across this sketch of developments and histories, copper per se is not a tool, but it is nonetheless an object shaped by organizational and technical processes that define social and economic imaginaries, transform bodies, and extend our sensory perception. In this regard copper can be understood as a technology. The many devices, objects, and infrastructures of which copper is a component part hold their own mediating properties and capacities. Yet no matter how much we may wish otherwise, objects do not speak. Even if they possess forms of what Katherine Hayles (2017) terms 'nonconscious cognition', objects remain elusive. While materiality is often operationalized within our idioms of intelligibility—whether through classification, modification, abstraction, and so forth—there nonetheless remains at a certain ontological level a stubborn refusal to inculcation.

How, then, to tune in to the silence of objects that is nonetheless underscored by the potential for force with transformative effects upon and within the world, on human subjectivity and ecological life? In recent years there has been a methodological predilection across the social sciences and humanities to 'follow the thing', as though revealing the movement and action of objects across networks lends an ontological substance brought about through the attribution of relations. There is undoubtedly some value in such a pursuit, but how to discern the hierarchy of relations—the politics—without succumbing to the depoliticized logic of 'flat ontologies' that prevails

---

[4] For a fascinating study of experiments in cybernetics and socialism in Chile's Allende government, see Medina (2011).

within the Latourian turn to assemblage theory? The history of copper modernity in Chile demonstrates that objects are never neutral. An elemental object as multivalent as copper traffics in confrontation.

The post-Kittlerian method of 'cultural technique' attentive to the material homologies between objects and their accompanying media-cultural techniques or practices makes possible a reordering of things beyond classificatory regimes that assume objects to conform to types or disciplinary systems, privileging instead the resonance and repetition of properties across otherwise dissimilar objects.[5] Given Kittler's own trajectory in his later years and the interest for researchers in cultural technique as a method that questions the relation between *physis* and *techné*, it should perhaps be no surprise that such a method returns us to the Greeks, and indeed perhaps always was Greek. Taking the object of copper, Aristotle made the case for understanding reality as 'a sequence of transitions from "matter" to "form" and from "form" to "matter"' (Asmus, 1979).

In his essay, 'Towards an Ontology of Media', Kittler (2009) maintains that the Aristotelian attention to matter and form neglects the relation of things to time and space. Perhaps buried in this insight of mediation between things, and thus organization, are the components for a critique of power and the techniques by which it operates through technologies of empire. Integrated into computer hardware and cable infrastructure, copper is an elemental media that also signals a regressive turn in making the Internet accessible to the Western world. In places where copper as infrastructure is absent, the leap to wireless and fibre optics can be made more easily. Copper can take us back to the Greeks, and it can turn us into 'German media theorists'. But above all else copper is elementary for colonialism. Copper is not an enabler, much in the same way that gold is not. Copper is a hoarder. It stores. And it switches and distributes.

REFERENCES

Anthony, Sebastian. 2016. 'The Secret World of Microwave Networks'. *Ars Technica*, 4 November. https://arstechnica.com/information-technology/2016/11/private-microwave-networks-financial-hft.

Asmus, Valentin F. 1979. 'Aristotle'. In *The Great Soviet Encyclopedia*, 3rd edn. New York: Macmillan. https://encyclopedia2.thefreedictionary.com/Aristotle+anomaly.

Bandmann, Oliver, Karl Heinz Wiss, and Stephen G. Kaler. 2015. 'Wilson's Disease and other Neurological Copper Disorders'. *Lancet Neurology*, 14(1), 103–13.

Burton, Mark. 2017. 'ANZ Stung as Paper Trail on Metal Loans Leads to Fakes'. *Australian Financial Review*, 5 July. http://www.afr.com/markets/commodities/metals/anz-stung-as-paper-trail-on-metal-loans-leads-to-fakes-20170705-gx4zp0.

Carey, James W. 1992. 'Technology and Ideology: The Case of the Telegraphy'. In *Communication as Culture: Essays on Media and Society*. London and New York: Routledge, 201–30.

---

[5] For an introduction to cultural techniques, see the collection of essays by some of the key proponents in a special issue of *Theory, Culture & Society* edited by Winthrop-Young et al. (2013).

Clark, Timothy David. 2013. 'The State and Making of Capitalist Modernity in Chile'. PhD dissertation, York University, Toronto.

Courtney, Michael. 2018. 'Blending Copper and Fiber'. *Datacenter Dynamics*, 19 February. http://www.datacenterdynamics.com/content-tracks/core-edge/blending-copper-and-fiber/99789.fullarticle

Cubitt, Sean. 2017. *Finite Media: Environmental Implications of Digital Technologies*. Durham, NC: Duke University Press.

Desai, Vishal and Stephen G. Kaler. 2008. 'Role of Copper in Human Neurological Disorders'. *American Journal of Clinical Nutrition*, 88, 855S–858S.

Dutta, Soumitra, Thierry Geigerand, and Bruno Lanvin (eds). 2015. *The Global Information Technology Report 2015*. Geneva: World Economic Forum. http://www3.weforum.org/docs/WEF_Global_IT_Report_2015.pdf.

Geman, Helyette and Matthias Scheiber. 2017. 'Recent Experiences of Copper in the Shanghai Futures Exchange: Some Lessons for Warehouse Monitoring'. *Resources Policy*, 54, 130–6.

Gethmann, Daniel and Florian Sprenger. 2014. *Die Enden des Kabels: Kleine Mediengeschichte der Übertragung*. Berlin: Kadmos.

Grappi, Giorgio and Brett Neilson. Forthcoming. 'Elements of Logistics: Under the Line of Copper'. *Environment and Planning D: Society and Space*.

Hayles, N. Katherine. 2017. *Unthought: The Power of the Cognitive Nonconscious*. Chicago: University of Chicago Press.

Hörl, Erich. 2017. 'Introduction to General Ecology'. In Erich Hörl with James Burton (eds), *General Ecology: A New Ecological Paradigm*. London: Bloomsbury Academic, 1–73.

Innis, Harold A. 1951. *The Bias of Communication*. Toronto: University of Toronto Press.

International Monetary Fund. 2015. *Chile: IMF Country Report*, no. 15/228, August. https://www.imf.org/external/pubs/ft/scr/2015/cr15228.pdf.

International Monetary Fund. 2016. *Chile: IMF Country Report*, no. 16/376, December. https://www.imf.org/external/pubs/ft/scr/2016/cr16376.pdf.

Jensen, Michael. 1995. 'Telematics for Development: Discussion Paper'. ITU, UNESCO, UNECA, Addis Ababa, 3–7 April. http://www.africa.upenn.edu/Padis/telmatics_Jensen.html.

Kittler, Friedrich. 2009. 'Towards an Ontology of Media'. *Theory, Culture & Society*, 26(2–3), 23–31.

'London Metal Exchange'. https://en.wikipedia.org/wiki/London_Metal_Exchange.

McLuhan, Marshall. 2008. *The Mechanical Bride: Folklore of Industrial Man*. Santa Rosa, CA: Gingko Press.

McLuhan, Marshall and Quentin Fiore. 1967. *The Medium is the Massage: An Inventory of Effects*. London: Penguin.

Martin, Reinhold. 2016. *The Urban Apparatus: Mediapolitics and the City*. Minneapolis: University of Minnesota Press.

Mattern, Shannon. 2017. *Code and Clay, Data and Dirt: Five Thousand Years of Urban Media*. Minneapolis: University of Minnesota Press.

Maxwell, Philip. 2004. 'Chile's Recent Copper-Driven Prosperity: Does it Provide Lessons for Other Mineral Rich Developing Nations?' *Minerals & Energy*, 19(1), 16–31.

Medina, Eden. 2011. *Cybernetic Revolutionaries: Technology and Politics in Allende's Chile*. Cambridge, MA: MIT Press.

Packer, James and Craig Robertson (eds). 2006. *Thinking with James Carey: Essays on Communications, Transportation, History*. New York: Peter Lang.

Parikka, Jussi. 2015. *A Geology of Media*. Minneapolis: University of Minnesota Press.

Peters, John Durham. 2015. *The Marvelous Clouds: Toward a Philosophy of Elemental Media*. Chicago and London: University of Chicago Press.

Rossiter, Ned. 2016. *Software, Infrastructure, Labor: A Media Theory of Logistical Nightmares*. New York: Routledge.

Sanderson, Henry. 2015. 'Shanghai Takes Command in Metals Markets'. *Financial Times*, 30 November. https://www.ft.com/content/4baa7aa2-94fd-11e5-bd82-c1fb87bef7af.

Sanderson, Henry and Neil Hume. 2017. 'Crunch Time for Increasingly Brittle London Metal Exchange'. *Financial Times*, 9 February. https://www.ft.com/content/3ab427e8-ed1c-11e6-ba01-119a44939bb6.

Stalder, Felix. 2018. *The Digital Condition* (trans. Valentine A. Pakis). Cambridge: Polity Press.

Tironi, Manuel and Israel Rodríguez-Giralt. 2017. 'Healing, Knowing, Enduring: Care and Politics in Damaged Worlds'. *Sociological Review Monographs*, 65(2), 88–109.

Vergara, Angela. 2008. *Copper Workers, International Business, and Domestic Politics in Cold War Chile*. University Park, PA: Pennsylvania State University Press.

Winthrop-Young, Geoffrey, Ilinca Iurascu, and Jussi Parikka (eds). 2013. 'Special Issue: Cultural Techniques'. *Theory, Culture & Society*, 30(6), 3–172.

Yardley, Matt. 2013. *Developing Successful Public–Private Partnerships to Foster Investment in Universal Broadband Networks*. Geneva: International Telecommunications Union. http://www.itu.int/ITU-D/treg/publications/SuccessfulPPPs.pdf.

CHAPTER 16

....

COPY MACHINE

....

MONIKA DOMMANN
TRANSLATED BY ERIK BORN

ON 19 September, 1959, the cover of *Business Week* featured Joseph C. Wilson, the CEO of Haloid Xerox, who stood beaming in front of an indistinct machine the size of large office furniture against a background of female office workers and stacks of paper (Figure 16.1; 'Out to Crack Copying Market', 1959). With the market launch of the Xerox 914 in 1959, the copier had finally cast off its historical ties to photochemical processing. The photostat process developed around 1900 involved a conventional camera with a lens fitted with a prism and made copies on a roll of bromide paper. The first photocopiers were marketed as efficiency and cost-saving devices in the 1920s. In 1937, Chester Carlson, a New York patent attorney with training in physics, patented a process he called 'electrophotography', which was based on the interaction of electrostatic charge, a photosensitive surface, and toner powder (Mort, 1989, 1994; Owen, 2004). Carlson's patent was not a success. It was not until 1947 that Haloid Xerox, a manufacturer of photo paper, would decide to invest in the technology. Thanks to Xerox's new electrostatic process, it was now possible to make copies on ordinary office paper instead of special photo paper. With this technical change, Haloid Xerox also attacked the dominant business model used by copier suppliers, which generated only 15 per cent of revenue through the sale of copy machines and the other 85 per cent through sales of special paper and copier accessories. Because the 914 model was expensive and could only be afforded by customers with extensive capital, it was brought to market with a novel rental system. As *Business Week* reported: '[I]t has set a monthly rental of $95, which works out to cost about 4 cents a copy for a 2000-copy-a-month user. Above a limit that will probably be around the 2,000 mark, a small charge per copy will be added to the rental. This charge will decline as the number of copies increases' ('Out to Crack Copying Market', 1959: 86). With the success of this model, Xerox quickly became synonymous with photocopying. A decade later, their copy machine would make history due to a famous act of whistleblowing.

The Pentagon Papers were released by Daniel Ellsberg, a RAND Corporation employee and former assistant in the Pentagon who had access to a secret 7,000-page,

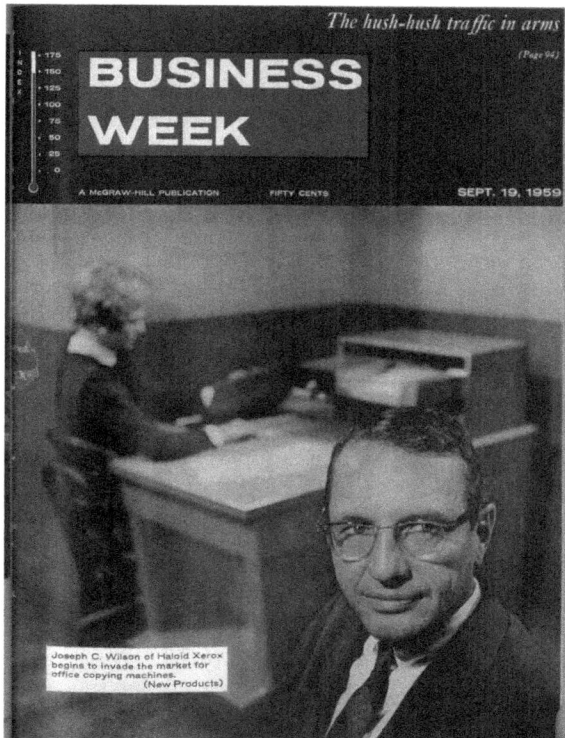

FIGURE 16.1 The businessman and the copier: Xerox makes the cover of *Business Week* following the launch of the Xerox 914, 1959

(*Source*: *Business Week*, September 19, 1959, Cover)

47-volume study of the historical involvement of US forces in Vietnam. Ellsberg had been part of the Vietnam Study Task Force, a group of thirty-six political analysts and historians tasked in 1967 with documenting and analysing the Vietnam War on behalf of Secretary of Defense Robert S. McNamara. The study, officially titled *Report of the Office of the Secretary of Defense Vietnam Task Force*, contained evidence of growing clandestine pressure well before the Gulf of Tonkin Resolution of August 1964. In October 1969, at the height of anti-war demonstrations in the United States, Ellsberg started copying the thousands of pages with the help of his trusted friend Anthony Russo (Douglas, 2008; Gitelman, 2014: 83–110). Every night, Ellsberg would make away with a single volume of the secret report in his briefcase and then copy it at the advertising agency where Russo's girlfriend worked. Several years later, Ellsberg leaked the completed copies to the American press. On 13 June 1971, the *New York Times* started publishing a series of articles informing the public about the contents of this secret 'Vietnam Archive' (Sheehan, 1971). If, in 1959, the Xerox machine had been hailed by *Business Week* as a device 'to crack copying markets', the copier's triumphant march into the office world, in the course of the 1960s, had started to challenge the powerful machinery of government. As Lisa Gitelman suggests: 'It seems clear in hindsight that copying copies was effectively the so-called killer app that Haloid Xerox had initially overlooked in developing and marketing the 914' (Gitelman, 2014: 92). In the Cold War society of secrecy, Haloid Xerox had inserted a potentially unruly medium of transparency and means of self-empowerment. Potentially, the operation of a copy machine turns organizations like courts, libraries, and administrations inside out, revealing their administrative core.

EMPOWERMENT AND CONTROL

At the margins of media and organization studies, the copy machine has occasionally appeared as an agent of media change in various forms: a technology of empowerment, a technology of control, an alternative technology to letterpress printing, and a transitional technology leading to the computer. In 1966, only seven years after the Haloid Xerox corporation introduced its first fully-automated model, Marshall McLuhan viewed the photocopier as a technology for empowering readers ('An Interview with Marshall McLuhan', 1966). For the Canadian media theorist, the copier was a medium for decentralizing and individualizing book production, insofar as the decentralized organization of copy culture provided a counter-model to the centralized organization of print culture. According to McLuhan, Xerox would transform readers from consumers of mass-produced books into producers of customized materials based on a different unit:

> The physical realities of xerography, let alone automation, mean that the book ceases to be a mass-produced package for the general public and becomes a personal, tailor-made custom item for the most exclusive personal needs and wishes.... In a few minutes or an hour a complete sheaf of Xerox material, in several languages, can be made available to suit its exact needs. This type of tailor made, custom-built servicing is characteristic of electrical circuitry. Instead of the mass-produced package, the immediate personal service is characteristic of electric technology. This is true even of the teaching machines, which are tutorial agents.
> ('An Interview with Marshall McLuhan', 1966: 68)

Less than five years later, a comment on McLuhan's vision of the individual as publisher would come from the publishing industry itself, especially American publisher William Jovanovich, whose medley of authors included Hannah Arendt, Milton Friedman, and Charles Lindbergh (Jovanovich, 1971; Eakin, 2001). For Jovanovich, the 'Universal Xerox Life Compiler Machine' and the period's autonomous social movements challenged the long-standing Renaissance and Enlightenment idea that the importance of a work consists in organizing the proliferation of knowledge: 'Indeed, property as a concept, although not necessarily a thing, is viewed suspiciously by the Young (as a caste), by the separatist Blacks, and by the Intellectuals who are able to subscribe to universities, foundations and government.... The radicals, the outsiders, in our society are less interested in the artifact, the object, than in the media' (Jovanovich, 1971: 252).

One of the main arguments in Jovanovich's defence of the traditional model of publishing was that a Xerox machine copied everything indiscriminately. Hence, the publisher considered Xerox to be a mere form of excessive mimesis (Balke, 2018) and not an artistic medium that would pose any serious, revolutionary threat to the publishing industry: 'The widespread use of xerography will not devalue literature, nor will it confuse the identity of the reader and writer, nor will it turn to antiquarianism the profession of publishing. It does not do these things because the act of copying is indiscriminate, unselective, uncompetitive. It is not a medium of art' (Jovanovich, 1971: 255).

The reception of the Xerox machine among the New Left has remained quite ambivalent since the late 1960s. Exemplarily, Hans Magnus Enzensberger's 'Constituents of a Theory of the Media' (1970) touched on the photocopier only in passing, classifying it as one of many 'electronic media', which interested him for their potential to mobilize people. For Enzensberger, the medium's subversive potential was manifest in the following example:

> The Soviet bureaucracy, that is to say the most widespread and complicated bureaucracy in the world, has to deny itself almost entirely an elementary piece of organizational equipment, the duplicating machine, because this instrument potentially makes everyone a printer. The political risk involved, the possibility of a leakage in the information network, is accepted only at the highest levels, at exposed switchpoints in political, military, and scientific areas. It is clear that Soviet society has to pay an immense price for the suppression of its own productive resources—clumsy procedures, misinformation, *faux frais*.   (Enzensberger, 1970: 16–17)

For Enzensberger, proof of the medium's danger to those in power could be found not only in the regulation of the copier in the East, but also in Xerox's capitalist business model based on the use of rental contracts:

> The technically most advanced electrostatic copying machine, which operates with ordinary paper—which cannot, that is to say, be supervised and is independent of suppliers—is the property of a monopoly (Xerox), on principle it is not sold but rented. The rates themselves ensure that it does not get into the wrong hands. The equipment crops up as if by magic where economic and political power are concentrated. Political control of the equipment goes hand in hand with maximization of profits for the manufacturer. Admittedly this control, as opposed to Soviet methods, is by no means 'watertight' for the reasons indicated.   (Enzensberger, 1970: 17)

In Xerox's rental model, which had been promoted in *Business Week* as a means of cranking up the copy volume with the stipulation of higher fixed costs, the German media critic saw the incarnation of a capitalist media system that sought to maximize profits in bourgeois property relations. As code for a new copy culture, Xerox would also be mobilized by left-wing publishers as an argument against copyright (Felsch, 2015).

Even as personal computers challenged the copy machine's status as a storage and transmission medium in the 1980s, Xerox continued to figure as code for both a fundamental shift in bureaucratic organization and a transitional technology in the nascent computer age. For communications scholar James R. Beniger, the photocopying technology developed by American companies like Photostat and Recordak in the 1930s is only a minor detail in a much longer list of techniques and technologies of information processing (Figure 16.2), which were put in the service of a social transformation he calls the

The One Machine that Performs Every Copying Requirement
... Makes Prints Size to Size ... Enlarges ... Reduces

FIGURE 16.2 Machines as office aids: Rectigraph brochure, USA, 1930s

(*Source*: Library of Congress, mm 78028043 Joint Committee on Material for Research, Box 13)

'control revolution' (Beniger, 1986: 395–6). From Beniger's perspective, the rise of copying techniques and technologies appears to be an organizational programme designed to counter the loss of control that had led to the disappearance of personal relationships in the course of the industrial revolution. Bureaucratic measures and information technologies were required to iron out the wrinkles left by the loss of control in the state and the economy. Hence, Beniger's concept of the control revolution remains bound up, in organizational terms, with an increasing tendency towards centralization and, in medial terms, with the ongoing development of new information technologies. In the late 1980s, Jean Baudrillard would still use the word 'Xerox' as a form of magic incantation in his commentary on the effects of intelligent machines on virtual people in the age of Minitel. However, 'Xerox' appeared only in the title of Baudrillard's commentary, 'Le Xérox et l'infini', serving as a distant echo of a time before the telematic people and the surface screen culture addressed in the text (Baudrillard, 1988).

KNOWLEDGE STORAGE AND DISTRIBUTION MEDIA

A crucial phase in the early history of the copy machine as a medium of organization came in the 1930s. Originally developed in the context of edition philology in the late nineteenth century, the photographic photostat method was mechanized by photography companies in the United States and Germany in the early twentieth century (Binkley, 1936: 161–76; Dommann, 2014: 134–69). Starting in the late 1920s, various American and European companies (e.g., Leitz in Wetzlar, Zeiss Ikon in Dresden, La Photoscopie in Brussels, Eastman Company in Rochester) sought to monetize the methods of 'bibliophotography', which had been invented in the context of the documentation movement. Further applications of bibliophotography included filming checks and customer policies for banks and insurance companies, and recording newspapers, books, and manuscripts for libraries (see Figure 16.3). Companies like Recordak and Siemens developed the first mechanized photocopiers, which worked faster and with fewer personnel, and automatically took care of developing the copies. One of the machine's main selling points was photocopies were produced in a quasi-automatic manner (see Figure 16.4), meaning mechanization might overcome threats to confidentiality, which had always been associated with copying. Delegating the task of reproduction to a mechanical apparatus was thought to eliminate any lingering concerns about copying a company's confidential files or an administration's secret documents:

> In such cases, the insertion of documents, the number of recordings, and the delivery of an equal number of reproductions can all be made and monitored by confidential representatives. Since the processing of photo prints occurs automatically after their exposure, the operator cannot gain any knowledge of the documents

FIGURE LXIX

Recordak copying camera

FIGURE 16.3 'Accounting by photography': filming checks with the Recordak camera, 1930s

(*Source*: Binkley, Robert C. Manual on Methods of Reproducing Research Materials. A Survey Made for the Joint Committee on Materials for Research of the Social Science Research Council and the American Council of Learned Societies, 172. Ann Arbor: 1936)

during the finishing stages. The absence of a negative in our process also guarantees that after delivery of the photograms no further copies can be made without authorization.   ('Siemens Reproduktions-Automat', 1936: 5)

The Berlin Patent Office started accepting copies as surrogates for patent documents for the first time in the early 1930s; and copies of excerpts from the civil registry were considered equivalent documents, by ministerial decree, because 'sufficient safeguards against counterfeiting are in place' ('Siemens Reproduktions-Automat', 1936: 5).

Libraries used photostat machines to produce white-and-black copies as a means of storing knowledge and distributing information. For instance, the League of Nations' library in Geneva sent copies of statistics and statutes to banks, libraries, ministries, and scientific institutes (Sevensma, 1931).

The photocopy combined the logic of two different systems: the reference library's site-specificity, which meant that books could not be lent out; and the League of Nations' internationality, which was intended to be a form of border crossing. After the First World War, new institutions like the International Council of Scientific Unions (ICSU) and the International Institute for Intellectual Cooperation (IIGZ) sought to promote

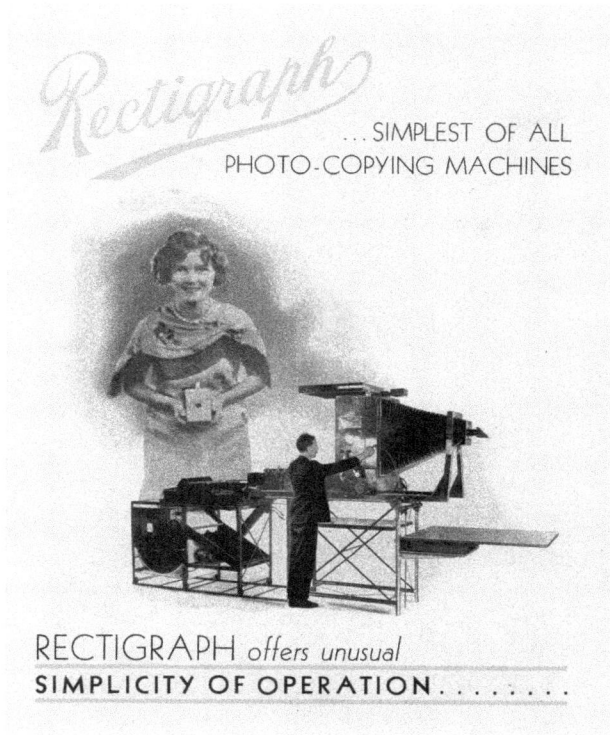

FIGURE 16.4 Photostat machines as alternatives to letterpress printing: Rectigraph brochure, USA, 1930s

(*Source*: Library of Congress, mm 78028043 Joint Committee on Material for Research, Box 13)

international scientific cooperation. In Europe, the international cooperation of libraries, the development of information sciences, and the application of new photographic reproduction techniques were all supposed to contribute to bringing the world closer together, exchanging knowledge, and integrating this collective knowledge into some common order. In the United States, the two decades after the First World War were also a period of change and internationalism for the scientific community. This was the context for the founding of the Joint Committee on Materials for Research under the aegis of the American Council of Learned Societies (ACLS) and the Social Science Research Council (SSRC). In August 1929, these organizations, which were financed by endowments from Carnegie and Rockefeller, commissioned historian Robert Binkley to study new alternatives to letterpress printing, especially in terms of collecting, disseminating, and storing research data in the social sciences and the humanities (Binkley, 1936).

Only libraries with demand for a high number of copies, such as the National Library of Medicine and the Library of Congress, were able to recoup the costs of renting the massive machine. 'Once you have it, you must keep it busy or else you're working for the Haloid Xerox', warned the head of copying services at Columbia University (Ballou, 1959: 88). Photographs depicting the almost exclusively black and female workforce, which

FIGURE 16.5 Frank Shifflett adjusting the Haloid Xerox machine

(*Source*: U.S. National Library of Medicine, Bethesda Md.)

can be found at the National Library of Medicine located just north of Washington, DC, show how much the operation of these machines was structured by categories of race and gender (see Figures 16.5 and 16.6).

The Xerox 914's rental model gave rise to a new copying craze in libraries, according to the librarian responsible for the reproduction department at the University of California, Los Angeles. The number of copies skyrocketed. Libraries soon acquired additional equipment and expanded the hours for their photocopying services to evenings and weekends. The fact that the large order volume for copies required libraries to adjust their organization and administration is also reflected in the fact that they started selling users a 'Xerox Credit Card': with a starting value of $10, the card would be punched for every copy; in doing so, libraries spared themselves the need for an accounting system (Crawford, 1963). Despite these optimizations, the copy machine itself was still prone to failure. The proverbial 'paper jam' was a stumbling block in the desired 'information flow', and copy machines disrupted normal operations by making noise, heating up rooms, and spreading the particular electrostatic odour that remains an olfactory characteristic of the process to this day.

The copy industry did business wherever there was a demand for copies—in libraries, offices, and busy public spaces like train stations. One thing was clear: the automation of the copy process, the trend towards self-service, and the diffusion of the machine

FIGURE 16.6 Haloid Xerox copy flo machine
(*Source*: U.S. National Library of Medicine, Bethesda Md.)

into the public sphere all made it impossible to make any definitive statements about who was copying what material and how often they were doing it. Still, the social usage of photocopiers can be partly reconstructed on the basis of studies initiated by science policy commissions and other associations in the context of copyright conflicts (e.g., Joint Libraries Committee on Fair Use in Photocopying, 1961; 'Survey of Copyrighted Material Reproduction Practices in Scientific and Technical Fields', 1963). Over 90 per cent of scholars responding to these surveys said they kept copies after using them; just as many researchers reported that they would pass copies on to others. Scholars clearly regarded photocopies as a distribution medium. However, they also complained that copies were causing them to lose track of their audience: 'You do not know who is interested in your work', one of the scholars surveyed reported ('Survey of Copyrighted Material Reproduction Practices in Scientific and Technical Fields', 1963: 84). The possibility of acquiring on-demand copies from libraries gradually made the practice of ordering of reprints directly from authors obsolete; and there was a corresponding drop in personal correspondence among scholars. If one of the characteristics of mass media is that they exclude the possibility of interaction between the transmitter and the receiver by imposing themselves into the communication channel (Luhmann, 2004: 10–11), the photocopier drove a particular wedge between the authors and readers of scholarly work.

REPAIRS AND REVOLTS

While photocopiers in libraries undermined authors' copyright privileges (i.e., the exclusive right to reproduction), those in state administrations challenged the internal power structures of bureaucracy. The copy machine's subversive potential, which had been addressed in Enzensberger's media theory, became reflected in organizational regulations that called for a central organization of the photocopier. In 1965, for example, the GDR's Institute for Administrative Organization and Office Technology published a booklet titled 'Copy—But How?' (Zieger, 1965). As a preventative measure against the paper flood, the Institute sought to centralize reproduction in terms of quality, efficiency, cost control, and material savings. In accordance with a regulation passed in 1959, however, the production of printed matter and Xeroxed products still required a government approval (i.e., a 'printing permission'). This tension indicates the political explosiveness and potential danger that came with the entrance of photocopiers into public administration and reveals the control measures state administrations developed in response to their potential loss of control. During the Cold War, the free flow of information commonly figured in the West as a counter-model to the control over information and copying in the East (though Enzensberger's thesis complicates this simplistic binary opposition). In 1957, UNESCO even put out an announcement about a meeting of government experts under the slogan 'Free Flow of Information' (UNESCO, 1957). In the early 1960s, the hypothesis of a rapid flow of information as a vehicle for social well-being would receive statistical support from the emerging research field of scientometrics. In *Little Science, Big Science*, for instance, American information scientist Derek de Solla Price presented captivating graphics, which did not merely explain the exponential growth of the 'scientific community' and its journal publications, but also the declining half-life of scientific 'papers' measured in terms of a publication's rapidly declining phase of citation (see Figures 16.7 and 16.8; Price, 1963).

The first institution in the federal administration of Switzerland to make a rental contract for a Xerox 914 was the Swiss National Library. In 1964, the Central Agency for Organizational Affairs authorized the library to rent a Xerox machine (Schweizerisches Bundesarchiv, 1963–70). The rental agreement created by the European subsidiary Rank Xerox was based on the exact same business model Xerox had used for the initial launch of the 914 copier in the United States five years earlier: the rental fee was now 120 francs per month, to be paid in advance; and each copy cost 37 rappen, with a minimum of 2,000 copies per month. According to the contract, Rank Xerox would provide free training for a National Library employee in the daily maintenance of the machine, as well as making special visits for repairs and revisions. The choice of paper, which had been entirely open at the launch of the 914, was now contractually restricted: if purchased from some other source than the machine's supplier, the copy paper would have to be inspected before use. The device remained the property of Xerox and could not be

Don't buy an office copying machine!

FIGURE 16.7 'Don't buy an office copying machine! Borrow ours!' Brochure for the Xerox 914, around 1962

(*Source*: Xerox Corporate Library)

subleased to another party. The rental contract reveals a whole set of control measures included by Xerox: determining their ownership of the equipment; monopolizing their technical know-how regarding repair and maintenance; guaranteeing their own right to permanent maintenance; and making external repair workers, who were always temporary employees, into troubleshooters for a technology that could never be fully controlled (Orr, 1996). Repairing and eliminating errors in the Xerox 914 became a fixed part of the schedule in many bureaucratic organizations (Suchman, 1985). One report on Xerox, published in *The New Yorker* magazine in 1967 under the apt title 'Xerox Xerox Xerox Xerox', testifies to both the fascination with human–machine interaction and the gender-specific connotations of maintenance (Brooks, 1967). The report describes the maintenance of the machine in the mode of erotic relationships (i.e., man–woman) and familial love (i.e., mother–child). The women responsible for refilling the toner and the paper and for cleaning the Selenium drum were of particular interest to the journalist:

> I spent a couple of afternoons with one 914 and its operator, and observed what seemed to be the closest relationship between a woman and a piece of office equipment that I have ever seen. A girl who uses a typewriter or switchboard has no interest in the equipment, because it holds no mystery, while one who operates a computer is bored with it, because it is utterly incomprehensible. But a 914 has distinct animal traits: it has to be fed and curried; it is intimidating but can be tamed; it is subject to unpredictable bursts of misbehavior; and generally speaking it responds in kind to its treatment.   (Brooks, 1967: 57; see Figure 16.9)

Borrow ours

The Xerox 914 Copier costs you nothing; we lend it to you. You only pay for copies ($95 a month for the first 2,000 copies). You'll be paying about 5¢ a copy—certainly no more and probably less than you pay now. And here's what you'll get: 1. Copies on ordinary paper. Copies that last as long as the paper lasts. 2. Fast copies. Seven copies a minute, automatically, by turning a knob and pushing a button. 3. Clear, precise copies. Anything you can see (colors, too), the 914 will copy in black and white. 4. No adjustments. No wet chemicals to add. The 914 is a dry machine. Bone dry. 5. Copies you can use. Every copy is a perfect copy. (In your present calculations, are you adding the cost of copies that wind up in your wastepaper basket?) Call your Xerox representative. Come in and see the 914 Copier perform. Xerox offices are in principal U.S. and Canadian cities. (If our arithmetic doesn't convince you, the copies will.) **XEROX** CORPORATION

FIGURE 16.8 'Don't buy an office copying machine! Borrow ours!' Brochure for the Xerox 914, around 1962

(*Source*: Xerox Corporate Library)

Far more than mere mechanical operation, maintenance is presented here as a caring activity. The journalist's description anthropomorphizes the machine, which has to be nurtured and cared for, contains secrets that need to be controlled, and generally threatens to disrupt the order of the office. At another point in the article, the journalist also ponders the birth of a new female profession out of the spirit of Xerox: 'The girls who

FIGURE 16.9 Brochure for the Xerox 914, 1961

(*Source*: Xerox Corporate Library)

operated the earliest typewriters were themselves called "typewriters", but fortunately nobody calls Xerox operators "xeroxes"' (Brooks, 1967: 57).

For an organization or a department to have its own copy machine meant more autonomy, and, in many organizations, photocopiers promoted an existing tendency towards decentralization. The high rental cost of Xerox machines, along with their supposed 'paper flood' (which is a driving force and phantom for any bureaucracy) presented administrations with a constant source of conflict over control and cost-cutting measures. 'Xerox' meant 'copy and pay'—namely, to a private company. Once touted for their ability to make states and companies more efficient, copy machines increasingly became a case for reorganization in the eyes of administrative business economists. By the 1970s, the same devices that had promoted communication and the coordination of an organization's goods, services, and information were viewed with suspicion by business economists trained in austerity measures. Equipped with the weapons of operational research methods, they advised reducing the amount of copies, the number of man-years required, and the personnel costs by centralizing reproduction and reviving courier services. Decentralized reproduction had become a luxury, in the eyes of operational research, that could no longer be afforded in the face of growing state and administrative tasks.

For the new social movements emerging after 1968, the copy machine became a medium for creating a 'counter-public sphere' (Negt and Kluge, 1972). While universities were at the centre of the student movements, many students were themselves reluctant to depend on the coin-operated copy machines that had been installed there. The real centres of the media revolution were the copy shops, which had flourished outside of libraries, universities, and state administrations (Eichhorn, 2016). Based on the Xerox rental system, small business owners established copy shops in public locations in various cities, while left-wing student organizations sought to lease and operate their own copy machines. In Zürich, the Copy-Quick copy shop put out advertisements in the journal of the student body and of the leftist student organization as early as 1974 (see Figure 16.10). The university's student body also owned their own Xerox Canon NP 70 since around 1975, and offered printing courses, according to McLuhan's motto that in the Xerox age every author (or, would-be author) could also be a publisher (see

FIGURE 16.10 Advertisement for the Copy-Quick copy shop, ca. 1975

(*Source*: UZH Archives, Zürich)

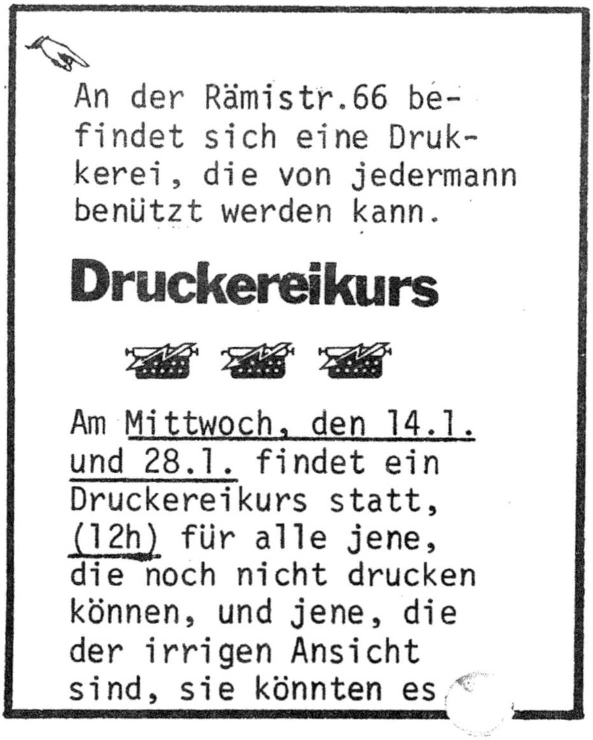

FIGURE 16.11 Advertisement for a printing course, Zürich, 1970s

(*Source*: UZH Archives, Zürich)

Figure 16.11). While the Zürich youth movement communicated primarily through the media of video and the small offset press, Xerox copies were also circulated as leaflets within the movement and on the street. In the University of Zürich's archive, there is a collection of thirteen leaflets for the Association of Zürich Students from June 1980. Appropriately called 'Copy Argument', the leaflets are numbered like a journal, and all of them contain a handwritten note requesting the reader to make further copies. If the revolution will not be televised, it must be Xeroxed! The students' leaflets present a copy-and-paste aesthetic in the tradition of John Heartfield's photomontages and other graphic artists of the 1920s and 1930s (see Figure 16.12). Nevertheless, their third leaflet, called 'Copy Argument no. 3', shows that the Xerox machine, the very medium that scientific policy and education foundations had once promoted as a beneficial medium for distributing edifying content, had been largely discredited in the student movement. For many students, Xerox ultimately became a symbol for the standardized, pre-packaged craving for knowledge. One flyer shows the Zürich students in rank and file—reading from Xeroxed pages! 'We don't need no education', Pink Floyd sang in 1979, and only a year later, the students' photocopies would be used in McLuhan's sense of a tailor-made, empowering medium for critiquing university training, which had become increasingly

FIGURE 16.12 'Copy Argument no. 3' against the standardized, prepackaged craving for knowledge, June 1980

(*Source*: UZH Archives, Zürich)

reminiscent of mass production. Despite various technical, political, and organizational measures, the copy machine was never brought under control. To this day, the small, dark, and stuffy copy room remains a risk factor for the state, the university, and the economy.[1]

REFERENCES

'An Interview with Marshall McLuhan'. 1966. *The Structurist*, 1 January, 61–9.

Balke, Friedrich. 2018. *Mimesis zur Einführung*. Hamburg: Junius.

Ballou, Hubbard W. 1959. 'Developments in Copying Methods—1958'. *Library Resources & Technical Services (LRTS)*, 3(2), 86–97.

Baudrillard, Jean. 1988. 'Le Xérox et l'infini'. *Traverses. Revue du Centre Georges Pompidou*, 44–45, 18–22.

[1] The author would like to thank Lucas Federer, Philipp Messner (UZH Archives), Stefan Nellen (Swiss Federal Archives), and Karin Schraner.

Beniger, James R. 1986. *The Control Revolution: Technological and Economic Origins of the Information Society*. Cambridge, MA: Harvard University Press.

Binkley, Robert C. 1936. *Manual on Methods of Reproducing Research Materials: A Survey Made for the Joint Committee on Materials for Research of the Social Science Research Council and the American Council of Learned Societies*. Ann Arbor, MI: Edwards Brothers, Inc.

Brooks, John. 1967. 'Xerox Xerox Xerox Xerox'. *The New Yorker*, 1 April, 46–90.

Crawford, Helen. 1963. 'Notes of a Librarian on Contemplating Her Xerox Machine'. *Bulletin of the Medical Library Association*, 51(3), 397–9.

Dommann, Monika. 2014. *Autoren und Apparate: Die Geschichte des Copyrights im Medienwandel*. Frankfurt am Main: Fischer.

Douglas, Martin. 2008. 'Anthony J. Russo, 71, Pentagon Papers Figure, Dies'. *The New York Times*, 8 August, A 17. https://www.nytimes.com/2008/08/09/us/politics/09russo.html [Accessed 16 July 2018].

Eakin, Emily. 2001. 'William Jovanovich, 81, Longtime Publishing Chief, Dies'. *The New York Times*, 6 December. https://www.nytimes.com/2001/12/06/business/william-jovanovich-81-longtime-publishing-chief-dies.html [Accessed 16 July 2018].

Eichhorn, Kate. 2016. *Adjusted Margin: Xerography, Art, and Activism in the Late Twentieth Century*. Cambridge, MA: MIT Press.

Enzensberger, Hans Magnus. 1970. 'Constituents of a Theory of the Media'. *New Left Review*, 64, 13–36.

Felsch, Philipp. 2015. *Der lange Sommer der Theorie: Geschichte einer Revolte 1960–1990*. München: Fink.

Gitelman, Lisa. 2014. *Paper Knowledge: Toward a Media History of Documents*. Durham, NC: Duke University Press.

Joint Libraries Committee on Fair Use in Photocopying. 1961. 'Report on Single Copies'. *Bulletin of the Copyright Society of the U.S.A.*, 9(1), 79–84.

Jovanovich, William. 1971. 'The Universal Xerox Life Compiler Machine'. *The American Scholar*, 40(2), 249–55.

Luhmann, Niklas. 2004. *Die Realität der Massenmedien*. Wiesbaden: VS Verlag für Sozialwissenschaften.

Mort, Joseph. 1989. *The Anatomy of Xerography: Its Invention and Evolution*, Jefferson, NC: McFarland.

Mort, Joseph. 1994. 'Xerography: A Study in Innovation and Economic Competitiveness'. *Physics Today*, 47, 32–8.

Negt, Oskar and Alexander Kluge. 1972. *Öffentlichkeit und Erfahrung: Zur Organisationsanalyse von bürgerlicher und proletarischer Öffentlichkeit*. Frankfurt am Main: Suhrkamp.

Orr, Julian E. 1996. *Talking about Machines: An Ethnography of a Modern Job*. Ithaca, NY: Cornell University Press.

'Out to Crack Copying Market'. 1959. *Business Week*, 19 September, 86–93.

Owen, David. 2004. *Copies in Seconds: How a Lone Inventor and an Unknown Company Created the Biggest Communication Breakthrough Since Gutenberg—Chester Carlson and the Birth of the Xerox Machine*. New York: Simon & Schuster.

Price, Derek John de Solla. 1963. *Little Science, Big Science*. New York: Columbia University Press.

Schweizerisches Bundesarchiv. 1963–70. E 3130A#1993/240#156*, Az 781.4, Xerox Schnellkopiergerät 914, 2. Xeroxapparat 720.

Sevensma, T. P. 1931. 'Die Bibliothek des Völkerbunds'. *Zentralblatt für Bibliothekswesen*, 48, 527–9.

Sheehan, Neil. 1971. *The Pentagon Papers as Published by The New York Times*. New York: Quadrangle Books.

'Siemens Reproduktions-Automat'. 1936. *I.I.D. Communicationes*, 3(1), Beilage.

Suchman, Lucy A. 1985. *Plans and Situated Actions: The Problem of Human–Machine Communication*. Palo Alto, CA: Xerox Corporation.

'Survey of Copyrighted Material Reproduction Practices in Scientific and Technical Fields'. 1963. *Bulletin of the Copyright Society of the U.S.A.*, 11(2), 69–124.

UNESCO. 1957. 'Free Flow of Information'. *Bulletin of the Copyright Society of the U.S.A.*, 6(4), 194–5.

Zieger, Gerhard. 1965. *Vervielfältigen—aber wie?* Leipzig and Berlin: Staatsverlag der Deutschen Demokratischen Republik.

CHAPTER 17

·······································

DATING APP

·······································

NANNA BONDE THYLSTRUP AND
KRISTIN VEEL

DATING apps offer unique insights into how we navigate and manage relationships through cultural techniques, and in a wider scope how we manage information in a productive oscillation between control and uncertainty under neoliberalism. Coupling media theories on hook-up and dating apps with cultural theoretical works on cultural techniques, we explore the ways in which dating apps use uncertainty as an organizational logic, addressing these apps as cultural technologies that take an active agency in organizing the patterns, logics, and identities of 'datable subjects' (Rosamond, 2018).

Work on dating and hook-up apps has been pioneered by the fields of health studies and in different branches of Internet sociology, including those informed by gender, critical race studies, and LGBTQ studies (see, e.g., Batiste, 2013; Stempfhuber and Liegl, 2016). While the work conducted by these fields takes on a wide variety of concerns and interests, it is also possible to identify recurrent themes across the board: questions of risk, uncertainty, and control (Handel and Shklovski, 2012; Albury and Byron, 2016; Brubaker et al., 2016), new forms of intimacy (Race, 2015; David and Cambre, 2016; Møller and Nebeling, 2017), and new patterns of mediated mobilities (Blackwell et al., 2015; Licoppe, 2015).

A particularly salient feature of dating apps is their reputation as risk-management technologies that help us organize life's uncertainties into controllable possibilities. They deploy calculative techniques to determine whether someone is a match, mitigating the many inherent risks of face-to-face communication with strangers, and giving off an air of algorithmic certainty and objectivity. A key aspect of this feature is that of warranting, i.e., the measures in place that enable a user to determine whether to trust that the person behind a particular online profile is actually who he or she claims to be (Stone, 1995; Walther and Parks, 2002). The risk-reducing strategies of dating apps in terms of matching people are haunted, however, by the security risks posed by the dating technologies themselves in the form of data leaks and other security breaches. From this perspective, dating apps emerge themselves as risky entities, threatening the user's privacy and safety.

Yet rather than reducing dating apps to risk or risk-elimination technologies, we instead suggest that they also organize around, and productively work with, *uncertainty*. In this regard, we follow recent theoretical calls for exploring uncertainty as logic that may be related to—but is not necessarily encompassed by—risk (Brashers, 2001; Appadurai, 2011; Schüll, 2012; Amoore, 2013; Nowotny, 2015; Samimian-Darash and Rabinow, 2015; Beckert, 2016; Corriero and Tong, 2016).

We here suggest that there are important organizational insights to be gained from focusing on the way dating apps 'represent a larger mode of organising and presenting information' (Morris and Elkins, 2015: 38) as well as shifting the discursive context of these apps from *control, security*, and *love* to the less goal-oriented notions of *uncertainty* and *flirting* (Phillips, 1996; Hoffman-Schwartz et al., 2015). On one level, dating apps incorporate ideas of serendipity through cultural techniques such as geo- and time-tagging. On another, they increasingly rely on habit-forming techniques such as flow to create small pockets of manageable uncertainty in which one can 'zone out' in a contained environment and forget about the larger uncertainties in life. Dating apps are thus designed not only to help users navigate safely but also to hook the users in a habitual pattern of usage in which the act of searching is itself a source of pleasure.

The need for dating apps to retain an element of uncertainty becomes especially obvious in the design of their infrastructure and interface, which in many cases employ elements of gamification. Thus, while most dating sites in some respects work to reduce uncertainties by turning them into statistically calculable risks, they also actively foster and play with uncertainties in ways that engage and even captivate their users. Uncertainty in the field of online dating thus emerges as a resource to invite, cultivate, and exploit, rather than as a liability to reduce, mitigate, or control. In this sense, the world of online dating increasingly aligns with other digitally mediated realms of uncertainty, such as gambling and finance. The following sections trace how uncertainty becomes an organizational force and a cultural technique in dating apps.

GRID, SWIPE, GEOLOCATION: THE ORGANIZATIONAL LOGIC OF DATING APPS

One specific challenge of researching dating apps as organizational technologies is that they belong to the regime of habitual update (Chun, 2017). The particularity necessary for careful analysis of specific apps and platforms and their modus operandi therefore runs the risk of outdating the analysis already at the time of writing. A further and related complication is the fact that the technological innovation of apps takes place within private commercial corporations, where access to an overview of the market is only obtained with difficulty. Dating apps therefore echo broader methodological questions about how to approach apps meaningfully in methodological terms (Morris and Murray, 2018). Yet, by approaching dating apps from the perspective of cultural technique and emphasizing their role as one example of a more generalized discourse on uncertainty and control, we may circumvent some of these obstacles.

On an infrastructural level, dating apps are organizational technologies that inscribe datable subjects in the logic of platform capitalism (Srnicek, 2016), embedding users in larger platform economies through data brokers, advertisers, and social media platforms. However, at the same time they also envelop users in complex algorithmic and technological infrastructures geared towards working with, and mitigating, uncertainty. For instance, Tinder uses ranking systems (e.g., Elo ranking score) developed for the world of sports to calculate the relative skill levels of players in uncertain zero-sum games. Furthermore, the infrastructural set-up of most dating apps can be regarded as networked software stacks layering other platforms such as Google Cloud Platform, Amazon Cloudfront, Zendesk, Docker, and Cloudflare. Dating apps emphasize these algorithmic and agile infrastructures as selling points, promising the user improved dating possibilities through their intermingling of virtual networks with traditional offline networks. It allows them to frame themselves as flexible organizational technologies that optimize dating processes, filter out unwanted intimacies, and speed up information-seeking behaviour.

Underpinning these large-scale infrastructures of uncertainty, however, are the much more intimate interfaces presented to users. Here, the organizational logic of uncertainty is mobilized not as infrastructure, but rather as interface. Focusing on how the significance of uncertainty as an organizing force in dating apps becomes visible to the user in the cultural techniques of the *grid*, the *swipe*, and *geolocation* we can situate dating apps as organizational objects that operate on the intimate level of the visual and haptic experience of the user. Moreover it also gives opportunity to gesture towards how these cultural techniques are inscribed in more fundamental discourses on uncertainty management as a condition of life under neoliberalism.

Grid

The grid is a classic way of organizing information, which has been used as a method for bringing order to chaos for millennia (Higgins, 2009). The Tibetan Book of Proportions, from the eighteenth century but developed as far back as the fourth century, delineates, in thirty-six ink drawings, precise iconometric guidelines for depiction of the Buddha and Bodhisattva figures. Medieval scribes used a grid system for composing papers that was later borrowed by German printer Johann Gutenberg to produce the first mechanically printed bible in 1455. Later yet, colonial powers deployed the gridded landscape as geographical violence (Blomley, 2003; Raymond, 2003). As Lewis Mumford notes, 'the standard gridiron plan in fact was an essential part of the kit of tools a colonist brought with him for immediate use' (Mumford, 1961: 192). Today, cartographers still use it as a way of plotting coordinates, architects convert mathematics into spatial arrangements, and it is essential to the work of typographers as well as computer engineers. Despite such variations, the grid has remained largely stable in structure and use.

Regardless of its ubiquity, the grid is not necessarily an intuitive mode of organization. Rather, as graphic artist Joseph Müller-Brockmann notes, grids are cultural techniques: 'one must learn how to use the grid; it is an art that requires practice' (Müller-Brockmann, 2003: 92).

Today, the cultural technique of the grid emerges as an organizing principle for how we view datable subjects in several dating apps. One example is Grindr, which organizes "datable subjects" along horizontal and vertical axes in a gridded structure that determines content, proportion, space, form, and temporality in a manner that bestows modular and systematic characteristics to the visual appearance.

Users learn how to navigate gridded structures that organize datable subjects in a visually, temporally, and spatially appealing way. The grid provides the user with an overview of datable subjects in thumbnail size, which can be scrolled through and clicked on if any of the images solicit interest. Dating apps thus appear to realize Kittler's central doctrine of 'information-theoretic materialism': 'Nur was schaltbar ist, ist überhaupt' (Kittler 1993: 182).[1] From this perspective, one might say that on dating apps organized around a gridded design, datable subjects exist as switchable nodes in much larger networks of desire.

While the grid technique appears in a wide variety of aesthetic and functional contexts (Krauss, 1979; Müller-Brockmann, 2015), it is also, as German media theorist Bernhard Siegert (2015: 97) remarks, a cultural technique, and its 'salient feature is its ability to merge operations geared toward representing humans and things with those of governance'. From this perspective, the grid view expresses the interaction between imaging technologies and mathematical, topographical, geographical, and governmental knowledge and speculation (Siegert, 2015: 97ff.). In the digital environment, we find the coming together of all these different forms of grids in GPS systems that overlay topographical and three-dimensional grids. For some, such systems represent lines of flight. For others, such as Siegert (2015: 98), this convergence is instead a sign of an increasing 'totalitarianism of the grid', from which hardly anything can escape.

The governance potential of the grid is obvious on sites such as Grindr. Displaying the potential dates in a gridded view offers the user a sense of overview and control. Yet as anyone who has experienced a gridded view in digital search environments such as Pinterest, Netflix, Google Image, and Facebook knows, the grid never displays a totality of information. Instead, as digital designer and theorist Mitchell Whitelaw (2011) notes, the grid configures computational technologies as potentially boundless: 'The array is a spatial device that says "and so on".' This sense of informational infinity is in fact a selling point for most digital apps, the very *raison d'être* of which is to offer limitless information for the user to not only find but also get lost in.

As art theorist Anna Munster notes, these new dynamic digital arrays preserve gridded infrastructures, while also folding them into new complex networks that intersect elements of biopower, vitality, urbanism, and aesthetics (Munster, 2013: 30ff.; see also

---

[1] John Durham Peters indeed translates this as 'Only that which can be switched (gridded) exists at all' (Peters, 2015: 320).

Parisi, 2013: 46ff.). Indeed, the new functions of the grid in its digital form add more complexity to its role as an organizational technique as it shifts from a primarily topographical to a topological technique (Lury et al., 2012). The grid thus re-inscribes itself in a complex network that fuses control and uncertainty.

We may therefore regard the gridded organization of dating apps as about more than the rigidities of geometrical compositions and rationalization of desire. Rather, and building upon art historian Jonathan Crary's discussion of Seurat's painting *Parade de Cirque*, we suggest that dating apps such as Grindr represent a modern ordering of desire and exchange 'in which any geometrical semblance of stability is a mask over a more powerful dispersal and circulation of affect' (Crary, 2001: 213). From this psycho-analytically attuned perspective, the cultural technique of the grid, endowed as it is with 'generative possibilities' and 'creative capacities', becomes not only an ordering principle but also manifests itself as part of the fantasy (Crary, 2001: 208ff.).

Swipe

While often overlooked as analytical objects of study, gestures are fundamental techno-cultural organizational tools, not only accompanying our handling of information, but also codifying meaning (Flusser, 2014). Marcel Mauss conceptualized bodily gestures as 'techniques of the body' (Mauss, 1973: 70). His observation was underpinned by an attentiveness not only to the psychology of gestures, but also their material dimension and how techniques of the body often presuppose an instrument from spades and chairs to stones and beds (Mauss, 1973: 79). These gestural techniques, then, rather than existing a priori, are shaped and institutionalized by normative patterns of behaviour formed by gendered, colonial, and economic structures (see Duffner, this volume).

Today, gestures have been incorporated as an important organizational element of dating apps. The most significant example being Tinder, which has introduced the swipe as the cultural technique of online dating par excellence.

How might we situate this gesture in the context of flirtation? The swipe may come across as an almost violent physical act, shoving another datable subject away or drawing them closer. Yet, swiping can also be read as a gesture of care and intimacy as a finger that caresses. None of these meanings were present when swiping was originally introduced as a feature in dating apps by Tinder, however. Rather, swiping was entangled in a topology of consumption and control through its association with turning on computers and swiping cards. Tinder introduced the swipe to 'make online dating fast, delightful, intuitive—and a little physical' (Pierce, 2016). In particular, the creators wished to transform online dating from a laborious cognitive task to a more embodied state of fun. The swipe thus adds haptic and gaming techniques to the digitally mediated flirt, just as it accelerates the pace of online dating by abandoning button interfaces in favour of gestural techniques. As such, the 'elemental binary meaning of yes or no' in Tinder's swipe (David and Cambre, 2016: 8) not only habituates the bodily gesture of the swipe, it also retrains the body, inserting its gestures and modes of attention, into the accelerated pace of digital capitalism. In this respect the swipe appears less as a playful gesture, and more like gestures of acceleration identified by Marcel Mauss, much like Charlie Chaplin's famous nervous tic in his assembly line choreography in the 1936 film *Modern Times* as he tries to keep up with the inhuman pace of capitalism.

The design ethos underlying the swipe as a cultural technique of acceleration is to simplify the user's experience by presenting fewer options. The simplicity of the swipe, with its binary yes or no option, defers all the nuances of a social encounter (from subtle hints to more direct signals) in favour of a simple gesture, which promises to release not only energy but also time (Werning, 2015). Yet as media theorist Stefan Werning notes, saving time does not necessarily mean that the user spends their time elsewhere and on other things. Indeed, as swiping is implemented and naturalized in digital interfaces, it often produces an urge in the user to keep swiping, insofar as the interface logic of the swipe activates the powerful cognitive modality described by behavioural psychologist B. F. Skinner in the 1950s as a 'variable ratio enforcement', which means users feel compelled to constantly swipe to find out 'what happens next'. As such, the swipe appears not only as an organizational device that optimizes and rationalizes dating processes but also an affective 'trigger' structure (Werning, 2015).

Uncertainty is central to this affective mechanism, both in terms of the uncertainty of serendipity and the sense of habitual control that dating apps induce by offering a relatively contained uncertain practice in which users can 'zone out' from the much more overwhelming state of uncertainty of the world. We may here draw a parallel to the work of Natasha Dow Schüll (2012), whose anthropological research on the underlying mechanisms of gambling identifies what she calls 'the machine zone', in which 'time, space, and social identity are suspended in the mechanical rhythm of a repeating process' (Schüll, 2012: 13). This zone and its gamified interaction, Schüll argues, should be viewed as a 'technologically contrived contingency management' that allows individuals to take the edge off the true contingencies in life (illness, death, poor social relations, failing economy, climate, political unrest, etc.) by engaging in a play of distilled risks and choices mediated by the digital interface. From this perspective, the gesture of the swipe departs from the intimate bodily techniques of intuition, violence, and care, becoming instead machinic cultural techniques of uncertainty in the accelerating system of social and infrastructural relations.

Geolocation

A third cultural technique to which we wish to draw attention is geolocation as an integrated part of many mobile dating apps, managing social interaction from the assumption that physical proximity is indicative of common ground for potential love interests. Following the trajectory from the conceptual grid to the tactile swipe, the embodied experience of the temporally coded mobile map inherent in the cultural technique of geolocation highlights ways in which uncertainty is configured as a spatio-temporal constellation in these apps. This speaks to the cultural history of maps as emotional geographies that give geographical form to affective sensibilities (Anderson and Smith, 2001; Davidson et al., 2005; Aitken and Craine, 2009). Yet geolocation can be seen as generating a particular configuration of the intertwinement between maps and their affects, one that possesses specific properties resonating from the technological possibilities afforded by dating apps and the way they organize and operationalize uncertainty.

Geolocation as a feature becomes visible and is employed in a variety of ways on mobile dating apps, ranging from notification of the proximity of a given profile in relation to one's own location, measured in miles or kilometres (Tinder, Grindr); to the general indication of a region, area, or city (Bumble); to an actual map locating where one has crossed paths with a potential match (Happn). The apps differ in the frequency with which they update the location—i.e., whether one needs to open the app to update the location or whether it tracks this silently. The various modes of embedding geolocation in the information conveyed about a profile are thus overlaid by different temporalities, which influence the affects produced by this information and consequently the ways in which uncertainty is configured.

In a trajectory from Georg Simmel's emblematic figure of 'the stranger' (1971 [1908]) as intrinsic to modern urban culture through to Stanley Milgram's 'familiar stranger' (1977) as an inspiration for the infrastructure of social networking sites, two central functions of the geolocation function of dating apps are, on the one hand, physical proximity as a matching technique that creates affinity between the user and potential partners and, on the other hand, geographic location as a warranting technique that a user may employ to determine whether to trust an online profile, insofar as the data points we see are perceived as less consciously performative than profile images and text (Veel and Thylstrup, 2018). It is in the vulnerable emotional space between desire for the unknown and anxiety as to the implications of the unknown that geolocation as a cultural flirtation technique manages uncertainty. It demarcates a continuum between seeking to identify a stranger with whom one wishes to become familiar, making sure that the stranger can be trusted and avoiding subjecting oneself to potential stalkers, and leaving enough space for performativity on the part of oneself and one's potential partner to allow for a sense of playful interaction. In other words, it permits the right amount of uncertainty to flourish to make the experience enticing while limiting the exposure to unwanted risks.

THE DATING APP AS ORGANIZING DEVICE AND ITS WIDER IMPLICATIONS

Dating apps are consequently to be regarded as organizational technologies that employ a wide range of cultural techniques. *Grid*, *swipe*, and *geolocation* can be seen as enabling and shaping the organization of intimacies in ways that manage uncertainty by allowing uncertainty to be contained at a level on which it might function as a productive generator for further interaction.

Dating apps hereby take part in wider optimization tendencies in terms of organizing various aspects of contemporary life, and they are often inscribed in tales of moral degeneration in which engagement with the *other* is becoming characterized by calculative commodification or superficial gamification.

One set of concerns focuses on how our intimate spheres become new sites of labour, subject to the same processes and ideals of rationalization and optimization as bureaucratic organizations and private corporations (Illouz, 2007). In this vein, dating apps may be understood as tools that assist in the task of dating as a laborious affair, in which responsibility for success is placed upon the individual, who can find support and advice in self-help literature that enforces a neoliberal doctrine of the individual's ability/responsibility to change his or her circumstances. This is dating as serious business, and the organizational structures offered by the dating apps support this logic as management tools. Another set of concerns regards these apps as transposing dating into a virtual game-like realm, removing it from the risk of emotional impact (Badiou, 2012). This may on the one hand relieve the tension of dating as a laborious affair, but it also raises its own set of concerns about modes of flow and addiction. As literature and film scholar Moira Weigel (2016) has argued, we find these laments of moral decline in dating culture throughout the twentieth century and often curiously aligned with critiques of technological innovation. However, they do not stand unchallenged. Building on the work of Niklas Luhmann (1987), cultural theorist and artist Lee MacKinnon (2016) has recently contested concerns about rationalization, arguing that love in pre-digital discourse was actually more deterministic than in the digital computational discourse: 'Love and intimacy no longer function to shield us from the "immense complexity and contingency of all the things, which could be deemed possible," but facilitate increasing access to complexity, contingency, and possibility. In an online context, love comes to be defined by novelty, differentiation, and incompatibility.'

Here we contend that placing these apps not in the discourse of love but instead in the cultural history of flirtation allows us to engage with dating apps and their affordances beyond the dichotomies of moral decay or incalculable liberation, approaching them instead as sites of uncertainty management that have specific organizational properties and implications. This allows us to regard dating apps as part of much more fundamental questions of the organization of life itself under neoliberalism, not least how to ensure a meaningful organization of information that productively balances control and uncertainty.

References

Aitken, Stuart C. and James Craine. 2009. 'The Emotional Life of Maps and Other Visual Geographies'. In Martin Dodge, Robert Kitchin and Chris Perkins (eds), *Rethinking Maps: New Frontiers in Cartographic Theory*. London: Routledge, 149–66.

Albury, Kath and Paul Byron. 2016. 'Safe on my Phone? Same-Sex-Attracted Young People's Negotiations of Intimacy, Visibility and Risk on Digital Hook-Up Apps'. *Social Media + Society*, 2(4), 1–10.

Amoore, Louise. 2013. *The Politics of Possibility: Risk and Security beyond Probability*. Durham, NC: Duke University Press.

Anderson, Kay and Susan J. Smith. 2001. 'Editorial: Emotional Geographies'. *Transactions of the Institute of British Geographers*, 26, 7–10.

Appadurai, Arjun. 2011. 'The Ghost in the Financial Machine'. *Public Culture*, 23(3), 517–39.

Badiou, Alan. 2012. *In Praise of Love*. New York: The New Press.

Batiste, Dominique P. 2013. '"0 Feet Away": The Queer Cartography of French Gay Men's Geo-social Media Use'. *Anthropological Journal of European Cultures*, 22(2), 111–32.

Beckert, Jens. 2016. *Imagined Futures: Fictional Expectations and Capitalist Dynamics*. Cambridge, MA: Harvard University Press.

Blackwell, Courtney, Jeremy Birnholtz, and Charles Abbott. 2015. 'Seeing and Being Seen: Co-Situation and Impression Formation using Grindr, a Location-Aware Gay Dating App'. *New Media & Society*, 17(7), 1117–36.

Blomley, Nicholas. 2003. 'Law, Property, and the Geography of Violence: The Frontier, the Survey, and the Grid'. *Annals of the Association of American Geographers*, 93(1), 121–41.

Brashers, Dale E. 2001. 'Communication and Uncertainty Management'. *Journal of Communication*, 51, 477–97.

Brubaker, Jed R., Mike Ananny, and Kate Crawford. 2016. 'Departing Glances: A Sociotechnical Account of "Leaving" Grindr'. *New Media & Society*, 18(3), 373–90.

Chun, Wendy H. K. 2017. *Updating to Remain the Same: Habitual New Media*. Cambridge, MA: MIT Press.

Corriero, Elana Francesca and Stephanie Tom Tong. 2016. 'Managing Uncertainty in Mobile Dating Applications: Goals, Concerns of Use, and Information Seeking in Grindr'. *Mobile Media & Communication*, 4(1), 121–41.

Crary, Jonathan. 2001. *Suspensions of Perception: Attention, Spectacle, and Modern Culture*. Cambridge, MA: MIT Press.

David, Gaby and Carolina Cambre. 2016. 'Screened Intimacies: Tinder and the Swipe Logic'. *Social Media + Society*, 2(2), 1–11.

Davidson, Joy, Liz Bondi, and Mick Smith. 2005. *Emotional Geographies*. Aldershot: Ashgate.

Flusser, Vilém. 2014. *Gestures*. Minneapolis: University of Minnesota Press.

Handel, Mark J. and Irina Shklovski. 2012. 'Disclosure, Ambiguity and Risk Reduction in Real-time Dating Sites'. In *GROUP '12 Proceedings of the 17th ACM International Conference on Supporting Group Work*, Sanibel Island, FL, 175–8.

Higgins, Hannah B. 2009. *The Grid Book*. Cambridge, MA: MIT Press.

Hoffman-Schwartz, Daniel, Barbara Natalie Nagel, and Lauren Shizuko Stone (eds). 2015. *Flirtations: Rhetorics and Aesthetics this Side of Seduction*. New York: Fordham University Press.

Illouz, Eva. 2007. *Cold Intimacies: The Making of Emotional Capitalism*. Cambridge: Polity Press.

Kittler, Friedrich. 1993. *Draculas Vermächtnis: Technische Schriften*. Leipzig: Reclam.

Krauss, Rosalind. 1979. 'Grids'. *October*, 9 (Summer), 50–64.

Licoppe, Christian. 2015. 'Seams and Folds, Detours, and Encounters with "Pseudonymous Strangers": Mobilities and Urban Encounters in Public Places in the Age of Locative Media'. *i3 Working Papers Series*, 15-SES-05.

Luhmann, Niklas. 1987. *Love as Passion: The Codification of Intimacy*. Cambridge, MA: Harvard University Press.

Lury, Celia, Luciana Parisi, and Tiziana Terranova (eds). 2012. *Topologies of Culture*. Special issue of *Theory, Culture & Society*, 29(4–5).

MacKinnon, Lee. 2016. 'Love Machines and the Tinderbot Bildungsroman'. *e-flux*, 74 (June). http://www.e-flux.com/journal/74/59802/love-machines-and-the-tinder-bot-bildungsroman [Accessed 28 September 2017].

Mauss, Marcel. 1973. 'Techniques of the Body'. *Economy and Society*, 2, 70–88.

Milgram, Stanley. 1977. 'The Familiar Stranger: An Aspect of Urban Anonymity'. In Stanley Milgram, *The Individual in a Social World: Essays and Experiments*. Reading, MA: Addison-Wesley, 51–3.

Møller, Kristian and Michael Nebeling Petersen. 2017. 'Bleeding Boundaries: Domesticating Gay Hook-up Apps'. In Rikke Andreassen, Katherine Harrison, Michael Nebeling, and Tobias Raun (eds), *Mediated Intimacies: Connectivities, Relationalities, Proximities*. London: Routledge, 208–24.

Morris, Jeremy and Sarah Murray (eds). 2018. *Appified: Culture in the Age of Apps*. Ann Arbor: University of Michigan Press.

Morris, Jeremy Wade and Evan Elkins. 2015. 'There's a History for That: Apps and Mundane Software as Commodity'. *The Fibreculture Journal*, 25. http://twentyfive.fibreculturejournal.org/fcj-181-theres-a-history-for-that-apps-and-mundane-software-as-commodity [Accessed 28 September 2017].

Müller-Brockmann, Josef. 2003. *Gestaltungsprobleme des Grafikers: Gestalterische und erzieherische Probleme in der Werbegrafik—die Ausbildung des Grafikers [= The Graphic Artist and his Design Problems: Creative Problems of the Graphic Designer, Design and Training in Commercial Art = Les probèmes d'un artiste graphique: typographie, dessin, photo, labels, couleurs, etc.]*. Sulgen: Verlag Niggli.

Müller-Brockmann, Josef. 2015. *Grid Systems in Graphic Design: A Visual Communication Manual for Graphic Designers, Typographers and Three Dimensional Designers*. Zürich: Niggli.

Mumford, Lewis. 1961. *The City in History: Its Origins, Its Transformations and Its Prospects*. New York: Harcourt, Brace and World.

Munster, Anna. 2013. *An Aesthesia of Networks: Conjunctive Experience in Art and Technology*. Cambridge, MA: MIT Press.

Nowotny, Helga. 2015. *The Cunning of Uncertainty*. Cambridge: Polity Press.

Parisi, Luciana. 2013. *Contagious Architecture: Computation, Aesthetics, and Space*. Cambridge, MA: MIT Press.

Peters, John Durham. 2015. *The Marvelous Clouds: Toward a Philosophy of Elemental Media*. Chicago: University of Chicago Press.

Phillips, Adam. 1996. *On Flirtation*. Cambridge, MA: Harvard University Press.

Pierce, David. 2016. 'The Oral History of Tinder's Alluring Right Swipe'. *Wired*, 28 September. https://www.wired.com/2016/09/history-of-tinder-right-swipe. [Accessed 28 September 2017].

Race, Kane. 2015. 'Speculative Pragmatism and Intimate Arrangements: Online Hook-up Devices in Gay Life'. *Culture, Health & Sexuality*, 17(4), 496–511.

Raymond, Mark. 2013. 'Locating Caribbean Architecture: Narratives and Strategies'. *Small Axe*, 17(41), 186–202.

Rosamond, Emily. 2018. 'To Sort, to Match and to Share: Addressivity in Online Dating Platforms'. *Journal of Aesthetics & Culture*, 10(3), 33–42.

Samimian-Darash, Limor and Paul Rabinow. 2015. *Modes of Uncertainty: Anthropological Cases*. Chicago: University of Chicago Press.

Schüll, Natasha Dow. 2012. *Addiction by Design: Machine Gambling in Las Vegas*. Princeton, NJ: Princeton University Press.

Siegert, Bernhard. 2015. *Cultural Techniques: Grids, Filters, Doors and Other Articulations of the Real*. New York: Fordham University Press.

Simmel, Georg. 1971 [1908]. 'The Stranger'. In Georg Simmel, *On Individuality and Social Forms*, ed. Donald N. Levine. Chicago: University of Chicago Press, 143–50.

Srnicek, Nick. 2016. *Platform Capitalism*. Hoboken, NJ: John Wiley & Sons.

Stempfhuber, Martin and Michael Liegl. 2016. 'Intimacy Mobilized: Hook-Up Practices in the Location-Based Social Network Grindr'. *Österreichische Zeitschrift für Soziologie*, 41(1), 51–70.

Stone, Allucquère Rosanne. 1995. *The War of Desire and Technology at the Close of the Mechanical Age*. Cambridge, MA: MIT Press.

Veel, Kristin and Nanna Bonde Thylstrup. 2018. 'Geolocating the Stranger: The Mapping of Uncertainty as a Configuration of Matching and Warranting Techniques in Dating Apps'. *Journal of Aesthetics & Culture*, 10(3), 43–52.

Walther, Joseph B. and Malcolm R. Parks. 2002. 'Cues Filtered Out, Cues Filtered In'. In Mark L. Knapp, John A. Daly, and Gerard R. Miller (eds), *Handbook of Interpersonal Communication*. Thousand Oaks, CA: Sage, 529–63.

Weigel, Moira. 2016. *Labor of Love: The Invention of Dating*. New York: Farrar, Straus & Giroux.

Werning, Stefan. 2015. 'Swipe to Unlock: How the Materiality of the Touchscreen Frames Media Use and Corresponding Perceptions of Media Content'. *Digital Culture & Society*, 1(1), 55–72.

Whitelaw, Mitchell. 2011. 'After the Screen: Array Aesthetics and Transmateriality'. http://teemingvoid.blogspot.dk/2011/04/after-screen-array-aesthetics-and.html [Accessed 28 September 2017].

CHAPTER 18

···

DESK

···

GIBSON BURRELL AND KAREN DALE

DURING the process of writing this chapter we have barely sat at a desk. Instead, the context of our text is a living room where we have sat on a sofa, with a laptop, and paper and books spread around us. We engage with the topic, if not the materiality, of 'the desk' at a time when the 'death of the desk' is being proclaimed (Hickey, 2014), when the 'loungification' of offices is being talked about (O'Doherty, 2017), and even a virtual desk is being trialled at technology fairs.[1] Why, then, would we focus on a piece of furniture which would seem to be in the process of becoming defunct? Our answer is because the mediating technology of the desk has had lasting effects, not only in the birth and (much vaunted though as yet somewhat exaggerated) death of bureaucracy—a rule by those who sit at desks—but also in a multitude of significant ways in which particular forms of knowledge, power, and social relations have become deeply embedded within the modern world. As we hope to show in this chapter, the desk is an assemblage intimately related to organization and organizing. It is a 'comprehensive architectonic of social order' (Albrow, 1992: 316), in that desks are central to regimes of social organizing within 'bureaucracy', within the 'Taylorism' of white collar work, in Braverman's analysis of clerical deskilling, and in Foucauldian accounts of the docile body in disciplinary society (Gerth and Mills, 1948; Braverman, 1974; Foucault, 1977). We will seek to show that a concentration upon this artefact reveals not a fixed contribution to modernity but offers an ever-changing grasp of what the desk signifies, for it has a strong adaptability and flexibility, both in its material forms and in its mediation of social relations.

In a material sense, why is a desk not simply a table? Indeed, the word 'desk' comes from the Latin for a table to write at, and upon (Conrad and Richter, 2013). Medieval illustrations which predate printing show furniture designed for the meticulous copying of texts. The detailed organization of materiality as a specific aid to writing is seen in the design of compartments within the desk for the ink pot, the blotter and the powder tray, and careful storage for pens or quills. Thus any desk is aided by other materialities—in and out trays, staplers, files of various sorts, and essentially paper and ink. The desk is

---

[1] 'Reimagining the Work Desk with AR', http://www.bbc.co.uk/news/av/40129729/meta-hopes-to-redesign-the-work-place-by-using-augmented-reality-to-organise-your-desk.

never an object without an assemblage, of both human and non-human actants. It is a focal mediating presence: materially, ideationally, and socially. Further, whilst the desk has a flat surface for writing, it also has drawers, compartmentalization of some sort and pigeonholes. Importantly, then, it combines writing with encoded classificatory organization (Figure 18.1). Different forms of desks have evolved to express both functional and social relations. The cylinder desk and its later relative, the rolltop desk, were designed with covering lids which fitted over the writing surface, and could be safely locked. These covers thus add security, privacy, and the appearance of tidiness to the desk. This perceived need for coverage clearly indicates the value and significance of the records made and then kept at the desk. A form of the cylinder desk, referred to as the 'bureau Kauntiz' after the Austrian ambassador who is said to have introduced it, became extremely popular in the French court in the eighteenth century. The significance of the secret paperwork held within the desk becomes obvious in the context of the burgeoning civil service in France, where in 1757 de Gournay coined the term 'bureaucracy' (Albrow, 1970: 16). Here the object itself—a desk—becomes the metonym for a whole system of organization. So 'bureau' represents successively the desk, the office, and the institutional form of administration. The desk becomes a major organizing force across the modern world (Mill, 1837).

As is well known, despite bureaucratic advocacy of rationality and expertise (Weber in Gerth and Mills, 1948)—or perhaps because of it—aristocratic objection to rule by those who sit at desks allowed the term 'bureaumania' to come into prominence at the time of the huge growth of the French civil service. Bureaucracy in fact has always been

FIGURE 18.1 The book-keeper's desk in the manager's office at Quarry Bank Mill, near Manchester, UK. Mill offices from ca. 1860 retain their original furniture

(Photo by the authors)

accompanied by the negativity of accusations of it being rule-bound and replete with dysfunctions (Merton, 1968; Clegg and Dunkerley, 1980). But for us the key point is that the desk becomes both materially and epistemologically productive of multiple consequences. It is perhaps *the* symbol of 'the control revolution of the early twentieth century' (Reed, 2011: 250). Desks are constitutive, therefore, of a major organizational force, replete with both positive and negative connotations.

Once sat in front of the desk, it is possible for a bureaucratic separation between the private individual and their public role to be enacted in socio-material form. The organizing impetus of the desk comes in major part from its link with stored records, and the recording often carried out upon the desk itself. The desk and its use facilitate the development of a rational-legal system whereby one can see what has been done through the concisely minuted recording of information, decisions, and actions, and can refer to the record to ensure consistency and an ongoing accountability (Wittfogel, 1957: 50–1). The desk gives rise in many places to the sense of being able to 'see' what has been carried out and why. The centrality of this object becomes apparent in the way it is used as a metonym. For example, 'the desk' on a newspaper or governmental agency (or indeed within many knowledge organizations) is the shorthand reference for a central collecting point or focus of intelligence and the production of outcomes from this arrangement. Thus the desk is not simply a material object but has epistemological implications. It enables particular ways of thinking and organizing, through sifting, categorization, and means–end rationality. The ways in which things are categorized in material form such as records and files create epistemic groupings and boundaries. The desk is irreducibly productive of ways of thinking and seeing the world.

Larger bureaucracies developed in the late nineteenth and early twentieth century, across not only state administration but in companies such as insurance firms, banks and other financial institutions, and very importantly, within mail order retailing (intimately associated with the expansion of railways which delivered these goods). The consequent growth of 'white collar' work required a huge expansion in the number of desks. The number of clerks in the United States increased tenfold between 1880 and 1920 (Leffingwell, 1925 in Graf Klein, 1982: 12). The mass production of desks both mirrors and facilitates the way in which the work techniques of mass production had been transferred into burgeoning white collar occupations. Compared to the black-coated male 'clerks' of Victorian times (Lockwood, 1958), these new clerical jobs were deskilled, with repetitive and routine paperwork the workers' focus of attention (Braverman, 1974: chapter 15). By 1920 offices were less cosy workspaces, and double pedestal desks had replaced the penchant for rolltop desks. These were more practical for the tasks undertaken, easier to mass produce and once cut down in height facilitated the overarching supervision of large numbers of clerks (Graf Klein, 1982: 20). In his *Office Management: Principles and Practice* (1925), Leffingwell noted that an abundance of drawers was no longer necessary. In his injunction we see the distinction drawn between the wooden executive desk, which maintains a craft appearance with its drawers and nooks for important, confidential papers, and the streamlined metal standardized clerical desk, expressing in obvious material form the hierarchy of the office.

In this expansion of mass white collar work, attention soon turned to the ergonomics of the desk, which were quickly related to Tayloristic principles of efficiency in the control and rationalization of workers' physical movements. New office machinery such as the typewriter occupied the desk, whilst the paperwork was moved off the desk to its own specific places in filing cabinets. Harry Braverman (1974: 320–6) speaks tellingly of the deskilling of clerical tasks in this period, where specific time values were given to each activity undertaken on or around the desk. The significance of this, as Foucault (1977: 153) notes, is that 'over the whole surface of contact between the body and the object, power is introduced, fastening them to one another'.

It is from this period that the idea of the 'desk job', and even more tellingly, the phrase 'chained to the desk' comes (Braverman, 1974: 336). In the Larkin Building, in Buffalo, New York State, where the company practically invented the business of mail order, the architect-designed desk for each person 'expressed the limited freedom allowed to employees', since the 'seats were cantilevered out and integral with the desks' (Duffy, 1997: 21). Desks themselves were often fixed to the floor. The clerical staff member was immobilized for most of the working day (Braverman, 1974: 310). As well as standardized tasks, the construction of the desk demands the standardized body and bodily movements as in, for example, assuming the norm of right-handedness and an able body (see Hjorth, this volume). This arrangement of desks also allows for particular forms of gendered behaviour and attitudes to be reproduced, primarily through the openness to surveillance and normalization of the male gaze. Wasserman's study of the open-plan offices of the Israeli Ministry of Foreign Affairs reveals this, showing the internalization of this gaze through:

> organisational expectations as to the required mode of speech (workers are expected to speak quietly, without getting excited), style of sitting ("I am expected to sit upright and well, not slouch over the desk even if I am tired"), appropriate dress ("I dress better now than previously because I am constantly on view") and full control of the body ("I cannot straighten my underwear, not even for one moment").
>
> (Wasserman, 2012: 19)

Thus the arrangement of desks within these white collar factories not only constructs the individual worker in a particular way, but also produces distinctive collective social relations. As well as the individual body effectively entrapped within the design of the desk, the collective body is entrapped within the wider organization of desks. The large open-plan design of offices became known as the 'bullpen', a term which exactly captures this notion. Again, this was based on manufacturing and Tayloristic forms of organization, resembling a production line with rows of desks which could be easily overseen by supervisors. Thus:

> The Bullpen office layout made it possible to standardise work activities and to supervise workers closely and easily. Employees were seated in open areas with no

partitions and no adornments and confined strictly to their jobs. Activities were rigidly segmented, so that typists typed and filing clerks filed. No other activity, including conversation, was permitted.[2]

The arrangement of people at desks expresses and makes visible a social ordering (Foucault, 1977). This is noticeable for example in the arrangement of children in a classroom to express their position in a hierarchy of how well they have done in a subject, or in the ordering of desks in a workplace where those closest to the windows are the highest ranking whilst those furthest away, in the centre of the bullpen, are in the least comfortable and thus least valued places in the hierarchy. From Foucault's analysis, we can also see the interlocking processes of this: *enclosure*, which he links with the development of specific defined spaces for labour; and *partitioning*, which is to locate individuals so as to facilitate some communications and prevent others, to keep their movements controlled and visible. Enclosure and partitioning mean that at every desk 'each individual has his own place; and each place its individual' (Foucault, 1977: 143). With individual desks then being grouped together into similar operations, *classification* across individuals is possible, so that supervision is 'both general and individual' (Foucault, 1977: 145) ensuring that workers doing the same task could be compared and classified according to speed and skill. This further enables a process of *ranking*, or placing individuals into a 'hierarchy of knowledge or ability' arranged in space (Foucault, 1977: 147). The desk becomes the mediating technology which allows these social and spatial organizing and disciplining relations to take place. It produces fixity, inherent in the notion of the 'work station', as well as enabling constant visibility. The 'grid of intelligibility' which Foucault talks about can be seen clearly in the spatial arrangement of individual desks in these large open-plan offices.

Leaving behind the rationalized rows of the Taylorized offices, and turning to the variety and messiness of everyday organizational life, 'desks are at the nexus of social life in offices' (Tyson, 1992: 31). Desks may be arranged in such a way as to create connections whereby teams are formed and maintained but also to create separations and enhance divisions. Desks are a way of constructing a little bit of personal territory, with the possibility of the expression of individuality and a life outside the office through, for example, family photographs (Tyler and Cohen, 2010). At the same time desks are both a private domain and a place of meeting, around which colleagues can gather. Shortt (2018) notes the tendency in modern offices of eating at one's desk in order to cut down on 'wasted' time in the lunch break. The desk becomes a meeting point for the contradictions of organizational life. A desk can form a place of refuge—one sits behind a desk, it can express a very material as well as symbolic barrier against the forces of organization—but at the same time it is public, and thus open to transgressions or invasions of this moral ordering. For example, how does it feel if someone else sits on or at your desk?

---

[2] https://web.archive.org/web/20080530151010/http://home.telkomsa.net/deycor/office%20planning.pdf.

In researching 'the desk as a social institution', Tyson points to the 'moral implicatedness' of desks. In interviewing employees he found that 'no-one could be entirely frank about what they did and how they used their desk since they all knew that what they do at the desk says a great deal about the social role an individual has, their commitment to work, attitudes to colleagues...and much more' (1992: 31). Desks themselves can be used as resistance to surveillance, to control, to transparency. Desks become complicit agents in the act of presenteeism and impression management, since the arrangements of objects on a desk can be used to imply the amount and pace of work being undertaken. The desk therefore can be enrolled as a theatrical stage in presenting one's presence in the organization. A central assumption of work, reflected in organizational rhetoric, is that when the employee is 'at their desk' they are involved in productive work. However, the desk can also lie about this presence, as in the example of car keys and jackets being left at the desk when the 'occupier' of the desk is themselves absent, perhaps socializing or at home. The mediating public presence of the desk can be used to tell different stories other than a rational objective account of work being performed. The supposedly silent object can 'say' a multiplicity of things.

This brings us back to the materiality of the desk (Conrad and Richter, 2013). Desks are constructed from materials as varied as mahogany, brass and bronze, pinewood, and MDF. These materials themselves incorporate other forms of social relations which become embedded into the desk. IKEA may sell thousands of desks per annum across the world but desks are also a key part of luxury markets and reflect Veblen's notion of 'conspicuous consumption' through elite aesthetic and symbolic qualities (Figure 18.2). Several elements are being consumed simultaneously in a high-end desk—space, rare materials such as walnut, and the craft skills built into the piece of furniture. Historically, the wealthy of Britain had been interested in their elaborate Chippendale and Gillow desks on which accounts were perused, letters written, and servants paid (Goodman, 2003: 79–81). But the doyen of all desks perhaps, made from a single piece of onyx, is to be found in the St Petersburg's Hermitage, and is currently occupied by the museum director. It is horseshoe shaped and is about 9 metres in diameter. It is lit from below to show off its patterning and colouration in the way the Tsar wished to see it.

Alongside conspicuous consumption, the powerful in modern times have used 'things to take your breath away' (Hitler quoted in Dovey, 1999: 55) and desks are no exception. Sudjic (2005: 13–18) recounts the journey of the Czech president, Emil Hacha, which ends in the 4,000 square feet of Hitler's 'study'. Hacha, with a weak heart, was to collapse twice under the pressure, having to be revived and forced to sign away Czechoslovakia into Nazi hands—on Hitler's desk. As Dovey explains, the architecture of President Hacha's journey 'needs to be understood as a form of symbolic choreography where the spatial structure operates to control the framing of a series of representational themes'. Similarly, Mussolini's office in Rome occupied two floors. In the main hall there was only his desk, his chair, and two chairs for visitors. The desk was placed 20 metres away from the door, requiring a long walk up to it, with the placing of the powerful figure seated behind it.

FIGURE 18.2 A luxury desk, demonstrating high-end design and opulent materials. Designed by Charles Rennie Macintosh, 1904, and made by Alex Martin, 1905, for Hill House, Helensburgh. Kelvingrove Museum, Glasgow, UK

(Author's own photograph)

The desk is therefore a key actant in the careful staging of power. Under the regimes of figures such as Hitler and Mussolini, the desk is symbolic of a centralized power: it is the focal point of domination. It is interesting to speculate as to why such prominence should be afforded to a desk. Is it because it signifies the ability to get things done; to have knowledge as well as power; to appropriate the appearance of rational-legal authority, or something of all of these? The socio-materiality of the desk becomes clear in such circumstances where impressive desks in impressive surroundings are meant to embody impressive human beings.

The desk then can be activated as a mechanism of domination over others, but its use may also be entirely rejected. Wordsworth saw the desk as an instrument of tyranny (Figure 18.3), chaining him to a piece of furniture whilst he wished to be out and about within the landscape. Yet, ironically, one of the Wordsworth tourist attractions in the English Lake District is his school desk in Hawkshead Grammar School with his and his brother's names carved into it. However, in refusing to use a desk himself, Wordsworth's wife and sister ended up being the ones to write down his poems for him, looking at him whilst he looked out, not at them but upon the landscape.

The refusal of the desk has also become a statement of elite position. In 1925, Leffingwell stated that 'an executive who is really a directing, thinking, planning head of

FIGURE 18.3 Wordsworth's father's desk. John Wordsworth (d. 1783) was a legal agent for the Earl of Lonsdale, also acting as bailiff and recorder at Cockermouth in the Lake District. Wordsworth House and Garden, National Trust, Cockermouth, UK

(Author's own photograph)

business doesn't need a desk with drawers unless for storing cigars and golf balls' (cited in Graf Klein, 1982: 98). The executive who was a believer in 'Management by Walking About' (Peters, 1987) had little need for stationary stationery. In the mid-1960s the Ford Foundation's new headquarters in New York used luxurious rich materials throughout, including for their office furniture. Their experimental desks combined the historical reference to a high prestige rolltop design with adaptations for new technology. However, it must be noted that this innovative technology was for the use of an exclusively female secretarial staff, while senior executives occupied themselves in 'a living room for the whole Foundation' (Graf Klein, 1982: 169). Repeating this perspective, an article in *The Wall Street Journal* of 1982 is entitled: 'Want Office Status? Remove All Papers from the Top of Desk...And then Remove the Desk: Very-Top Bosses Favor Living-Room Atmosphere' (Bralove, 1982).

In modern organizing practices, the desk is more generally taking a decentred position. From a technology of fixity, to one associated with mobility; and from a material solidity to a virtual graphic or augmented reality interface, the desk is once again a locus of changing social, technological, and organizational relations. Contrasted to the fixity of the desk in the white collar factories which rendered visible the presence of employees, many contemporary organizations have become conscious of the lack of efficiency and cost effectiveness of having one desk assigned to each employee. 'Hot-desking' has been a feature for decades (and of course 'hot bunking' is a much older concept from the Royal Navy) though it is hard to find academic research which puts a positive gloss on the experience of those having to employ it. For this temporary location of one's desk is transgressive of the moral order. To take your desk away from you, to have no 'home' or sense of belonging, may well lead to employee anomie. Senior management's thinking about hot-desking in terms of cost and efficiency, rather than social relations, leads to unintended consequences. Although hot-desking is often presented in social terms—that is to say, flexibility, greater communication, teamworking, and openness are said to accompany such changes—these rhetorical flourishes often obscure the economic rationale of architectural change. The movement to have desks merely as functional objects, decoupled from their personal and social attachments, is sold as an apparatus of positive change management. It is rarely experienced as such by those on whom it is imposed.

One of the conditions of possibility for this shift is that the 'desktop' itself has been encoded into other forms, so central is it to our ways of organizing. Where once the desktop referred to the material writing surface of the desk, as computing expanded it came to refer to a personal computer which fitted on the desktop of an individual worker, facilitated by the development of the microprocessor. It then moves on further as a term to describe the graphical representation of information management within the IT systems on PCs. This extensive usage of the term and its interpretive flexibility shows the importance of 'desk' as an object to think with as well as to work at.

Early studies for the development of the human–machine interface for the desktop computer considered how employees used their actual desks for organizing their work. One such piece by Malone, at Rank Xerox's Palo Alto site, characterizes desks as 'personal information environments' (1983: 100). Malone uses his studies and interview data to conclude that: 'Two of the most important units of desk organization are *files* and *piles*. Both files and piles are ways of collecting groups of elements into larger units' (1983: 105). Files assume a categorizing and classifying rationale, while piles are unordered collections of documents. He also notes that desk organization is involved with both finding and reminding: 'Much of the information that is visible on top of the desks and tables in most offices is there to *remind* the user of the office to do something, not just to be available when the person looks for it'(1983: 106). This formative research into how desks enable organization became incorporated into the representation of digitized information *within* the computer, as well as into the ways that that information is made accessible and retrievable. Although the 'desktop' form of the computer has proliferated into more portable devices which themselves do not need the physical desktop itself—as the very word 'laptop' indicates—the processes and routines associated with the desk remain embedded in our organizing.

As a result of this research in Silicon Valley not only were computers adapted to be used on desks, but associatedly, desks became adapted for use with computers.

For example, desks began to appear with designed compartments to fit the 'tower' form of computers, holes to enable the wires to be connected, space for printers and so on. But as time progressed in the IT world, the desktop PC was seen to have severe disadvantages:

> The inert desktop is as aptly called a 'personal computer' as a grandfather clock is called a 'personal time-keeping device.' It's always at your service—until you leave the room.[3]

At a time when liquidity (Bauman, 2000) and mobility (Urry, 2007) have become conceptualized as crucially important, the desktop as well as the desk itself are seen as a drag. Fleming (2015: 107) relates such societal shifts to our changing relationship to work. Employees are constantly facing 'an injunction to produce' and as a result are becoming their own individual 'micro-enterprises' (Foucault, 2008: 242). This process of the corporate 'enlisting of life' means that they (perhaps 'we'?) cannot just be associated with one desk in our workplace, but need to have multiple 'desks' in order to be constantly 'at work'. These might consist of a desk in our home office, the table of a café, the tray of an aeroplane seat, as well those 'desktops' incorporated into the mobile technologies that facilitate this dromomania (Virilio, 1986) and the always 'switched on' employee, who has now become an 'entrepreneur of the self' (Gordon, 1987: 300; Foucault, 2008: 226). Thus there is not just a quantitative increase in work (longer hours, often facilitated by technology), but a qualitative change in the relationship between individuals and their jobs. The Italian autonomists call this phenomenon the 'social factory' since it is not bounded by the physical workplace as in the traditional factory or office, and work is no longer external to the person doing it. Therefore, 'work is transformed into something we *are* rather than something we simply do among other things' (Fleming, 2015: 37).

Despite, or maybe because of this, the desk, conceived as a fixed piece of furniture in a dedicated office, remains crucially important. New houses across the UK are sold by estate agents as having a specific room designated as a 'study', and each show-home sports a desk. If we live in post-bureaucratic times, as some maintain (Clegg et al., 2011), and if the laptop has replaced the desktop because our laps and not our desks are always with us, it is not surprising that obituaries for the desk are being written. But the desk has enframed the social and material relations of organizing, shaping the world in which we live. So it would be a foolish commentator to fully accept the rhetoric of the death of the desk, an object placed at the very centre of material and epistemological concerns for centuries. And in case you are wondering, we returned to the desk to finish this final draft because, over an extended period, the laptop was far too hot and the sofa was far too uncomfortable.

---

[3] Randall Stross, 'The PC doesn't have to be an anchor', *The New York Times*, 18 April 2009, http://www.nytimes.com/2009/04/19/business/19digi.html.

REFERENCES

Albrow, Martin. 1970. *Bureaucracy*. London: Pall Mall.

Albrow, Martin. 1992. 'Sine ira et studio—or Do Organizations Have Feelings?' *Organization Studies*, 13(3), 313–29.

Bauman, Zygmunt. 2000. *Liquid Modernity*. Cambridge: Polity Press.

Bralove, M. 1982. 'Want Office Status? Remove All Papers from the Top of Desk…And then Remove the Desk: Very-Top Bosses Favor Living-Room Atmosphere'. *The Wall Street Journal*, 15 January, 1.

Braverman, Harry. 1974. *Labor and Monopoly Capital*. New York: Monthly Review Press.

Clegg, Stewart and David Dunkerley. 1980. *Organization, Class and Control*. London: Routledge & Kegan Paul.

Clegg, Stewart, Martin Harris and Harro Höpfl (eds). 2011. *Managing Modernity: Beyond Bureaucracy?* Oxford: Oxford University Press.

Conrad, Lisa and Nancy Richter. 2013. 'Materiality at Work: A Note on Desks'. *Ephemera*, 13(1), 117–36.

Dovey, Kim. 1999. *Framing Places: Mediating Power in Built Form*. London: Routledge.

Duffy, Francis. 1997. *The New Office*. London: Conran Octopus.

Fleming, Peter. 2015. *The Mythology of Work*. London: Pluto Press.

Foucault, Michel. 1977. *Discipline and Punish*. Harmondsworth: Penguin.

Foucault, Michel. 2008. *The Birth of Biopolitics: Lectures at the Collège de France, 1978–1979* (trans. G. Burchell). Basingstoke: Palgrave Macmillan.

Gerth, H. H. and C. Wright Mills (eds). 1948. *From Max Weber*. London: Routledge & Kegan Paul.

Goodman, Dena. 2003. 'Furnishing Discourses: Readings of a Writing Desk in Eighteenth-Century France'. In Maxine Berg and Elizabeth Eger (eds), *Luxury in the Eighteenth Century: Debates, Desires and Delectable Goods*. Basingstoke: Palgrave Macmillan, 71–88.

Gordon, Colin. 1987. 'The Soul of the Citizen: Max Weber and Michel Foucault on Rationality and Government'. In Sam Whimster and Scott Lash (eds), *Max Weber, Rationality and Modernity*. London: Allen & Unwin, 293–316.

Graf Klein, Judy. 1982. *The Office Book*. London: Quarto.

Hickey, Shane. 2014. 'Death of the Desk: The Architects Shaping Offices of the Future'. *The Guardian*, 14 September.

Leffingwell, W. H. 1925. *Office Management: Principles and Practice*. Chicago: A. W. Shaw.

Lockwood, David. 1958. *The Black Coated Worker: A Study in Class Consciousness*. London: Allen & Unwin.

Malone, Thomas W. 1983. 'How Do People Organize Their Desks? Implications for the Design of Office Information Systems'. *ACM Transactions on Information Systems (TOIS)*, 1(1), 99–112.

Merton, Robert K. 1968. *Social Theory and Social Structure*. New York: Simon & Schuster.

Mill, John Stewart. 1837. 'Armand Carrel, His Life and Character'. *Westminster Review*, 28(6), 7.

O'Doherty, Damian P. 2017. *Reconstructing Organization: The Loungification of Society*. London: Palgrave Macmillan.

Peters, Tom J. 1987. *Thriving on Chaos: Handbook for a Management Revolution*. New York: Harper & Row.

Reed, Michael. 2011. 'The Post-Bureaucratic Organization and the Control Revolution'. In Stewart Clegg, Martin Harrisand, Harro Höpfl (eds), *Managing Modernity: Beyond Bureaucracy?* Oxford: Oxford University Press, 230–56.

Shortt, Harriet. 2018. 'Cake and the Open Plan Office: A Foodscape of Work through a Lefebvrian Lens'. In Sytze Kingma, Karen Dale and Varda Wasserman (eds), *Organization Space and Beyond: The Significance of Henri Lefebvre for Organization Studies*. London: Routledge, 207–34.

Sudjic, Deyan. 2005. *The Edifice Complex*. London: Allen Lane.

Tyler, Melissa and Laurie Cohen. 2010. 'Spaces That Matter: Gender Performativity and Organizational Space'. *Organization Studies*, 31(2), 175–98.

Tyson, P. J. 1992. *The Desk as a Social Institution*. Technical Report. Cambridge: EuroPARC Rank Xerox.

Urry, John. 2007. *Mobilities*. Cambridge: Polity Press.

Virilio, Paul. 1986. *Speed and Politics: An Essay on Dromology* (trans. Mark Polizzotti). New York: Semiotext(e).

Wasserman, Varda. 2012. 'Open Spaces, Closed Boundaries: Transparent Workspaces as Clerical Female Ghettos'. *International Journal of Work Organisation and Emotion*, 5(1), 6–25.

Wittfogel, Karl. 1957. *Oriental Despotism*. New Haven, CT: Yale University Press.

CHAPTER 19

··

ELEVATOR

··

ANDREAS BERNARD
TRANSLATED BY ERIK BORN

THE elevator is a means of transport that organizes space. After its emergence in the mid-nineteenth century, this architectural element began to play a constitutive role in the organization of the multi-storey building. For its facilitation and automation of movement, which represents its active, formative mediality, the elevator was not a mere technical aid to be integrated into the existing containers of the home or office. Rather, the apparatus was responsible for a much more fundamental change in the organization, perception, and imagination of multi-storey buildings. In the following, I will describe this transformation process, which took place from roughly 1860 to 1930 in the United States and with a slight delay in Europe through the combination of elements from the history of architecture, law, medicine, and technology.

A starting point for an approach to the elevator from the perspective of media and organization studies might consist in the question of its influence on the 'order of the imagination' for living and working spaces: how did the collective imagination of multi-storey office and apartment buildings change with the rise of the elevator in the decades around 1900? What effects did the technical apparatus have on what was thinkable and sayable about the events occurring inside a building, about the distribution of people and the division of spaces? These structures are equally visible in building codes and hygiene handbooks, in architecture manifestos and the organization of space in the modernist city novel.

From this perspective, the early history of the elevator is of particular interest because it resides, as it were, in the gap between an old and a new organization of space, and registers its very points of transition. Furthermore, a cultural-archaeological perspective on the history of the elevator represents a clear departure from the conventional forms for presenting the history of technology, which have been long dominant in museums and encyclopaedias. In these contexts, the appearance of an apparatus is still frequently presented as a triumphant chronicle with a definite origin followed by a continuous series of refinements and innovations. The problematic aspects of this approach are manifest, in the case of the elevator, in the careless and unopposed attribution of its origin to a singular event, a quantifiable date for the 'invention' of the apparatus. Almost every encyclopaedia article on the elevator starts with Elisha Otis' demonstration of a security elevator at the Exhibition of the Industry of All Nations in New York City in 1854. However, the contemporary reception of this demonstration reveals that this allegedly historical event occurred without attracting the notice of the public, and was only circulated as an origin story by the Otis Elevator Company in retrospect. In fact, the formation of the elevator in the mid-nineteenth century involved many (now-forgotten) engineers, who contributed to its various functional and aesthetic elements. In other words, the 'origin' of a technical apparatus refers less to an event that actually took place and more to the narratological necessities and economic power structures in the conventional staging of the history of technology (Bernard, 2014: 1–13).

To reconstruct the elevator's medial and organizational power, we can identify four main problem areas, which can be captured under the keywords 'breaches', 'hierarchies', 'controls', and 'interiors'.

BREACHES

The elevator organizes the vertical. In this respect, one of its characteristics is of particular significance: the elevator shaft cuts a vertical swath straight through a building. The implications of this action are apparent in the contrasting architectural layouts of the first buildings equipped with elevators and those built before the emergence of this means of transport. In the traditional German office building, known as a Kontorhaus, like the one in Gustav Freytag's novel *Debit and Credit* (1855), the vertical was not yet a structural axis, nor did the vertical play an integral role in early accommodations for the masses in the working-class districts of European metropolises. Instead, the interiors of these mid-nineteenth-century structures were characterized by an assortment of inter-mediate floors, secondary staircases, and corridors with dead-ends, which frustrated any attempt to draw a clear distinction between each spatial unit or to determine straightforward relationships among them. In the mass housing quarters in Berlin and Hamburg, for instance, there were already four levels of living space stacked on top of each other, but there was no category of the floor in architecture or building management (Wischermann, 1997). We need to consider this very aspect of the elevator—its creation of a better way of conceiving of multi-storey spaces. When an early essay about the floorplans of modern office buildings in New York City speaks of their intended 'simplicity of arrangement' and emphasizes that 'from the point where the elevators deliver in each story, the door of every office on that floor should be visible' (Robinson, 1891: 194), it implicitly highlights the extent of the elevator's breach of the building. The elevator not only enabled the construction of high-rises with more than six or seven storeys. It also domesticated the entire floorplan, eliminated organic vertical growth, and reinforced the previously amorphous category of the floor, since the vehicle would only stop on whatever was defined as the first, second, third, or fourth storey.

FIGURE 19.1 Elevator bank

(*Source*: From Englert and Englert, *Lifts in Berlin* (Jovis, 1998))

As a result, the infrastructure of buildings changed dramatically: the stairwell, which had been the traditional principle for organizing the vertical, was downgraded, within the span of only a few decades in the United States, to a mere escape route. Henceforth, all routing would proceed through the elevator lobby (Figure 19.1).

The significance of the elevator's breach of the building is even more vivid against the backdrop of other architectural straightening projects from the mid-nineteenth century. The swath the elevator cuts through the vertical contrasts with more visible horizontal breaches in the form of channels, avenues, boulevards, railroad tracks, or subway tunnels. One might even claim that the practice of breaching was an expression of the project of modernity itself. Ever since the late nineteenth century, there have been many studies of the cultural and political consequences of this practice—above all, in Haussmann's renovation of Paris (Engels, 1962 [1872]; Giedion, 1992 [1941]; Benjamin, 1982; Sennett, 1997 [1994]). The redesign of multi-storey houses can be described in terms of a similar change from the proliferation of space to its canalization—a kind of vertical Haussmannization. Once again, the most crucial factor in this development was the elevator's influence on the category of the floor, which eliminated any notion of organic branching or in-betweenness. The elevator was unrelenting in its division, partitioning, and organization of space, which is substantiated in a related historical detail: the doorbell panel, which would eventually be installed at the front door of every multi-storey residence in Europe, gained acceptance at the same time as the new transport channel. In the mid-nineteenth century, it would have been difficult to visualize the corresponding addresses for the organic growths of intermediate floors and secondary staircases. (What would a doorbell panel for one of these houses even have looked like?) In the early twentieth century, on the other hand, the orderly doorbell panel must have seemed like a microcosm of the interior order that had been brought about by the elevator.

Thus, the elevator can be understood as a disciplinary element in the history of architecture with one important caveat—its relentless organization of the building into individual floors, its breach of the vertical, simultaneously obscured this same dimension. The elevator fragmented the multi-storey building and transformed it into a series of separate platforms. The vehicle's discontinuous stopping points, its limitation of accessible space to the first floor, second floor, and so on, made each level of the house appear to be a discrete unit. Whatever may have been located between these units no longer existed. In the terminology of media studies, the transition from the stairwell to the elevator can be described as one from the analogue to the digital, the implications of which are particularly apparent in the history of literature. At the end of the nineteenth century, the multi-storey building was frequently the central setting for the plot of city novels, such as Émile Zola's *L'Assommoir* (1877) and *Pot-Bouille* (1882), where the shared stairwell makes the tenement house into an interconnected microcosm and motivates the intrigues and romantic affairs of its inhabitants. In the age of the elevator, a novel about the network of relations in a multi-storey building has become less possible, and there are rather few representatives of this genre from the late twentieth century in either American or European literature.

Hierarchy

In addition to its organization of the vertical axis, the elevator made a further intervention in the structure of multi-storey houses, re-coding the traditional hierarchies of living and working spaces. Between 1890 and 1930, at different rates and with different consequences in Europe and the United States, there was a fundamental change in the meaning of the upper floors, which not only informed the architectural standards governing offices and tenement houses, but also influenced growing concerns about public hygiene and shaped the settings of the period's literature. As a technical apparatus, the elevator was responsible for massive changes at the nexus of the semantics of space and the history of technology, which were particularly apparent in the development of the grand hotel around 1900. Within a short span of time, the most desirable part of the grand hotel changed from the lowermost to the uppermost floors, as is evident in both the common topography of the hotel novel and the classifications and price breakdowns found in Baedeker travel guides. Ultimately, the elevator marked the end of the era of the *bel étage* and the start of the era of the penthouse.

In this respect, it can be revealing to compare the connotations of living spaces located near the top of buildings before the invention of the elevator. In the mid-nineteenth century, the code for this region of the building was primarily based on an image of the attic room, paradigmatically represented in Carl Spitzweg's famous painting *Der arme Poet* ('The poor poet', 1837) and late-Romantic stories like Ludwig Tieck's 'Des Lebens Überfluß' ('Life's Luxuries', 1986 [1839]). At the time, the only conceivable image of a room located under the roof was of an isolated space removed from the world. The contemporary status of the upper floors is also particularly evident in the widely contested triumph of the mass tenement house in German-speaking cities. In the last third of the nineteenth century, the emergent hygiene movement, then negotiating new forms of housing in urban areas, was concerned with the vertical expansion of buildings. Before the creation of zoning laws, the construction of multi-storey tenement houses sparked a debate among hygienists, above all in Germany and France, about the detrimental health effects of certain apartment levels; besides the basement, their main target of criticism was any apartment located above the third floor. The hygiene movement developed an outright 'pathology of the upper stories' (Bernard, 2014: 91) using statistics about increased mortality rates on the upper floors and increased still-births on the top floor (Bernard, 2014: 82–93). In the jargon of the hygienists, there was talk, starting around 1870, of 'anomalous tenements' above the third floor. As a reaction to this persistent criticism, amendments were made to German housing laws in the 1880s prohibiting the construction of more than four storeys for residential buildings. However, all of these hygienic and juridical excuses—and this is the decisive point—are only conceivable in an age before the establishment of the elevator. At the start of the twentieth century, the highest floor of a multi-storey building was still perceived, at least in Europe, as a demonic space. The European aversion to these upper storeys can also be seen in the literary imagination, where the attic was the central setting for four significant modernist texts from around 1900: Henrik Ibsen's *The Wild Duck* (1888), Robert Musil's *Young Törless* (1906), Gerhart Hauptmann's *The Rats* (1911), and Franz Kafka's *The Trial* (1914). In each case, the attic setting suggested the shape of an uncanny, inaccessible, and illegitimate sphere.

In the early twentieth century, there was a complete reversal of the decades-long tendency to demonize the top floor, which eventually came to be glorified with the emergence of new types of spaces: the public roof garden, as it evolved in New York; the penthouse in hotels and apartment buildings; and the executive floor in multi-storey office buildings. The semantic connotations of these upper storeys no longer conveyed stuffiness, inaccessibility, and illegitimacy, but rather transparency, repre-sentation, and power. Hence, in the age of the elevator, the building regulations cre-ated to restrict the height of buildings were quickly repealed. The eminent historicity of interior space can be read, above all, in the glass-enclosed executive floor, which has been burned into our imaginations through countless films and television series about lawyers. Today, this type of space may appear to be a natural, ahistorical alli-ance of absolute power and the total gaze. However, what made its realization pos-sible in the first place was the elevator.

CONTROLS

The development of control systems played a significant role in the history of the elevator, reflected in the transition from the elevator operator to automatic push-button controls. During the early history of the elevator, the attendant's authority and responsibility were most clearly manifest in the fact that those interested in the position were required to pass a comprehensive examination. In the elevator aptitude test, aspiring operators were required to prove their mastery of cable-based, lever-based, and hand-cranked control systems, all of which were used in early hydraulic and electric elevators. Thus, the elevator operator's tasks included the autonomous acceleration of the cabin and its deceleration at the right point in time before reaching a floor. This procedure demanded a high amount of skill and finesse, especially in the case of cable control systems, which were in use for decades and did not include any labels showing the elevator's starting position. Indeed, what the virtuosic handling of the cabin at the end of the nineteenth century was able to achieve can be seen in Thomas Mann's *Confessions of Felix Krull* (1955). In Mann's celebrated novel, the lift boy's skills at operating the elevators of the Parisian Hotel Saint James and Albany, which were equipped with a lever-based control system, propelled his career as a confidence man.

However, the unreliability of many other lift boys and the regularly occurring accidents they were responsible for increasingly called attention to the necessity of an automatic control system. Push-button controls were introduced for the first time in electric elevators in 1893 in the United States and 1903 in Germany. Following a short transitional period, they made the elevator operator obsolete. To a certain extent, they also democratized access to the awe-inspiring machine: any child could operate an elevator with the push of a button. The paragraphs in elevator regulations stipulating the presence of a trained lift boy were qualified, bit by bit, and ultimately repealed in the 1920s. Thus, the push-button played a decisive role in what we might call, adapting Wolfgang Schivelbusch's work on the railroad (1977: 198), the 'cultural assimilation' of the apparatus into buildings. The emergence of previously unknown controls in the elevator cabin around 1900 also supplied a condition of possibility for a novel perception of the elevator as something strange and disconcerting. In the sequence of technical processes, the push-button was associated with a radically new relationship between visibility and invisibility, cause and effect. The mere gesture of pushing a button took the place of a continuous movement, perceptible throughout the duration of this process, as had been the case with an elevator equipped with cable controls. What happens between the push of a button and its desired result—the arrival of the elevator cabin—plays out in the dark. This transition from *creating* a movement to *triggering* one signals a decisive break in the history of the elevator (Blumenberg, 1981 [1963: 35]). A push-button control system makes the cabin appear to function on its own, thereby providing the technological underpinning for a standard horror topos in late twentieth-century films and novels—the imagination of the uncontrollable, murderous elevator.

INTERIORS

The interior space of the elevator cabin was perceived, from the start, as an irritating crossroads in urban modernity where the greatest possible intimacy met with the greatest possibility anonymity. In the elevator, the line between public and private spheres was never completely clear. Similarly, the shared stairwell in tenement houses, another frequent target of hygiene debates, was never definitively classified as either the continuation of the street or that of the building's residential units, which contributed to the delayed acceptance of apartment houses in the United States. While the multi-storey residential building had provided the first test-site for the difficult integration of precarious intermediary spaces like stairwells and corridors into the modern floorplan, this conflict intensified once again with the emergence of the elevator. The primary challenge involved organizing the meeting of people, who barely know each other, in the smallest space imaginable. Against this background, it comes as no surprise that the elevator was perceived during its early years as a massive foreign body, which, like some inhabitants, required decades to settle in. Over the course of this process, there were many deadlocks. At first, it was unclear whether the elevator, initially called an 'ascending room' in England and the United States, was a vehicle or a space. In William Howell's popular play *The Elevator* (1960 [1884]), for instance, there were lengthy arguments, reflecting a common debate, about whether to take off one's hat in an elevator, as in other kinds of rooms, or to leave it on, as in other modes of transportation. The fact that elevators were always switched off during the evening and at night shows that people did not yet completely trust the machine at the turn of the twentieth century. Eventually, elevator-specific diseases even started cropping up, such as the short-lived 'elevator sickness', which was rampant in New York City in the 1890s (Anonymous, 1890). Unlike other forms of motion sickness, the modern pathology was supposed to have originated in the abrupt stopping of the cabin, which overtaxed the nervous system.

The elevator only came to be taken for granted as the natural core of multi-storey buildings in the early twentieth century. By this time, its effect on building interiors extended well beyond its mere transport function, and affected the distribution of occupants and affected their social order. There is something like a politics of the elevator, and one might even claim that the vehicle conveys not only passengers but a particular image of humanity itself. Elevator politics are particularly apparent when one asks why a concept like the servant's elevator, which was introduced before the First World War to replace the widespread servant's stairwell, never took off. Without ignoring historical evidence like the decline in the employment of servants during this period, the failure of the servant's elevator can still be attributed to some purely architectural factors, insofar as the vertical swath cut out by the elevator shaft creates a priori structural conditions for an increasing homogeneity of a building's traffic flow. To some extent, the elevator made building access more egalitarian. For many contexts in the early twentieth century, the elevator's egalitarian function was particularly consequential for the late phase of

European monarchy. Amazingly, the historical documents agree that the new means of vertical transport filled monarchs with a fundamental dread. In 1913, for instance, the Russian Tsar's refusal to use the elevator in the Hotel Adlon led to a complicated revision of the entire court protocol and dozens of guests had to be removed from and to lower floors (Adlon, 1955: 39–41). The emergence of the elevator threatened the traditional ceremonial order of the monarchy, which had ascribed the most significance to vertical passages, like imposing staircases, for the spatial representation of the monarch. Interrupting the necessary continuum of visibility, the elevator constituted a dark spot in ceremonial space. Future studies should not wholly discount the elevator's effect on the demise of the European monarchy.

ELEVATOR STORIES

As an interior space, the elevator cabin has performed diverse dramaturgical functions in literary, filmic, and promotional narratives, though most elevator stories tend to focus on a potential crisis between intimacy and anonymity for which three factors are of particular significance. First of all, the elevator is a space of contingency, which organized the plot of many city novels. If the constitution of modernity has largely been based, ever since Baudelaire's famous statement, on the irritating multiplicity and ephemerality of metropolitan encounters, then the elevator appears to be the paradigmatic space of this constellation. An elevator ride plays a pivotal role in many stories set in hotels, office buildings, or tenement houses, often signalling a turning point in the narrative. The combination of movement and concentration, of a freely-accessible means of transport and the form of a hermetic capsule, makes the elevator cabin into a decisive crossroads for both characters with differing biographies and different strands of the plot. As a contingent space, an elevator can set a story in motion, as happens in Vladimir Nabokov's first novel, *Mary* (1926), or it can connect different narrative strands occurring simultaneously, as is the case with Arthur Hailey's *Hotel* (1965) and other bestsellers of the hotel novel genre. In each case, the elevator generates coincidental but momentous encounters.

Second, the elevator is a space of transformation. A ride on an elevator functions as such an effective narrative turning point because the cabin interior is completely hidden from view. As a result, individuals and couples are given to believe, with absolute but usually misguided certainty, that they are unobserved for the duration of their passage, whereby the elevator becomes the literary or filmic scene of secret crimes and love affairs. Even more conspicuously, the elevator makes an appearance as a refuge and a safe changing room for figures with precarious identities, such as cross-dressers, con-men, and comic book superheroes. In twentieth-century Hollywood cinema, whenever there was a story about someone leading a double life in some metropolis, there would almost inevitably be an elevator. (Admittedly, the elevator has slowly been losing its status as a hidden refuge over the last ten to fifteen years due to the increasing installation of security cameras.)

Third, the elevator is a space of truth. In books and movies, elevators break down with a frequency hardly corresponding to that of real glitches, and they always seem to get stuck between floors. The dramaturgical popularity of this incident can be explained by the fact that an interruption in service breaks down the distance between passengers, which required effort to maintain even during failure-free trips, once and for all. An elevator malfunction pierces passengers' protective armour and creates a communicative dynamic, which ends with a predictable confession of long-kept secrets or some meditation on basic questions about life. Remarkably often, the elevator becomes a profane confessional.

The greater the confinement, the more authentic the speech—an alliance made clear in the genesis of the Christian confessional, from a free-standing stool in the Middle Ages to a sealed up cabin in the early modern period (Figure 19.2). In the twentieth-century metropolis, on the other hand, the stuck elevator eventually took over the function of the confessional, and provided a uniquely hermetic space within a large city shaped by the ceaseless flow of traffic.

Fig. 162.

FIGURE 19.2 Early elevator cars often also looked like confessionals

(*Source*: From Simmen, Jeannot, and Uwe Drepper. *Der Fahrstuhl: Die Geschichte der vertikalen Eroberung* (Prestel, 1984))

In recent years, the elevator's confessional function has remained evident in the novel ritual of the 'elevator pitch', where some simple employee takes advantage of a coincidental elevator ride with the company boss to persuade him of a powerful business idea in a matter of seconds. This new urban myth brings together many prominent elevator motifs: the contingency of the encounter in the smallest space imaginable; the brief absence of hierarchy and social class; the intimacy of a space capable of transforming this unforeseen encounter into a close cooperation; the matter of timing due to the precisely calculable duration of this short ride; and even the cabin cum confessional's 'truth effect', which promises a radical change of someone's career through the mythic chance of the successful elevator pitch. If nothing else, the moments of crisis and triumph triggered by the elevator reveal what any analysis of this inconspicuous yet influential apparatus will show: the elevator needs to be understood as a paradigmatic space of both modernity and social organization.

REFERENCES

Adlon, Hedda. 1955. *Hotel Adlon: Das Haus, in dem die Welt zu Gast war*. Munich: Kindler.

Anonymous. 1890. 'Elevator Sickness'. *Scientific American*, 12 July, 17.

Benjamin, Walter. 1982. *Das Passagen-Werk*, vol. 1. Frankfurt am Main: Suhrkamp.

Bernard, Andreas. 2014. *Lifted: A Cultural History of the Elevator* (trans. David Dollenmayer). New York: New York University Press.

Blumenberg, Hans. 1981 [1963]. 'Lebenswelt und Technisierung unter Aspekten der Phänomenologie'. In Hans Blumenberg, *Wirklichkeiten in denen wir leben: Aufsätze und eine Rede*. Stuttgart: Reclam, 7–54.

Engels, Friedrich. 1962 [1872]. 'Zur Wohnungsfrage'. In Karl Marx and Friedrich Engels, *Werke*, vol. 18. Berlin: Dietz, 211–87.

Freytag, Gustav 1855. *Soll und Haben*. Munich and Zurich: Droemer. http://www.gutenberg. org/files/19754/19754-h/19754-h.htm (flawed 1858 English translation).

Giedion, Sigfried. 1992 [1941]. *Raum Zeit Architektur: Die Entstehung einer neuen Tradition*, 5th edn. Zürich: Birkhaeuser Verlag.

Hauptmann, Gerhart. 1929. 'The Rats' (trans. Ludwig Lewisohn). In *The Dramatic Works of Gerhart Hauptmann*, vol. 2: *Social Drama*. New York: Viking.

Howells, William. 1960 [1884]. 'The Elevator'. In William Howells, *The Complete Plays*. New York: New York University Press, 300–13.

Ibsen, Henrik. 1978. *The Wild Duck* (trans. Rolf Fjelde). In Henrik Ibsen, *The Complete Major Prose Plays*. New York: Farrar, Straus & Giroux, 387–490.

Kafka, Franz. 1998. *The Trial* (trans. Breon Mitchell). New York: Schocken Books.

Mann, Thomas. 1955. *Confessions of Felix Krull, Confidence Man (The Early Years)* (trans. Denver Lindley). New York: Knopf.

Musil, Robert. 1964. *Young Törless* (trans. Eithne Wilkins and Ernst Kaiser). New York: New American Library.

Robinson, John Beverly. 1891. 'The Tall Office Buildings of New York'. *The Engineering Magazine*, 1, 185–202.

Schivelbusch, Wolfgang. 1977. *Geschichte der Eisenbahnreise: Zur Industrialisierung von Raum und Zeit im 19. Jahrhundert*. Munich: Carl Hanser.

Sennett, Richard. 1997 [1994]. *Fleisch und Stein: Der Körper und die Stadt in der westlichen Zivilisation*. Frankfurt am Main: Suhrkamp.

Tieck, Ludwig. 1986 [1839]. 'Des Lebens Überfluß'. In *Schriften*, vol. 12: *Schriften 1836–1852*, ed. Uwe Schweikert. Frankfurt am Main: Deutscher Klassiker Verlag, 193–249.

Wischermann, Clemens. 1997. *Mythen, Macht und Mängel: Der deutsche Wohnungsmarkt im Urbanisierungsprozeß*. In *Geschichte des Wohnens*, vol. 3: *Das bürgerliche Zeitalter, 1899–1918*, ed. Jürgen Reulecke. Stuttgart: Wüstenrot-Stiftung, 333–502.

....

CHAPTER 20

EXECUTIVE DASHBOARD

....

ARMIN BEVERUNGEN

'UNDERSTAND the past, predict the future, and drive immediate execution.' 'Predict, simulate and analyse all your enterprise information from one source.' The voiceover in a promotion video for SAP's Digital Boardroom lays out SAP's (2015) vision for the 'next-generation boardroom experience' as the camera zooms into the executive dashboard, surrounded by white-collar executives: we are at the heart of the Digital Boardroom. The executive dashboard—the SAP version is one amongst others—partakes in a long history of computing in management, from early forms of electronic data processing and information systems to contemporary enterprise resource planning and business intelligence systems, while its specificity is constituted by the way in which the computational systems it makes available to the executive managers is enhanced by contemporary advances associated with big data, cloud computing, the Internet of Things, and machine learning. The executive dashboard bridges the gap between computation and human understanding through particular forms of data visualization and analysis, and in doing so augments and formats human decision making in computational terms as data-driven, algorithmic decision making. The decision making capacities of human managers remain underdefined and constitute a residuality in relation to the computational prowess made accessible through the executive dashboard. The executive decision, then, is being separated and reasserted as an exception to computation and the operational functioning of the system, yet the scope of the separation remains limited by the way in which decision making capacities are distributed in the computational system, both to humans and machines.

THE EXECUTIVE DASHBOARD
AS INTERFACE

....

SAP's Digital Boardroom, piloted as Boardroom Redefined in early 2015 and rolled out in late 2015, is an example of contemporary developments in business intelligence and analytics. Competitors in business intelligence and enterprise resource planning software such as Oracle or IBM provide similar application suites. SAP was founded in 1972 by five ex-IBM employees. Its enterprise resource planning software—its primary product—was particularly successful compared to Oracle and other competitors because it enjoyed a first-mover advantage in the 1980s, SAP was early in adopting a client–server architecture for its R/3 system, and it benefited greatly from the re-engineering hype of the 1990s (Campbell-Kelly, 2003: 191–7). It became so central to organizing large corporations that Campbell-Kelly notes: 'If overnight R/3 were to cease to exist...the industrial economy of the Western world would come to a halt, and it would take years for substitutes to close the breach in the networked economy' (2003: 197). The Digital Boardroom complements the new generation of SAP, S/4HANA, and builds on the post-relational database HANA.

According to SAP's own history (recounted, for example, in Plattner and Zeier, 2011), the Digital Boardroom springs from a vision of Hasso Plattner, one of the founders of SAP, from the early 2010s. It picks up Bill Gates' 1990 vision of 'information at your fingertips' in which he conceives the computer as a 'general and complete, that is universal information technology, which allows its users to receive all information they might desire' (Burkhardt, 2015: 12; my translation). More immediately, it wants to translate the logic of the search engine, particularly its instantaneity in query response and its accessibility to big data, to enterprise resource planning software. In Plattner's vision of the 'management meeting of the future' (Figure 20.1), we thus see a number of executives, partly sat at a desk and partly virtually present via video conferencing, surrounded by a number of screens, some of them integrated in the desk at the centre of the scene, and all presumably touch-enabled, displaying various tables and charts, thereby making company information available at the executives' fingertips. The scene recalls a history of

FIGURE 20.1 Vision of the 'Management Meeting of the Future'

dashboards and control rooms (cf. Deane, 2015), and infuses it with contemporary technologies of touch and data visualization.

The Digital Boardroom offers 'information that is timely, reliable and insightful'; more specifically, it promises 'business leaders can monitor, simulate, and foster change' while 'using large, triple-interlinked touch screens'—the material bearer or hardware of the dashboard as interface—to achieve 'total transparency', 'real-time, data-driven insights', and 'simplified boardroom processes' (SAP, 2017a). The immediate hardware, in this case the 'large, triple-interlinked touch screens', is of negligible interest here, since apart from its capacity for touch, which is key for interaction with the interface and for the provision of 'information at your fingertips', these screens are precisely meant to disappear as the dashboard itself comes into view (Figure 20.2).

What the Digital Boardroom as interface reveals though, is what it puts into relation. As Hookway (2014: 5) notes, the interface is constituted by, as much as it produces, its elements, which include the system and the subject. In this case, the dashboard as interface puts SAP's computational system in relation with the executive managers conceived as its users. Hookway defines the interface as:

> that form of relation which is defined by the simultaneity and inseparability of its processes of separation and augmentation, of maintaining distinction while at the same time eliding it in the production of a mutualism that may be viewed as an entity in its own right, with its own characteristics and behaviors that cannot be reduced to those of its constituent elements.   (2014: 4)

The interface both separates and augments: it separates the machine from the human, and it augments both in their relation. In this case, the executive dashboard as interface separates the machine that is the enterprise resource planning system from the human that is the executive manager, in the way it distributes capacities for action, decision, and control between them, and regulates their interaction. It augments both the machine and the human in their relation, in that the decisions taken by managers are meant to complement the capacities of the enterprise resource planning system, and executive decisions are enhanced in data-driven decision making supported by computational capacities such as predictive analytics.

Since the interface puts human and machine into relation, like any other human–machine system the interface 'is not only defined by but also actively defines what is human and what is machine', through communication and contestation (Hookway, 2014: 12). For the interface 'the questions of what is human and what is machine are only posed operatively' (Hookway, 2014: 139). The executive dashboard and its history is marked by this contested relation and operational definition of human and machine, and it demonstrates how management, decision, and control are conceived. In that sense, the dashboard as interface serves as an entry point for an inquiry into the complex system and its various elements—its cloud infrastructure, its non-relational database, its learning algorithms, its sensorial apparatus—that is made accessible to the executive managers here, and into the executive managers who are constituted as subjects of decision and control through the dashboard in their relation to the computational system.

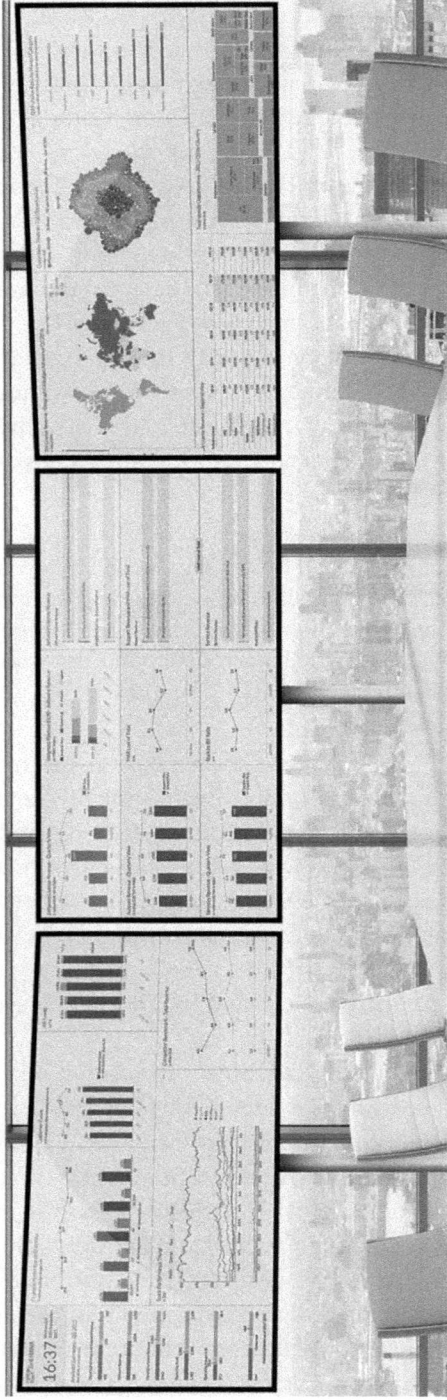

FIGURE 20.2 SAP's Digital Boardroom

What does the dashboard reveal about the capacities of the computational system to deliver 'total transparency' and 'real-time, data-driven insights', and how do these relate to human capacities for decision making in this organizational scene?

From Decision Devices to Enterprise Resource Planning

As the executive boardroom is conceived primarily as a space of managerial decision, we can position the executive dashboard in a long history of a plethora of decision devices or media of decision (see Conradi et al., 2016) meant to assist and to partially replace executive managers in taking decisions. Some of these media include graphs, charts, and tables and its various carriers from paper to PowerPoint. Hoof (2016) recounts how graphical methods and visualization devices, such as Gantt charts enabling a form of 'decision-making at a glance', produced a 'new visual culture of managerial decision-making' around the turn of the twentieth century, which reverberates in the executive dashboard today (see also Yates, 1985). The French *tableau de bord*—i.e. dashboard or 'boardroom charts'—from the middle of the twentieth century, was used primarily for accounting and reporting, but also a 'new mode of representing' company activities and intervening in them, and consisted simply of a series of paper printouts of graphs (Pezet, 2009). Pias (2009) explores how PowerPoint was initially not designed as a simple presentation tool, but its prehistory as 'electronic overheads' shows its conception as a kind of groupware which would enhance collective decision making. Banking on the widespread dislike and limitations of PowerPoint and other media of decision, the Digital Boardroom promises to liberate the executive manager from them:

> By eliminating the multiple Microsoft Excel spreadsheets, reports, and static Microsoft PowerPoint presentations you once needed for each meeting, SAP Digital Boardroom reduces meeting preparation time and effort. No longer must senior business analysts collect information and produce presentations in advance. Information is delivered live from your back-end transactional applications, as you need it.   (SAP, 2017a)

The most immediate context in which computational media are developed specifically for managerial decision making is the field of decision support systems or executive information systems, designed to assist managers with semi- or unstructured decisions and to help them deal with information overload (see Power, 2007). These systems already rely on 'the individually tailored access to the broader, more detailed sweep of data that only computers can provide', and are built on 'a common core of data'—a 'data cube' characterized by 'the sheer breadth of its cross-functional sources and the depth of its detail', promising executives 'inclusive information at their fingertips' (Rockart and Treacy, 1982).

The history of decision support systems claims many of the key developments in computing for itself: it considers the Semi-Automatic Ground Environment (SAGE) air defence system the 'first data-driven DSS [decision support system]', and lists Engelbart's oN-Line System and Bush's memex among precursors for 'communications-driven' and 'document-driven' decision support systems respectively (Power, 2007). Ensmenger recalls how computing developed from electronic data processing to information technology, and how already in the 1960s a vision of total management information systems emerged (2010: 137–61), which partly translated 'total systems' approaches from the war room to the boardroom (Haigh, 2002). The executive dashboard falls easily into this narrative of computing development. When SAP and others introduced 'fully integrated business applications software' they named it enterprise resource planning software to differentiate it from earlier, less integrated, and more partial software systems (Campbell-Kelly, 2003: 169). Yet in many ways enterprise resource planning software constitutes a revival of the total management of the 1960s (Haigh, 2003: 813–23). Even dashboards for managers were already conceived in the 1950s, and by 'the late 1990s, the idea of a "digital dashboard" was widely promoted'; it was 'intended to supply a single screen of easy to read gauges and dials to permit each manager to see at a glance how well his or her operations were performing' (Haigh, 2003: 819).

There is nothing conceptually novel about the executive dashboard, then, it is only its sheer computational capacities that make a difference. Firstly, the amount of data gathered for detailed analysis and scrutiny has expanded significantly. Earlier decision devices relied on specific, limited forms of capture, i.e., the gathering of data as part of the tracking of human activities and physical objects in organizations (cf. Agre, 1994), whereas now capture is extensive. With data gathered in the industrial Internet of Things, for example from smart sensors in the supply chain, SAP promises to optimize operations, for example in enabling companies to 'track, monitor, analyze, and maintain all moving assets, wherever they are in the network' (SAP, 2017c). Secondly, the post-relational database HANA makes this data available in one database rather than a number of separate databases and partially incompatible datasets. HANA offers 'real-time'—meaning 'under the eight seconds of maximal human attention span' (Plattner and Leukert, 2016: 114)—access to a unified source of data, since all business applications in use are meant to feed data into one database that can be installed on the premises or in the cloud, and allows parallel, in-memory processing of data (SAP, 2017b). Thirdly, data analysis is enhanced by machine learning and artificial intelligence, such as predictive analytics, which for example allows for scenario planning.

Augmented Decisions and Human Residuality

As media, the executive dashboard largely conceals rather than reveals how the machine works, but nonetheless invites the human to partake in part of its operation. The visualization that is part of the dashboard is a technique for dealing with the distance between computation and human understanding. Visualizations 'imply that not all computational processes can be fully automated and left to run themselves', and that human decisions are still called for (Wright, 2008: 79). Visualization here involves 'the language of the act of translation between a complex world and a human observer' (Halpern, 2014: 22); the visualizations coded for the executive dashboard both reduce the distance between computational processes and human understanding and more specifically make the vast amounts of data amenable to human exploration and inquiry augmented through analytics. Visualizations here play a key role in the practices of depicting and modelling which produce certain forms of attention for the executive managers (cf. Halpern, 2014: 21–7).

In that way the executive dashboard as interface supplements and augments the agency of the executive managers, constituting a form of subjectification which produces a kind of 'fragmented and augmented subjectivity' (Hookway, 2014: 17). Fragmented, because human intelligence here is incomplete without its machinic supplement; augmented, because it can operate at a higher level of effectivity. The encounter which here takes place through the executive dashboard as interface between computational system and executive manager 'is an introjection of machine intelligence into human selfhood, as well as a projection of human intelligence onto the machine' (Hookway, 2014: 46), providing a different kind of business intelligence. Notwithstanding the computational powers of the enterprise resource planning system, the executive dashboard supplements and augments it with human intelligence, as much as human intelligence is augmented with the kinds of data analytics that the executive dashboard makes available through visualizations. One particular feature of the Digital Boardroom is the ability to access not just aggregate data but to look at different aggregates as well as particular data points. Plattner and Leukert suggest: 'Playing with business data interactively will improve awareness of facts and developments, assist reactions to changes, and enhance projections into the future' (2016: 121).

The executive dashboard, in mediating the computational capacities associated with data and analytics, transforms managerial decision making. McAfee and Brynjolfsson note that the kinds of analyses based on big data are different in terms of the volume of information available for analysis, the velocity insofar as real-time analysis provides particular advantages, and the informational variety that databases such as HANA can make available in the same dataset (2012: 62–3). The consequence is for McAfee and Brynjolfsson quite clear: 'Data-driven decisions are better decisions…Using big data enables managers to decide on the basis of evidence rather than intuition. For that reason it has the potential to revolutionize management' (2012: 63). Even if there is ample criticism of this position with regards to the limits of data-driven decision making (e.g., Martin and Golsby-Smith, 2017), and to the use of big data in strategy more generally (e.g., Constantiou and Kallinikos, 2015), it is precisely this formatting of decision making in the image of data and analysis which is inherent to the executive dashboard. Whereas Cortada (2011) in his reflections on information and the modern corporation, for example, maintains a clear hierarchy between data, information, knowledge, and wisdom, these distinctions are undermined by the way in which the executive dashboard operatively imposes computational categories on managerial decision making. Data is likened to truth, for example when SAP suggests that HANA provides a single source of truth: 'SAP Digital Boardroom provides visibility across your entire management chain using a single source of truth. Decision makers all see the same information and have answers they can trust' (SAP, 2017a).

While the executive dashboard, then, calls for managerial involvement in decision making, how human decision making here is conceived mirrors machine logics: decision making becomes a question of data gathering and algorithmic analysis. Even though data does not speak for itself but requires context and interpretation (e.g., Boyd and Crawford, 2012), the executive dashboard does not define a positive role for human cognition that is clearly distinguished from its machine counterpart. The human becomes an increasingly residual category. Whereas Herbert Simon, for example, in a reflection on the automation of management in the 1980s, surmised that one of the comparative advantages of humans over computers is 'the use of his [sic] brain as a flexible general-purpose problem-solving device' (1960: 31), it is precisely the computer as universal machine that today promises these general capacities, not only in the move towards general artificial intelligence. And whereas Simon understands decision making to involve first 'detecting the occasions for decision', and second 'developing possible problem solutions' (1960: 39), it is precisely in these areas where computational capacities such as pattern recognition (e.g., discovering outliers in data) and simulation (plotting different scenarios) come into play. The executive dashboard shows how far decision making has been taken over by computers, and how far even human decision making has been coded in its image.

Managerial Exceptionalism
and Distributed Decisions

What remains, then, of the human in the human–machine relations constituted by the executive dashboard? The executive dashboard partakes in a history of interfaces in which the boundaries between human and machine are continually redrawn. In this history, despite the tendencies described above, the human never disappears, since as Hookway notes:

> For even if the interface suggests in its provision of augmentation the possibility of leaving behind once and for all the weight and drag of humanness, it also operates continually upon that very humanness, dividing it, shaping it, and bringing it to light, finding within it a territory of limitless expansion.   (2014: 134)

The 'limitless expansion' of humanness associated with the executive dashboard in particular can be likened to a certain kind of managerialism: a managerialism in which the human manager's executive powers and decision making capacities are continually reasserted, even as they are undermined. The executive dashboard operationally reaffirms human decision making, beyond its coding as data-driven and algorithmic, as exception to computation. Von Foerster already noted that only those questions which are principally undecidable can be decided, since otherwise they are always already decided by the framework of the question and the rules that lead to an answer (von Foerster, 2003: 291–5; cf. Conradi et al., 2016: 14). The computational system to which the executive dashboard as interface gives access already provides a very clear framework and rules for decision making. Furthermore, machine learning today does not operate on the basis of right decisions but of likelihoods (e.g., the likelihood that a certain image is of a dog). In that sense even beyond the formal definition of von Foerster the decision making capacities of computers are expanding, since where answers are not immediately available the system provides simulations. Nonetheless, it is not necessarily immediately clear what constitutes a decidable or undecidable, or which simulation is most likely, and this is where the executive manager can operate and make decisions. And managers can always overrule; even if data-driven decision making questions managerial intuition, it does not crowd out the space for decision making that executive managers can assert. The executive manager's decision, insofar as it exceeds or defies the algorithmic logic of the system provided through the executive dashboard, constitutes an exception in which the executive manager operates.

Here the human exception to computational decision making marked by the limits of decidability coincides with a certain managerialism which is also part of the history of the development of managerial media of decision making. In this history it comes to the fore in conflicts between managers and the agents, such as programmers or consultants, who administer computation in organizations. Ensmenger, for example, recounts how managers revolted against the status given to programming and programmers in early electronic data processing, asserting their own capacities for problem solving and decision making (2010: 158–61). In a similar way, Haigh (2001) retraces the history of the 'systems men' who, on the basis of computers understood as managerial information systems, sought to establish their authority as key for managerial decision making and control. And while the systems men ended up sidelined, today's enterprise resource planning systems—and the swathes of SAP consultants that accompany their implementation—can be considered their heirs. That the executive dashboard is nonetheless strongly marked by a certain managerialism becomes apparent in comparison to one example from computing history which regularly gets excluded from the historical accounts of computing in management (e.g. Power, 2007): Project Cybersyn. Tkacz (2015) suggests that we might look towards the operations room of Cybersyn 'for inspiration as to how contemporary dashboards can be designed to support and indeed reimagine a future politics beyond the neoliberal mode of capitalism'.

Project Cybersyn, a project to computationally manage the socialized economy of Chile under Salvador Allende in the early 1970s, whose principal architect was management cybernetician Stafford Beer, demonstrates a different kind of relation between computing and executive management (see Medina, 2014). The centre of Cybersyn, much like the executive dashboard for SAP's system, is constituted by the operations

FIGURE 20.3 'Operations Room' of Project Cybersyn

(© Gui Bonsiepe. Used with permission)

room or 'Opsroom' (Figure 20.3), a hexagonal room in which executives would sit in seven swivel chairs equipped with a number of operational controls surrounded by walls featuring a number of screens displaying key economic data and further operational controls. As much as in the executive dashboard the decision makers are central to the set-up here, even though the system made available through the interface of the Opsroom was far less sophisticated, consisting of a few mainframe computers and a telex system. Yet the power of the executives in relation to the computational system is here also conceptually restrained. Designed with Beer's 'viable systems model' in mind, a model of an adaptive organization in which different lower system levels would by and large recursively regulate themselves (see Beer, 1972), it was assumed that the executives would here only need to be involved in the highest levels of systems management. This is in contrast to the Digital Boardroom, which promises full data transparency and a plethora of algorithmic analytic capacities, granting the executive a kind of omniscience and omnipotence not limited to strategy but deeply involved in operations.

Nonetheless, the executive dashboard as interface also undermines the managerialism that it portrays. It gives the executive managers access to data about the organization, but it does not provide any feedback loops to the organization. Managers can dabble with data, but ultimately the executive dashboard is only a reporting and accounting device—it enables and augments decision making, but it does not allow for the exertion of control (which in Cybersyn was at least conceived, even if it never worked). In the meantime, the enterprise resource planning system on which it relies to generate all the data that it draws on operates independently at least of the organizational space constituted by the executive dashboard. Equipped with the capacities to track and trace both human activities and physical objects, the enterprise resource planning systems can effectively predict, monitor, and control organizational processes independently of executive involvement. As Rossiter notes: 'Who really needs a manager when decisions become computational calculations? The world increasingly becomes coded vanilla' (2016: 125).

DASHBOARDING HUMAN–MACHINE INTELLIGENCES

Are we moving towards a system in which smartness reigns, in which human reason and cognition are reconfigured in the image of the computer? The executive dashboard is far from visions of automation in which humans are completely replaced by computers; the managerialism encoded in the executive dashboard assures as much. Yet the executive dashboard is one interface in which these developments in computing and in human–machine relations play themselves out. Perhaps they are also a good example of how human–machine relations will develop in the coming years. As many, for example Brynjolfsson and McAfee (2016), note, while the cognitive capacities of computers have developed significantly, with computers beating humans in games such as chess and Go, and many other tasks, usually it is humans *with* machines which outsmart either machines or humans on their own. In a review of how artificial intelligence will impact business, Brynjolfsson and McAfee (2017) note: 'Over the next decade, AI won't replace managers, but managers who use AI will replace those who don't.' At the same time, intelligence is reconceived as a kind of distributed smartness (Halpern et al., 2017), with many of the devices surrounding us turned into smart devices and capacities for sensing, gathering, and processing data as well as controlling environments, thereby further displacing human intelligence. As dashboards have become 'a generalized but infinitely customizable template for acting in the present' (Tkacz, 2015), the executive dashboard might also reveal some more general features of the dashboarding of human–machine intelligences prevalent in digital cultures.

REFERENCES

Agre, Philip E. 1994. 'Surveillance and Capture: Two Models of Privacy'. *The Information Society*, 10(2), 101–27.

Beer, Stafford. 1972. *Brain of the Firm*. New York: McGraw-Hill.

Boyd, Danah and Kate Crawford. 2012. 'Critical Questions for Big Data: Provocations for a Cultural, Technological, and Scholarly Phenomenon'. *Information, Communication & Society*, 15(5), 662–79.

Brynjolfsson, Erik and Andrew McAfee. 2016. *The Second Machine Age: Work, Progress, and Prosperity in a Time of Brilliant Technologies*. New York: W. W. Norton.

Brynjolfsson, Erik and Andrew McAfee. 2017. 'The Business of Artificial Intelligence'. *Harvard Business Review*, July. https://hbr.org/cover-story/2017/07/the-business-of-artificial-intelligence.

Burkhardt, Marcus. 2015. *Digitale Datenbanken: Eine Medientheorie im Zeitalter von Big Data*. Bielefeld: transcript.

Campbell-Kelly, Martin. 2003. *From Airline Reservations to Sonic the Hedgehog: A History of the Software Industry*. Cambridge, MA: MIT Press.

Conradi, Tobias, Florian Hoof, and Rolf F. Nohr (eds). 2016. *Medien der Entscheidung*. Münster: Lit.

Constantiou, Ioanna and Jannis Kallinikos. 2015. 'New Games, New Rules: Big Data and the Changing Context of Strategy'. *Journal of Information Technology*, 30(1), 44–57.

Cortada, James W. 2011. *Information and the Modern Corporation*. Cambridge, MA: MIT Press.

Deane, Cormac. 2015. 'The Control Room: A Media Archaeology'. *Culture Machine*, 16. https://www.culturemachine.net/drone-culture/the-control-room.

Ensmenger, Nathan. 2010. *The Computer Boys Take Over: Computers, Programmers, and the Politics of Technical Expertise*. Cambridge, MA: MIT Press.

Haigh, Thomas. 2001. 'Inventing Information Systems: The Systems Men and the Computer, 1950–1968'. *Business History Review*, 75(1), 15–61.

Haigh, Thomas. 2002. 'Lost in Translation: "Total Systems" from War Room to Boardroom'. Presented at the Annual Meeting of the Society of the History of Technology, Toronto, November.

Haigh, Thomas. 2003. 'Technology, Information and Power: Managerial Technicians in Corporate America, 1917–2000'. PhD dissertation, University of Pennsylvania, Philadelphia.

Halpern, Orit. 2014. *Beautiful Data: A History of Vision and Reason since 1945*. Durham, NC: Duke University Press.

Halpern, Orit, Robert Mitchell, and Bernard D. Geoghegan. 2017. 'The Smartness Mandate: Notes toward a Critique'. *Grey Room*, 68, 106–29.

Hoof, Florian. 2016. 'Medien managerialer Entscheidung: Decision-Making "at a Glance"'. *Soziale Systeme*, 20(1), 23–51.

Hookway, Branden. 2014. *Interface*. Cambridge, MA: MIT Press.

McAfee, Andrew and Erik Brynjolfsson. 2012. 'Big Data: The Management Revolution'. *Harvard Business Review*, 90(10), 60–68.

Martin, Roger L. and Tony Golsby-Smith. 2017. 'Management is Much More than a Science: The Limits of Data-Driven Decision Making'. *Harvard Business Review*, 95(5), 128–35.

Medina, Eden. 2014. *Cybernetic Revolutionaries: Technology and Politics in Allende's Chile*. Cambridge, MA: MIT Press.

Pezet, Anne. 2009. 'The History of the French *tableau de bord* (1885–1975): Evidence from the Archives'. *Accounting, Business & Financial History*, 19(2), 103–25.

Pias, Claus. 2009. '"Electronic Overheads": Elemente einer Vorgeschichte von PowerPoint'. In Wolfgang Coy and Claus Pias (eds), *PowerPoint: Macht und Einfluss eines Präsentationsprogramms*. Frankfurt am Main: Fischer, 16–44.

Plattner, Hasso and Bernd Leukert. 2016. *The In-Memory Revolution*. Cham: Springer International.

Plattner, Hasso and Alexander Zeier. 2011. *In-Memory Data Management*. Berlin: Springer.

Power, Daniel J. 2007. 'A Brief History of Decision Support Systems'. *DSSResources.com*. http://DSSResources.COM/history/dsshistory.html.

Rockart, John F. and Michael Treacy. 1982. 'The CEO Goes On-Line'. *Harvard Business Review*. https://hbr.org/1982/01/the-ceo-goes-on-line.

Rossiter, Ned. 2016. *Software, Infrastructure, Labor: A Media Theory of Logistical Nightmares*. New York: Routledge.

SAP. 2015. *Digitize Your Boardroom Experience*. Walldorf: SAP. https://www.youtube.com/watch?v=qdBmodcVxks.

SAP. 2017a. *Reimagine Your Boardroom for Today's Digital Economy*. Walldorf: SAP. https://www.sap.com/products/cloud-analytics.html.

SAP. 2017b. *Rethink the Possible with the SAP HANA Platform*. Walldorf: SAP. https://www.sap.com/documents/2016/04/ac1e84d4-697c-0010-82c7-eda71af511fa.html.

SAP. 2017c. *Connect Internet-Enabled Things to Business Processes for Innovative Outcomes*. Walldorf: SAP. https://www.sap.com/documents/2017/10/de5eefe8-d67c-0010-82c7-eda71af511fa.html.

Simon, Herbert. 1960. 'The Corporation: Will It be Managed by Machine?' In M. Anshen and G. L. Bach (eds), *Management and Corporations, 1985*. New York: McGraw-Hill, 17–55.

Tkacz, Nate. 2015. 'Connection Perfected: What the Dashboard Reveals'. Keynote address presented at the Digital Methods Initiative Winter School, Amsterdam.

von Foerster, Heinz. 2003. *Understanding Understanding: Essays on Cybernetics and Cognition*. New York: Springer.

Wright, Richard. 2008. 'Data Visualization'. In Matthew Fuller (ed.), *Software Studies*. Cambridge, MA: MIT Press, 79–86.

Yates, Joanne. 1985. 'Graphs as a Managerial Tool: A Case Study of Du Pont's Use of Graphs in the Early Twentieth Century'. *Journal of Business Communication*, 22(1), 5–33.

CHAPTER 21

....

FILTER SYSTEM

....

WENDY HUI KYONG CHUN

FILTER systems are software tools which survey, select, prioritize and predict information, items, and users. The most prevalent form is the personalized recommendation system, which rose to prominence in the mid-1990s with the emergence of e-commerce. Personalized recommendation systems have been deployed by sites such as Amazon and Netflix to amplify sales, build user loyalty to their sites, and to understand user wants (Ricci et al., 2015, p. 5).

Such filter systems, however, do not benignly optimize to user experience; they also serve the needs of sellers, whose interests do not always coincide with those of their buyers. Through their content and form, they shape behaviour by collecting individuals and items into similarity-based neighbourhoods. They limit choice and amplify past trends in the name of efficiency and desire, i.e., by using historical data to anticipate the 'user wants'. The impact of these systems also goes far beyond e-commerce: recommendation systems and their algorithms have been crucial to the evolution of search engines and increasingly intrusive methods of data-mining, as well as to the fracturing of the World Wide Web into echo-chambers.

Recommendation systems were initially promoted to help establish e-commerce as a viable competitor to bricks-and-mortar shopping. Since e-commerce sites could offer direct contact with commodities, they sought to compensate via 'value added' curating based on analyses of past purchases across users and items. Most positively, these systems are described as democratic technologies: economic ways of giving advice to those 'who cannot afford to or are not willing to pay for high-quality advice by experts', where high-quality means personalized (Jannach et al., 2011, p. xiv). Less positively, they are framed as helping users incapable of competently evaluating 'the potentially overwhelming number of alternative items'. They organize information by reducing the possible items to a ranked listing, thus alleviating the 'misery-inducing tyranny' of choice (Ricci et al., 2015: 1–2). These systems differ from other filter systems, such as the widely available 'editor's choice' format, because they tailor their advice to individual user needs. This tailoring, however, is based not simply on users' individual actions, but rather actions of users determined to be 'like them'. In this sense, the term 'personalized' is a misnomer, since recommendation systems are built on the organizing principle of homophily, the idea that similarity breeds connection, that 'birds of a feather flock together'(McPherson et al., 2001); they assume that similar people like the same objects and/or that individuals desire similar items. They amplify the twentieth-century trend towards market segmentation by providing micro-segments—called neighbourhoods or clusters—and/or by considering individual items and users as sums of various weighted manifest or latent factors. Rather than simply displacing identity, they create new categorizations that refine and perpetuate older notions of race, gender, class, and sexuality (Gandy 2009).

Recommendation systems are divided into five basic categories, based on how they 'predict' missing links: content-based, collaborative-filtering, community-based, demographic, and knowledge-based (Ricci et al., 2015: 11–14). These systems make recommendations by either focusing on user–user relations and histories (homophily between users); on attributes of items (homophily between items); or through user questionnaires or detailed requests (for rarely bought items, such as cars). These different approaches, however, employ similar methods, such as nearest neighbour clustering, and they often bleed into each other: most systems use a hybrid of these approaches. Recommendation systems are also divided into 'memory-based' or 'model-based' systems, although both systems rely on past interactions. The difference between the systems speaks to the more technical difference between memory and storage: memory-based systems hold all data (or most) 'in memory' and use them directly to generate recommendations. This is computationally intensive—especially at run time; in contrast, model-based systems pre-process the data and then use the 'learned' model to make predictions. These model-based systems, as elaborated upon later, usually 'decompose' matrices into various factors.

COLLABORATIVE FILTERING AND ORGANIZATION OF TIME/SPACE

These filtering systems immediately raise the question: What are the ramifications of similarity? More simply: How do we measure similarity? How are items and users determined to be alike?

A common measure of similarity between users is the Pearson correlation coefficient, which is:

Covariance (X,Y) / (standard deviation of X*standard deviation of Y)

This coefficient measures the covariance between items, that is, how they increase or decrease in value together and divides it by the product of their standard deviations, so their strength can be more easily compared to other measured coefficients. Thus, a score of 1 would indicate a strong correlation (the variables perfectly coincide), 0 would reveal no correlation (the variables are independent) and −1 would reveal a negative correlation (the variables are polar opposites). After having computed this coefficient, these systems—such as the early user-based nearest neighbour collaborative filters—identify *peer users* or *nearest neighbours* for any given user. They basically assign a probability to all other items the user has not rated or seen based on ratings by algorithmically determined 'peer users' (Jannach et al., 2011: 13).

Information is organized by reducing the future to the past. That is, the filtering systems assume a flat relationship between past, present, and future: they assume that users' tastes do not change over time. They are further riddled by false positives: it is impossible to determine absolutely if users chose certain items because of the recommendation or if they would have chosen these items regardless. More insidiously, they do not simply presume this relationship; systems may also construct this relationship through the recommendations they make. By focusing the user on certain items and by hiding others, they strengthen certain correlations. They can also deploy well-known priming tactics to make certain purchases or items more attractive, such as adding 'irrelevant (inferior) items in an item set [to] significantly influence the selection behavior', placing items at the beginning or end of a list to emphasize their importance, introducing items earlier, etc.; or they incorporate psychological theories or personality type-analysis (Jannach et al., 2011: 234–52). That is, although they are based on the principle of homophily rather than contagion or imitation, their effectiveness is arguably linked to imitative behaviour.

These collaboratively-based systems also carve networks into affectively intense neighbourhoods. They place users into neighbourhoods, based on how they fall from the norm, the mean, the common denominator. 'Liking' something like *Harry Potter*, for instance, creates a neighbourhood too vast to be useful. Thus, 'liking' *Harry Potter* has to be supplemented by another measure that restricts one's neighbours more adequately: what matters are the moments users 'think out of the box' with others. Not surprisingly, the Pearson coefficient is often supplemented by a 'case amplification' function that weights agreement in accordance with controversy, for 'an agreement by two users on a more controversial item has more "value" than an agreement on a generally liked item' (Jannach et al., 2011: 16). These transformations reveal the extent to which controversy is key to predictability—and the ways in which controversy is fomented in the name of predictability. Users are more predictable the more they fall from the norm with others. In these affectively charged zones, users presumably fall prey to confirmation bias because these zones are zones of belief, that is, 'authenticity'.[1] In other words, these are moments and areas in which users are most aware of the fact that their own views are 'controversial', and thus—no matter how dominant they may be (it could be a 50–50 split)—these views feel 'subversive' or 'resistant' to hold. The point is to find triggers that ensure predictable user reactions and that can be used to delineate the boundaries between neighbourhoods.

---

[1] For more on this, see Chun (2019).

MODEL-BASED SYSTEMS AND
ORGANIZATION VIA LATENT FACTORS

Collaborative-based systems are assumed to be the most accurate filtering systems because they draw on actual data. They are, however, as noted earlier, computationally intense. Since they presume a static relationship between past–present–future, they also fall victim to the 'cold start' problem (it is impossible to come up with weighted ratings for new users or items) and they elide the possibility of serendipitous purchases and change. In response to these difficulties, model-based filter systems that compute predictions offline have been developed. These models often further simplify calculations by using Principal Component Analysis or related matrix factorization techniques, such as Singular Value Decomposition, to 'decompose' the database. The goal is to identify the most important manifest or latent factors or classes. Each item or user is then considered to be a weighted sum of these vectors.

The efficacy of matrix factorization models was proven via the Netflix Prize (2006–9), during which Netflix.com offered a large chunk of its database and a significant cash prize to whoever could improve its recommendation system by 10 per cent. Many of the teams published their results during the competition. Within a year of the competition, the importance of matrix factorization methods to detect 'weak signals' had become clear (Bell and Koren, 2007: 75). In particular, latent factor models, which measured the agreement of users and moves across a series of features that are 'algorithmically learned from the data' (Bell and Koren, 2007: 77) became 'central to predicting causal relationships and interpreting the hidden effects of unobservable concepts' (Anandkumar et al., 2012: 1). Users and/or movies thus became weighted sums of these latent factors (see Figure 21.1). Importantly, these matrix factorization models do not need actual ratings: they can take other implicit signals such as viewing time, etc., all made available after Netflix turned to streaming films. Further, they can add factors/weightings that account for time, since movies' popularity and users' preferences change with time.

Not surprisingly, Netflix is not using the winning prize algorithm with all its complexity and hybridity (it was a mix of neighbourhood and matrix factorization models) (Chen, 2011), but rather relying heavily on metadata and latent factors in its recommendation systems and to determine its 'original' programming (Madrigal, 2014). These latent factors and nearest neighbour analyses allegedly show the 'coarsity' of factors such as gender, race, and class, since these neighbourhoods and factors do not simply correspond to these identity categories. Indeed, Hallinan and Striphos in their in-depth analysis of the Netflix Prize competition and the competing algorithms produced argue: 'the parameters of human cultural identity stretch beyond the human, all too human to include "prepersonal" or "incorporeal" aspects perceptible to machines (Guattari, 1995: 9). These emerging aspects of cultural identity contain profoundly ambivalent potentialities, and their relationship towards existing modes of personal and cultural identity is far

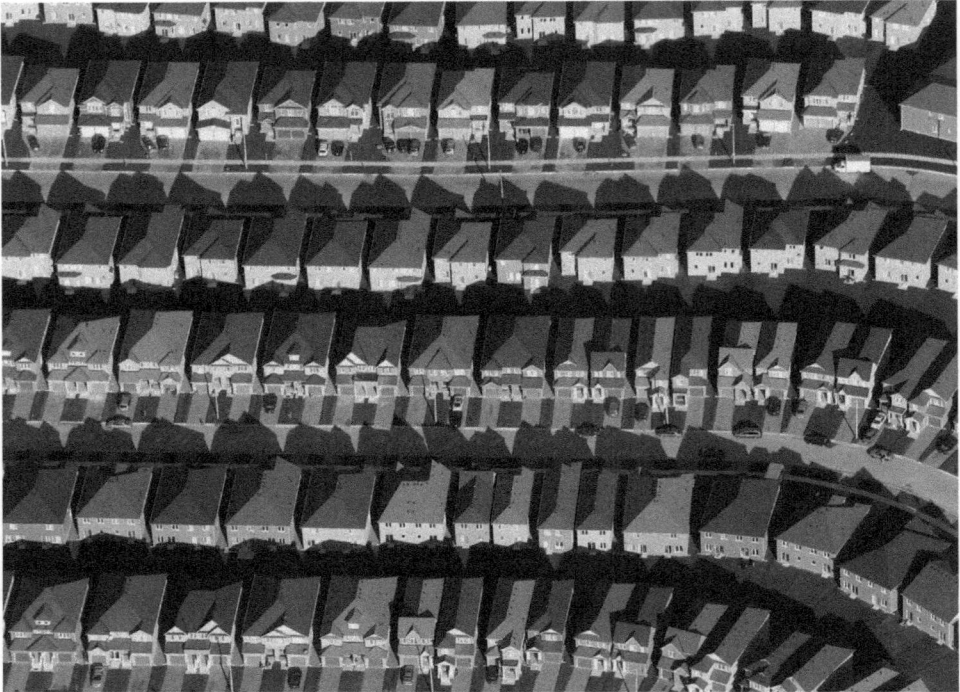

FIGURE 21.1 An aerial view of housing developments near Markham, Ontario
(*Source*: Photo by IDuke, November 2005, https://commons.wikimedia.org/wiki/File:Markham-suburbs_id.jpg)

from determined. Will the latent categorizations complement or eclipse extant human understandings?' (Hallinan and Striphas, 2016: 127).

Identity categories, however, are not discarded. Most blatantly, they are key to solving the 'cold start' problem. As mentioned previously, one of the biggest challenges facing recommendation filter systems is sparseness of data: How can recommendations be made when there is nothing or little 'in memory'? How does one make recommendations for new users or items? To resolve these problems, machine learning is deployed, along with demographic information:

> One straightforward option for dealing with this problem is to exploit additional information about the users, such as gender, age, education, interests, or other available information that can help to classify the user. The set of similar users (neighbors) is thus based not only on the analysis of the explicit and implicit ratings, but also on information external to the ratings matrix. These systems...which exploit demographic information—are, however, no longer 'purely' collaborative, and new questions of how to acquire the additional information and how to combine the different classifiers arise.   (Jannach et al., 2011: 22–6)

In the United States, zip codes are particularly 'useful' in determining the characteristics of users, since they reveal intersectional identities: not just race, but also class. As Cathy O'Neill explains, many systems 'draw statistical correlations between a person's zip code or language patterns and her potential to pay back a loan or handle a job. These correlations are discriminatory, and some of them are illegal' (O'Neil, 2016: 17–18). They are key to the kinds of discrimination and creation of 'uber users' described by Marion Fourcade and Kieran Healy (Fourcade and Healy, 2017). Pointedly, the direct consideration of factors of gender, age, and education (note here that race is implicitly, rather than explicitly factored in) is marked here as the 'limit' of collaborative filtering systems: such explicit calculations are no longer 'purely' collaborative.

This desire to place these factors as outside collaborative filtering methods reveals the uneasy relationship between identity and collaboration: such direct appeals to categories are to be avoided (if possible), even if they are easily inferable from the data. As Faiyaz Al Zamal et al. have shown in their analysis of Twitter, latent attributes, such as age and political affiliation, are easily inferred via a user's 'neighbours' (Al Zamal et al., 2012). Kosinski et al. have similarly shown that race, gender, and other private traits and attributes are predictable from publicly available Facebook likes (Kosinski et al., 2013). The Donald Trump administration's use of country of origin to refuse entry to the United States was initially blocked by US courts because discriminating on the basis of religion is unconstitutional. This overt ban arguably revealed the inadequacy of US Intelligence models in dealing with the 'cold start' problem. Banning this ban, however, did not 'solve' the problem of religious profiling. The question is not simply whether or not race, gender, sexuality, and so on are directly used to determine recommendations, but rather how they figure as 'latent factors'. As Kate Crawford and Jason Schultz have argued, these Big Data machine learning algorithms compromise privacy protections afforded by the US legal system by making personally identifiable information about 'protected categories' legible (Crawford and Schultz, 2014). These systems have thus been critiqued for perpetuating discriminatory practices in policing, hiring, and school selection (O'Neil, 2016).[2]

To return to the example of the Netflix algorithm, the winning prize used SVD to determine latent factors. Intriguingly, Netflix also employs humans to create 'metatags' for its movies and it has, as Alexis Madrigal has revealed in her remarkable reverse engineering of Netflix's 'reverse engineering of Hollywood', created thousands of micro-genres (Madrigal, 2014). What this arguably reveals is a fascinating back and forth between machine and human learning, in which latent factors are identified by matching them to humanly-produced categories.

LIMITATIONS AND FURTHER CONSEQUENCES

In order for these filtering systems to work, users must be authenticated and considered to be operating authentically. These systems can be and are routinely gamed in order to bias results. Recommendation systems are thus part of an overarching push of filtering

---

[2] Historically, the development of statistical correlation and boundary detection techniques is linked to the rise of eugenics, in particular biometric eugenics. For more on this see Chun (2019).

for Real Names and unique identifiers in order to track users across systems, key to the emergence of Big Data. The rise of authenticity and branding within contemporary politics and marketing, however, undermines any simplistic understanding of authenticity or identity, and authenticity has been historically linked to dramatic performance (Trilling, 1972; Bernstein, 2007; Fleming, 2009; Banet-Weiser, 2012; McCarthy, 2014).

Filtering systems are part of larger efforts to reorganize consumers and citizens through micro-segmentation, based on historical relations and actions. They are thus updating, and perhaps exacerbating, the organizational consequences of media technologies. 'Media organize', after all, and the question of organization becomes one of the scripts and performances of socio-technical ordering (Martin, 2003: 15; Beyes et al., 2019). Although it is outside the purview of this chapter to consider fully the historical relationship between these methods and older historical methods to segregate populations, it is important to acknowledge this link. The notion of homophily is drawn from the work of Paul Lazarsfeld and Robert Merton, who coined the term while studying segregation within two US housing estates (Lazarsfeld and Merton, 1954); work on network neighbourhoods draws from Thomas Schelling's agent-based analysis of segregation during the Civil Rights period (Schelling, 1971). The concepts of correlation and linear regression were developed by Karl Pearson and Sir Francis Galton as a way to provide a quantitative basis to their biometric theories of eugenics, in particular to determine the impact of previous generations on future and current ones (Pearson, 1901, 1920). Pattern recognition emerged from R. A. Fisher's work on linear discrimination and his desire to determine mathematically boundaries between races and species (Cortes and Vapnik, 1995). In these earlier mathematical models, the past determined the present. These systems are thus part of attempts to use perceived similarities to shape future actions and preferences.

REFERENCES

Al Zamal, Faiyaz, Wendy Liu, and Derek Ruths. 2012. 'Homophily and Latent Attribute Inference: Inferring Latent Attributes of Twitter Users from Neighbors'. Paper presented at the Proceedings of the Sixth International AAAI Conference on Weblogs and Social Media.

Anandkumar, Animashree, Daniel Hsu, Adel Javanmard, and Sham M. Kakade. 2012. Learning Topic Models and Latent Bayesian Networks under Expansion Constraints. UC Irvine. https://escholarship.org/uc/item/38c8m71n.

Banet-Weiser, Sarah. 2012. Authentic TM: Politics and Ambivalence in a Brand Culture. New York: New York University Press.

Bell, Robert M. and Yehuda Koren. 2007. 'Lessons from the Netflix Prize Challenge'. ACM SIGKDD Explorations Newsletter, 9(2), 75–9.

Bernstein, Elizabeth. 2007. Temporarily Yours: Intimacy, Authenticity, and the Commerce of Sex. Chicago: University of Chicago Press.

Beyes, Timon, Lisa Conrad, and R. Martin. 2019. Organize. Minneapolis: University of Minnesota Press.

Chen, Edwin. 2011. 'Winning the Netflix Prize: A Summary'. http://blog.echen.me/2011/10/24/winning-the-netflix-prize-a-summary.

Chun, Wendy H. K. 2019. 'Discriminating Data: Individuals, Neighborhoods, Proxies'. Manuscript in progress.

Cortes, Corinna and Vladimir Vapnik. 1995. 'Support-Vector Networks'. *Machine Learning*, 20(3), 273–97.

Crawford, Kate and Jason Schultz. 2014. 'Big Data and Due Process: Toward a Framework to Redress Predictive Privacy Harms'. *Boston College Law Review*, 55(1), 93–128.

Fleming, Peter. 2009. *Authenticity and the Cultural Politics of Work: New Forms of Informal Control*. New York: Oxford University Press.

Fourcade, Marion and Kieran Healy. 2017. 'Seeing Like a Market'. *Socio-Economic Review*, 15(1), 9–29.

Gandy, Oscar. 2009. *Coming to Terms with Chance: Engaging Rational Discrimination and Cumulative Disadvantage*. New York: Routledge.

Guattari, Félix. 1995. *Chaosmosis: An Ethico-Aesthetic Paradigm*, trans. Paul Bains and Julian Pefanis. Bloomington, IN: Indiana University Press.

Hallinan, Blake and Ted Striphas. 2016. 'Recommended for You: The Netflix Prize and the Production of Algorithmic Culture'. *New Media & Society*, 18(1), 117–37.

Jannach, Dietmar, Markus Zanker, Alexander Felfernig, and Gerhard Friedrich. 2011. *Recommender Systems: An Introduction*. New York: Cambridge University Press.

Kosinski, Michal, David Stillwell, Thore, Graepel. 2013. 'Private Traits and Attributes are Predictable from Digital Records of Human Behavior'. *Proceedings of the National Academy of Sciences*, 110(15), 5802–5.

Lazarsfeld, Paul and Robert Merton. 1954. 'Friendship as Social Process: A Substantive and Methodological Analysis'. In M. Berger, T. Abel, and C. Page (eds), *Freedom and Control in Modern Society*. New York: Van Nostrand, 18–66.

McCarthy, E. Doyle. 2014. 'Emotional Performances as Dramas of Authenticity'. *La Critica Sociologica*, 48(190), 25–39.

McPherson, Miller, Lynn Smith-Lovin, and James M. Cook. 2001. 'Birds of a Feather: Homophily in Social Networks'. *Annual Review of Sociology*, 27, 415–44.

Madrigal, Alexis C. 2014. 'How Netflix Reverse-Engineered Hollywood'. *The Atlantic*, 2 January.

Martin, Reinhold. 2003. *The Organizational Complex: Architecture, Media, and Corporate Space*. Cambridge, MA: MIT Press.

O'Neil, Cathy. 2016. *Weapons of Math Destruction: How Big Data Increases Inequality and Threatens Democracy*. New York: Crown.

Pearson, Karl. 1901. 'LIII. On Lines and Planes of Closest Fit to Systems of Points in Space'. *The London, Edinburgh, and Dublin Philosophical Magazine and Journal of Science*, 2(11), 559–72.

Pearson, Karl. 1920. 'Notes on the History of Correlation'. *Biometrika*, 13(1), 25–45.

Ricci, Francesco, Lior Rokach, and Bracha Shapira (eds). 2015. *Recommender Systems Handbook*, 2nd edn. New York: Springer.

Schelling, Thomas C. 1971. 'Dynamic Models of Segregation'. *The Journal of Mathematical Sociology*, 1(2), 143–86.

Trilling, Lionel. 1972. *Sincerity and Authenticity*. Cambridge, MA: Harvard University Press.

CHAPTER 22

HIGH HEELS

MIKE ZUNDEL

HIGH heels attract attention—even in their absence. Nicola Thorp, an accountant at PWC in London, was sent home from work without pay for not wearing at least two-inch heels, her refusal sparking global media attention and a petition to outlaw such demands (Medland, 2017). Equal outrage was caused when Danish film producer Valeria Richter was stopped four times attending a gala premiere at the Cannes film festivals, each time having to explain why she was not wearing heels: 'Four times of Show and Tell. Four times of explaining that with a missing toe and a partly amputated foot, heels are not an option. Then again, I shouldn't have needed a reason; I shouldn't have needed to explain at all' (Richter, 2015). A similar sense of scorn drives a 'Jurassic Park High Heels edition', a spoof depicting the film's protagonists—women, men, reptiles—stumbling through the film-set on absurdly high stilettos (Lerf, 2015). The spoof takes aim at actor Bryce Dallas Howard's portrayal, in 'Jurassic World', of the Dino-Park's 'senior asset manager' wearing 3.5-inch nude Sam Edelman heels throughout the entire action-packed film. Asked why, Howard explains her film character needed to be 'somebody who looks like she belongs in a corporate environment' (Warner, 2015). Heels mean business; the higher the office, the higher the stilettos. This brings us to another caricature, this time provided by the now iconic image of 'FLOTUS' Melania Trump striding towards Air Force Once for a flight to Texas to survey the devastation and catastrophic flooding caused by Hurricane Harvey—sporting 5-inch Malono Blahnik stilettos (Friedman, 2017). Immortalized by *Sex and the City*'s Carrie Bradshaw, Blahniks retail upwards of £500 and in concert with the Louboutins, Jimmy Choos, Puccis, Givenchys, and Yves Saint Laurents, have come to epitomize the fetishization of commodity (Debord, 1967: 18). They are expensive, impractical, damaging to health and floors, and at the same time objects of desire: productive of exaggerated gender identity; glamour, magic and attraction; vulnerability and danger; and status and power (Rabinowitz, 2001: 55).

Heels, and especially stilettos—Italian for 'dagger' but also stylus, the pen—have a long association with power. Louis XIV, of France, wore dazzlingly red heels, lifting him above the people (Semmelhack, 2008). Today, heels signify power in the boardroom, nowhere more so than in technology-intensive contexts. Vogue (Weisberg, 2013) pictures Melissa Mayer, then CEO of Yahoo, reclining on a garden lounger, legs pointing up, resting her Yves Saint Laurent heels where her head should be—all in apparent contrast not just to conservative pin-striped boardroom fashion, but also to the traditionally male-dominated appearance of the geek culture (Miller, 2012). Cheryl Sandberg, COO of Facebook, posed for a cover of *Time* in stilettos and a subtitle stating '... her mission to reboot feminism' (Time, 2013). But then again, Mayer and Sandberg purport rather selective and restrictive versions of feminism; the latter's project to reignite the women's revolution appearing much like a self-marketing activity by what Dowd (2013) called a 'PowerPoint Pied Piper in Prada ankle boots'.

LOTUS

Shoe designer Louboutin comments: 'The core of my work is dedicated not to pleasing women but to pleasing men' (Collins, 2011). John Berger suggests that '[m]en survey women before they treat them' (Berger, 1972: 46). High heels are part of the apparatus of such ways of seeing: 'supernormal stimuli' exaggerating elements of gait associated with the feminine: pelvic rotation, vertical motion in the hip, and shorter strides (Morris et al., 2013). Some see in this primal forms of messaging, where heels issue 'come-hither' signals as they 'thrust out the buttocks and arch the back into a natural mammalian courting—actually, copulatory—pose called "lordosis." Rats do it, sheep do it ... lions do it, dogs do it. It is a naturally sexy posture that men immediately see as sexual readiness' (Fisher, in Brown, 2017). While such claims are not just primitive but also dangerous (Valenti, 2011), high heels remain tied to the projection of a predominantly male fantasy of 'looking-at' and the residual, passive role of the female figure as sexual object or erotic spectacle (Mulvey, 1975: 11, 46), often with a view towards commercialization. Louise Brooks was one of the first movie actresses to unveil ankles and high heels in the provocative moving picture *Pandora's Box* (Parmentier, 2016), while Marilyn Monroe's famous 'wiggle' in her custom-made Ferragamo heels 'masquerades excessive femininity' precisely because the performances are so artificial, uninhabitable, and in excess of a woman's body.

This link between commodification and deformation invokes eerie connections to the practice of foot binding which, similar to high heels, produced alterations in a woman's posture: the ideal of slender feet and curved arches effecting 'the shifting of her centre of gravity [that] produced a mincing gait ... eulogized as "lotus steps" in poetry' (Ko, 2005: 136). Once widespread in China, and lasting for a millennium, foot binding involved painfully compressing women's feet often until bones broke, resulting in permanent deformities. Taking years to complete, foot binding impacted on mobility and balance, limiting women's ability to conduct work and social engagements beyond the house. And while its reasons are complex, there are at least some economic connections. Textile production, typically performed by women from home (Bossen et al., 2011), required strength and skill in the hands. While foot binding did not limit this substantial economic contribution of women it diminished their social and political influence, 'mystifying female labor' (Ko, 2005: 2–3). And as with foot binding's atrociously painful procedures and disabling effects, the high heeled shoe is responsible for conditions including hallux vagus, corns, calluses, metatarsalgia, Achilles tendon tightness, planar fasciitis, and Haglund's deformity. High heels are more slippery on most floors, inviting domestic and occupational accidents (Manning and Jones, 1995), and long-term wearing of heels leads to lower-extremity joint dysfunction, inelasticities in muscle fibres, chronic muscle shortening, soft tissue issues, back pain (Zollner et al., 2015), and the list goes on. The injuries caused by heels and narrowly constructed shoes leave many women unable to walk without pain, with surgery often being the only partial treatment of 'hobbled feet' (Linder, 1998).

GLAMOUR

The high heel is obstinate—no longer needed to grip a saddle's stirrup to balance a rider (Semmelhack, 2013), nor accelerating movement or stability. Its physical elevation comes at the price of discomfort and instability. When objects take on value beyond any actual use they become enchanted. The 'secular magic' (Thrift, 2008: 14) of useless objects is 'glamour', instilling in them a life of their own. On billboards or in glossy adverts, high heels captivate; let us glimpse an imaginary ideal; alternate versions of 'me'. They do so effortlessly, instilling their magic through our envy and anxiety (Thrift, 2008: 15); a promise of fulfilment to be realized in the purchase of the commodity (Myers, 1982).

The high heeled object itself, however, speaks to a higher, 'spiritual' need. For Thorstein Veblen (2007: 104), the divorce of fashionableness from the mechanical service yields a 'pecuniary culture'. These objects have to be expensive while simultaneously lowering a 'subject's vitality'. Rendering their wearers unfit for work, they become displays of reputability, showing wasteful expenditure and 'exemption from, or incapacity for all vulgarly productive employment' (Veblen, 2007: 110). This, Veblen suggests, goes especially for women who 'have been required not only to afford evidence of a life of leisure, but even to disable themselves for useful activity' (Veblen, 2007: 111). High heels may thus signal the growth of the middle classes, of service and intellectual industries, of wealth and mass production and the decline of manual labour. Like priests whose vestments are testament to their servile status and vicarious life, the high heel indicates refrain from useful effort, coupled with excess. As one of Oscar Wilde's dandies tells us: 'Nothing succeeds like excess'; so high heels represent something strange, a fantasy-world of the spectacular in the commodity (Debord, 1967: 18).

High heels worn at work or worn by workers indicate complex processes of cultural consumption. Enstad (1999) details how American immigrant working women in the late nineteenth century took to wearing French heels. These cheap, flimsy, three-inch constructions with paper-thin soles were utterly unsuited for the rugged and muddy streets of US industrial cities. And yet, Enstad suggests, they allowed working women to express multiple identities. For immigrants eschewing traditional clothing they signalled 'Americanization'; as signifiers of 'ladyhood' they provided flamboyant counterpoints to the masculinity associated with 'workers' and to the tedium of often repetitive and dirty work; as symbols of a generational struggle they signalled the growing demands of working women to spend some of their wages on themselves. The popularity of high heeled shoes, along with colourfully adorned hats, elaborate dresses, and fine undergarments as central parts of workplace culture, was related to the growing popularity of fiction, first in the form of 'dime' novels and later the apparatus of the movie theatre. Dime novels were frequently read in work breaks and discussed at work fanning the collective development of imagined futures centred in commodities (Enstad, 1999: 68). Their typical chronotopic 'adventure-romance' pattern involved a protagonist facing a series of challenges which had to be overcome to reap returns and attain valorized social positions: 'When working-girl heroines encountered challenges, their task was not simply to endure, or to wait to be rescued by someone. On the contrary, they provided their bravery as much as their virtue, and often saved themselves…"really" [being] ladies by accomplishing dramatic and daring physical feats' (Enstad, 1999: 74). All this, crucially, without changing the heroine herself; without denouncing her identity as a worker or attempting to rectify the system that so puts her to the test.

Fast-forward a hundred years and we find a contemporary revival of such accounts in the self-help genre peddled by the likes of Sandberg. Her 'lean in' book, self-described as a 'sort of feminist manifesto' (Shade, 2014), falls into a genre that 'largely manifests a studiously structurally unconscious relation to patriarchal constraints or…that these can be transcended through self-belief and perseverance' and through the 'authenticity of the fully realized female self'; the murmured mantra being: 'Go along to get ahead' (Negra, 2014: 278, 280). Current day 'heroines' of the tech-world equally don heels as codes for a form of femininity that valorizes bravery and virtue by overcoming their 'internal obstacles' (Sandberg, 2013: 10). However, the ultimate success of Sandberg or Mayer as some of the world's richest business people signifies but a utopian promise of transformation as, in presenting their own biographies as blueprints for other women to follow, they mask the immense privilege that first affords many of their advances. While private jets (see Dowd, 2013) are less easy to obtain, what can—and is—being copied in the increasingly glamorized tech-entrepreneurship culture (Negra, 2014: 277) are elements of fashion such as high heels which specifically code femininity (Enstad, 1999: 77); signets of a 'neoliberal feminism' with no chance of fulfilling the utopian dreams of the many as any collective desire to critique or change the status quo is eclipsed by the individual's desire to reach the top herself (Rottenburg, 2013).

FETISH

Like the glamorous object that derives its enchantment from its removal from everyday use, the fetish object is isolated from its surroundings. It stands for something else; a supernatural sign releasing certain behaviours; a 'simulacrum [that] is often more potent than the object...that it stands for' (Sebeok, 1989). This objectification in altocalciphilia—shoe fetish—is preserved through an optical casting of a perceptual frame that makes a 'cut' from the field; immobilizing that cut within a static framework (Bryson, 1980: 97), so as to produce a redundancy: the material dead object and its referent, which it presents (Berkeley, 1989). For Freud, the fetish is 'a substitute for the woman's (the mother's) penis that the little boy once believed in and...does not want to give up' (quoted in Rabinowitz, 2001: 57). In opposition to glamour, the high heel as fetish performs a distinction: the sexual difference through the masculinization of the female body. The stiletto, the spike-heeled dagger, long, hard, and phallic, is often paired with shoe openings revealing equally phallic, visible peep toes: 'a simulated rendition of [a watcher's] own masculine possessions [becoming] part of the female's bodily text' (Berkeley, 1989: 84). The harmful discovery as the result of the child's exploration—the castrated female body—is neutralized through the fetish: like a 'frozen, arrested, two-dimensional image, a photograph to which one returns repeatedly to exorcize [these] dangerous consequences' (Deleuze, 1991: 31).

Examples abound. The UK's *Daily Mail* (2013), in stereotypical sleaziness, calls the heels in Mayer's *Vogue* shot 'daring'; the same tabloid printed a picture of Scottish First Minister Nicola Sturgeon next to UK's Prime Minister, Theresa May, both in heels, asking who won 'legs-it'. Columnist Sarah Vine opined that Sturgeon's legs were 'altogether more flirty, tantalisingly crossed...a direct attempt at seduction' (Malkin, 2017). Where male politicians like UK Labour's Jeremy Corbyn are being picked up on their appearance, for instance when then Prime Minister Cameron asked him to 'put on a proper suit' (BBC, 2016), such comments are typically less venomous and do not invoke particular kinds of gender role expectations. The substitution of a fetish, be it in form of the high heeled shoe or through the represented figure itself, marks the transformation of something dangerous into something reassuring. The fetishized product is perfect; powerful politicians looked at and cast in passivity; a spectacle created for the gaze of a (male) audience (Mulvey, 1975).

SUCCESS

The fetishized high heel masculinizes the female body (Berkeley, 1989). Take the following *New York Times* special describing the posse of the then Libyan leader: 'Aloof from all this were the women who serve Colonel Qaddafi's as bodyguards, part of a larger contingent that sometimes accompanies him at home. Some wear pistols and high-heeled shoes, one sports a black jump suit, high heels and webbing holster for her Tokarev. Three, in desert camouflage, wear revolvers on Sam Browne belts that match wedge-heeled patent leather shoes. They are an object of some admiration and much comment, but they are not all glamour' (Cowell, 1983). Stilettos puncturing, stabbing, and injuring just like a gun; even the Tokarev's 4.57 inch barrel length is a play on a heel's height. High heels: 'a woman's power tool' (Brokman, 2000); a 'bullet case, sheathing the woman's foot and hardening it against the concrete pavement she traversed in her search for desire and power'; elevating women to greater height; commandeering air and sound-spaces through the rhythmic clattering of heels against floors (Rabinowitz, 2001: 63).

The Spice Girls' sky-scraping platforms signalling 'girl power', and perhaps more than others, 'Barbie', have contributed to what Collins (2011, in D'Angelo, 2016: 115) calls a 'body project': the possibilities prescribed by images such as Barbie for women's careers other than motherhood on the basis of physical appearance—the 'glittering pink aesthetic' that has come to define many women's vision of self, of success, and, relatedly, of beauty (Baker, 2017: 52). Barbie portrays a 'model of female success and beauty [in form of] a tiny waist, long blonde hair and [...] couture' (Toffoletti, 2007: 59)—with tip-toed feet permanently moulded to fit high heels. A smart suit, shedding one's accent, and a 'good pair of pumps' have come to be key elements of female mobility (Rabinowitz, 2001: 65), helping to overcome the shame of the unmodified body's perceived failures (Anders, 2010: 36). In the 1988 film *Working Girl*, Melanie Griffith plays Tess, a working-class secretary who assumes her boss's identity to move up the career ladder at a New York brokerage firm. The script (2018) begins:

> *FADE IN: STATEN ISLAND HEIGHTS—A SERIES OF SHOTS—EARLY MORNING. CLOSE ON FEET, one pair of them, padding quickly down a hill in well-worn, rain-soaked running shoes.*

Twenty-five minutes in, on the eve of her transformation, the script places her in her absent boss's apartment accompanied by her friend Cyn, getting dressed in her boss's clothes for an important social meeting:

> Tess: *it's important for me to start interacting with people, you know, not as a secretary....*
> Cyn: *but as a total impostor. Right.*
> Tess: *slips into the dress. A little sexy, a little plunging.*
> Cyn: *It's not maybe a little...much?*
> Tess: *Nah. It's elegant, simple, and yet...makes a statement. Says to people, confident. A risk-taker. Not afraid to be noticed. Then you hit 'em with your smarts.*
> Tess: *Shoes. I need shoes.*
> *She selects a pair of pumps and screws them on, with difficulty.*
> Tess: *(wincing) God, she's got small feet.*
> *She strikes a smart pose for the mirror, attitude in place. Her right eye starts twitching, comically, like a wink gone out of control.*
> Tess: *Damnit. There goes my eye.*
> Cyn: *Little antsy?*
> Tess: *I guess.*   (Kevin Wade, *Working Girl*, 1988)

The script places high heels at the centre of the transformation; her old secretarial life symbolized by worn-out trainers; her new executive one by uncomfortable high heels that are not her own. Her feet will be in pain for the rest of the evening and for most of the charade. Dressing for success made the high heel and the clutch bag symbols of high-powered women executives (D'Angelo, 2016: 127): the sexual suggestiveness; the play on gender and dominance; violence and vulnerability; and on presence and Freudian absence; a continuous battle between age-old concerns about women's 'propensity for economic folly' and (the fear of) their reclamation of power and success (Semmelhack, 2008: 63).

Prosthesis

Perhaps we should therefore not only consider questions of artificial bodies, dieted, heeled, surgically modified, and made up, versus versions of natural bodies; but also the labour that goes into the aesthetics of 'femininity' (Baker, 2017: 52). Swift (2012: 105), in an auto-ethnographic essay on her experiences as a queer femme and a sex worker, suggests that there are 'under-explored feminist possibilities in discomfort. Femmes and sex workers have something to gain by explicitly embracing discomfort, symbolized in our high heels, as we are too familiar with the ways in which comfort in fashion is not simply about how one's body fits into their clothes, but how one's clothes enable their body to interact in the world, and that therefore comfort is not as self-evident a goal as it may seem.' Comfort, Holliday (1999) notes, means different things in different contexts: flat, comfortable shoes may embody resistance to the hegemony discourses of femininity; but comfort may also act as a sedative to lull groups into passive spectatorship, making active resistance too uncomfortable (Sennett, 1994). Barbie equally carries 'transformative figuration' (Toffoletti, 2007: 68) precisely because of her plasticity and the exaggeration and artifice that goes into the creation of hyperfeminine attributes. Barbie and Ken are frequently re-appropriated as queer symbols. Just like high heels, they point to the artifice and transformability of gender itself (Toffoletti, 2007: 74). In the 1970s, heels were taken up by glam rock bands and punks in deliberate 'spectacular plays' (Catalani, 2015); by drag acts and dominatrixes crossing, masquerading, or rewriting gender and power relations (Morgan, 1989; Preciado, 2003: 39).

High heels insert incongruity into the otherwise fluid processes of (organized) life. Like a broken hammer, they stand in the way of achieving projected ends, making journeys on foot arduous; inviting dangerous slips; requiring a continuous balancing act; even the act of standing still turns into continuous little movements; into 'continued variation'—an un-balancing releasing of new potentiality, difference, and variation (Murray, 2009: 206). They punctuate the pleasing comfort of the 'satisfied equilibrium' (Moretti, 1985: 122) and therefore the sameness and neat alignment of things and purposes. They render obstinate that with which we are primarily concerned (Heidegger, 1962 [1927]: 102). Louboutin has a matching story about a client who, having bought her first pair of his heels, was forced to slacken the pace of her morning walk. 'She began to notice the little details of her neighborhood for the first time' (Collins, 2011).

FIGURE 22.1 Jeffery West, men's boots: 'Muse', 2.5 inch heel

This physical un-balancing is accompanied by what Swift (2012: 101) calls 'reconnais-sance': 'the act of going in disguise to recover that which is "lost"'. Heidegger's (2002: 13) description of what he thought was a pair of boots belonging to a woman peasant farmer in a van Gogh painting emphasizes the disappearance of the object into the work world: 'The peasant woman wears her shoes in the field. Only then do they become what they are. They are all the more genuinely so the less the peasant woman thinks of her shoes while she is working, or even looks at them, or is aware of them in any way at all. This is how the shoes actually serve.' Stilettos are not farmers' boots; their obtrusive presence pierces the reliability especially of the work world, imposing difference and, with this, the continuous imposition on wearers and spectators to negotiate these distinctions of gender; power and vulnerability; sexual availability and exploitation; class—and perhaps beyond all: organization, as they resist the comfort and fluidity of an already determined set of relations. Perhaps it's time for men to again wear them as well. For this, I submit exhibit one (Figure 22.1).

REFERENCES

Anders, Günther. 2010. 'Promethean Shame'. In Christopher J. Müller (ed.), *Prometheanism: Technology, Digital Culture and Human Obsolence*. London: Rowman & Littlefield, 29–96.

Baker, Sarah E. 2017. 'A Glamorous Feminism by Design?' *Cultural Studies*, 31(1), 47–69.

BBC. 2016. 'Cameron "put on a proper suit" jibe at Corbyn at PMQs'. http://www.bbc.co.uk/news/uk-politics-35651000.

Berger, John. 1972. *Ways of Seeing*. London: Penguin.

Berkeley, Kaite. 1989. 'Reading the Body Textual: The Shoe and Fetish Relations in Soft and Hard Core'. *American Journal of Semiotics*, 6(4), 79–94.

Bossen, Laurel, Wang Xurui, Melissa J. Brown, and Hill Gates. 'Feet and Fabrication: Foot Binding and Early Twentieth-Century Rural Women's Labour in Shaanxi'. *Modern China*, 37(4), 347–83.

Brokman, Elin S. 2000. 'A Woman's Power Tool: High Heels'. *The New York Times*, 5 March. http://www.nytimes.com/2000/03/05/weekinreview/a-woman-s-power-tool-high-heels.html.

Brown, Lauretta. 2017. 'Newsweek Goes After Trump Women for... Wearing Heels?'. *Townhall. com*, 11 August. https://townhall.com/tipsheet/laurettabrown/2017/08/11/newsweek-goes-after-trump-women-forwearing-heels-n2367422.

Bryson, Norman. 1980. 'The Gaze in the Expanded Field: Vision and Visuality'. In Hal Foster (ed.), *Vision and Visuality: DIA Art Foundation, Discussions in Contemporary Culture*, vol. 2. Los Angeles: University of California Press, 86–116.

Catalani, Anna. 2015. 'Fashionable Curiosities: Extreme Footwear as Wearable Fantasies'. *Fashion Theory*, 19(5), 565–82.

Collins, Lauren. 2011. 'Sole Mate'. *The New Yorker*, 28 March. https://www.newyorker.com/magazine/2011/03/28/sole-mate.

Cowell, Alan. 1983. 'Qaddafi Bodyguards in Patent-Leather High Heels'. *The New York Times*, 8 June. http://www.nytimes.com/1983/06/08/world/qaddafi-bodyguards-in-patent-leather-high-heels.html.

Daily Mail. ' "I'm shy and like to code": Yahoo boss Marissa Mayer poses for Vogue and insists she's "just geeky"...but her husband says otherwise'. *Daily Mail*, 19 August. http://www.dailymail.co.uk/news/article-2397426/Not-shy-thinks-Marissa-Mayer-sits-stylish-Vogue-interview-insists-shes-just-geeky-husband-says-otherwise.html#ixzz54RbGq5JD.

D'Angelo, Francesca. 2016. 'Standing Tall: The Stiletto Heel as Material Memory: A Contemporary Cross-Cultural Look at Perceptions of the Stiletto Heel'. PhD dissertation, York University, Toronto.

Debord, Guy. 2005 [1967]. *The Society of the Spectacle*. New York: Zone Books.

Deleuze, Gilles. 1991 [1967]. *Masochism: Coldness and Cruelty & Venus in Furs*. New York: Zone Books.

Dowd, Maureen. 2013. 'Pompom Girl for Feminism'. *The New York Times*, 23 February. http://www.nytimes.com/2013/02/24/opinion/sunday/dowd-pompom-girl-for-feminism.html.

Enstad, Nan. 1999. *Ladies of Labor, Girls of Adventure: Working Women, Popular Culture, and Labor Politics at the Turn of the Twentieth Century*. New York: Columbia University Press.

Friedman, Vanessa. 2017. 'Melania Trump, Off to Texas, Finds Herself on Thin Heels'. *The New York Times*, 29 November. https://www.nytimes.com/2017/08/29/fashion/melania-trump-hurricane-harvey-heels-texas.html?_r=0.

Heidegger, Martin. 1962 [1927]. *Being and Time*, trans. John Macquarrie and Edward Robinson. Oxford: Basil Blackwell.

Heidegger, Martin. 2002. *Off the Beaten Track*, ed. Julian Young and Kenneth Haynes. Cambridge: Cambridge University Press.

Holliday, Ruth. 1999. 'The Comfort of Identity'. *Sexualities*, 2(4), 475–91.

Ko, Dorothy. 2005. *Cinderella's Sisters: A Revisionist History of Foot Binding*. Berkeley: University of California Press.

Lerf, Matthias. 2015. 'Dinos mit High Heels'. *Berner Zeitung*, 30 July. https://www.bernerzeitung.ch/kultur/kino/Dinos-mit-High-Heels/story/15938876.

Linder, Marc. 1998. 'High Heels to Blame'. *The New York Times*, 8 March. http://www.nytimes.com/1998/03/08/style/l-high-heels-to-blame-908878.html.

Malkin, Bonnie. 2017. Daily Mail "Legs-it" front page criticised as "sexist, offensive and moronic"'. *The Guardian*, 28 March. https://www.theguardian.com/media/2017/mar/28/daily-mail-legs-it-front-page-sexist.

Manning, D. P. and C. Jones. 1995. 'High Heels and Polished Floors: The Ultimate Challenge in Research on Slip-Resistance'. *Safety Science*, 19(1), 19–29.

Medland, Dina. 2017. 'High Heels and Workplace Dress Codes: Urgent Action Needed, Say U.K. MPs'. *Forbes*, 25 January. https://www.forbes.com/sites/dinamedland/2017/01/25/high-heels-and-workplace-dress-codes-urgent-action-needed-say-u-k-mps/#4f1d74b47731.

Miller, Claire C. 2012. 'Techies Break a Fashion Taboo'. *New York Times*, 3 August. http://www.nytimes.com/2012/08/05/fashion/in-silicon-valley-showing-off-their-louboutins.html?pagewanted=all.

Moretti, Franco. 1985. 'The Comfort of Civilization'. *Representations*, 12, 115–39.

Morgan, Thais E. 1989. 'A Whip of One's Own: Dominatrix Pornography and the Construcxtion of a Post-Modern (Female) Subject'. *American Journal of Semiotics*, 6(4), 109–37.

Morris, Paul H., Jenny White, Edward R. Morrison, and Kayleigh Fisher. 2013. 'High Heels as Supernormal Stimuli: How Wearing High Heels Affects Judgment of Female Attractiveness'. *Evolution and Human Behavior*, 34, 176–81.

Mulvey, Laura. 1975. 'Visual Pleasure and Narrative Cinema'. *Screen*, 16(3), 6–18.

Murray, Timothy. 2009. 'Like a Prosthesis: Critical Performance à Digital Deleuze'. In Laura Cull (ed.), *Deleuze and Performance*. Edinburgh: Edinburgh University Press, 203–31.

Myers, K. 1982. 'Fashion 'n' Passion'. *Screen*, 3(4), 89–98.

Negra, Diane. 2014. 'Claiming Feminism: Commentary, Autobiography and Advice Literature for Women in the Recession'. *Journal of Gender Studies*, 23(3), 275–86.

Parmentier, Marie-Agnès. 2016. 'High Heels'. *Consumption Markets & Culture*, 19(6), 511–19.

Preciado, Paul B. 2003. *Kontrasexuelles Manifest*. Berlin: b_books.

Rabinowitz, Paula. 2001. 'Barbara Stanwyck's Anklet'. *Lectora*, 7, 53–79.

Richter, Valeria. 2015. 'As an amputee, I can't believe that I was stopped for wearing flats at Cannes'. *Independent*, 20 May. http://www.independent.co.uk/voices/comment/as-an-amputee-i-cant-believe-that-i-was-stopped-for-wearing-flats-at-cannes-10264889.html.

Rottenburg, Catherine. 2013. 'The Rise of Neoliberal Feminism'. *Cultural Studies*, 28(3), 418–37.

Sandberg, Sheryl. 2013. *Lean In: Women, Work, and the Will to Lead*. New York: Alfred Knopf.

Sebeok, Thomas A. 1989. 'Fetish'. *American Journal of Semiotics*, 6(4), 51–66.

Semmelhack, Elizabeth. 2008. *Heights of Fashion: A History of the Elevated Shoe*. Pittsburgh, PA: Periscope.

Semmelhack, Elizabeth. 2013. 'A Delicate Balance: Women, Work and High Heels'. *The New York Times*, 1 November. https://www.nytimes.com/roomfordebate/2013/11/01/giving-stilettos-the-business/a-delicate-balance-women-work-and-high-heels.

Sennett, Richard. 1994. *Flesh and Stone: The Body and the City in Western Civilization*. London: Faber and Faber.

Shade, Leslie R. 2014. 'Give us Bread, but Give us Roses: Gender and Labour in the Digital Economy'. *International Journal of Media & Cultural Politics*, 10(2), 129–144.

Swift, Jayne. 2012. 'Life's Too Short to Wear Comfortable Shoes: Femme-ininity and Sex Work'. In Shira Tarrant and Marjorie Jolles (eds), *Fashion Talks: Undressing the Power of Style*. New York: New York University Press, 99–117.

Thrift, Nigel. 2008. 'The Material Practices of Glamour'. *Journal of Cultural Economy*, 1(1), 9–23.

Time. 2013. 'Don't hate her because she's successful'. *Time*, 18 March. http://content.time.com/time/covers/0,16641,20130318,00.html.

Toffoletti, Kim. 2007. *Cyborgs and Barbie Dolls: Feminism, Popular Culture and the Posthuman Body*. London: I. B. Tauris.

Valenti, Jessica. 2011. 'SlutWalks and the Future of Feminism Online'. *Washington Post*, 3 June. http://www.washingtonpost.com/opinions/slutwalks-and-the-future-of-feminism/2011/06/01/AGjB9LIH_story.html.

Veblen, Thorstein. 2007 [1899]. *The Theory of the Leisure Class*, ed. M. Banta. Oxford: Oxford University Press.

Warner, Kara. 2015. '"Jurassic World" Star Bryce Dallas Howard Thinks Heelgate Was Feminist'. *Cosmopolitan*, 20 October. http://www.cosmopolitan.com/entertainment/movies/q-and-a/a47996/bryce-dallas-howard-jurassic-world-interview/.

Weisberg, Jacob. 2013. 'Yahoo's Marissa Mayer: Hail to the Chief'. *Vogue*, 16. August. https://www.vogue.com/article/hail-to-the-chief-yahoos-marissa-mayer.

Zollner, A. M., Pok, J. M., McWalter, E. J., Gold, G. E. & Kuhl, E. 2015. 'On high heels and short muscles: a multiscale model for sarcomere loss in the gastrocnemius muscle'. *Journal of Theoretical Biology*, 365: 301–10.

CHAPTER 23

···

INTERFACE

···

NISHANT SHAH

. . . is as interface does

I<small>N</small> the network societies that we live in, our lives are organized by and around electronic networks of information management and delivery. However, the network is a nebulous thing—impossible to reduce to visual representations (Chun, 2016), and when translated into maps, false (Lovink, 2011). The ineffability of the digital network demands a 'face', allowing it to become visible and tangible (Galloway, 2012). The interface, especially the Graphical User Interface, has become this 'face' of the digital, being variously posited as the site of human–machine encounter (Wood, 2012), as a space of human–machine transaction (Gil and Garcia, 2006), and as facilitator of transfer and circulation of resources (Becker et al., 2017). This is especially so in Human Computer Interaction and User Experience Design, in which fields the 'black mirrors' of our digital screens (Papagiannis, 2017) have emerged as the de facto reference point to understand how we live and cope with ubiquitous digitalization. Screen studies has evolved as an entire body of inquiry, trying to examine how human engagements with digital interfaces are shaping social behaviour, interpersonal relationships, and connected environments (Yue, 2003). In almost all this focus on the interface, the interface is positioned as a fiercely personal experience of the individual (Warwick, 2014), who is conceived as engaging with both customized algorithmic networks and the other nodes the interface finds a visual and human connection with. This is as true for the earliest studies of the WIMP visualization (that helped massification of the personal computer) as it is of the contemporary post-WIMP studies of on-screen driven cultures (Van Dam, 2012) where blinking devices continue to embody the connected lives we live.

The interface thus performs a dual job: it simultaneously seeks to describe the network while hiding its mechanisms and materialities. The interface is 'on' and 'off' at the same time. Nodes lurk, are activated through calculated randomness, making it difficult to touch or name a network, but always offering the network as existing in potential. We don't just live in interface realities but we use the interface to imagine, understand, organize, and govern our societies.

In this chapter, I argue the interface needs to be studied as performing dual organizational tasks that often go unacknowledged in inquiries of how the interface organizes our lives. Its first task is as a site of intimate connections between the individual device and the user, a one-to-one site that belies the larger connected networks that lurk behind. The second task is controlling the conditions of interaction and informationality, one which performs the job of visualizing complex systems and big data flows, but obfuscates the individual and the lived experiences of these networked mechanisms. Studying these two aspects of the interface simultaneously, as a site (noun) and controlling (a verb), allows us look at the individual and collective mechanisms of computational networks, and offers a more critical and political view of the lived human and machine experiences of networked societies.

THE INTERFACE AS A NOUN

In Human Computer Interaction studies, where the interface is probably the most fetishized space, giving form to the design impulses, solutionist imperatives, and governance ambitions that shape the system, the interface has long been established as the site where the human and the computational meet. The interface is positioned as the reified cybernetic space which gives meaning and completes the feedback loop between the computing system and the computed human being (Hookway, 2014). In this discourse, the interface serves the primary function of consolidation, allowing for an idealized, seamless interaction between the computational and human paradigms. The quest for human language computing, for gesture and haptic interaction models, for smart technologies that inform the Internet of Things, and for embedded and visible technological implants, all consider the interface as a benign, passive, and neutral site[1] where the user and the digital come together.

In this proposed notion of the interface as a passive noun, there is a naturalized separation of the user from computation. It produces two deterministic narratives of the human and the computational as discrete, separate, and coming in touch only when validated by the visual testimonies of the interface. The idea that the human and the computational only come in contact strategically and sporadically, produces a narrative that treats the digital technologies as instrumental, functional, and largely transactional. In this idea of the interface as noun, the user reigns supreme—able to activate the digital transaction for personal needs, able to disconnect from the digital network, and choosing to engage with the digital logics and logistics through personal agency. This idea of an empowered and agential human user—the user in action—enables a techno-utopic

---

[1] For example, in enterprise network systems, passive interface commands are introduced to 'stop both outgoing and incoming routing updates' which ensures that the networked routers cannot communicate with each other, and are thus not allowed to be 'neighbors any more' (Cisco, 2005).

narrative to emerge where any critical reflection on the invisible mechanisms and insidious practices of digital conditions becomes both unnecessary and impossible.

Take the metrification system of social media engagement as an example. On popular social media platforms like Facebook and Twitter, all the material that is produced is presented as fiercely personal and extremely customized. The algorithms that measure friendship and followers determine both the context within which the user generated content appears and also the content that the user would consume. However, this content is not merely within the context of friendship and sociality. It is continually tracked and measured by numbers that nudge us towards creating more actions and guiding our interactions with different content types. Facebook creates a complex metrification system where every interaction—human or algorithmic—is open to interaction, engagement, and measurement (Fuller, 2012). Every post you see, every page you visit, every person you encounter, every advertisement that is presented to you, is followed by a range of emotional, affective, numerical, and personal endorsements that make you believe that your conversations and engagements are purely personal.

So strong has been this idea of social media as personal, that for the longest time, Facebook's privacy policies were all geared at protecting your individual data from other predatory users, hackers, or sniffers who might be looking at creating identity theft. Facebook's help pages were also about users protecting who gets to see their posts in their extended networks, and how to deploy multiple filters so that the user feels in control of the information that they produce. It is only with the latest high profile controversies around Cambridge Analytica exploiting Facebook data to create user profiles for emotional and political manipulation, which reminded the users that the biggest predator of their private data is Facebook itself. The easy GUI, the metrics of emotional engagement, and the flatness of Facebook, which does not have any secondary layer in its user design, all engineer a false sense of safety and intimacy that blinds the users to the idea that Facebook algorithms are tracking, profiling, and shaping their behaviour in unexpected and opaque ways.[2]

Thus, the users, under the impression of personal ownership and engagement with the interfaces of their devices are not always aware of the ways in which data is leaked, information is intercepted, correlations are made, movements are tracked, and new cultures of surveillance and monitoring are legitimized by the digital infrastructure. This idea of the interface as the noun often puts into danger activists on the ground who use digital devices for coordination, organization, and communication under authoritarian and military regimes of oppression. The naturalization of the interface as neutral also produces a false sense of security for younger users who put themselves in precarious positions by sharing data and information, images and videos that could be used to bully, shame, and abuse them through vicious phenomena like slut shaming and doxing

[2] Ben Grosser's Facebook Demetricator project that adapts the platform's interface so that the numerical data that it foregrounds is stripped from the user experience is a remarkable study in how quantified interfaces shape and nudge user behaviour. The Facebook Demetricator can be found at https://bengrosser.com/projects/facebook-demetricator/.

(Shah, 2015). The benign interface often hides the malign intentions of those who remain invisible in the control of data, information, code, and protocols that remains both invisible and hidden in our interfaced interactions in everyday practice (Bunz and Meikle, 2017).

THE INTERFACE AS A VERB

It would seem that the solution to the presumed benign nature of the interface, its static neutrality as a noun, would be to look at the interface as a verb. The idea of delineating the different layers of operation that support the interface—Interaction Design scholar Steven Heim (2008) calls it 'the resonant interface'—making it transparent, and showing the various processes of data and information consolidation and circulation, is well intentioned in its impulse. This active state of the interface almost as an action, as an active state of negotiation which allows us to perform the various tasks of aggregation, consolidation, intersection, connection, and collection of data makes it more accessible in understanding the logics and logistics of the network. The interface as a digital verb can be understood as performing systems of fragmentation, distribution, and paradigmatic precariousness of computation as a democratic space of negotiations.

This idea of the interface as active reinforces the idea that the haptic, cybernetic, reciprocal loop of interactivity between the human user and the technological system is a closed loop. However, even when the interface is described as an action and actively performing different techno-human tasks, the narrative of responsibility and the onus of culpability still lie with the human subject. In Computer Science, for instance, two of the most popular and foundational acronyms that are taught in early years are GIGO and WYSIWYG.

GIGO stands for 'Garbage In, Garbage Out'. For Computer Science students, it is drilled down that an error in the final outputs of computational processing is always human error (Van Riper, 2002). If a computational system fails to deliver the intentions of a program, and is unable to execute the code that has been written, it necessarily means that the code writer or the programmer has introduced errors in it. The machine is never wrong, and the computer cannot make a mistake. If at the end of the program, what you get is garbage, it is because your input had garbage in it, making the code impossible to render. It is possible to reverse engineer code, to do forensic investigations into the errors that might have been committed—often-called bug fixing—and making the entire process of the various application layers of the program transparent. However, in this bug fixing and error correction, there is never the questioning of the hardware and the protocols of network logic that might be equally responsible for the problem. When the interface becomes active, the presumed neutrality is pushed further down into the system, so that the hardware is presumed to be benign, and scanning of the possible backdoors, vulnerabilities, and designed flaws that are maintained in the hardware is not a common thing. Even though it has now been shown over and over again that machines introduce errors, and more importantly, that larger institutions which regulate and control machines can train them to amplify errors and reinforce them, the first suspect in times of a stack overflow or a system crash is always the user.

WYSIWYG, an acronym that declares that with computing, 'What You See Is What You Get', offers similar reassurances about the infallibility of the computational machine (Murray, 2011). WYSIWYG, especially with the rise of the GUI has been established as a way of championing the values of transparency and visibility that interfaces offer. The doctrine confirms that there is a direct, unmediated, and transparent correlation between the input data and the received results. The various layers of computation like machine translation, processing, error correction, database rendering, parsing, and visualizing are all made invisible as we put our faith in the device as essentially benign; merely a conduit for the transfer of human information. This is further amplified by the making invisible of various layers of interception, data capture, storage, sharing and circulation that remain unacknowledged in the fetishization of the interface. As a verb, the interface continues to divert attention, create protocols of traffic that can be visualized, and determines the directionality of our informational transactions while continually hiding the complex, and often exploitative conditions that our digital devices put us into.

The interface, by producing a huge amount of interaction possibilities at the level of the surface, produces a condition where the human user never gets to own the logics and logistics of the digital network, but only gets reduced to being a user. Wendy Chun argues that this naturalization of the human as the user converts the digital networks into 'opaque metaphors' (2011). The transparency of personal engagement on the interface, according to Chun, belies the opaqueness of the transactions and quantifications of the digital networks. Chun declares that 'the more our devices become transparent, the more they become opaque', and this condition of opacity is about hiding the processes of alienation and data extraction that remain outside the scope of the attention and screen time of the visual interface.

THE INTERFACE AS A PROCESS

The seeming duality of the interface and the way these are often treated as either-or, stem from the fact that the interface is generally presented to us a thing. Be it the hardware and material object of computational processing or the software and the fluid spaces of engagement, the interface is always considered as static, and as something that can be easily pinned down and understood. This idea of the interface as thing leads to the narrative that the interface organizes human intention, action, thought, and transactions by giving it visual order and designing spaces where the encounters for all of these can happen. The interface, in computational theory, is referred to as a compiler or consolidator. In more popular imaginations the interface is seen as a translator or equipped to render and visualize without any additional input or interference. The interface then is seen as the representation of a system, an extension of the human user, a visualization of complex processes which can be reduced to simple actions and user stories, and a site of correlation and causal encounters to emerge.

In all of these descriptions of the interface—conceptual, theoretical, or popular—the interface is seen as a prosthesis and not as an organizing force. It is seen as a vehicle of presentation or edge of transfer but not as defining or creating the logics, logistics, and mechanics of computational networks and user engagement. In order to move away from this benign idea of the interface as thing, I would propose that we look at the interface as a process. As a digital process, the interface is no longer studied for itself, and its appearance is more symptomatic of a variety of other transactions and uses which need to be analysed. Especially for thinking through organization, the interface as a process gives us three distinct entry points into understanding how the presence of the interface leads to the construction of new techno-social practices, identities, and politics in digitally networked societies.

The first sites of inquiry are the thresholds of access and information transfer. The interface does not just visualize but also determines the different thresholds that information has to pass through, in order to reach the final visual destination. It is important to remember that even the most cursory layers of software, hardware, application layer, and information management systems all work through prescribed protocols and determined algorithms in order for the interface to come to life. The interface does not just show informational engagements, it also hides those which might be corrupt, disruptive, or unassimilable in the logic of the interface. From voices and expressions to people and communities, structures, objects, and people that do not make themselves legible to the logic of these different layers and do not have the currencies (social, political, economic, or digital) to cross the thresholds are often made invisible through the capacities of the processes that activate the interface. Non-standard languages, modes of expression that might override terms of services, dark web networks that remain in the shadow lands of digital anonymity are overridden as illegible and hence not allowed to cross the thresholds of digital information circulation. The interface, in its attempts at transparency, even visualizes the death of this information, or the ban of access through cute and playful means—like the 404 Error page or the Twitter Whale—thus exercising the organization of content in public though still opaque ways.

The second site of inquiry is to look at the intentionality of the interface. The interface as a process makes it clear that the interface is not just a destination, it is a gatekeeper and it has clear intentions of what it allows through, and what remains hidden. Recognizing the GUI is not just a medium of making all things visual, but that it hides, manipulates, and restricts certain kinds of processes and transactions, is important. Thus, when we look at the individual and personal user–GUI relationship, we need to be aware that the larger network mechanisms are being hidden from that particular instance. Also, the intentions of the same GUI can be different. Thus, even as the network mechanisms can be hidden from the user, the same mechanisms might be shown to a system administrator or a server manager, whose levels of access clearance and privilege in roles would produce a different dashboard of information. Understanding the GUI as the end point of multiple intentions helps map out the different profiles, diverse inequity in access to information, and the various ways in which the same information gets circulated and visualized for different kinds of user profiles.

The third site of inquiry is to position the interface as a measurement. Digital processes are essentially measures of the validity, veracity, velocity, and volume of the information that is being transferred through the system. Positioning the interface as a part of this system of measurement immediately brings to the fore questions of politics, of authority, and of control. The interface then is not just a pre-programmed structure that shows and hides particular data. It is, in fact, a system by which various practices are rewarded when they measure up to the expectations of the interface design. Thus, conditions like mindjacking, continued engagement, push and pull based notifications, systems of garnering attention and keeping the user in a state of distraction economies become critical question to ask of the interface. This approach also questions the taken-for-granted nature of the haptic, gestural, visual, and affective designs of interfaces and helps unravel who profits from these particular designs and forms that have been massified through the standardization of the GUI.

As we move from the interface as thing to interface as process, it becomes a way by which we reconcile its noun and verb forms, bringing together a complex narrative of identities, practices, and politics that expand the scope of interface studies. It establishes the interface not as a description of the organization principles but an active process that intervenes, interferes, and invents new forms of organization. At a personal level this might be in the ergonomics of human form as we develop twitter thumbs and hunched shoulders. At a collective level, this might be in the neighbourhoods of proximity and echo chambers of intolerance that we unwittingly find ourselves occupying. At the level of social organization this identifies the interface not merely as a screen or a site of encounter but as a process of organization that selectively hides and reveals various layers of computation to various human and non-human actors. At the level of political organization, this opens up the interface as betraying the rights and entitlements of the user as well as supporting the insidious intrusion of surveillance actors that exploit user data and shape new collective imaginaries. The interface as a process stops the fetishization of organizing around the interface and instead charts out the ways in which the interface organizes, beyond and around its appearance, a series of processes that need careful planning, design, questioning, and critique. The interface, as a process, is a thing in the making, thus not taken as a naturalized default but as an iterative space that can be variously owned, publicly distributed and can also be deployed towards the counter-cultural alternatives and resistances that mark the landscape of digital cultures. The interface as process gives us new organizational principles and also the promise of recalibration of structures and powers to question the future scripts of the digital worlds we want to live in.

REFERENCES

Becker, Valdecir, Daniel Gambaro, and Thais S. Ramos. 2017. 'Audiovisual Design and the Convergence between HCI and Audience Studies'. In Masaaki Kurosu (ed.), *Human–Computer Interaction: User Interface Design, Development and Multimodality*. Cham: Springer, 3–22.

Bunz, Mercedez and Graham Meikle. 2017. *The Internet of Things*. Cambridge: Polity Press.

Chun, Wendy H. K. 2011. *Programmed Visions: Software and Memory*. Cambridge, MA: MIT Press.

Chun, Wendy H. K. 2016. *Updating to Remain the Same: Habitual New Media*. Cambridge, MA: MIT Press.

Cisco. 2005. 'How does the passive interface feature work in EIGRP?' https://www.cisco.com/c/en/us/support/docs/ip/enhanced-interior-gateway-routing-protocol-eigrp/13675-16.html [Accessed 10 May 2018].

Fuller, Matthew. 2012. 'Don't Give me the Numbers—an interview with Ben Grosser about Facebook Demetrication'. *Rhizome*. https://www.cisco.com/c/en/us/support/docs/ip/enhanced-interior-gateway-routing-protocol-eigrp/13675-16.html [Accessed 14 May 2018].

Galloway, R. Alexander. 2012. *The Interface Effect*. Cambridge: Polity Press.

Gil, A. B. and F. J. Garcia. 2006. 'Recommender Systems in E-Commerce'. In Claude Ghaoui (ed.), *Encyclopaedia of Human–Computer Interaction*. London: Idea Group Reference.

Heim, G. Steven. 2008. *The Resonant Interface: HCI Foundations for Interaction Design*. Boston: Pearson/Addison Wesley.

Hookway, Branden. 2014. *Interface*. Cambridge, MA: MIT Press.

Lovink, Geert. 2011. *Networks Without a Cause: A Critique of Social Media*. Cambridge: Polity Press.

Murray, J. Katherine. 2011. 'WYSIWYG, Recognition Memory & Computing Machines'. Chapter 4 of 'Breaking Boundaries: A Study of Human-Mobile Interaction'. Dissertation submitted to Stanford University.

Papagiannis, Helen. 2017. *Augmented Human: How Technology is Shaping the New Reality*. Sebastopol, CA: O'Reilly Books.

Shah, Nishant. 2015. 'Sluts "r" Us: Intersections of Gender, Protocol and Agency in the Digital Age'. *First Monday*, 20(4). http://firstmonday.org/ojs/index.php/fm/article/view/5463 [Accessed 14 May 2018].

Van Dam, Andries. 2012. 'Post-Wimp User Interfaces: The Human Connection'. In Rae Earnshaw, Richard Guedj, Andries van Dam and John Vince (eds), *Frontiers of Human-Centred Computing, Online Communities and Virtual Environments*. London: Springer, 163–78.

Van Riper, A. Bowdoin. 2002. *Science in Popular Culture: A Reference Guide*. London: Greenwood Press.

Warwick, Kevin. 2014. 'A Tour of Some Brain/Neuronal–Computer Interfaces'. In Gerd Grübler and Elisabeth Hildt (eds), *Brain–Computer Interfaces in their Ethical, Social, and Cultural Contexts*. New York: Springer, 131–46.

Wood, Aylish. 2012. *Digital Encounters*. New York: Routledge.

Yue, Audrey. 2003. 'Paging "New Asia": Sambal is a Feedback Loop, Coconut is a Code, Rice is a System'. In Chris Berry, Fran Martin, and Audrey Yue (eds), *Mobile Cultures: New Media in Queer Asia*. Durham, NC: Duke University Press, 245–66.

CHAPTER 24

....

MIND TRACKER

....

ALEKSANDRA PRZEGALINSKA

FIGURE 24.1 Mind tracker

(Photo: Albert Zawada / Agencja Gazeta)

USING tracking methodologies has a long tradition. Personal informatics that we know today started with so-called life-logging—tracking data generated by various behavioural activities in the late 1980s (Calvo and Peters, 2014). Nonetheless, already way before that there existed analogue 'wearable' technologies such as the abacus ring, a seventeenth-century Chinese implement that allowed bean counters to perform mathematical tasks by moving tiny beads along nine rows (da Costa and de Sá-Soares, 2016). Moreover, in the 1950s, long before the development of the Internet of Things Edward O. Thorpe and Claude Shannon created a smart shoe that could fairly accurately predict where the ball would land on a roulette table. These devices were, however, not widespread and their availability was limited to experimentation (Guizzo, 2003).

Since we are going to focus here on a particular kind of tracking device, namely those that track activities of the mind, it is also important to mention that the field of brain-tracking (or, more precisely, neuroimaging) and its diverse applications has a long history behind it. Richard Canton's discovery in 1875 of electrical signals in animal brains initiated this technique (Teplan, 2002). As one of the first common uses of brain computer interface technology, EEG neurofeedback has been in use for several decades. In the 1960s and 1970s various advancements were made and the explorations in neuroimaging were enriched by a context of expanding the potential of the mind through insight. Mass production of the microchip in the 1980s brought with it the potential to create smaller and lighter computers than ever before. Steve Mann, a researcher and inventor specializing in electronic photography, began his EyeTap project in the early 1980s. At the time it was clunky and made the wearer look like a motorcyclist who had experienced a pretty unfortunate event with a television, but the idea itself was astounding. Over the past few decades Mann has gradually whittled the device down to a sleek and simple headset which looks rather a lot like the Google Glass. The explosion of portable computing in the early 1990s resulted in lots of stabs at wearables with varying degrees of success. Key to this early wave of wearables was Reflection Technology's Private Eye (Mann, 2000), a head-mounted display which used a vibrating mirror to create a display directly in the wearer's field of vision. The year 1998 marked a significant development in the field of brain mapping when researcher Philip Kennedy implanted the first brain computer interface object into a human being. John Donoghue and his team of Brown University researchers formed a public traded company, Cyberkinetics, in 2001. The goal was to commercially design a brain computer interface, the so-called BrainGate. The company has come up with NeuroPort (Normann, 2007), its first fully commercial product.

Nonetheless, before the current sensor revolution the cost and expertise needed for working with large-scale datasets and visualizations limited access to most kinds of trackers mainly to professionals and scientists. Tracking first occupied the health sector and then became visible in wellness and recreational sports activities. With biomarker testing, health metric tracking was traditionally an expensive one-on-one process ordered by physicians for patients in response to specific medical risks (Swan, 2009). However, the proliferation of mobile digital devices has seen life-logging tools break out of research labs and move to the hands of the masses. This is related to the fact that the costs of these technologies have decreased significantly. Furthermore, improvements in tools have made data collection and manipulation more available to the individual.

Monitoring, measuring, and recording elements of one's body and life as a form of self-improvement or self-reflection have been discussed since ancient times. However, the arrival of digital technologies, particularly biosensing ones, has created a new space for communities organized around self-tracking activities (Swan, 2012). One should mention at least two such communities formed in the second half of 2008 to explore, brainstorm, and share their self-tracking experiences: Quantified Self, established in the San Francisco area, and HomeCampInt, from London (2015). Soon, other groups like PatientsLikeMe or DIYgenomics followed. These communities coined and developed the idea of 'n = we' understood as experimenting with tracking devices on the individual level and coming together in collaborative health-oriented communities that made their n = 1 discoveries less anomalous, and statistically significant.

These groups advocated data-sharing, and a more proactive health self-management and responsibility-taking that could be performed in a playful style. At the beginning they were mainly making use of very simple trackers that measured calories or steps taken, but over the years started demanding devices that would measure more intangible states and activities. Thus, increasing numbers of mind-related wearable devices have become commercially available. Some of these devices are just passively monitoring user's activity, whereas others actually influence alterations of physical cognitive states (Calvo and Peters 2014; Kopeć et al., 2015; Mazurek and Tkaczyk, 2016). Among currently accessible trackers, most popular are those that allow users to foster increased innovative thinking, creativity, or work effectiveness, and/or reduce anxiety and stress. These technologies seek to foster human well-being, and aspire to be classified as 'positive computing' devices. Clearly, on a more structural level, both creativity and stress are considered conditions that can be introspected, managed, and altered through technology. The ultimate goal of self-tracking communities is to smoothly integrate technology with the human body and make it a part of people's daily life. Within this context the human body becomes the central element of Human–Computer Interaction through moving away from the desktop-based interaction towards mobile and wearable applications (Bordegoni et al., 2012). An underlying assumption here is that data is an objective resource that can quickly bring visibility and information to a situation, and that psychologically it should entail an element of empowerment, control, and fun. The goal is not only to gain access to data, but also to build a motivational system that helps with removing harmful habits from daily routines. We should note that collaborative wearables may introduce control where consolidated communities like Weight Watchers emerge. Their members may rely more on the feedback their trackers provide than on doctors' and dieticians' advice. This is an emerging phenomenon, in which computers are perceived as rational, unbiased, and fully objective, whereas humans are perceived by other humans as failing to provide objective judgement and fallible. We can call this process the deprofessionalization of those disposing expertise in a given field. The expertise of trainers, diet specialists, even doctors is currently questioned by virtual support communities of non-experts and their trackers. This new digitized trust consisting of familiarity built over time with the online community and 'real', data-driven knowledge represented by the device can sometimes be deceitful and potentially harmful.

Among current biosensors there is a particular class that attempts to encode something that for a long time was not transparent outside the laboratory context and intangible for individual users. Mind-tracking devices or mind trackers allow for monitoring brain activity and affective states as well as influence them with biofeedback properties (Figure 24.1). Mind trackers should be associated with affective computing: bridging emotions and computers. And even though the term 'affective computing' was coined already in 1995 by Rosalind Picard (Picard, 1995; Nissan, 1999; Picard and Wolf, 2015), it had to wait two decades for material manifestations. Today we know that the machine can measure non-verbal communicative traits and affective/mental states (which could be considered rough approximations of emotions) in different ways: studies of physiological parameters, using text analysis, observing behavioural patterns, facial expressions, and recognizing emotions in voice or body posture of users. Concerned by the capacity for new technologies, such as constant notification systems, to produce cognitive overload, for instance by notification systems, such as overstimulation, a group of researchers turned to the development of what they call 'attentive user interfaces' (Vertegaal, 2003). Particularly precise measurement of facial expressions is gaining popularity, because results of this type of tracking can be sold to marketing departments and advertising content producers. By tracking the user's gaze an interface can adapt highlighting urgent issues while backgrounding those of less importance. Further on, new applications in affective computing for reflection and for mental health appeared. The era of personalized 'mind tracking' has long been pronounced upon and it is just beginning to take shape. It is true that users of self-tracking devices and applications still mainly track and support their physical rather than mental activities, but this can change in the future, as they move on to more sophisticated devices. On the market, there are a lot of devices designed to gather physiological data, while there is a scarcity of tools and methods to gather and sensibly analyse psychophysical or mental data. It is clear that a growing number of self-tracking individual consumers express vivid interest in tracking more refined aspects of their overall state than calories or steps. Significant numbers of users express their interest in tracking mood, memory, or affects. Several users who were previously tracking medical or health-related parameters decided they needed more data on their mental states and affects. Thus, one could carefully assume that the need for quantification scales up from measurement of simple processes to more sophisticated ones. Also, technology seems to have rapidly adapted to this change and accelerated it, too.

This is where the current mind trackers step in. Nowadays, many portable EEG devices are consumer-grade, low-cost devices that are targeted for lifestyle applications. Vendors like InterAxon, NeuroSky, or Emotiv (Güneysu and Akin, 2013; Tabakcıoğlu et al., 2016) and Versus provide some of these off-the-shelf, fairly inexpensive devices for consumers. The number of electrodes on these devices is limited (i.e., 2–14 electrodes) compared to the clinical grade devices (i.e., 16–32), their resolution is lower, and the electrodes are usually focused on a specific portion of the brain. Many of these portable EEG devices are, on the other hand, far simpler to set up; they connect via Bluetooth to a smartphone, a computer, or a microcontroller, where data can be analysed directly. Dry electrodes used in most of these devices do not require a time-consuming, cumbersome preparation process. These portable, cost-effective devices may also be used with related available EEG tools and open-source platforms (such as BCI2000, OpenEEG, or MuLES). These changes have helped evolve EEG applications in both novel and established fields. More importantly, however, they allowed consumers lacking professional expertise to explore and use devices previously reserved solely for medical or scientific purposes. They could access their brainwave patterns, 'hear' their brain through biofeedback, and try to assess their degrees of focus and stress.

These products also often rebrand EEG data with the simpler, easily understood term 'neurofeedback', understood as a type of biofeedback that uses real-time displays of brain activity—most commonly EEG—to teach self-regulation of brain function. It is clear that from self-tracking of simple and easily quantifiable activities we are moving to more collaborative and sophisticated, even though scientifically unsatisfactory forms of tracking. Trackers such as Muse enter a different level of interaction with users and bring about profound changes in how the role of a tracker is perceived. They are more personalized and usable anytime and anywhere. Despite the fact that they provide feedback on very complex activities, they are also highly portable.

Mind trackers that aspire to foster human physical and mental potential are often called 'positive computing' devices (Calvo and Peters, 2014). And mind trackers indeed bear the claim to be able to provide information such as one's state of mind, sleep pattern, emotional spectrum, and the strength of the emotions. Furthermore, their producers declare the devices can assist in improving focus, reduce stress, and increase attention. In the past, tracking has been focused on looking at exceptions and variations of parameters (such as detecting diseases). Currently, it focuses more on improving achievement as well as higher level needs such as reaching work–life balance and emotional stability. There is a clear demand for precise and context-aware data responding to complex processes tracked. Also, new reasons for tracking become more and more significant. Among the most prominent ones one could enumerate increasing work effectiveness combined with better work–life balance and the urge for a sort of 'posthumanist' technology-enabled transgression where current physical and mental capacities are significantly enhanced.

This brings us to the issues of reliability of mind trackers which actually can be questioned. Muscle movement and other electronic devices in the vicinity also produce signals that are very similar to EEG and therefore may negatively affect the measurement of actual brain waves. The inability of mind trackers to cancel out such 'noise' may generate less reliable results providing rather a mere indication of our brain activity. This reliability could be improved by including more sensors in order to get a better signal; however that increases costs, weight, and design of the product making it less attractive for consumers to purchase (Chuah et al., 2016). What is more, the devices currently on the market are still struggling with devising proper and individualized algorithms that would be able to produce reasonable output based on psychophysiological data.

The reliability of devices and the accuracy of data provided on such refined activity as affective states are an important problem when one considers the level of trust that is allocated in them. Trust used be attributed to relationships between people and it can be demonstrated that humans have a natural disposition to trust and to judge trustworthiness that can be traced to the neurobiological structure and activity of the human brain. However, now one of the key current challenges in the social sciences is to rethink how the rapid progress of wearable technologies impacts constructs such as trust. This is generally true for all information technology that dramatically alters causation in social systems, and specifically for wearables that enter strong and complex relationships with their users. The attribution of trust is a matter of dispute. We do observe, however, that users tend to trust the data and follow instructions provided in apps attached to wearable gadgets, treating them frequently as experts in the field of wellness, while at the same time they are afraid that their privacy will be breached through uncontrollable circulation of sensitive data. It is interesting to look at this phenomenon through Luhmann's classic work on trust and distrust (*Mißtrauen*). According to Luhmann (1968), distrust is not just a sole opposite of trust, but a functional equivalent. For Luhmann, trust's and distrust's main social function is the reduction of social complexity. With trust, this is done through positive expectations. With distrust, however, this reduction is done using negative strategies (such as defining the other as the enemy, building up emergency reserves, attacking, etc.). What is more, reductions of complexity/uncertainty are generally more complex when based on negative expectations than on positive expectations. Trusting the device as your doctor or expert that will manage your health and wellbeing properly can be easily achieved. Whenever the user gets a signal that the strategy optimized for the user by the device is working, he or she may think that the device knows him/her well and through its objectivity (and non-humanity) can actually find the best solutions. However, when it comes to data collection, it is not necessarily the device or its producer that should be distrusted. It is perhaps the community or other companies that are trying to obtain that data.

Quite frequently, it is inner motivation that drives self-tracking, but in many cases it is the need to follow others. As mentioned before, in several contexts people are encouraged, 'nudged', obliged, or even, in some cases, coerced into using digital devices to produce personal data which is then used by others. In a perplexing manner, self-tracking practices do resemble some form of post-Taylorism where self-optimization practices intersect with coercive efficiency improvement discourses. In the light of such reflections one can put forward the thesis that future, advanced versions of mind trackers will become highly personalized self-management devices and consequently producers of Self—the kind of Self that is striving (or compelled to strive) for a precisely measurable perfection, defined by quantitative body and mind data. Obviously, self-management through tracking can take many different shapes. As Deborah Lupton (2016) notices, there have been quite a few modes of self-tracking that have emerged recently: private, communal, pushed, imposed, and exploited. On a more abstract level, one could try to distinguish what self-management means in the context of using wearable devices, both active and passive ones. Self-management can be understood in diverse ways. It can be understood, on an private level, as self-care: individual control of health care. However, it can also be understood—in a professional context—as a form of organizational management based on self-directed work processes. In this latter sense, in the context of wearable technologies, it would indicate an emancipatory dimension of self-tracking as maximizing potential and being in control of one's achievement as well as autonomy in setting out productivity goals. However, self-management of that kind is not necessarily self-imposed and this is where several problems emerge. It is clear that from self-tracking of simple and easily quantifiable activities we are moving to more collaborative and sophisticated, even though scientifically unsatisfactory and somewhat deceiving forms of tracking. It is clear that on the individual level self-tracking (if not becoming addictive) frequently becomes either boring or frustrating over time (Nafus and Tracey, 2002; Nafus and Sherman, 2014; Lupton, 2016). Being part of something larger than oneself can surely have an empowering and motivating effect. Exploring the potential of highly personalized, open to collaboration data has, however, several problematic aspects. Self-tracking is not an individual action even though its name may imply that. It may take the form of collaborative tracking and discovering, but it may just as well take the form of 'other tracking', exercising control and surveillance. The change towards even more meticulous and personalized tracking, combined with efforts to create consolidated tracking communities is of crucial importance, as tracking of parameters that correlate best with various processes, and the evolution of context-aware systems, can bring about a profound change in our understanding of collaboration: either empowering or disempowering the community and either redefining top-down solutions of modern corporations and institutions or consolidating them. The producers know it, too, and this is why their efforts are to bring trackers into wellness programmes of organizations and corporations and make them become transparent companions of everyone's routines. This is why, despite potentially promising collaborative dimension of tracking, there is a concern that identification of the boundaries between a self-/community-driven collaboration and abject commercial exploitation becomes harder and harder. Perhaps such a polarization is artificial and counterproductive as the communality of tracking does not have to be compromised or destroyed by its commercial dimension, but bearing in mind the sensitivity of data and the particular organizing power of these devices it is legitimate to ask questions about communities in the context of a possible social contract here. Who is legitimate to view others data, whose data is transparent and whose not? What rules are governing the relationship between the technological system, the community that uses it, and the individual that feeds it with data and is supposed to benefit from it somehow?

What is more, self-tracking movements can promote addiction to technological gadgets and data. Various measurements of this kind feed into circuits of reproduction, making performances visible and thus reproducible. Users track their own 'progress', and the 'progress' of other users and compete against each other for better results that they can publicly display. This, in itself, is not bad. But at the Quantified Self meet-ups various q-sers presented themselves at meet-ups as 'downgraded' or 'mediocre' without a tracker, experienced depression whenever a tracker was not displaying their progress, and felt the urge to migrate to ever newer tracking software and hardware pieces without any clear purpose. Also, they perceived other members of self-tracking communities as elite clubs of people who have better access to themselves and generally are better informed than others. This, again, links with the problem of deprofessionalizing experts in the field.

It seems that with the advent of the Internet of Things and ubiquitous technologies, problems that humanistic management was addressing have become even more complex. The current state of technological development does not clarify what will be the next stage of affairs and what sort of use we will make of those technologies that are either replacing people or opening up a new, radically deeper level of machine–human interaction and interdependency. At the same time, critical and humanistic management approaches (Alvesson and Willmott, 2002) that claim to strongly oppose any form of Taylorism claim that management as a practical science should offer people assistance in how to live better, and how to achieve social progress understood as increased welfare and general well-being. It requires further diligent research to understand what will be the role of wearable tracking technologies in this dynamically changing landscape. What we already know is that the shift towards mind tracking is of crucial importance Of course, one can rightfully ask whether that shift has a qualitative character. As mentioned at the beginning, both problems—self-quantification and accessing the mind—have been prominent themes of scientific and non-scientific practices for several centuries. Experiments related to these fields date back to the late nineteenth century, especially in the 'Würzburg School' (Ogden, 1951; Hoffmann et al., 1996; Kusch, 2001) of experimental psychology. Wilhelm Wundt was trying to create a methodology bridging inner states of mind with external, objectivized data (*Bewußtseinszustände*). Thus, the status of these devices as truth-bearers and truth-revealers is not relevant even if the new, more accessible and more accurate devices claim to give full access to the mind and reveal its secrets (the ancient concept of truth as aletheia—uncovering reality). This important shift happens elsewhere. People do perform facial expressions in order to be recognized as experiencing a certain affect. Trackers work by elevating the hitherto invisible rhythms of the body to the status of quantified patterns. Their users want to see themselves in the devices and are seeking hidden patterns formed from the collected data that could lead to novel insights about themselves.

What is more, tracking of parameters that correlate with various mental processes, and the evolution of context-aware systems, can bring about a profound change in our understanding of productivity, creativity, or expertise. Trackers in general also work by managing us. Mind trackers in particular manage (or aspire to manage) the complexity of our mental activities by way of organizing thoughts through patterns (which is in a way a thoughtless way of organizing thoughts). The direction of this thought management is still unknown, and we also know very little about the 'bigger picture' that is supposed to emerge as its product.

To sum up, quite frequently, researchers and users tend to think about self-tracking as either good, empowering, and bringing people closer to work on resolving issues that they share or as essentially evil, highly addictive, and enslaving. Frankly speaking, this is a kind of dichotomy that relies on subjective judgement and is thus unsolvable in any near future. There is a meta-level question related to trackers, though. Future, advanced versions of trackers may become highly personalized self-management devices. Trackers and tracking apps essentially are, like many other technologies, producers of the Self. Trackers shape our image of ourselves and others and thus they change us. The kind of Self they shape is one that is striving (or is compelled to strive) for a precisely measurable perfection. What is more, trackers may become even more pervasive producers of Self than many other devices, because of the nature of their relationships with their user. It is a close relation, built on trust in the device. The device becomes an important channel of contacting both oneself (displaying data that we are unaware of) and others (data sharing and exchange).

REFERENCES

Alvesson, Mats and Hugh Willmott. 2002. 'Identity Regulation as Organizational Control: Producing the Appropriate Individual'. *Journal of Management Studies*, 39(5), 619–44.

Bordegoni, Monica, Secil Ugur, and Marina Carulli. 2012. 'When Technology Has Invisible Hands: Designing Wearable Technologies for Haptic Communication of Emotions'. *Proceedings of the ASME 2012 International Design Engineering Technical Conferences and Computers and Information in Engineering Conference*. American Society of Mechanical Engineers, 581–9.

Calvo, Rafael A. and Dorian Peters. 2014. *Positive Computing: Technology for Wellbeing and Human Potential*. Cambridge, MA: MIT Press.

Chuah, Stephanie H. W., Philipp A. Rauschnabel, Nina Krey, Bang Nguyen, Thurasamy Ramayaha, and Shwetak Lade. 2016. 'Wearable Technologies: The Role of Usefulness and Visibility in Smartwatch Adoption'. *Computers in Human Behavior*, 65, 276–84.

da Costa, Filipe and Filipe de Sá-Soares. 2016. 'Authenticity Challenges of Wearable Technologies'. In Andrew Marrington, Don Kerr, and John Gammack (eds), *Managing Security Issues and the Hidden Dangers of Wearable Technologies*. Hershey, PA: IGI Global, 98–130.

Guizzo, E. M. 2003. 'The Essential Message: Claude Shannon and the Making of Information Theory'. MS thesis, Massachusetts Institute of Technology. https://dspace.mit.edu/handle/1721.1/39429?show=full [Accessed 7 March 2018].

Güneysu, Arzu and H. Levent Akin. 2013. 'An SSVEP Based BCI to Control a Humanoid Robot by Using Portable EEG Device'. *Proceedings of the 35th Annual International Conference of the IEEE Engineering in Medicine and Biology Society (EMBC)*, 6905–8.

Hoffmann, J., A. Stock, and R. Deutsch. 1996. 'The Würzburg School'. In J. Hoffmann and A. Sebald (eds), *Cognitive Psychology in Europe: Proceedings of the Ninth Conference of the European Society for Cognitive Psychology*. Lengerich: Pabst Science Publishers, 147–72.

Kopeć, J., K. Pacewicz, A. Przegalińska, M. Smoleń, J. Wencel, M. Kominiarczuk, and S. Wróbel. 2015. 'Gamification: Critical Approaches'. Manuscript, Faculty of 'Artes Liberales', University of Warsaw. https://depot.ceon.pl/handle/123456789/8013.

Kusch, Martin. 2001. 'The Politics of Thought: A Social History of the Debate between Wundt and the Würzburg School'. In Liliana Albertazzi (ed.), *The Dawn of Cognitive Science: Early European Contributors*. Dordrecht: Springer, 61–88.

Luhmann, Niklas. 1968. *Vertrauen: ein Mechanismus der reduktion sozialer Komplaxitat*. Stuttgart: Ferdinand Enke Verlag.

Lupton, Deborah. 2016. *The Quantified Self*. Cambridge: Polity Press.

Mann, W. Steve G. 2000. 'Comparametric Transforms for Transmitting Eye Tap Video with Picture Transfer Protocol (PTP)'. In K. Rao and P. Yip (eds), *The Transform and Data Compression Handbook*. Electrical Engineering & Applied Signal Processing Series, vol. 1. Boca Raton, FL: CRC Press, 79–116.

Mazurek, Grzegorz and Jolanta Tkaczyk (eds). 2016. *The Impact of the Digital World on Management and Marketing*. Warsaw: Poltext.

Nafus, Dawn and Jamie Sherman. 2014. 'Big Data, Big Questions—This One Does Not Go Up To 11: The Quantified Self Movement as an Alternative Big Data Practice'. International Journal of Communication Systems, 8, 1784–94.

Nafus, Dawn and Karina Tracey. 2002. 'Mobile Phone Consumption and Concepts of Personhood'. In James E. Katz and Mark Aakhus (eds), *Perpetual Contact: Mobile Communication, Private Talk, Public Performance*. Cambridge: Cambridge University Press, 206–22.

Nissan, Ephraim. 1999. 'Review of Rosalind W. Picard, *Affective Computing*'. *Pragmatics & Cognition*, 7(1), 226–39.

Normann, Richard A. 2007. 'Technology Insight: Future Neuroprosthetic Therapies for Disorders of the Nervous System'. *Nature Clinical Practice: Neurology*, 3(8), 444–52.

Ogden, R. M. 1951. 'Oswald Külpe and the Würzburg School'. *American Journal of Psychology*, 64(1), 4–19.

Picard, Rosalind W. 1995. *Affective Computing*. Cambridge, MA: MIT Press.

Picard, Rosalind and Gary Wolf. 2015. 'Sensor Informatics and Quantified Self'. *IEEE Journal of Biomedical and Health Informatics*, 19(5), 1531.

Swan, Melanie. 2009. 'Emerging Patient-Driven Health Care Models: An Examination of Health Social Networks, Consumer Personalized Medicine and Quantified Self-Tracking'. *International Journal of Environmental Research and Public Health*, 6(2), 492–525.

Swan, Melanie. 2012. 'Sensor Mania! The Internet of Things, Wearable Computing, Objective Metrics, and the Quantified Self 2.0'. *Journal of Sensor and Actuator Networks*, 1(3), 217–53.

Tabakcıoğlu, M., Çizmeci, H., & Ayberkin, D. 2016. 'Neurosky EEG Biosensor Using in Education'. *International Journal of Applied Mathematics, Electronics and Computers*, 4, 76–8.

Teplan, Michal. 2002. 'Fundamentals of EEG Measurement'. *Measurement Science Review*, 2(2), 1–11.

Vertegaal, Roel. 2003. 'Attentive User Interfaces'. *Communications of the ACM*, 46(3): 30–3.

CHAPTER 25

OFFICE PLANT

STEFAN RIEGER
TRANSLATED BY ERIK BORN

THE OFFICE AS NATURAL HABITAT

THERE are currently various attempts to approach the office plant in an artistic and documentary manner. One of the main attractions for them consists in depicting this natural organism within the dreary landscape of office space, entangled with an assemblage of other office equipment. As a natural relic, the office plant stands out among storage places and communication media, telephones and pencil sharpeners, staplers and fax machines, hole punches and computer monitors, three-ring binders and hard drives, stamp pads and sticky notes, envelope moisteners and document portfolios, index card boxes and hanging file folders, coffee machines and personalized mugs, paperclips and colour markers, desk pads and mouse pads, erasers and typewriters. The office plant is supposed to liven things up.

The news magazine *Der Spiegel* juxtaposes two projects on office plants in workplaces that impressively demonstrate this phenomenon. It seems to have been the case for a certain period of time—and the series of images support this finding for the long run—that the office always has been the natural habitat for plants. The office plant has always and in a very unobtrusive way been a medium of organization. As such, it not only makes material demands on the arrangement and perception of an office space, but also permits, in its affective dimension, a characterology of those who work there. The first project in the *Spiegel* is a series of photographs, taken by the artist Saskia Groneberg for her thesis project, that capture the demure charms of office plants in a series of pictures (Groneberg, 2013). The other project in the *Spiegel*, a series by photographer and media artist Frederik Busch, is also dedicated to the everyday life of office plants (Hielscher, 2017). Both photo series capture atmospheres and negotiate stereotypes.

There are not particularly many types that could be attributed to an organization-theoretical category of the office plant. Rubber trees, elephant ears, bowstring hemp, money plants, dragon trees, Flaming Katys, cactuses, and other low-maintenance succulents announce a veritable invasion of plants capable of causing anarchy in the otherwise dull office routine. Depending on its species and habitat, an office plant can structure spaces, change sight lines, create partitions or disruptions, maintain its distinction as something solitary (like the rubber tree) or fall in line with others of its kind in a floral assemblage. According to this logic, a cactus loves company. Plants create counterpoints; their growth can mimic stacks of files or mountains of data, as in the case of ivy, which will grow right along a row of folders. As Groneberg notes, the mailroom appears to be the most privileged site of floral excess. But office plants catch one's eye in other places, too. For instance, one opinion piece in *Der Freitag* presents an A to Z office list, which contains the following for the keyword 'Greenery':

> The true heroes of our times are green and, in most cases, ugly. They are named dragon tree, ficus, Schleffera, or money plant. Their reason for being: to convert the dust and sweat of many millions of workaholics, their bad breath, their gloomy thoughts and badly-fitting suits into positive energy. Office plants work under the most adverse conditions: little light, hardly any water, never any fertilizer. However, if one of them ever gets caught losing its composure, its position is terminated without notice.

> Naturally, the common office plant has no business in any more exclusive atmosphere, in the roomy lofts of the creative and financial class. There, hydrocultures concocted by botanical artists purify the air of even the smallest pathogens. Greenery comes from far away and acts like something superior. However, the proletariat is in the majority and, like always, does the dirty work.   (Stöhr, 2012)

Such journalistic impressions have also attracted the attention of some scholars, and the role of office plants has landed on the research agenda of empirical sociology as well as labour economics. Articles in specialized journals such as *HortScience* and *HortTechnology* discuss the benefits of plants for an 'office climate' in every possible sense of the term, and report on their effects on American office environments, often with a greater claim to their representative status. For instance, the starting point for a study of the 'Reported Impacts of Interior Plantscaping in Office Environments in the United States' is the following intuition: 'Most people intuitively feel that contact with plants and nature is restorative to the human spirit. This common belief in the positive effects of plants on people is evidenced by the widespread use of interior plantscaping in office environments, particularly in those of large businesses' (Pearson-Mims and Lohr, 2000: 82).

There is no doubt that there are psychological benefits to office plants, whether in claims about the absolute 'Psychological Benefits of Indoor Plants in Workplaces' (Bringslimark et al., 2007) or 'The Relative Benefits of Green versus Lean Office Space' (Nieuwenhuis et al., 2014). The latter study, which has attracted attention in diverse fields since its publication, situates the phenomenon among various philosophies of office organization, a branch of general workplace design that has passed through various phases since the publication of Frederick Winslow Taylor's *Principles of Scientific Management* in 1911 during the heyday of classical modernism (see Sundstrom and

Sundstrom, 1986). Office management is no longer only a matter of architectural layout (e.g., cubicles vs. open-plan office), but also of the very possibility of plants and floral decorations as an example of the functionality of various office styles and philosophies of working environments (see Szyperski et al., 1982; Petendra, 2015; and especially for the current discussion, Haslam and Knight, 2010). As a medium of organization, the seemingly-unimportant plant organizes the entire order of the office.

OFFICE AND OUTER SPACE

The article on 'The Relative Benefits of Green versus Lean Office Space' opens with a little gallery piece about the British prime minister David Cameron, one thorn in whose side was the country's expenditures on office planting and one stated aim of whose governance was a corresponding cost-cutting policy. According to this matter-of-fact politician, the country should be led on a small budget and in a businesslike atmosphere. However, empirical studies often interfere with ideological presuppositions and historical categories like the standardized austerity of Taylorism.

> Principles of lean office management increasingly call for space to be stripped of extraneous decorations so that it can flexibly accommodate changing numbers of people and different office functions within the same area. Yet this practice is at odds with evidence that office workers' quality of life can be enriched by office landscaping that involves the use of plants that have no formal work-related function. To examine the impact of these competing approaches, 3 field experiments were conducted in large commercial offices in The Netherlands and the UK. These examined the impact of lean and "green" offices on subjective perceptions of air quality, concentration, and workplace satisfaction as well as objective measures of productivity. Two studies were longitudinal, examining effects of interventions over subsequent weeks and months. In all 3 experiments enhanced outcomes were observed when offices were enriched by plants.   (Nieuwenhuis et al., 2014: 199)

In subsequent years, the study's hypothesis of a roughly 15 per cent increase in job performance in floral milieu has persisted with a certain tenacity. The factors reportedly responsible for this increase are based on a familiar topos about the benefits of spending time outdoors: plants in office spaces are supposed to counteract the symptom of 'nature deficit disorder', to provide for calm and mindfulness, regeneration and personal wellness (Louv, 2005; Fletcher, 2017). In doing so, they contribute to the maintenance of a good working environment among office employees, and thus manage the 'atmosphere' in every respect. But there are also scientists who take the discourse of needing to improve the climate in a non-metaphorical sense; proponents of office plants have received additional support from a completely different source—namely, NASA. The American space agency was responsible for a series of large-scale experiments documented in a report published under the title 'Interior Landscape Plants for Indoor Pollution Abatement' (Wolverton et al., 1989). The widely-publicized final report covers the fact, the reasons, and the extent to which plants are capable of minimizing pollutants and thus beneficial to an indoor climate, as well as the methods and protagonists involved in bringing these findings to light (see Stutte, 2012).

The principal investigator for the programme, Bill Wolverton, took a convoluted path to arrive at the space agency; his involvement was as unusual as the role played by NASA in providing the foundational research for what would become the policy of office plants. A common starting point for Wolverton and NASA was the 'sick building syndrome' (SBS), a condition caused by inhabiting interior spaces with atmospheric contamination. The list of health risks associated with SBS is long and ranges from asthma, headaches, skin irritations, and difficulty concentrating, across a spectrum of diverse allergies, to cancer. The primary culprit for indoor air pollution, along with the use of certain building materials, was taken to be the lack of air circulation, which was directly linked to another factor—thermal insulation. Hiding behind these factors, in turn, were energy politics, the first oil price crisis in 1973 and the second in 1979, and the period's corresponding energy-saving measures. What may have been good for an office climate— namely, maximum air circulation—proved to be bad for the environment and bad for an economy suffering from an oil crisis.

> It was during this period of the early 1970's and 1980's when the issues associated with Sick Building Syndrome were gaining attention that the United States National Aeronautics and Space Administration (NASA) became an unlikely leader in identifying biological solutions to the problem of poor indoor air quality. NASA had been supporting work using biological systems for atmospheric regeneration since the 1950's, with the emphasis on using photosynthetic systems for the removal of carbon dioxide and regeneration of oxygen as part of a life support system.
>
> (Stutte, 2012)

In this situation of general conflict, the space agency came into play. The investigators behind real spaceships, which were inspired by the Cold War, recognized that there was a tried-and-tested solution to the problem of manned space travel. For 'spaceship earth', to use Richard Buckminster-Fuller's grand and memorable formulation, a form of travel could be found in nature's power of regeneration:

> The National Aeronautics and Space Administration (NASA), faced with the task of creating a life-support system for planned moon bases, began extensive studies on treating and recycling air and wastewater. These studies led NASA scientists to ask a very important question. How does the earth produce and sustain clean air? The answer, of course, is through the living processes of plants. With this basic knowledge, NASA scientists began to study the development of sustainable, closed ecological life-support facilities. Working toward these goals, scientists at NASA's John C. Stennis Space Center in southern Mississippi discovered that houseplants could purify and revitalize air in sealed test-chambers.   (Wolverton, 1997: 7)

To achieve this objective, technological atmospheres were created, e.g., in the form of 'biodomes'; due to their self-containment, every conceivable factor could be measured using a large amount of sensor technology. The results were a series of findings about which plants made which contributions to an improved indoor climate, and which pollutants they were able to filter out of the air in which amounts. The same findings can also be found in the handbook *How to Grow Fresh Air: 50 House Plants that Purify Your Home or Office*, whose author was none other than the study's principal investigator. Behind Wolverton's handbook, which at first glance seems like a special issue of *Better Homes and Gardens*, lurks a chapter from the history of science, which Wolverton (1997) specifically calls to mind in the Acknowledgements with a reference to the environment for his work: 'My sincere appreciation to NASA and, in particular, the John C. Stennis Space Center, for allowing me to study nature amidst a world of electronics and rockets' (Wolverton, 1997: 4).

OFFICE PLANT RANKINGS

What grounded NASA's ambitious project was by no means an ecological position, but quite simply the political climate of the Cold War. After the Cold War the results were aimed at promoting the health not only of astronauts but of all people: Wolverton's guidebook presents the results of his research at NASA in the form of a series containing many of the same plants found in the artistic documentations mentioned at the start of this article—not as elements of artwork, but rather as the result of a formidable ranking system. Wolverton's guidebook contains fifty plants, which are described in terms of their ecological benefits, e.g., for purifying pollutants. It lists each plant's name, taxonomical location, a portrait image, tips for cultivation, potential pests, and other information like transpiration rates. The order is determined not by the plant's taxonomy, nor even the alphabet, as would be the case in the genre of the florilegium, but rather each plant's total score. Hence, the house plants are ranked from the areca palm (*Chrysalidocarpus lutescens*) with an overall score of 8.5 to the Flaming Katy (*Kalachoe blossfediana*) with a score of 4.5. Additionally, the rubric 'medium' provides information about which form of cultivation is suited for each plant—the potential use of hydroculture and the avoidance of organic soil fall under this point. The origins of hydroculture have less to do with the 'botanical artists' mentioned in the lexicon entry above, and more with the designation 'hydroponics', which was used for the accommodation of passengers on Pan American Airways. The site for Pan-Am's experiments was Wake Island air base, located on an atoll in the Pacific Ocean and used for stopovers on long-distance flights (Taylor, 1939); it also played a role for the provisioning of American soldiers in the Second World War (Wolverton, 1997: 34–5).

The results obtained in a certain biosphere or in the so called 'Wolverton house' can be transferred to the office environment, and are thus being researched there—e.g., in field studies of office plants' ability to reduce harmful volatile organic compounds (VOCs) (Wood et al., 2006). The following caption, presented below a wonderful diagram from everyday office life, applies in miniature to those neighbourhoods of land parcels predefined through open-plan office architecture: 'Plants influence air quality within a personal breathing zone' (Wolverton, 1997: 29). Keeping plants in offices is thus part of the cultural history of the house plant—or, more precisely, the cultural history of its cultivation including selections among media of acculturation and fantasies about the use of greenhouses or the switch to hydroculture (Rieger, 2014). Last but not least, it is a history of dealing with plants in interior spaces—both of their cultivation indoors and of the agents of office plant welfare. An affinity for plants and the ability to care for them are two virtues that are also relevant to everyday office work. The house gardener has long been an ideal for the employee (Waller, 1821). Caring for a plant, as a medium of organization, can also provide a subliminal indicator of employee character, allowing an individual to be read as reliable or perhaps even rooted in the business.

Multispecies–Computer Interaction

Nevertheless, the demands made by plants on those who work in offices have changed to some extent over the years. The amount of maintenance has been minimized through the introduction of hydrocultures, and responsibility for their life or death has been delegated further to technical infrastructures in the form of automated indoor climate controls commonly found in smart homes and smart offices. Smart gardening, too, has long since reached the office: the Parrot Pot and other smart flower pots are able to keep plants alive on their own; and plants are able to communicate their own needs using a protocol known as Botanicalls, which was created 'to provide a new way for plants and people to interact to develop better and longer-lasting relationships, regardless of the physical or genetical distance' (Kobayashi, 2015a: 101). With a circuit board printed in the form of a plant leaf and an unconventional imaging system reminiscent of the artistic office plant works mentioned at the start of this essay, Botanicalls made a fitting addition to MoMA's permanent collection.[1] Even without direct human–plant communication, auxiliary devices can simplify plant care. For instance, smart lamps regulate light and temperature on their own, without the competency of a green thumb (Salamone et al., 2016). In the smart office, plants no longer need to be watered by hand. However, the act of watering only becomes completely unnecessary in the case of virtual plants, which are taking over the position previously occupied by real plants in the office system (Koles and Nagy, 2014). In a wonderful functional equivalence, there has even been at least one study of virtual plants' purported contribution to an office's psychological welfare. Ironically, given the context of its presentation at an event dedicated to

[1] Available at https://www.botanicalls.com [Accessed 23 July 2018].

'Universal Access in Human–Computer Interaction', the study coupled high-tech solutions with chauvinistic ideas about gender stereotypes in the modern office. Taking the virtualization of plants seriously, the title of the study, 'Design Research of Augmented Realty [sic] Plant to Depressurize [sic] on Office Ladies', called a spade a spade (Hsieh et al., 2013). According to these Taiwanese researchers, displaying virtual plants with design variations will minimize occupational stress, above all, among the class of worker known in Japan as an 'office lady'.

These kinds of interactions are the object of a multispecies computer interaction, which increasingly works at freeing the use of computers from the constraints of human–computer interaction and transferring it to representatives of other species. Here, one might mention Clara Mancini's (2011) attempts to create animal–computer interaction, and Fredrik Aspling's (2015) interest in understanding multispecies computer interaction, which also accounts for plant–computer interaction. Aspling's clear goal consists in nothing less than bringing media to life: 'Other nonhuman species such as plants and trees and their possibilities in computer-mediated interactions has [sic] even more recently started to gain interest in the margins of the HCI community, a topic sometimes framed as "living media".'

However, the colourful operationalization in these kinds of products should not deceive us about their implications for mass applications in the future. What Mancini and Aspling have in common is that they do not get lost in the details of bringing these kinds of products to the shelf, nor in their proximity to art, but rather emphasize their theoretical relevance with great verve, especially their entanglement with the concept of *multispecies communities*. In one table, Aspling compares 'categories of motivation for plants in computer systems, design and real life' (Aspling et al., 2016). Along with another reference to displays, which are again purported to counteract depression among female office employees, Aspling also mentions a gadget called Plantio, whose makers describe their product as 'an interactive pot to augment plants' expressions' (Kuribayashi et al., 2007). With their platform for inter-species communication, the makers of Plantio inadvertently touch on one of the foundational stories in the history of plant communication—a legendary effect that sparked interest in plant communication like no other (Rieger, 2009). As a strange sort of punchline, it was a common office plant, the dragon tree, that was responsible for the infamous effect, which had been diagnosed by American polygraph specialist Cleve Backster.

ORGANIZATIONAL TRUTH AND PLANTS OF OTHER KINDS

Appearing in the international *Journal of Parapsychology*, Backster's attention-grabbing study 'Evidence of a Primary Perception in Plant Life' (1968) presented one of the most noted contributions on the language of plants. What formed the condition of possibility for the Backster Effect was the convergence of individual curiosity with the availability of a certain apparatus. The media coverage of the work's origins presented a narrative that was typical of encounters with other species in its mixture of composure, feigned scepticism, and the accumulation of seemingly objective details (Tompkins and Bird, 1973). In Germany, the tone for this reception was set by science journalists Dagny and Imre Kerner's influential book *Der Ruf der Rose. Was Pflanzen fühlen und wie sie mit uns kommunizieren* (The Fame of the Rose: What Plants Feel and How They Communicate with Us, 1992). The Kerners' book owed its wide reception in Germany to a unique situation, involving equal parts fascination with their technique of setting up the experiments and openness to a diffuse spirituality whose main representative could be found in the New Age movement. Under the heading 'Der Drachenbaum, der Gedanken lesen kann' (The Dragon Tree that Can Read Minds), one encounters a familiar narrative, though it is about an American scientist who works not just anywhere but on the cutting edge of modern interrogation and truth-finding methods, and thus already seems to be immune to the reproach of gullibility. As a polygraph specialist, Backster founded a lie detector school for the CIA and developed procedures for diverse American government agencies, especially the army. In his own school in New York City, he trained law enforcement and security officials from around the world in polygraph truth-finding methods. Backster's discovery of plant feelings and communication came about thanks to an ordinary dragon tree, which stood in his office for customary aesthetic reasons. Backster himself depicts the first impulse for his research with a mixture of sheer curiosity and excitement about the available technology:

> RESEARCH ON A PRIMARY PERCEPTION in plant life was triggered by curiosity about a common botanical function. On February 2, 1966, immediately following the watering of an office plant, the author wondered if it would be possible to measure the rate at which water rose in a plant from the root area into the leaf. [...] The pair of instrument electrodes could be attached to a leaf of the plant, and hypothetically, by so using the wheatstone bridge circuitry involved, the relative decrease in the plant leaf's electrical resistance—due to the expected increase in its moisture content— should be indicated by an upward trend of the ink tracing on the chart recording.
> (Backster, 1968: 329)

Backster's encounter with his dragon tree follows a pattern similar to the constellation with Wolverton, where plant research occurred in the environment of rockets and computers: in the Kerners' words, 'He had just developed a new lie-detector procedure for the American army, when, in the middle of the night—or, put better, in the early morning hours of February 2, 1966—he looked at the dragon tree in his office and came upon the idea of hooking up the plant to this device, so that he could see how long after watering it would take for the water to reach the leaves' (Kerner and Kerner, 1992: 46). What came out of this encounter with office plants and then-advanced equipment for truth-finding, is a far-reaching story. Even if its protagonists sit on the same windowsill, their story would require turning over another leaf... of paper (Rieger, 2015).

With changing forms of organization, which increasingly play out in the virtual, the anachronistic charm of office plants today seems untenable and obsolete (Putnik and Cunha, 2008). These relics of office culture may have outlived their usefulness for the logic of cooperation among virtual teams, which is often independent of both time and space. Apparently, office plants have little to do with the reality of virtual organizations or their virtual problems, and are only good for documenting the remains of the real working world (Hughes et al., 2001; Shekhar, 2016). At the same time, they remain present indirectly: casuistically, as floral ornaments on screensavers or virtual plants on computer monitors; and systematically, in the modalities of natural interface design, which bet on user-friendly strategies like 'ecological interfaces' and 'ecovisualization' for gaining acceptance. Naturally, office plants are also present as agents of communication and interaction, which occur under the enlistment of other, non-human agents and under the abandonment of habitual communication systems. Even in a biosphere and in the post-human, highly technologized mode of multispecies communication, they ensure that the connection to nature is maintained or can be established by technical means (Kobayashi, 2015b).[2]

REFERENCES

Aspling, Fredrik. 2015. 'Animals, Plants, People and Digital Technology: Exploring and Understanding Multispecies-Computer Interaction'. Lecture. ACE '15: Proceedings of the 12th International Conference on Advances in Computer Entertainment Technology.

Aspling, Fredrik, Jinyi Wang, and Oskar Juhlin. 2016. 'Plant–Computer Interaction, Beauty and Dissemination'. Lecture. ACI '16: Proceedings of the Third International Conference on Animal-Computer Interaction, Article No. 5.

Backster, Cleve. 1968. 'Evidence of a Primary Perception in Plant Life'. *International Journal of Parapsychology*, 10(4), 329–48.

Bringslimark, Tina, Terry Hartig, and Grete Grindal Patil. 2007. 'Psychological Benefits of Indoor Plants in Workplaces: Putting Experimental Results into Context'. *HortScience*, 42(3), 581–7.

Fletcher, Robert. 2017. 'Connection with Nature Is an Oxymoron: A Political Ecology of "Nature-Deficit Disorder"'. *The Journal of Environmental Education*, 48(4), 226–33.

Groneberg, Saskia. 2013. Interview with Saskia Groneberg by Maria Huber. 'Pflanzen im Büro: Kunststudentin fotografiert deutsche Arbeitsplätze'. *Spiegel Online*, 26 April. http://www.spiegel.de/karriere/pflanzen-im-buero-kunststudentin-fotografiert-deutsche-arbeitsplaetze-a-896282.html [Accessed 23 July 2018].

Haslam, S. Alexander and Craig Knight. 2010. 'Cubicle, Sweet Cubicle: Why Some Office Spaces Alienate Workers, While Others Make Them Happier and More Efficient'. *Scientific American Mind*, 21, 30–5.

[2] On Kobayashi's individual projects in Human–Computer–Biosphere Interaction, see his homepage, available at http://hhkobayashi.com [Accessed 23 July 2018].

Hielscher, Ines. 2017. 'Büropflanzen: Ein Leben neben Aktenstapeln'. *Spiegel Online*, 20 September. http://www.spiegel.de/karriere/bueropflanzen-ein-leben-neben-aktenstapeln-a-1168704. html [Accessed 23 July 2018].

Hsieh, Jei-Che, Chang-Chang Huang, and Hwa-San Kwan. 2013. 'Design Research of Augmented Realty [sic] Plant to Depressurize [sic] on Office Ladies'. In Constantine Stephanidis and Margherita Antona (eds), *Universal Access in Human–Computer Interaction: Design Methods, Tools, and Interaction Techniques for eInclusion (UAHCI 2013)*. Berlin: Springer, 297–303.

Hughes, John A., Jon O'Brien, Dave Randall et al. 2001. 'Some "Real" Problems of "Virtual" Organisation'. *New Technology, Work and Employment*, 16(1), 49–64.

Kerner, Dagny and Imre Kerner. 1992. *Der Ruf der Rose. Was Pflanzen fühlen und wie sie mit uns kommunizieren*. Köln: Kiepenheuer & Witsch.

Kobayashi, Hill Hiroki. 2015a. 'Human–Computer–Biosphere Interaction: Toward a Sustainable Society'. In Anton Nijholt (ed.), *More Playful User Interfaces: Interfaces that Invite Social and Physical*. Singapore: Springer, 97–119.

Kobayashi, Hill Hiroki. 2015b. 'Research in Human–Computer–Biosphere Interaction'. *Leonardo*, 48(2), 186–7.

Koles, Bernadett and Peter Nagy. 2014. 'Virtual Worlds as Digital Workplaces: Conceptualizing the Affordances of Virtual Worlds to Expand the Social and Professional Spheres in Organizations'. *Organizational Psychology Review*, 4(2), 175–95.

Kuribayashi, Satoshi, Yusuke Sakamoto, Maya Morihara et al. 2007. 'Plantio: An Interactive Pot to Augment Plants' Expressions'. Lecture. 4th International Conference on Advances in Computer Entertainment Technology/ACE 2007, 13–15 June. Salzburg, Austria.

Louv, Richard. 2005. *Last Child in the Woods: Saving Our Children from Nature-Deficit Disorder*. Chapel Hill, NC: Algonquin Books.

Mancini, Clara. 2011. 'Animal–Computer Interaction (ACI): A Manifesto'. *Interactions*, 18(4), 60–73.

Nieuwenhuis, Marlon, Craig Knight, Tom Postmes et al. 2014. 'The Relative Benefits of Green versus Lean Office Space: Three Field Experiments'. *Journal of Experimental Psychology*, 20(3), 199–214.

Pearson-Mims, Caroline H. and Virginia I. Lohr. 2000. 'Reported Impacts of Interior Plantscaping in Office Environments in the United States'. *HortTechnology*, 10, 82–6.

Petendra, Brigitte. 2015. *Räumliche Dimensionen der Büroarbeit. Eine Analyse des flexiblen Büros und seiner Akteure*. Wiesbaden: Springer VS.

Putnik, Goran D. and Maria Manuela Cunha. 2008. *Encyclopedia of Networked and Virtual Organizations*. Hershey, NY: Information Science Reference.

Rieger, Stefan. 2009. 'Wacholder'. In Benjamin Bühler and Stefan Rieger (eds), *Das Wuchern der Pflanzen. Ein Florilegium des Wissens*. Frankfurt am Main: Suhrkamp, 277–90.

Rieger, Stefan. 2014. 'Treibhaus'. In Stefan Rieger and Benjamin Bühler (eds), *Kultur. Ein Machinarium des Wissens*. Berlin: Suhrkamp, 232–46.

Rieger, Stefan. 2015. 'What's Talking? On the Nostalgic Epistemology of Plant Communication'. In Patrícia Vieira, Monica Gagliano, and John Ryan (eds), *The Green Thread: Dialogues with the Vegetal World*. Lanham, MD: Lexington Books, 59–79.

Salamone, Francesco, Lorenzo Belussi, Ludovico Danza et al. 2016. 'An Open Source "Smart Lamp" for the Optimization of Plant Systems and Thermal Comfort of Offices'. *Sensors*, 16(3), Article 338.

Shekhar, Sandhya. 2016. *Managing the Reality of Virtual Organizations*. New Delhi: Springer.

Stöhr, Markus. 2012. 'Grünzeug'. *Der Freitag*, 'Gummibaum und Diddl-Tasse: A–Z Büro', 10 April. https://www.freitag.de/autoren/der-freitag/gummibaum-und-diddl-tasse [Accessed 23 July 2018].

Stutte, Gary W. 2012. 'Phytoremediation of Indoor Air: NASA, Bill Wolverton, and the Development of an Industry'. *NASA Technical Reports Server* (NTRS), Document Number 20120003454.

Sundstrom, Eric and Mary Graehl Sundstrom. 1986. *Work Places: The Psychology of the Physical Environment in Offices and Factories*. Cambridge: Cambridge University Press.

Szyperski, Norbert, Erwin Grochla, Klaus Hering et al. (eds). 1982. *Bürosysteme in der Entwicklung. Studien zur Typologie und Gestaltung von Büroarbeitsplätzen, Braunschweig*. Wiesbaden: Vieweg & Sohn.

Taylor, Frank J. 1939. 'Nice Clean Gardening'. *The Rotarian* (July), 14–15.

Tompkins, Peter and Christopher Bird. 1973. *The Secret Life of Plants: A Fascinating Account of the Physical, Emotional, and Spiritual Relations between Plants and Man*. New York: Harper & Row.

Waller, Carl Alexis. 1821. *Der Stubengärtner, oder Anweisung die schönsten Zierpflanzen in Zimmern und vor Fenstern zu erziehen und auf eine leichte Art zu überwintern*, 3rd edn. Sondershausen und Nordhausen: Bernhard Friedrich Voigt.

Wolverton, B. C. 1997. *How to Grow Fresh Air: 50 Houseplants that Purify Your Home or Office*. New York: Penguin.

Wolverton, B. C., Anne Johnson, and Keith Bounds. 1989. *Interior Landscape Plants for Indoor Pollution Abatement*, 15 September. https://ntrs.nasa.gov/archive/nasa/casi.ntrs.nasa.gov/19930073077.pdf [Accessed 23 July 2018].

Wood, Ronald A., Marget D. Burchett, Ralph Alquezar et al. 2006. 'The Potted-Plant Microcosm Substantially Reduces Indoor Air VOC Pollution: I. Office Field-Study'. *Water, Air, and Soil Pollution*, 175, 163–80.

CHAPTER 26

OVERHEAD PROJECTOR

CLAUS PIAS

TRANSLATED BY VALENTINE A. PAKIS

'IN ancient history, overhead projectors were used to give presentations. They involved static pages of see-through "paper" on which black images were printed. Historians suggest ancient presenters even used marking pens to write on these transparent pages.'[1] This, at any rate, is how the triumphant history appeared on the homepage of a now defunct projector dealer in 2009. Because technical media change so rapidly, their brief histories often quickly become obscure. Yet, if the Nobel Prize-winning physicist Edward M. Purcell is to be believed, the overhead projector represents nothing less than 'the greatest invention since chalk'.[2] As in the case of other media that continue to survive in niches, it is first and foremost the fans and nerds, the devoted dilettantes and scrupulous antiquarians, who perpetuate an expert knowledge of the present through its transition to the construction of scientific and historical memory. The Austrian artist collective *monochrom*, for instance, not only collects images of overhead projectors but has also composed a farewell song for them,[3] and in Denmark an *Overhead Festival for Forgotten Media* was held as early as 2005.[4]

My considerations below will not exclusively be concerned with the overhead projector, but rather I wish to point out an historical constellation of various technologies, practices, and concepts of knowledge, cooperation, and presentation that made it possible for something like PowerPoint to arise and become ubiquitous. This list of circumstances is sure to be incomplete (a history of the projector, for instance, has yet to be written), and it is just as certain that a phenomenon such as PowerPoint (and similar software) cannot be entirely explained by examining its precursors or conflating it with their histories. Nevertheless, the historical elements that came together in a highly

---

[1] http://www.projectorwarehouse.com.au/glossary.asp [Accessed 10 January 2009].
[2] https://www.harvard.edu/about-harvard/harvard-glance/honors/nobel-laureates [Accessed 19 July 2018].
[3] http://www.monochrom.at/farewell-overhead/ [Accessed 19 July 2018].
[4] http://rhizome.org/community/34108/ [Accessed 19 July 2018].

peculiar mixture continue to affect that which can be known, said, and demonstrated through slide presentation.

Bowling for Interdisciplinarity

Little is known about James Earl Bancroft, who seems to hold the honour of patenting the first overhead projector.[5] His early years as an inventor were devoted to the prosaic matter of opening elevator doors, and only his so-called 'vehicle response mechanism'— a device that registers a car's arrival in the driveway and opens the garage door as it approaches—displays a certain degree of futuristic charm, reminiscent of the golden age of Detroit, the 'Motor City' in which Bancroft lived.[6] Yet amid all these transport-related devices—submitted at the beginning of the Second World War and accepted during the year that the United States entered it—Bancroft patented a device with the simple title 'projector' (Figure 26.1).[7]

Even though it could be inferred from the name of his employer—The American Bowling and Billiard Corporation—that this projector had meant to serve the sole purpose of presenting highly legible bowling scores, Bancroft seemed to have greater ambitions for it. 'It will be understood,' as he wrote on the first page of his patent application, 'that my invention is not to be limited by such illustrations or language or disclosures', such as those imagined for the projection of game scores. Beyond the tedious style of 'thickly' describing ostensibly obvious things, which is essential to the poetology of patent applications, Bancroft's introduction suddenly flashes with the promise of a new type of visibility: 'This invention relates [...] to projectors [...] in which a true image [...] of an object situated in front of an observer is formed on a screen situated in the field of vision of said observer, so that he can see both the object and the image without shifting his position and merely by shifting his gaze.' Rather than transport people to objects, objects would be transported to people, visually.

The instruction to shift one's gaze is not without philosophical appeal. For, unlike the desk, the place of writing is no longer the only place where the writing can be seen. The writing place is now split into an intimate space and a public space. The writer writes, then repeatedly has to move away from the projection in order to engender the presence of the textual image. Conversely, moreover, he has to lift his gaze and suspend the act of writing in order to be a viewer of the image among the others present. As one image always and simultaneously causes another to disappear, Bancroft's projector creates a

---

[5] Bancroft is never quoted in the histories of the overhead projector that usually start more generally with the seventeenth or nineteenth century (https://en.m.wikipedia.org/wiki/Overhead_projector [Accessed 19 July 2018]. The decisive point for me to include lies in his concept of using a projector for writing and presenting simultaneously.

[6] Vehicle Responsive Mechanism, US Patent No. 1836058, filed 9 August 1926.

[7] Projector, US Patent No. 2250174, filed 16 August 1939; also see: Projector, US Patent No. 2310273, filed 18 June 1943.

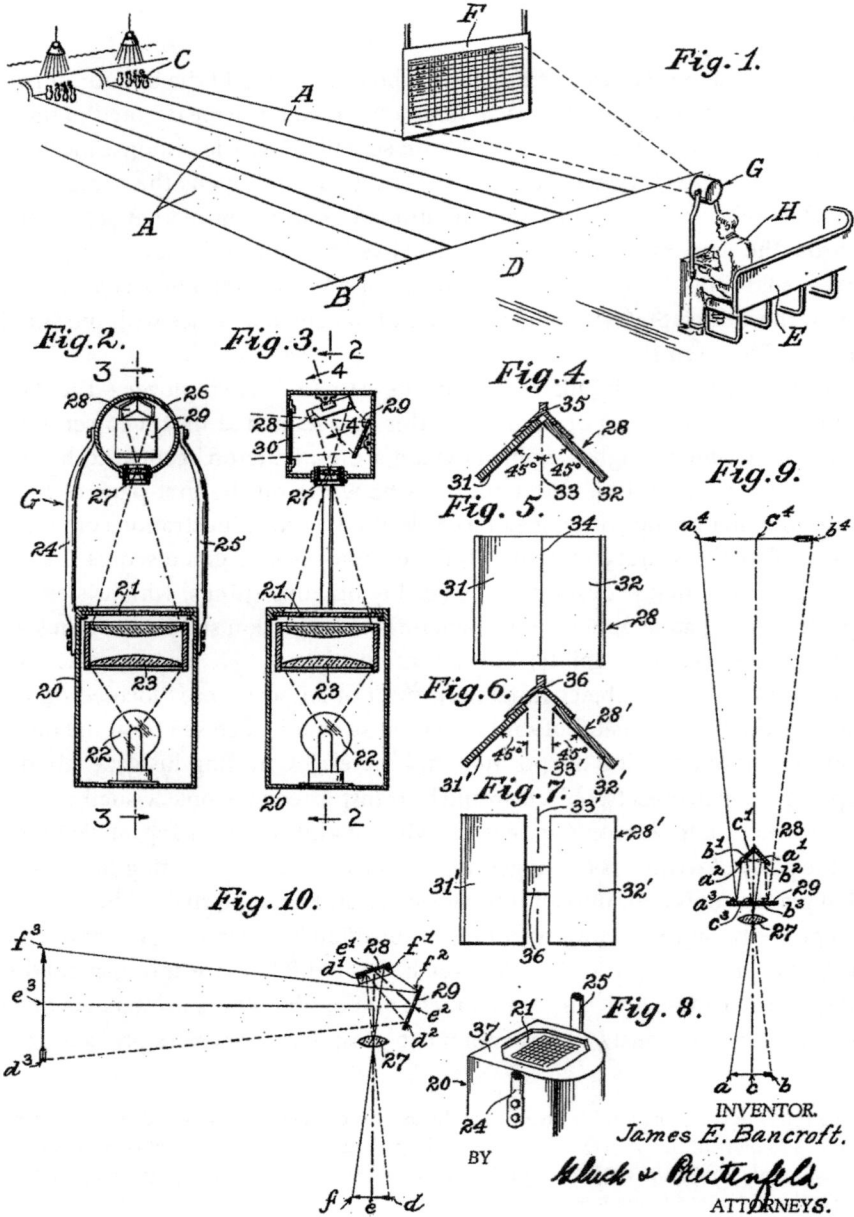

FIGURE 26.1 James E. Bancroft, patent for a projector to be used in a bowling alley (1939)

(US Patent No. 2250174, Aug 16, 1939)

loop between presence and absence, a delay in vision that is structurally similar to that which, as described by Derrida (1993) in his *Memoirs of the Blind*, is experienced by someone producing a self-portrait. The overhead projector creates a public sphere for writing, where nothing finished is presented. Rather, something commonly visible is produced 'live'.

Bancroft's 'projector' was not really 'invented' out of thin air. It rather derives from other apparatuses such as the epidiascope, which had enjoyed popularity since the nineteenth century, above all in art history lecture halls, because it could manage with incident light (unlike the sciopticon, for instance).[8] Thus it was possible to project onto the wall anything that could be written or printed on paper, but it was not possible to write simultaneously *while* projecting. This was to be the decisive difference with the overhead projector.

It is also worth noting that transparent folios had long been widely used in military contexts. In 1908, Farrand Sayre published *Map Manoeuvres* based on his lectures at Fort Leavenworth (Sayre, 1908; Wilson, 1969: 17). This was a 'free' American adaptation of the *Kriegsspiel* (war game) whose special feature found the player playing the referee, who managed and controlled the enemy troops. The game introduced the first transparencies made of celluloid, upon which manoeuvres and information could be written without ruining the valuable maps. The simple conclusion that was later drawn from this was that, if war games are to be played with multiple participants, it would be practical to combine the epidiascope with transparencies. And in order to be rid of the need of having to remove the transparencies over and over again, it would be helpful to illuminate the platform of the game from below instead of from above. The military need for such things seems so evident that we might assume the overhead projector was developed simultaneously in different war rooms during the Second World War, in which media-technically savvy and cooperative work was taking place that would provide the basis for the later CSCW (Computer Supported Cooperative Work). Thus around this time, for instance, the Naval Air Warfare Center Training Systems Division (NAWCTSD) boasted of having developed 'new approaches for military training': 'Some of these approaches were adopted for public education, including the overhead projector, which was originally developed for navigation training.'[9]

So far, however, all that can be demonstrated with evidence is a civilian line of development from Bancroft to Harold G. Fitzgerald, who, beginning in 1947, registered several patents that correspond quite precisely to today's conception of an overhead projector.[10] Fitzgerald's achievement was to free the overhead projector from dedicated contexts such as bowling or war games and to generalize it so that it became an abstract media entity with which several people could either follow a piece of writing accompanied

---

[8] On the use of projection in art history, see Meyer (1883), Schmid (1894), Ratzeburg (1998), Grimm (1897), and especially the groundbreaking essay by Dilly (1975).

[9] http://www.federallabs.org/labs/profile/?id=1359 [Accessed 19 July 2018].

[10] Projector for Handling Transparent Plates, US Patent No. 2564057, filed 20 August 1947. See also: Projector, US Patent No. D164293, filed 30 April 1951; Overhead Projector Apparatus, US Patent No. 2828666, filed 9 January 1956.

by a talk without having to see the writer, or see and hear a speaker standing before his prepared written material. The overhead projector became an oscillator between presence and presentation. Fitzgerald thus imagined various ways of using a light and transportable device 'in connection with lectures, speeches, educational work, sales promotional work, etc.'. In this sense, the overhead projector turned out to be perfect for the booming organizational form of 'meetings' in the 1950s, that is, for small interdisciplinary conferences and for collaborative or 'creative' decision-making in heterogeneous fields of knowledge.

DRAWING THINGS TOGETHER

However, in order for the overhead projector to be implemented widely in schools, lecture halls, and conference rooms and deeply change the culture of teaching and presentations, it first needed a few additional developments. This cultural 'invention of invention' (in Lewis Mumford's terms) can be attributed to a convergence of various conditions and technologies—a convergence in which the company 3M played a prominent role (3M, 2002: 57ff.). The first essential component was the transparency, which emerged from the field of photocopying. In 1951, with its Model 17 Secretary, 3M had commercialized the world's first dry copying system, based on a method by Carl Miller (1948) and reformulated by Carl Kurmmeyer, a machine maker from Minnesota. One of the many problems Miller faced was the mirror-image inversion of material created by the copying process. This he attempted to offset by means of transparent pages. What thus originated as a simple by-product of copying were transparencies with images or texts for which there was not yet any use. At least not until a young engineer Roger Appeldorn had the playful idea of putting these transparencies on one of the heavy and expensive Beseler Vu-Graph overhead projectors and was disappointed to realize that, unfortunately, they appeared only 'dim and brown' on the wall (3M, 2002: 57).[11] More light was needed, and that meant better transparencies, and hardware with greater light intensity.

In the development and production of transparencies (by then 3M's core business), the company was tried and tested, for there was already one big buyer: to play through the possible strategies in its war room, the Strategic Air Command alone used around twenty thousand transparency pages per month (Figure 26.2).

The second decisive component was thus a method for producing high-quality, Fresnel lenses out of material that was considerably cheaper, stronger, and lighter than glass. Through a higher light efficiency, moreover, the economy of lamps would increase (the bulb of a Beseler Vu-Graph, for instance, had just a ten-hour lifespan), and the

---

[11] Because the sheets were transparent, the mirror-image inversion was no issue (as it would be with paper).

FIGURE 26.2 A map supervisor on the overhead projector of 'Theaterspiel' (1962)

(*Source*: Pias and Coy 2009: 25)

devices could be built more compactly and cheaply.[12] On 15 January 1962, Appeldorn's team demonstrated the first overhead projector of 3M's own making with the new Fresnel lens, and this was put into production in the summer of 1963. As early as 1966, the light efficiency had increased to such an extent that J. W. Lucas submitted a patent that promised to install the lamp in the head of the projector.[13] In the meantime, moreover, the printed transparencies were no longer being treated as by-products but were being made especially for transportable copiers.

Historically, all of this came together with the first national Federal Aid to Education Program, initiated in 1957: sales shot through the proverbial roof and the overhead projector conquered the classroom. This led to comprehensive debates about the mediatization of education, which might be valuable to an archaeology of the present discussions about the epistemic effects of PowerPoint and e-learning. It is enough to name just a few book titles from the time: *The Teacher and Overhead Projection, A Survey of Overhead Projectors: Educational Foundation for Visual Aids, We Learn with the*

[12] The morphology of early overhead projectors can be reconstructed by historical sales catalogues for light bulbs such as http://www.donsbulbs.com [Accessed 19 July 2018].
[13] Overhead Projector, US Patent No. 3459475, filed 14 November 1966.

Overhead Projector, or the *Handbook of Overhead Projection: Contributions to the Didactics and Methods of Overhead Projection in Instruction and Education* (Schultz, 1965; Crocker, 1969; Bauer and Fischer, 1978; Allendorf and Wiese, 1977). Just as valuable as historical evidence are the educational materials for instructional units available in the form of prefabricated sets of transparencies for the broadest variety of topics. Whether the issue is modularized learning units, which today are treated as the salvation or downfall of the university, or bullet points, which Edward R. Tufte (2003: 23) believed to cause 'detectable intellectual damage to 80%', or whether the *interactivity* of students should be promoted by calling them to the head of the class to fill out transparencies or whether their *interpassivity* is more important, which is purported to foster decontextualized and non-narrative knowledge (Tufte, 2003: 6, 10)—all of these discussions were already underway, and this is not even to mention the topic of increased effectiveness in the classroom.

Already in these historical developments of the overhead projector, it is easy to recognize a division. In the case of war games, the precise goal was to visualize the information of an ongoing and cooperative work process for everyone involved and, with the help of these representations, to reach decisions. The transplantation of overhead projectors into classrooms and conference rooms led to them being used in entirely different ways. Contexts opened up in which no intermediate results could be shown

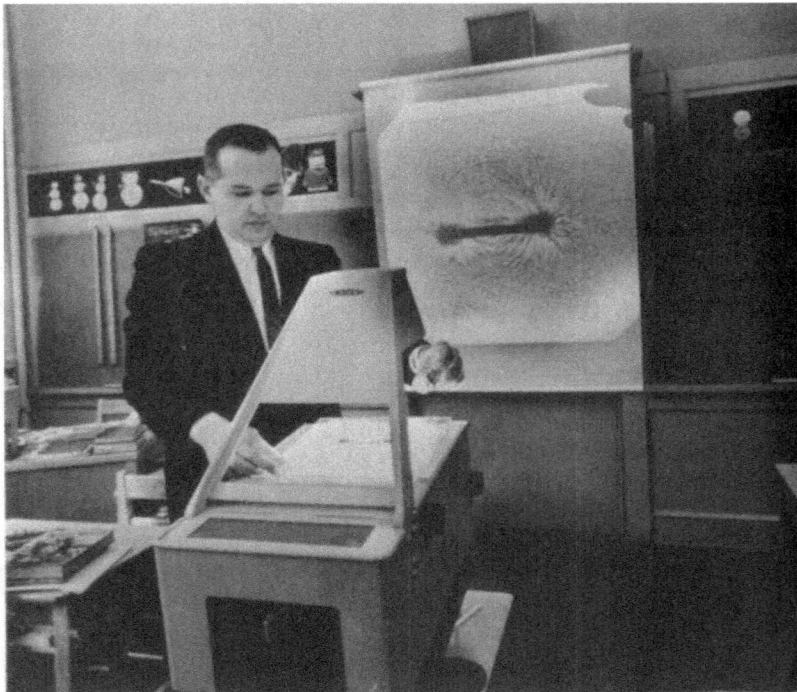

FIGURE 26.3 The overhead projector as a table for experiments (1965)

(*Source*: From Pias and Coy 2004, 29)

(as in bowling scores) and in which there was nothing to be decided (as in the war room), but rather in which the goal was simply to *display* something that was prefabricated. Nevertheless, both things are possible: the overhead projector can imitate the slide projector or the table, or it can also open up entirely new possibilities. The literature that accompanied its increasing popularity explored all of these possibilities: be it that one can reveal information in a step-by-step manner; be it that one can create bar charts with adhesive tape; be it that one can spread iron fillings on the projector to visualize magnetic fields; or be it that one can create a chronological graphic by overlapping multiple transparencies (Figure 26.3).

MEDIA OF DECISION-MAKING

In opening up new possibilities, for some, Edward Tufte amongst them, the overhead projector was also responsible for closing things down. Tufte's particular gripe was the bullet-point which had spread rapidly onto educational transparencies. Tufte saw in the black dots the elimination of all narrative context—the very context required for knowledge to make sense. In his estimation, a 'deeply hierarchical single path structure' serves 'as model for organizing every type of content' (Tufte, 2003: 4), even though not every type of content is suitable for such a structure. To the extent the bullet point remains, users come to assume the world is of a form conducive to point by point summary. Yet as much as this (somewhat overblown) argument makes sense within the framework of a 'poetology of knowledge' (Vogl, 1999), the production and display of transparencies or slides has a computer-related history that does not coincide with that of PowerPoint and similar software. Just as the overhead projector had its beginning in the context of games and war, communal presentations and visualizations with computers also began in specialized work environments. This parallel development to the overhead projector, which took place during the 'axial period of computing' in the 1960s (to borrow Hans Dieter Hellige's term), is at least worthy of a cursory examination. Two pertinent scenes from the history of the computer come to mind.

While working on his 'oN-Line System' (NLS), Douglas Engelbart (1962) had formulated a theory of cultural techniques that culminated in the so-called H-LAM/T System. In the latter, intelligence is not treated as something purely human but rather as a 'heterogeneous collective' of humans (H) with languages (L), artefacts (A), and methods (M) whose interaction can be trained (T). The underlying assumption is that the objective of every 'augmentation' (which was Engelbart's term for interaction) was to dissect complex problems so they become solvable. Even more: every thought process is composed of sub-processes (or sub-routines) from which the process as a whole can be fully explicated. There are seemingly no scaling problems involved with this. Every H-LAM/T System, according to Engelbart, is a collection of routines, a sum of 'process capabilities'. According to the analogical argumentation of cybernetics, human thinking is conceptualized in the model of a program of sub-routines and conditional jumps and is thus itself, in turn, a model for programs with sub-routines and conditional jumps. Here we are dealing with a profoundly Cartesian point of view, to the extent that problems seem to be solvable principally spatially (*res extensia*) by being parsed, in which case it is believed that the sum of optimal partial solutions will also yield an optimal total solution. Accordingly, for instance, the texts that are meant to be written for Engelbart's NLS are to be organized hierarchically and can be opened and closed as a tree structure by those who wish to understand it more precisely. Arguments of Tufte's sort, which maintain that such a nested system of points and sub-points is not applicable for every type of content, could be countered by *pointing* out that Engelbart's system was not conceived to account for everything. His prediction of a 'tremendous increase of augmentation' by means of the external and automated manipulation of symbols was intended for a working group (of engineers, for instance) that primarily had to solve the sort of problems that could be solved by being broken down and for which the optimality principle was valid.

This idea can be clarified by looking at the work of Joseph Licklider (Licklider and Taylor, 1968), which was done around the same time. Licklider begins his argument by claiming that Shannon's theory of communication is insufficient because humans do not exchange information like devices and because communication is not simply a loss-free transfer of information. Rather, communication should be understood as a process, within which 'mental models' are adjusted to each other. This causes problems because it is impossible to read someone's mind, because mental models are not static but are rather constantly changing, because they cannot be repeated exactly, and because they are governed by emotions. A social context is created, according to Licklider, by medially externalizing and reconciling such models. In this regard, the computer opened up entire new possibilities: as a 'plastic or moldable medium that can be modeled [as] a dynamic medium', it is especially suited for presenting to others how someone imagines something and, likewise in a unique way, it also allows for a common idea of something to be developed (see Skoglund, this volume).

The model case for this was, again, the technical project meeting at which different people with different ideas (models) come together in a small group and, by the end of the meeting, are supposed to have agreed upon a common idea. An illustration in one of Licklider's texts from 1968, for instance, shows a group of bridge builders (although the text itself is concerned here with tactical combat) (Figure 26.4).

One of the participants is giving a presentation on a large screen, while the others are sitting at networked computers, where they can work on things, look things up, make improvements, or also simply play around. Such activity, he believed, would not only bring about higher 'creativity', which was of the utmost important to survival at the time of the Cold War and its 'arms race', but rather the computer could also take on the role of a 'switch' or 'interactor' and become a medium that revolutionizes collaborative work with models. In Licklider's work, the computer forms the centre of a communication utopia: within his networked, graphical work environment, 'models in mind' became somewhat comprehendible, visible, and communally processable. In this regard, the interactive presentation in the working group turns out to be an emblem for the

FIGURE 26.4 An illustration from a text by Joseph Licklider: 'At a project meeting held through a computer, you can thumb through the speaker's primary data without interrupting him to substantiate or explain'

(*Source*: From Pias and Coy 2004, 33)

temporal structure of technical modernity, which is defined as a future horizon of problems and their potential solutions. A presentation such as this, moreover, is entirely different from a PowerPoint presentation today in front of a silently listening audience.

Two things can be gathered from this: first, the early computer-supported work environments were thematized within an epistemological framework that emerged from a simplifiable, hierarchical, and recursive order of knowledge and from structural phenomena that paid little attention to narration. In the 1960s, the practices of small group conferences, which had been established since the 1940s, and those of small (interdisciplinary) working groups influenced the conception of software in such a way that its structure is isomorphically similar to this style of working.[14] And perhaps it could be said that, here, this programme of cognitivism (which was closely associated with cybernetics) was implemented as software and that this software spread, lives on today, and continues to have unpredictable effects even after the epistemological expiration of cognitivism itself.

[14] A similar isomorph is mentioned by Tufte: 'the metaphor behind the PP cognitive style is the software corporation itself. That is, a big bureaucracy engaged in computer programming (deeply hierarchical, nested, highly structured, relentlessly sequential, one-short-line-at-a-time) [...] To describe a software house is to describe the PowerPoint cognitive style' (Tufte, 2003: 11).

Second, presentations were primarily thought of in terms of decision processes. They were thus imagined less as a representation of static bundles of knowledge than as a flexible medium for interactive processes of cooperative model discovery. It was from this basis that the problematic transition could take place, for instance, into the pedagogical context. Here, on the one hand, a different sort of content was brought into a format that had historically made sense only in certain contexts and, on the other hand, it was not treated interactively for the sake of reaching decisions but rather only as something to be communicated and learned. It was not for nothing that the ambitious pedagogical projects of the computer age were less concerned with presentation techniques than with interactivity and programming (Oelkers, 2008). Seymour Papert's famous 'Mindstorms: Children, Computers, and Powerful Ideas' from 1980, for instance, went out of its way to avoid the topic of 'mere' presentation entirely. He only mentions the television once—as an obsolete model of teaching that is only good for providing 'explanations' of existing 'material' within fixed curricula. In contrast to that, he believed that the computer (and particularly in the mode of programming and debugging) would be able to undermine Piaget's separation of 'concrete' and 'formal' knowledge and turn children into epistemologists. In Papert's work, the pedagogical project of the information age was to create a framework for 'thinking about thinking' and 'learning about learning'. The project would involve using the computer as an 'object-to-think-with' (Papert, 2003: 416) and it would have no use at all for prefabricated transparencies and the communication of 'content'.

CULTURES AND ECONOMIES OF PRESENTATION

Perhaps at this point it would beneficial to examine the ways in which business communication changed during the 1960s and 1970s, that is, how suddenly (it is assumed) people had to talk to others who spoke different languages, how people were recruited from other departments under the imperative of 'teamwork', and the extent to which all of this led to a greater trust in the visual. Simply for reasons of cost, the first victory was not won by Engelbartian or Lickliderian variants of 'co-thinking', symbiotic, or collegial machines but rather by stupid pointing devices and overhead projectors. In the year 1975 alone, according to Ian Parker (2001), fifty thousand overhead projectors were sold in the United States. In 1985, the victory parade was still going on, and the number of sales reached one hundred twenty thousand. Their advantage over 35mm slides, which were still being used in ambitious presentations at the time, is obvious: anything that could be put on a photocopy machine, which Xerox had been installing in businesses and agencies through its leasing model since the early 1970s, could be turned within seconds into an overhead transparency, though its aesthetic quality might leave something to be desired.

Help with aesthetics came in the form of a hacker solution in 1981, when the mathematician and cryptographer Whitfield Diffie, who was then an employee of Bell Northern Research (BNR), wrote a program that could generate and print out multiple boxes with text that could then be turned into presentation slides. This inspired Bob Gaskins, who likewise worked at BNR and happened to come from a family involved with the industrial photography business (including overheads). The rest of the story is well known: Gaskins founded Forethought, Inc. and developed a program called Presenter, which Microsoft bought up in 1987 and introduced to the market as PowerPoint 1.0 for the Apple Macintosh (Figure 26.5).[15]

Today's ubiquitous laptops and projectors make it easy to forget, however, that PowerPoint was at first not a presentation program at all but rather a tool for making transparencies for overhead projectors (either by conventionally printing and copying or by using Apple's Laserwriter printer, which had been on the market since 1984).

And, with this, an entirely different prehistory leaps into the concept of PowerPoint, namely that of the professional production of overhead transparencies.[16] In the early 1970s—if the corporate history of Genigraphics can be believed—NASA issued a large contract to General Electric with the instruction to develop a system that could produce, with little effort, thousands of graphics for the training of astronauts. To fulfil this task,

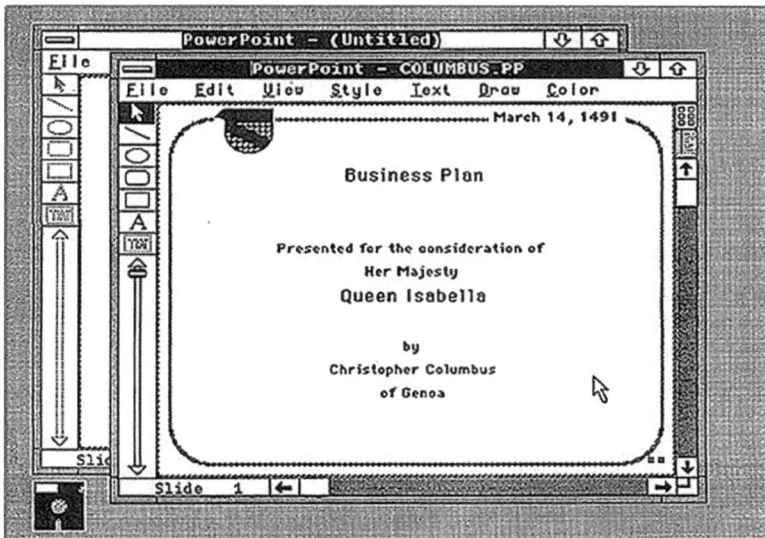

FIGURE 26.5 Design for PowerPoint 1.0 for Windows 2.0, October 1987

(*Source*: Pias and Coy 2009: 37)

[15] Told by himself here: http://www.robertgaskins.com/#ppt [Accessed 19 July 2018].

[16] For example, Douglas Engelbart used 35mm colour slides to present his invention in contexts like fundraising presentations, which is in every sense the complete opposite of the ideas his famous NLS from 1968 stand for (see http://sloan.stanford.edu/MouseSite/gallery/slides2/basic.html [Accessed 19 July 2018]).

General Electric constructed a combination of efficient graphic workstations and filmsetters; in 1979, it outsourced this product line under the name Genigraphics. By the middle of the 1980s, Genigraphics was operating twenty-three production facilities with professional designers who created slides and transparencies for nearly every Fortune 500 company.[17] And somehow it happened that Genigraphics developed the first templates for Bob Gaskins and his future PowerPoint (Endicott, 2000). I should not fail to mention that, in 1988, Genigraphics was charging its clients $240 an hour. It was probably for this reason that 1988 was also the year when PowerPoint 2.0 came out, which in addition to 'send to back', 'bring to front', and 'import postscript', had one highly interesting option among its features, namely 'send to Genigraphics'. At the time, Microsoft logically advertised that now anyone could have Genigraphics produce his or her own self-designed 35mm slides (in colour!) or overhead transparencies. This (like so many things) was not entirely true. For a full year, the designers at Genigraphics simply copied the submitted transparencies as accurately as possible, which clashed somewhat with the artistic honour of their profession. It was not until 1989 that an automated solution was implemented.

Thus PowerPoint's metaphorics and mode of operation have become evident: that it was never concerned with developing anything visually for a ludic or scientific event (such as for bowling or for a war game or for a discussion between engineers), and that it also never had anything to do with enabling someone to think aloud while writing—this much is written in the historical investment documents of PowerPoint, which was not even conceived as a presentation tool but rather simply as a front end for making transparencies for overhead projectors, with the aim of saving Genigraphics and other companies a bit of money. Bob Gaskins' first proposal of the business idea, which is dated 14 August 1984, leaves no room for doubt in this regard: it says that Presenter, as PowerPoint was then called, is for 'people who make presentations to others' and who produce, thanks to a 3.5 billion dollar industry, around 520 million 35mm slides and 380 million overhead transparencies per year, of which about 60 per cent (and soon 100 per cent) are being generated by computers because they usually involve no more than text, diagrams, spreadsheets, and corporate logos.[18] With the computer, this could be done more effectively, quickly, and cheaply, above all if the company worked together with large overhead producers such as 3M, Beseler, Bell & Howell, or Elmo, each of which is explicitly named as part of a potential joint venture.

Yet not even Gaskins himself was able to foresee that the overhead metaphor would be taken even further by a combination of computer graphics and projectors. His only concern was for which markets transparencies could be made with Presenter. These markets, however, were so fundamentally different that the decision came down to a particular ideology. In 1986, Gaskins had sorted out the situation with a comprehensive market analysis and extensively discussed the media-historical and epistemological

[17] http://www.ge-genigraphics.org/pages/about.html [Accessed 19 July 2018].
[18] http://www.gbuwizards.com/files/gaskins-original-powerpoint-proposal-14-aug-1984.pdf. For oral history see also: https://blog.indezine.com/2012/08/powerpoint-at-25-conversation-with.html [Accessed 19 July 2018].

differences between slide projection, overhead projection, and video projection (Gaskins, 1986: 16–23). According to his analysis, the 35mm slide possessed a certain prestige because its production costs were such that only large companies could afford them. Its presentation in large, darkened rooms was suitable for lectures in which discussions could only be held at the end or (better yet) not at all. And it also had a clearly higher entertainment value than the overhead transparency because one could admire the colourful 'artistry' of the slides and follow their 'fancy transitions'. Overhead projection, on the contrary, was suitable for small group meetings that take place in small, bright (and probably enlightened) rooms in which discussions may (and should) be held during the presentation and in which there are no fancy transitions to distract anyone's concentration from the information. These presentation styles are so different, both aesthetically and historically, that video projection, he noted, would have to decide between being a 'video replacement for overhead' or a 'video replacement for 35mm'. Thus, for Presenter, Gaskins had to make his own clear (and political) decision, which was that his program would not be concerned with 'artistic slides' or eye candy or with competing against Genigraphics but rather with producing 'electronic overheads' as a collaborative tool for work. 'Our target is overhead style' (Gaskins, 1986: 17, 22) and thus did Gaskins—a computer scientist who was certainly aware of Engelbart's and Licklider's work—express his commitment to the small, fact-based, and communal working group that thought things through together in a well-lit room. As to when and how this business idea not only gloriously ascended but also came to develop an unexpected waywardness that allowed PowerPoint itself to become a presentation tool with 'fancy transitions', these facts are ironically tied to the balance sheet of Genigraphics, which stood on the edge of ruin in 1994, when it was not coincidentally rescued by a company that produces LCD screens. One of the new features that were introduced in PowerPoint 3.0, which came out in the autumn of 1992, was the ability to *project* one's presentation (instead of simply printing it).[19]

REFERENCES

3M. 2002. *A Century of Invention: The 3M Story*. Minnesota: 3M Company. http://multimedia.3m.com/mws/media/171240O/3m-century-of-innovation-book.pdf [Accessed 19 July 2018].

Allendorf, Otmas and Johannes Gerhard Wiese. 1977. *Taschenbuch der Overhead-Projektion*. Beiträge zu einer Didaktik und Methodik der Overhead-Projektion in Unterricht und Ausbildung. Cologne: Interorga.

Bauer, Ingrid and Dieter Fischer. 1978. *Wir lernen mit dem Overhead-Projektor*. Neues Lernen mit Geistigbehinderten. Würzburg: Vogel-Verlag.

Crocker, A. H. 1969. *A Survey of Overhead Projectors*. Educational Foundation for Visual Aids. London: National Committee for Audio-Visual Aids in Education.

[19] This chapter is based on my introduction to Pias and Coy (2009: 16–44). In the meantime, this historical research has been continued by Knoblauch (2013: 26–49) and Robles-Anderson and Svensson (2016). Also Robert Gaskins (2012) published a book on his invention.

Derrida, Jacques. 1993. *Memoirs of the Blind: The Self-Portrait and Other Ruins*. Chicago: University of Chicago Press.

Dilly, Heinrich. 1975. 'Lichtbildprojektion—Prothese der Kunstbetrachtung'. In Irene Below (ed.), *Kunstwissenschaft und Kunstvermittlung*. Gießen: Anabas, 153–72.

Endicott, Jim. 2000. 'Growing up with PowerPoint'. *Presentations* (February), 61–6.

Engelbart, Douglas C. 1962. *Augmenting Human Intellect: A Conceptual Framework*. Summary Report AFOSR-3223 under Contract AF 49(638)-1024, SRI Project 3578 for Air Force Office of Scientific Research, Stanford Research Institute.

Gaskins, Robert. 1986. 'Presenter: Product Market Analysis', 27 June. Sunnyvale, CA: Forethought, Inc.

Gaskins, Robert. 2012. *Sweating Bullets: Notes about Inventing PowerPoint*. San Francisco and London: Vinland Books.

Grimm, Herman. 1897. 'Über die Umgestaltung der Universitätsvorlesungen: Über Neuere Kunstgeschichte durch die Anwendung des Skioptikons'. In *Beiträge zur Deutschen Culturgeschichte*. Berlin: W. Herts, 276–395.

Knoblauch, Hubert. 2013. *PowerPoint, Communication, and the Knowledge Society*. Cambridge: Cambridge University Press.

Licklider, Joseph C. R. and Robert W. Taylor. 1968. 'The Computer as a Communication Device'. *Science and Technology*, 76, 21–31.

Meyer, Bruno. 1883. *Glasphotogramme für den kunstwissenschaftlichen Unterricht: Erstes Verzeichnis (Nr. 1-4000). Mit einer Einleitung und reich illustrierten Abhandlung über 'Projektionskunst'*. Karlsruhe: Self-Published.

Oelkers, Jürgen. 2008. 'Kybernetische Pädagogik: Eine Episode oder ein Versuch zur falschen Zeit?' In Michael Hagner and Erich Hörl (eds), *Die Transformation des Humanen. Beiträge zur Kulturgeschichte der Kybernetik*, Frankfurt am Main: Suhrkamp, 196–228.

Papert, Seymour. 2003. 'Mindstorms: Children, Computers, and Powerful Ideas'. In Noah Wardrip-Fruin and Nick Montfort (eds), *The New Media Reader*. Cambridge, MA: MIT Press.

Parker, Ian. 2001. 'Absolute PowerPoint: Can a Software Package Edit our Thoughts?' *The New Yorker*, 28 May, 76–87.

Pias, Claus and Wolfgang Coy. 2009. *PowerPoint. Macht und Einfluß eines Präsentationsprogramms*. Frankfurt am Main: Fischer.

Ratzeburg, Wiebke. 1998. 'Die Anfänge der Photographie und Lichtbildprojektion in ihrem Verhältnis zur Kunstgeschichte'. MA thesis, Berlin.

Robles-Anderson, Erica and Patrik Svensson. 2016. '"One damn slide after another": PowerPoint at every occasion for speech'. *Computational Culture*, 15 January. http://computationalculture.net/one-damn-slide-after-another-powerpoint-at-every-occasion-for-speech/ [Accessed 19 July 2018].

Sayre, Farrand. 1908. *Map Maneuvers*. Springfield, IL: Staff College Press.

Schmid, M. 1894. 'Über Lichtbilder-Apparate im kunsthistorischen Unterricht'. In *Offizieller Bericht über die Verhandlungen des kunsthistorischen Kongresses zu Köln*, October. Nürnberg: Stich, 1–3.

Schultz, Morton J. 1965. *The Teacher and Overhead Projection*. Englewood Cliffs, NJ: Prentice-Hall.

Tufte, Edward R. 2003. *The Cognitive Style of PowerPoint*. Cheshire: Graphics Press.

Vogl, Joseph (ed.). 1999. *Poetologien des Wissens*. Munich Fink.

Wilson, Andrew. 1969. *The Bomb and the Computer: Wargaming from Ancient Chinese Mapboard to Atomic Computer*. New York: Delacorte Press.

CHAPTER 27

··

PAPER SHREDDER

··

ALICE COMI

INTRODUCTION

A relic of the analogic era, the paper shredder is largely neglected in the everyday life of organizations. It is often relegated to hidden and somewhat 'dirty' spaces of modern organizations, such as office corners, print shops, and waste rooms. Yet this unexciting object played a prominent role in many episodes that marked our history of organizations and organizing. One infamous example is the Enron scandal: the accounting firm Andersen was found 'guilty of shredding' tons of documents (Wastell, 2002) after the US Security and Exchange Committee launched a probe into Enron. The film *Enron: The Smartest Guys in the Room* (2005) depicted Andersen employees toiling around the clock to operate fork lifts and throw bales of paper into large shredders. The shredding operation extended for over two weeks, as dozens of trucks filled with Enron papers were sent to Andersen's offices in Houston, Chicago, London, and Portland for destruction (Market Watch, 2002).

A far larger shredding operation of recent history took place in the former German Democratic Republic, across district offices of the Ministry for State Security (Stasi) (Vismann, 2008). In the dying days of the East German regime, Stasi employees were feeding their shredding machines in a systematic attempt to withhold secret documents from posterity. A large-scale project is currently underway to piece back together some 17,200 sacks of shredded paper. While software played a role in reconstructing some of the destroyed documents, work will have to proceed manually as scanning technology is not sophisticated enough (Oltermann, 2018). This testifies to the power of the information destruction that is performed by the paper shredder, and its tendency to 'resist' even the most sophisticated technologies of information production.

The paper shredder also featured in the 1986/7 Iran–Contra scandal, when Lieutenant Colonel North was convicted for shredding documents that would have revealed his involvement in illegal activities, such as the sales of arms to Iran, and the diversion of proceeds to the anti-Sandinista Contras in Nicaragua. Though the then president

Ronald Reagan escaped implication, this was not the case for President Nixon whose re-election committee was found to have used a bulky shredder to destroy potentially incriminating documents during the Watergate scandal (Berke, 1986). As Richard Berke wrote in the columns of the *New York Times* in late 1986: 'In this capital city of paper and secrets, one thing that often comes between the most classified of documents and the eyes of a curious public is the paper shredder. Name a national crisis and, almost inevitably, one of these trusty office machines that transform paper into confetti is lurking behind the scenes.'

In the shift to the digital era, the paper shredder still plays a role. In the UK, the Data Protection Act of 1998 mandates that 'personal data processed for any purpose or purposes shall not be kept for longer than is necessary for that purpose or those purposes' (Great Britain, 1998: 48). This implies that all files containing names, addresses, financial and legal details are securely destroyed when they are no longer necessary. A quick Google Maps search limited to the Greater London area returned approximately one hundred companies operating in the business of 'information destruction'. The range of shredding services has been extended to include destruction of paperless media such as hard drives, memory sticks, and X-rays; as well as destruction of branded products such as badges, uniforms, and counterfeits. In contemporary organizations, the shredding

FIGURE 27.1 A shredding machine placed next to a shredding console and bags

(Photo by the author)

machine has been replaced by, or placed next to, shredding consoles and bags, which store the items that will be shredded off-site by specialized companies (Figure 27.1).

In this chapter, I ask the questions: How does the paper shredder enable and shape practices of organizing? In what lies its organizational force, and what are its organizational effects? To address these, I will be interweaving theories of organization with empirical observations on the paper shredder.

A Brief History of the Paper Shredder

The practice of destroying documents is as old as the practice of writing on paper. At least since information has been recorded on paper, there is a need to destroy documents containing inaccurate or sensitive information (Reichmann, 2015). In ancient times, documents on papyrus, parchment, or vellum were ripped or burnt—as exemplified by the legendary fire of the Royal Library of Alexandria, Egypt. Here, ancient classics (deemed to be sacrilegious) were destroyed—with historians debating whether responsibility should be attributed to the Christian Church or to Muslim Caliphs (Furlani, 1924). It is further retold that, during the Middle Ages, amanuensis monks used to censor ancient classics, by throwing out the originals and making amendments in their transcriptions.

Since then, the practice of information destruction has taken giant steps. The invention of the modern paper shredder is attributed to Abbot Augustus Low of New York, who patented a 'waste-paper receptacle' in 1909 (Reichmann, 2015). Low's invention was designed 'for use in offices and other places where not only the collection and storage of waste paper is desirable, but also its cancellation or mutilation in such manner as to render it unavailable or unintelligible for re-use or for information' (Low, 1909: 1). However, the waste-paper receptacle was never manufactured, due to Low's premature death a few years after his filing of the patent (Shred All, 2015).

The first paper shredder was manufactured in Germany in 1935 (Reichmann, 2015). Finding inspiration in the design of a hand-cranked pasta maker, Adolf Ehinger produced a hand-cranked paper shredder that was large enough to handle one sheet. Apparently, the inventor had been interrogated about anti-Nazi leaflets found in his bin, and created a shredding machine to destroy documents that could have got him into trouble. He subsequently added an engine to automate the device, and started selling his patented shredders to governmental agencies and institutions.

Ehinger's shredder became increasingly common during the Cold War. In the 1960s, his company (EBA Maschinenfabrik) sold the first cross-cut shredder, which cuts paper into tiny confetti rather than just strips (so as to offer increased security) (Reichmann, 2015). The cross-cut shredder grew in popularity after the American Embassy in Tehran was occupied by militant students, who seized confidential documents—of which some were already shredded. As paper shredders at the Embassy were only strip-cut (rather than cross-cut), the strips could be pieced back together to expose US activities in Iran (Povey, 2016).

Yet another model of paper shredder was developed in Germany—more precisely, in the former German Democratic Republic, right before the fall of the Berlin Wall (Vismann, 2008). Because the paper shredders at the district offices of the Ministry for State Security (Stasi) were glutting from destroying tons of classified files, more sophisticated machines were sought. This resulted in the development of a 'wet shredder' (*Verkollerungsanlagen*) that mashed snippets of paper with water (Probst, 2018). Some of the shredded documents (including papers, micro-films, and audio-tapes) were mixed with oil, and transformed into lumps of cellulose (and other materials) that look like stones (Schwartz, 2016).

Before the 1980s/1990s, paper shredders were used primarily by government agencies. Along with the spread of privacy concerns, identity thefts, and dumpster diving, paper shredders have been adopted also by businesses and individuals (Shred Instead, n.d.). Their use in the UK, for example, saw a big upsurge with promulgation of the Data Protection Act of 1998 (Taylor, 2016a). They have become widespread also in the US after the Supreme Court decision in *California v. Greenwood* of 1988 (Shred Instead, n.d.). This established that garbage left on the street for collection does not constitute private property; therefore its search and seizure does not infringe the law (MacDonald, 1994).

As document production has shifted from paper to digital, shredding machines have been improved to handle also non-paper media such as CDs, credit cards, and passports. Since Ehinger's cross-cut shredder, modern machines have grown even better in their shredding capacity and power. Industrial shredders can safely destroy thousands of documents in less than a minute (Reichmann, 2015). This is further sustained by diffusion of companies that offer shredding and recycling services for the private and public sectors alike.

RETRACING THE ORGANIZATIONAL FORCE AND EFFECTS OF THE PAPER SHREDDER

To retrace the organizational force and effects of the paper shredder, I draw on the 'tricks of the trade' offered by Susan Leigh Star (1999: 377) to study material infrastructures and other mundane objects. This approach is largely based on the practice turn in organization studies (Nicolini, 2012), which shifted attention away from neoclassical themes of 'structures' and 'systems' to explore what people actually *do* in organizations. Here, practices are conceived as 'embodied and materially mediated arrays of human activity organized around shared practical understanding' (Schatzki et al., 2001: 2). This implies that organizational activity is inscribed in the human body, is dependent on learning of shared know-how, and is interwoven with ordered constellations of non-human actors—i.e., objects.

The paper shredder can be ascribed to the category of *material infrastructures* (Nicolini et al., 2012), together with other artefacts that make up everyday life in organizations—such as chairs, desks, and wires. As compared to other types of objects such as boundary objects (Star and Griesemer, 1989), material infrastructures perform invisible and silent work (Star, 1999). They are not at the centre of organizational practices, but provide the sub-structures that support such practices. They are uncontested, taken for granted, and remain in the shadow; becoming visible only in case of breakdowns. Such objects are rarely given any attention in themselves, and instead are studied as assemblages that constitute large-scale infrastructures (Star and Ruhleder, 1996) or 'media infrastructures' (Parks and Starosieski, 2015) whereby information is distributed—such as Internet data centres, satellite earth stations, and mobile phone towers. The paper shredder is a type of material infrastructure often sunk within broader infrastructural arrangements.

Susan Leigh Star (1999: 377) offers 'tricks of the trade' to study material infrastructures, objects, or technologies that typically are 'mundane to the point of boredom'. These tricks include *identifying master narratives*, *surfacing invisible work*, and *understanding the paradoxes of infrastructure*. Applied from within a practice-based perspective, Star's 'tricks of the trade' carry a promise to unveil the dramas inherent in apparently dead objects. They aim to bring material infrastructures back to life, making the researcher sensible to the historical context, politics, and authorship (Nicolini et al., 2012: 624) embedded in such objects. In what follows, I use these 'tricks of the trade' as a way to problematize the paper shredder, make it stand out from wider infrastructure, and retrace its organizational force and effects. I then discuss the role of the paper shredder in organizations and organizing.

Identify Master Narratives

Many material infrastructures employ 'what literary theorists would call a master narrative, or a single voice that does not problematize diversity. This voice speaks unconsciously from the presumed centre of things' (Star, 1999: 384). For example, an application form that lists 'heterosexual' and 'homosexual' as alternative forms of gender orientation speaks from within a dominant perspective on gender, which neglects other forms of gender orientation (e.g., gender fluid). A number of literary devices can be used to foreground master narratives; including the *global actor*, *passive voice*, and *personification*—i.e., 'making a set of actions into a single actor with volition' (Star, 1999: 385). Using personification, I identified two competing tales around the shredding machine. The first, dominant in the business press, portrays the paper shredder as an accomplice in corporate frauds. The second, proposed by the shredding industry, portrays the paper shredder as a guarantor of the confidentiality of information.

Surfacing Invisible Work

Material infrastructures embed the history, politics, and authorship of organizational work (Nicolini et al., 2012). Such work is not visible in the shape of material infrastructure, but rather is enmeshed in its constituting artefacts such as computers, cables, and networks.

Surfacing the invisible work inscribed in material infrastructures involves following the traces left behind by a range of organizational actors. For example, to find invisible work in information systems, it is necessary to document the work processes of coders, designers, and users of such systems (Star, 1999). According to Star (1999: 385), this may require 'going backstage, in Goffman's (1959) terms, and recovering the mess obscured by the boring sameness of the information represented'. Applied to the paper shredder, this 'trick of the trade' provides an invaluable frame of analysis to foreground its organizational force and effects. The shredding machine, in fact, emerges as *the* device that sets the boundaries between visibility and invisibility in organizations. It organizes the social practices by which secrecy is manufactured in organizations, and information is concealed from the public. To surface invisible work involves retracing the consequences of the information that is being destroyed, and the social processes that unfold around the destruction of such information.

Understanding the Paradoxes of Infrastructure

In studying material infrastructures such as information technologies, Star (1999: 386) questioned why 'the slightest small obstacle often presents a barrier to the user of information technologies'. She mentioned the example of an extra button to push on the keyboard, recounting the stubbornness of this small barrier, and the seemingly insurmountable challenges that it triggers. Going beyond the visible interaction between the user and the system, she then foregrounds 'the process of assemblage, the delicate, complex weaving together of desktop resources, organizational routines, running memories of complicated task queues (only a couple of which really concern the terminal or system) and all manner of articulation work performed invisibly by the user' (Star, 1999: 386–7). She argues that by describing both the production task and the hidden articulation of work, we can achieve a comprehensive understanding of why some systems work and others do not work. Applied to the paper shredder, this 'trick of the trade' involves looking at breakdowns, disruption, and errors that make the object become more problematic and hence visible in practices of organizing.

ORGANIZATION AND THE PAPER SHREDDER

The paper shredder contributes to aspects of confidentiality, transparency, and secrecy in organizations; often in a paradoxical tension between legal and illegal practices.

On the one hand, its use might protect the confidentiality of information, therefore enforcing laws whereby sensitive information (e.g., medical records) is kept for no longer than is necessary (Great Britain, 1998). On the other hand, the paper shredder might be mobilized to destroy traces of unethical practices, whose disclosure might compromise (or challenge) the continuity of the organization. It therefore plays a double role, serving as either a zealous enforcer of law, or a trusted accomplice in crime. In the Enron case, for example, paper shredders were mobilized to cover up bank frauds, wire frauds, and false statements.

The shredding machine, therefore, mediates the relationship between the organization and its public. It performs the silent work of safeguarding the confidentiality of personal information, becoming visible only in corporate scandals. Yet the paper shredder organizes aspects of confidentiality not just towards the publics, but also *within* the organizations themselves. By destroying any visible traces of organizational secrets, the paper shredder draws boundaries between actors who are trusted to know (in-group), and actors who are prevented from knowing (out-group). This effectively creates a 'circle of secret-holders' (Costas and Grey, 2014: 1440), envied by organizational actors who know of its existence and yet are excluded. Such circles enjoy a privileged status (Ashforth, 2016), regardless of whether the motives for organizational secrecy are virtuous (e.g., protecting trade secrets) or vicious (e.g., engaging in political manoeuvres).

By stepping into the relationship between organizational actors, the paper shredder not only protects organizational secrets, but also organizes secrecy—intended as the socio-material processes through which secrets are maintained (Costas and Grey, 2014). Through the mediating force of the shredding machine, organizational secrets shift from being visible traces on paper to become invisible traces in the memories of actors. Unless a breakdown occurs in socio-material processes of organizing secrecy, this shift remains invisible to the eyes of organizational actors who are not in the know. Such breakdowns create an opportunity for insider threats such as whistleblowing, where confidential information is apprehended and leaked by disgruntled members of the organization.

One such breakdown is nicely depicted in the film *Erin Brockovich* (2000), in which a former employee of the Pacific Gas and Electric Company (PG&E) kept internal papers he was supposed to shred, and handed them over to Erin Brockovich. More recently, this happened in the WikiLeaks case, in which classified documents found their way to the press (rather than ending up in the shredding machine). As published in a blog of paper and data shredding:

What [WikiLeaks] revelations did do, was bringing to the forefront how much information we actually carry with us. Be it in paper form or digitally. And how much at risk that information can be. We and others within the industry found an immediate spike in the number of people wanting their hard drives shredded and their computers destroyed after Edward Snowden spoke out. The more articles came out about him, the more we got calls. People had suddenly become very conscious about leaving sensitive information on old phones or computers.   (Taylor, 2016b)

The controversial force of the paper shredder lies in its capability of destroying, rather than creating information (for good or bad). It is an antagonistic force in the 'society of information', where all activities revolve around the production of information. Showing no mercy, the paper shredder devours information that might have taken years to collect, manage, and organize. Its action marks the difference between information that is transparent, and information that becomes invisible. The shredding machine produces a loss—sometimes an irreversible one. It stands in opposition to mediating technologies that become implicated in the production of transparency, such as the *disclosure devices* (e.g., rankings) that work to make knowledge accessible (Hansen and Flyverbom, 2014). It does so in a somewhat anachronistic way, relying on analogic technology to destroy information on paper, hard drives, and memory sticks.

The power of the paper shredder, however, lies exactly in the physicality of its machinery of information destruction. Even in the digital era, shredding a hard disk is probably the safest way to make sure that the stored information is gone for good. It is true that the availability of 'shredding software', the digital counterparts of the shredding machine, carries the promise to actually delete files by overwriting their position on the machine. Yet its effectiveness remains questionable in the face of the increasing sophistication of data recovery software available for both licit and illicit uses (e.g., forensic investigation vs. industry espionage) (Benny, 2013).

Unlike its digital counterpart, the shredding machine makes the product of its information destruction visible, as confetti shreds, or scraps of hard drive. The material traces of information destructions send a reassuring message to the human agent. Of course, the paper shredder neglects the possibility that the information stored on hard drives was passed on to another machine. Yet, the same risk occurred in the analogic world, as information on paper might have been photo-copied, hand-copied, or even micro-filmed before its destruction. This leads to a quest between information diffusion and destruction, with the paper shredder materializing contradictory aspirations to transparency and confidentiality in modern organizations.

FROM THE PAPER SHREDDER TO A THEORY OF INFORMATION DESTRUCTION?

Albeit appearing as a 'boring object' (Star, 1999), the paper shredder performs important work in organizations and organizing: it becomes implicated in organizational secrecy, contributes to protect confidentiality, and organizes transparency towards external audiences. It further produces secrecy, sets the boundary between confidential and public information, and participates in both virtuous and vicious practices. Even in the shift towards a paperless world, the shredding machine continues to exert an organizational force—with its physicality remaining the safest method of destroying information. The visible traces of hard disks reduced to scraps of metal and plastic confirm that confidential or secret information is gone for ever.

This chapter has implications for organizational theories of material infrastructures and mundane objects (Star and Ruhleder, 1996; Star, 1999; Nicolini, 2012). A brief review of the literature has revealed that material infrastructures are generally defined as technologies that are implicated in the *production* of information. Our theories of material infrastructures have thus far neglected technologies that are implicated in the *destruction* of information. This is surprising, since the production of information unavoidably involves its manipulation and destruction. A more comprehensive theory of material infrastructures should look into the tensions between machineries of information production and destruction. It should also articulate the socio-material practices of information destruction, and how objects (or technologies) mediate the production of secrecy in organizations. The concept of *concealing devices*, which acts as an alternative to Hansen and Flyverbom's (2014) concept of *disclosing devices*, might offer an initial step into this investigation.

It is also interesting to note that current research on confidentiality, transparency, and secrecy in organizations has ignored material infrastructures, in spite of their organizing force and effects. Even in research that acknowledges the materiality of business ethics (Hansen and Flyverbom, 2014), material infrastructures tend to 'disappear' within the practices in which they are embedded—such as the performance of due diligence, the production of rankings, or the analysis of big data. I believe that our theories of organizations would benefit from a closer inspection of material infrastructures (i.e., boring objects such as binders, printers, and scissors) and their role in corporate accountability and social responsibility. It seems that such practices have been abstracted from the material infrastructures in which they are embedded. By reconnecting sociality with materiality, we will gain a richer understanding of concepts such as confidentiality, transparency, and secrecy.

References

Ashforth, Blake E. 2016. 'Jana Costas and Christopher Grey: Secrecy at Work: The Hidden Architecture of Organizational Life'. *Administrative Science Quarterly*, 61(4), NP44–NP46.

Benny, Daniel J. 2013. *Industrial Espionage: Developing a Counterespionage Program*. Boca Raton, FL: CRC Press.

Berke, Phillip R. 1986. 'Shredders: Better than Bonfires'. *The New York Times*, 4 December. http://www.nytimes.com/1986/12/04/us/washington-talk-shredders-better-than-bon-fires.html.

Costas, Jana and Christopher Grey. 2014. 'Bringing Secrecy into the Open: Towards a Theorization of the Social Processes of Organizational Secrecy'. *Organization Studies*, 35(10), 1423–47.

Furlani, Giuseppe. 1924. 'Sull'incendio della biblioteca di Alessandria'. *Aegyptus*, 5(3), 205–12.

Goffman, Erving. 1959. *The Presentation of Self in Everyday Life*. Garden City, NY: Doubleday.

Great Britain. 1998. *Data Protection Act*. London: The Stationery Office.

Hansen, Hans Krause and Mikkel Flyverbom. 2014. 'The Politics of Transparency and the Calibration of Knowledge in the Digital Age'. *Organization*, 22(6), 872–89.

Low, Abbot A. 1909. US Patent No. 929,960. Washington, DC: US Patent and Trademark Office.

MacDonald, Gordan J. 1994. 'Stray Katz: Is Shredded Trash Private?' *Cornell Law Review*, 79(2), 452–90.

Market Watch. 2002. 'Andersen Shredded 'Tons' of Paper'. *Market Watch* website, 14 March. http://www.marketwatch.com/story/andersen-shredded-tons-of-enron-documents-doj-says.

Nicolini, Davide. 2012. *Practice Theory, Work, and Organization: An Introduction*. Oxford: Oxford University Press.

Nicolini, Davide, Jeanne Mengis, and Jacky Swan. 2012. 'Understanding the Role of Objects in Cross-Disciplinary Collaboration'. *Organization Science*, 23(3), 612–29.

Oltermann, Philip. 2018. 'Stasi Files: Scanner Struggles to Stitch Together Surveillance State Scraps'. *The Guardian*, 3 January. https://www.theguardian.com/world/2018/jan/03/stasi-files-east-germany-archivists-losing-hope-solving-worlds-biggest-puzzle.

Parks, Lisa and Nicole Starosielski. 2015. *Signal Traffic: Critical Studies of Media Infrastructures*. Urbana: University of Illinois Press.

Povey, Tara. 2016. *Social Movements in Egypt and Iran*. London: Palgrave Macmillan.

Probst, Robert. 2018. 'Zerreißprobe'. *Süddeutsche Zeitung*. http://www.sueddeutsche.de/politik/stasi-unterlagen-zerreissprobe-1.3811221.

Reichmann, Alex. 2015. 'A Brief History of the Paper Shredder'. *iTestCash* blog, 25 March. https://www.itestcash.com/blogs/news/a-brief-history-of-paper-shredders.

Schatzki, Theodore, Karin Knorr Cetina, and Eike von Savigny. 2001. *The Practice Turn in Contemporary Theory*. London: Taylor & Francis.

Schwartz, Robert. 2016. 'Art Made of Destroyed Stasi Documents Displayed in Berlin'. *Deutsche Welle*, 27 April. http://www.dw.com/en/art-made-of-destroyed-stasi-documents-displayed-in-berlin/a-19215813.

Shred All. 2015. 'Let's Walk Down the History Lane of the Paper Shredder'. *Shred All* blog, January. http://shredall.ca/news/2015/01/lets-walk-down-the-history-lane-of-the-paper-shredder.

Shred Instead. N.d. 'Paper Shredding Facts'. *Shred Instead* website. http://www.shredinstead.com/paper-shredding-facts.

Star, Susan Leigh. 1999. 'The Ethnography of Infrastructure'. *American Behavioral Scientist*, 43(3), 377–91.

Star, Susan Leigh and James R. Griesemer. 1989. 'Institutional Ecology, "Translations" and Boundary Objects: Amateurs and Professionals in Berkeley's Museum of Vertebrate Zoology, 1907–39'. *Social Studies of Science*, 19(3), 387–420.

Star, Susan Leigh and Karen Ruhleder. 1996. 'Steps Toward an Ecology of Infrastructure: Design and Access for Large Information Spaces'. *Information Systems Research*, 7(1), 111–34.

Taylor, Rupert. 2016a. 'Understanding the Data Protection Act of 1998'. *Total Shred* blog, 25 September. http://www.total-shred.com/the-data-protection-act/.

Taylor, Rupert. 2016b. 'Edward Snowden Made Shredding Companies Even Busier'. *Total Shred* blog, 11 October. http://www.total-shred.com/shredding-companies/.

Vismann, Cornelia. 2008. *Files*. Stanford, CA: Stanford University Press.

Wastell, David. 2002. 'Andersen Guilty of Shredding'. *The Telegraph*, 16 June. http://www.telegraph.co.uk/finance/2765455/Andersen-guilty-of-shredding.html.

CHAPTER 28

....

PEN

....

DANIEL HJORTH

AFTER the opening brief attempt to historically locate the pen and see it as a product of a desire to write, I try to arrive at it via the pencil. This is merely to anchor it in a personal time of memory, learning, and school, where also a desire to write intersects with the demands and responsibilities that come with knowledge. This binds this entrance into the study to the companionship of desire between pen, writing, and expression.

The pen, Jotter, Bic, Biro, ball-point pen, roller-ball pen, and fountain pen. This is not exactly an example of the wisdom that we have many names for the things we love, for these are not all 'just pens', they are also different technologies for solving the problem of how to swiftly get lasting marks onto a surface you can bring and file, using an instrument that you can carry on you. A desire to write probably emerges—around 3100 BC— to provide support for memory (Spar, 2000; Lundqvist, 2001). The pen makes a lasting mark, it scratches into a receptive surface that holds and keeps the mark as a record, like with the reed pen used in ancient Egypt around 3000 BC to write on papyrus. The pen that looks and works like something closer to how we know it today is co-emergent with the need to make a longer note, to communicate and control at a distance, beyond a marking, that holds some form of narrative. At what point in history we can speak of proper writing is of course a matter of heated debate amongst historians (Spar, 2000; Fischer, 2001). However, if we imagine that we are not only scratching but telling; that we are not 'simply' counting but describing; if we are not just helping memory but also expressing—then we need an instrument that can serve us during a full sentence, that can host or carry affect. Several sentences that together hold a message, address a reader, convey a view, generate affect would, in effect, be a sign of writing. The earliest coherent texts are dated to 2600 BC (Houston, 2004). Writing would then be characterized by a certain performativity, a text that does what it should by being written (and read); no longer a record to support a counting or to keep track of things needed for an act to be properly conducted. This locates the pen in a different relationship to the system of signs; it becomes a pen proper in relationship to writing. When the pen is engaged in writing, it has a more direct relationship to the 'said' as an equipment used to perform: like the instrument for the musician. Like with the musician, we can develop a sensual, tactile, aesthetic, visceral, and thus personal relationship to the pen. A particular pen affects me in a particular way and my power to produce can be increased accordingly. The pen is a vast potentiality that, when used wisely, has changed the world.

FROM PENCIL TO PEN

One cannot get to the pen without passing the pencil. This is not only because most of us learn first how to handle the pencil, before we are trusted with the pen. It is also because, as technology, it is 'first' in the sense of being more simple and direct—not the pencil as we today know it (which arrived in the late eighteenth century), but as a pin that marks a surface, as something that spares the finger. The pencil's erasable impact on the world is probably the reason for its status as 'first' of writing technologies for many of us as kids. Writing, as something we learn in school, is a process of becoming comfortable with grasping and using the pencil. The pencil, in Winnicott's (1971) terms, is our transitional object that brings us from the writerly mute to the writing (and not only written) 'I'. We discover that, as writers, we can express ourselves without raising our voices, without entering the spotlight. We learn we can scream in a text, and only the readers can hear us. We learn we are writing *and* written (Gendron, 2004; Pullen, 2006; Barthes, 2012; Heehs, 2013). Writing is first a letter that should look good and stand on a line. Then a word that sounds like several letters that are forced to collaborate, yet should still look like a composition of letters that actually belong together. Thus far, it is more drawing than writing (Figure 28.1). The sentence is the next hurdle and it places additional demands on us as writing bodies: capital letter at the start, full stop at the end. They are as much spelled as written. In sheer excitement over that the imagined can be told and the told can be written, my daughter 'writes a book' during these years, before spelling, composition, and grammar have become a concern at all.

Some, like Steinbeck, cannot get past the pencil at all. He is said to have used everything from fifty up to several hundreds of pencils completing his book manuscripts. It is unique for the pencil that it vanishes the more you write. It will vanish equally quickly also when the writing hits the rubbish bin. It gets blunt and needs to be sharpened. It is literally ground down. The pencil can also be broken in many pieces, and all the parts would work equally well as pencils. It works by friction, like all pens, but requires a bit more effort from you to make a mark, to scrape off some of it onto the surface. Nothing flows by the nature of being fluid, as with ink, it is an exercise of friction.

FIGURE 28.1 Drawing words

There is something about the fact that pencil-writing works with coal, with a graphite tip. Coal that is one of the most abundant substances on the planet, also as hydrocarbon compounds. Graphite is the most stable form of carbon under standard conditions, the books tell us. This is a fantastic contrast to what it has done and can do in the form of a graphite tip on a pencil. The most stable form of the most common mineral/atom/compound is implicated in some of the most upsetting deeds mankind has seen. The visceral sensation of working with a pencil is indeed quite different from working with a pen, but they are both aiding the process of inscribing, as in adding to the world, by marking it.

Marking the world using a pencil is first noted by Conrad Gesner (1516–65), a Swiss natural scientist, who had purchased such a writing instrument in England. An English coalmine in Cumberland (now Cumbria) was the source of fine graphite pencils (Derwent, an early company). But it was expensive. German (Nürnberg) methods refined this. A French method (Nicholas Jacques Conté; Conté is a company that is now in the Bic corporation group) revolutionized the making of pencils as graphite and clay could now be used and the scale of hardness was then established—from 8B to 6H (the latter is hardest). Faber-Castell (founded in 1761) is now, by far, the world leader in pencil making, followed by Staedtler (also German).

The history of pencils and pens is also a history of entrepreneurs, patents, market fights, competition, and the gradual emergence of a really uncostly and reliable writing instrument. There is a US patent from 1888 of a ball-point pen of an early type invented by John Loud. The Biro brothers then—after several patents in the United States, England, France, Germany, and Czechoslovakia—would represent the next significant step when they introduced a significantly better construction for which they filed a patent in 1943. Paper Mate was the next step—during the 1950s—with an enhanced solution, and the company, later acquired by Gilette (1955), was successful thanks to a superior refiller function. In 1953 Parker introduced the Jotter (John F. Kennedy's favourite), and in 1953, the French baron Marcel Bich started to produce a refiller with built-in ball: the price could be decreased significantly and the Bic pen was born. Also the Swedish Ballograf, starting in 1947, made recognized ball-point pens.

If we inquire into what writing instrument has drawn the most affect throughout the history of writing, it is probably the fountain pen. Already in the year 953 there is a mention of a pen (owned by the Caliph Mu'izz of Egypt) with a sack of ink. Notes of a fountain pen are also found from Johann Mathesius, Germany, from 1663 when a Mr Coventry, England, writes about receiving such a gift, and in 1748 Catherine the Great of Russia is said to have used an 'eternal feather' to write her memoirs. The history of the modern fountain pen properly starts in 1884 when the US insurance company agent Lewis Edson Waterman registere a patent for a functioning fountain pen. The US Parker Pen company enhanced the construction of the fountain pen in the 1890s and the golden age of the fountain pen is considered to be 1925–40, and includes brands such as Waterman (founded 1884, USA), Parker (founded 1888, USA), Montblanc (founded 1906, Germany), and Sheaffer (founded 1912, USA), after which the ball-point pen achieved success and took over as an ink-based writing instrument (Petroski, 1989; Gostony and Schneider, 1998; Lambrou, 1998).

PEN AND WRITING

Now it is difficult to think of the pen as separate from writing. Writing, of old Germanic origin—*rizan*, meaning to write, scratch, tear, and *reissen*, meaning to tear, pull, tug, sketch, draw, design, or drag—is in Old English introduced as *wrïtan*, meaning to score, form, to carve, to outline, or draw the figure of (something). Old Norse—and still present in the Scandinavian languages—is *rita*, meaning to draw, sketch, or outline. In Indo-European languages to write originally means to carve or scratch or cut.

Hand–ink–pen is an assemblage that organizes the desire to write. With a pen the scratching is refined and enhanced by ink. Ink has been used for writing with 'pens' or brushes since the twenty-third century BC in China. More modern versions of ink appear in China 256 BC. But several ancient cultures independently developed the technology to write with ink, including South India and Rome. In the fifteenth century, notably, a new type of ink was invented to enable Johannes Gutenberg's revolutionary printing press. This is also when, properly speaking, we left the monopoly of manuscripts—literally inscriptions made by the hand, *manus* (Latin *manu scriptus*, written by hand)—and entered the machine era of multiplication, copying, and mass reproduction (cf. Benjamin, 2008). With that, the spread and circulation of ideas accelerate and the change of the world with it, and the pen took a hit. The text does not need to scratch itself into the world to enable the flow of thought to continue into the mind of the reader. Thoughts in the shape of ideas, expressed by letters in words, make their job equally well as unscratched yet read. Today, the paper is no longer necessary for this flow to continue. The keyboard, a display, a screen, a monitor is all you need to publish and circulate ideas, to add to the world, to make it change. However, the flow of ink is thereby interrupted in this physical-material sense of a liquid form giving an outline to the letters that writing uses. The art of using the pen to let the ink express what you want to 'say' in a text is perhaps withering. Handwriting is only to some extent a priority in schools today. Developing a handwriting style is of minor concern to young people. Style is more a question of fonts that others have created; that one can chose in a software. The clarity of the image—the pen is mightier than the sword—has somewhat declined. The message is of course still the same, and correct. The prominence of the pen as a writing instrument has been taken over by the mobile phone that has the Internet as its fluidity, as its ink.

'Pen' comes from the Latin *penna* (a feather or plume). The pen and the feather have perhaps flight in common. The feather allows the bird to take flight. The pen (*penna*) allows thought to take flight. What the birds' flight means is for the inaugurer to interpret: *inaugurare* (Latin) means to take omen from the flight of birds. When you are taken up in the community of professors (typically overrepresented in the history of writing), you are inaugurated, you are acknowledged as someone that should know how to take omens from the flight of birds; someone that understands what the signs mean. You are supposed to act as an augur (*in* + *augurare*). To be a reader—Old English *rædan*, Old Norse *rada*, Old High German *ratan*—also means to forebode and to give advice or counsel.

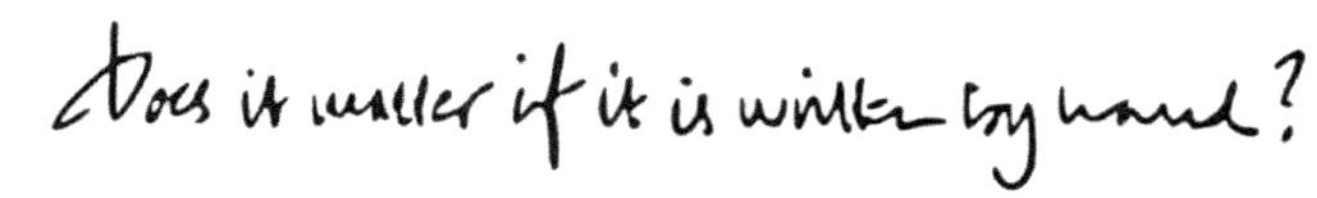

Does it matter if it is written by hand?

FIGURE 28.2 Writing by hand

Does it matter if it is written by hand? (Figure 28.2). I believe so. There is not so much 'me' in the Helvetica, Book Antiqua, or Courier. There is no guarantee of this being the case with the handwritten of course. Whether there *is* personal style or not is perhaps a question of the choice of words rather than how they look. Written by hand though, with a pen, the opportunities for personalization seem greater. Or, to put it differently, you have to avoid a number of contingencies that would—taken together—mix into a rather unique form if a template were not pressed upon them. Gutenberg's press, towards the end of the 1430s, literally press a standard upon writing that limits the 'I's' graphical individuality. The mass-producible copy is released from the waiting room of technologies. Editing is more difficult when writing progresses by hand, and you think one more time before you put pen to paper. Graphologists can thus trace 'you' in comparison with others, based upon the curves and slopes and bindings. The irony of the 'handwritten' example in Figure 28.2 is of course that it is now infinitely digitally reproducible. It is written by a digital stylus onto a tablet screen, into the world of ones and zeroes—a perfect digitalization that perhaps makes the idea of an original obsolete? And yet, the pleasure I get from the tactile experience of picking up my fountain pen and pressing its nib onto a sheet of paper, releasing the ink, is quite different from working on an ever so well-balanced keyboard. Apart from the beauty of the object, the feel of it in my hand, there is something about the movement of my hand leaving a rather precise trace that expresses content, a message, words. Movements become words recording a particular body's movement. We can understand Gutenberg's press, from the perspective of Kittler's statement 'Media determine our situation' (1999: xxxix), as distancing writing from its dependence on the moving hand that grasps the pen. For expression, the new situation was precisely the independence from *situ* or original place from where it needed to be copied. Ideas could now travel at greater speed, to mass audiences via the democratized aesthetics of standard types.

Pen–Writing–Expression/Style is an assemblage that organizes the pen-as-threshold, where the nib enters expression into the world. The pen writes as if to give personal style. Style, the only way becomings can be expressed (Deleuze and Parnet, 2007: 3), is the source of writing, just as charm is the source of life (Deleuze and Parnet, 2007: 5). 'Style is the economy of language' (Deleuze, 1988: 113). Style is here thought as a deviation from a norm, which is how it efficiently reaches an expression.

> After all, what we came to understand by style, regardless of its definitional vagaries
> [...] is already a deviation from a more originary meaning, a metonymic slippage
> from the stylus or metal tip used to engrave or impress in a wax tablet to the
> manner of reworking what had been thus incised or expressed, as well as a spelling

deviancy—stile becoming style—due to the erroneous association with Greek stulos: column, whereas its reconstructed Indo-European root would be "-sti": to pick, hence sting, stimulus.   (Milesi, 2013: 219)

Towards the end of this quote, Milesi arrives at emphasizing writing-style similar to how we have described it above; noticeable as the nib's pecking at the world. It is not merely a scratching, but a pecking (nib), a touch that wants to add something for the purpose of changing. Style that gives expression a sting, that makes it stand out and deviate from the norm. Style irritates in the Latin sense of the word—to excite, to provoke, or to annoy (cf. Nietzsche's style; Deleuze, 2006). To annoy is also to make impatient, to potentialize in the sense of compressing a spring—the pen's nib—that is thus provided energy that makes movement or action incipient. To provoke would then be to call forth or to challenge, which is what we have associated with writing above—to call for an answer that confirms I'm not alone, an answer that of course changes everything.

Writing is the physical ownership of style. Traditionally writing was about moving an instrument over a surface that registered and preserved that movement. Writing, as Rhodes and Brown so eloquently put it (with reference to Clifford, 1986), '...is an ethics of choice rather than an ethics of rules, where to write is always about making representational decisions—decisions which are apparently ethical because they construct others, and embalm transient events into a textual permanency' (Rhodes and Brown, 2005: 479). The pen is this instrument that aids your ambition to add to the world your textual expression that, when received, is also recorded as a trace of your signature, your mark—you have become responsible. This is not necessarily a choosing subject that represents the other and is therefore, as author, responsible for an entity/text (Kristeva, 1986). There is desire to write, and expression is an affect that finds its words, and both desire and expression need a body that—in the process of writing—becomes written as a subject of expression, as an affected body that affects. The school is a place for learning conversations, and the scholar is one that engages in such conversations because of a desire to learn. The writing scholar seeks to capture the transient and reach beyond the immediate audience and moment. The pen is the technology that inaugurated this possibility.

The Pen and Media, Technology, and Organization

Kittler's prophesy that it is we who adapt to the machine, and not the other way around, can help us explain how the bureaucracy flows out of the pen. Writing's desire to flow smoothly onto the surfaces of the world, and management's desire to control the organized world finds an empowering technology in the bureaucracy. 'Organization studies' is therefore a great site for inquiring into the role and impact of the pen. Central to the emergence of the modern organization are the principles of bureaucracy that Weber has distilled and formulated: what he called 'modern officialdom' is operating in a jurisdictional area governed by rules, laws, or administrative regulations. Official duties of officials (that qualify for their jobs only when fulfilling the requirements specified in rules) are regulated when it comes to the authority to give commands that these duties require, and the methods for exercising the corresponding rights demanded to fulfil these duties. Office hierarchy is also regulated in a specified system of super- and subordination. The management of the office is based upon written documents. 'The body of officials working in an agency along with the respective apparatus of material implements and the files makes up a *bureau*...' (Weber, 1978: 957, emphasis in original). Office management then presupposes training and specialization, and full-time devotion to the task, which require that the officer '...follows general rules, which are more or less stable, more or less exhaustive, and which can be learned. Knowledge of these rules represents a special technical expertise which the officials possess. It involves jurisprudence, administrative or business management' (Weber, 1978: 958). Writing is crucial for this bureaucracy to work. And the *bureau* of bureaucracy is literally the desk that supports this writing. More specifically, the bureau originally described a cloth covering for a desk, all there to support or to make writing smoother. Efficiency based on the reduction of friction, resistance, and the manager as bureaucracy's pen, recording, registering, reducing friction. The pen can be understood as implicated in one of the sources of this desire to write smoothly. A desire to control is of course the source of writing in the case of bureaucracy, although not simply a control that seeks power-over (Hosking et al., 1995, as in the common description in conventional bureaucracy-bashing popular management literature, Osborne and Gaebler, 1992). It is also a question of controlling when the recruitment process secures the meritocratic basis of the organization and when an ethos of the office secures an impartial running of the administration (du Gay, 2000). As both MacIntyre (1981) and Bauman (1989) have argued, meritocratic and fair/impartial employees in organizations still have to face the question of how far duty associated with one's role as bureaucrat can be prioritized when ethical decisions need to be taken. What is it that prevents bureaucracies from becoming utilitarian machines? We cannot pull out our clipboard and check a box on a form without asking what this means for the other whose life is affected by that box/form. Bluntly put—a societal telos suggests that the other human rules, it is not the rule that rules. The bureau does not determine the pen.

Nietzsche stressed that there is a constant danger that the ethical question of good and bad is smothered over by the moral question of good and evil (Spindler, 2009). We would have to think of both our assemblages: hand–ink–pen and pen–writing–style/ expression. It is in those assemblages, related, that we would have to seek the 'pen that wants'. There is desire to control and to express that hold these assemblages together, and although we would now have gluttonous excel software and savvy spreadsheets that crave ever-more data, the assemblages have not changed much. We desire writing as an expression and as a control, and we adapt to the pen, the press, the typewriter, the word processor. The moral good of doing what the procedures require from you and the spreadsheets crave, can thus be the bad of ethical passivity, neglecting to resist protocol or abstaining from preventing mechanical reproduction. Losing sight of the human other because of authority or routine, Bauman says, seems deceptively easy when an apparatus of 'beautifully working' rationalization wants to further itself into its incipient future, its 'natural' nextness. This can be the processual dynamics that leads up to the dropping of the bomb, the pressing of the button, the sanctioning of 'more of the same'. Bureaucracies, Bauman says, have a tendency to shift focus from the ends (writing) of the outcome to the means (bureau) of the process. Again, the pen is implicated; the bureau has merely given it an even smoother surface.

We can be organized by our pens (budgets, protocols, regulations, spreadsheets, business plans) rather than organize by our pens. Organizations are always more than the people through which they are relationally constituted. There are also the relationships and the situation of people being related. As Massumi reminds us: 'Is it an affront to objectivism to say that there is, in addition to the ingredients, their interaction and its effect?' (2002: 221). You and your pen, your relationship and its effect constitute a particular bloc of becoming in the history of organizations. But this bloc necessarily opens its unique creation process, each time. Bureaucracy can, following Weber, account for the general conditions for organization to emerge as a series of events, but it cannot provide a theory for predicting the eventness of the events. Massumi again: 'The singular, contingent ingredients give it [the event] its uniqueness, its stubbornness in remaining perceptible itself in addition to being a member of its class—its quality. The event retains a quality of "this-ness," an unreproducible being-only-itself, that stands over and above its objective definition' (2002: 222). This is also why the bureaucracy, the procedures, the regulations, cannot exhaust the general conditions for organizing. The pen's way over the sheet of paper is always open as long as it is held by a trembling hand. It is never the template, *the* way (as Nietzsche said), the bureau, but always the situational uniqueness of the event of expression that provides the conditions for the ethical relationship to the other. There and then is where the question of good and bad requires an answer, a responsibility, a standing by the mark you have just added to the world, with the nib of your pen, the point in which you touch, flow into the world (Ink, Ich, I). If organizations can be cont-rolled, it also means they can be pro-rolled (Hjorth, 2012), i.e., its constant relational-situational capacity for becoming different, its incipient rolling into unique expressions, can be affirmed rather than negated. This is the Nietzschean nib pecking/ hammering at the organized world. Foucault referred to it in his inaugural speech, when he said: 'But what is dangerous about people speaking? In that their discourse continuously multiplies? Where is the danger?' (1982: 221). The pen is not there for the bureau. The bureau is there for the pen. But they are always related. Style is the mobilization of politics by expressing a deviation from the norm. When you hold 'the pen', there is always a choice, because you are in the midst of the assemblages: hand–ink–pen–writing– expression, and you can leave a mark rather than just use a template. That choice is always also ethical.

References

Agamben, Giorgio. 2009. *The Signature of All Things*. New York: Zone Books.

Barthes, Roland. 2012 [1953]. *Writing Degree Zero*. London and New York: Macmillan.

Bauman, Zygmunt. 1989. *Modernity and the Holocaust*. New York: Cornell University Press.

Benjamin, Walter. 2008. *The Work of Art in the Age of Mechanical Reproduction*. London: Penguin Books.

Clifford, James. 1986. 'On Ethnographic Allegory'. In James Clifford and George E. Marcus (eds), *Writing Culture: The Poetics and Politics of Ethnography*. Berkeley: University of California Press, 98–121.

Deleuze, Gilles. 1986. *Essays Critical and Clinical*, trans. Daniel W. Smith and Michael A. Greco. London and New York: Verso.

Deleuze, Gilles. 1988. *Bergsonism*. New York: Zone Books.

Deleuze, Gilles. 2006. *Nietzsche and Philosophy*, trans. Hugh Tomlinson. New York: Columbia University Press.

Deleuze, Gilles and Claire Parnet. 2007. *Dialogues II*. New York: Columbia University Press.

Du Gay, Paul. 2000. *In Praise of Bureaucracy*. London: Sage.

Fischer, Steven R. 2001. *A History of Writing*. London: Reaktion Books.

Foucault, Michel. 1982. 'The Subject and Power'. *Critical Inquiry*, 8, 777–95.

Gendron, Sarah. 2004. 'A Cogito for the Dissolved Self: Writing, Presence, and the Subject in the Work of Samuel Beckett, Jacques Derrida, and Gilles Deleuze'. *Journal of Modern Literature*, 28(1), 47–64.

Gostony, Henry and Stuart Schneider. 1998. *The Incredible Ball Point Pen: A Comprehensive History and Price Guide*. Atglen, PA: Schiffer Publishing.

Heehs, Peter. 2013. *Writing the Self: Diaries, Memories, and the History of the Self*. New York: Bloomsbury.

Hjorth, Daniel. 2012. 'Organisational Entrepreneurship: An Art of the Weak?' In Daniel Hjorth (ed.), *Handbook of Organizational Entrepreneurship*. Cheltenham: Edward Elgar, 169–92.

Hosking, D. M., H. P. Dachler, and K. J. Gergen (eds). 1995. *Management and Organisation: Relational Perspectives*. Aldershot: Ashgate.

Houston, Stephen D. 2004. *The First Writing: Script Invention as History and Process*. Cambridge: Cambridge University Press.

Kittler, Friedrich A. 1999. *Gramophone, Film, Typewriter*, trans. and intro by Geoffrey Winthrop-Young and Michael Wutz. Stanford, CA: Stanford University Press.

Kristeva, Julia. 1986. *The Kristeva Reader*, ed. Toril Moi. New York: Columbia University Press.

Lambrou, Andreas. 1998. *Fountain Pens*. London: Classic Pens Ltd.

Lundqvist, Ingemar. 2001. *Pennan—fetisch och skrivdon* [The Pen—Fetish and Writing Instrument]. Laholm: Byggförlaget Kultur.

MacIntyre, Alasdair. 1981. *After Virtue*. Notre Dame, IN: University of Notre Dame Press.

Massumi, Brian. 2002. *Parables of the Virtual: Movement, Affect, Sensation*. Durham, NC and London: Duke University Press.

Milesi, Laurent. 2013. 'St!le-in-Deconstruction'. In Ivan Callus, James Corby, and Gloria Lauri-Lucente (eds), *Style in Theory: Between Literature and Philosophy*. London and New York: Bloomsbury Academic, 217–48.

Osborne, David and Ted Gaebler. 1992. *Reinventing Government*. New York: Penguin Press.

Petroski, Henry. 1989. *The Pencil: A History of Design and Circumstance*. New York: Alfred A. Knopf.

Pullen, Alison. 2006. 'Gendering the Research Self: Social Practice and Corporeal Multiplicity in the Writing of Organizational Research'. *Gender, Work & Organization*, 13(3), 277–98.

Rhodes, Carl and Andrew D. Brown. 2005. 'Writing Responsibly: Narrative Fiction and Organization Studies'. *Organization*, 12(4), 467–91.

Spar, Ira. 2000. 'The Origins of Writing'. *Heilbrunn Timeline of Art History*. New York: Metropolitan Museum of Art.

Spindler, Fredrika. 2010. *Nietzsche*. Göteborg: Glänta Produktion.

Weber, Max. 1978. *Economy and Society: An Outline of Interpretive Sociology*, 2 vols., ed Guenther Roth and Claus Wittich. Berkeley: University of California Press.

Winnicott, Donald W. 1971. *Playing and Reality*. London: Tavistock.

CHAPTER 29

....

PLANNING TABLE

....

LISA CONRAD

INTRODUCTION

PLANNING tables are ubiquitous and standard tools of organizing. Yet, they look very different depending on the material they are made of. They can be drawn on chalkboards or whiteboards. They can be filled in with magnets, Post-its, cards, plastic chips, or any other thinkable tokens of information (Figure 29.1).[1] A planning table can even be made of Lego. Planning tables are also a standard component of enterprise software; hence its electronic form is very prominent today (Figure 29.2). On top of that, planning tables are not always called this way. They can have other names such as planning bulletin, scoreboard, Kanban board, or simply calendar.

Planning tables dispose of two central capacities uniting the many different variants. Quite conveniently, these two capacities relate to the terms *table* and *planning*. Firstly, planning tables have a tabular structure. Its basic elements are rows, columns, their respective labels and values, and cells. The tabular structure affords clear and unambiguous attribution: an item can go into one cell only. It cannot appear in two different cells, and two items cannot go into one cell. Next to clarity, the structure of the table generates table-specific information by the way it distributes and relates items across the surface (Gregory, 2013: 313). Secondly, their use as a tool for planning implies they can accommodate different drafts of a plan as well as changes to its points of departure. This again implies that planning tables dispose of a certain handiness or

---

[1] I use the term 'token' from Schmandt-Besserat's study on prehistoric accounting and its use of tokens in different shapes. The concept of the token underlines the 'one-to-one correspondence' that is also relevant for planning tables: 'One jar of oil was shown by one ovoid, two jars of oil by two ovoids, and so on' (Schmandt-Besserat, 1992: 6). In the case of planning tables, too, one token (a card, a magnet, a Post-it, a ticket, etc.) usually corresponds to one item of work, be it a customer, an order, a task, or an idea.

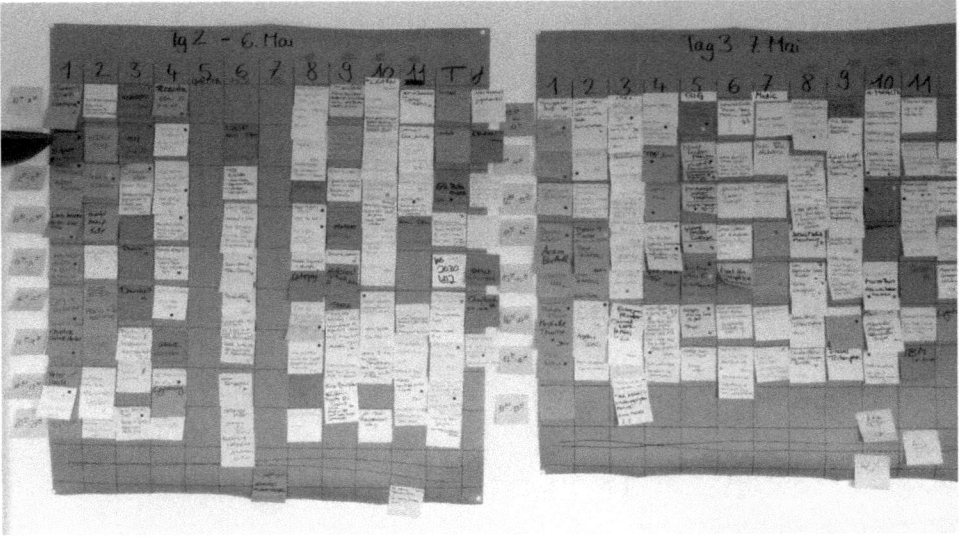

FIGURE 29.1 Paper-based planning at re:publica, an annual conference on digital society

(received upon request (CC BY-SA 2.0))

FIGURE 29.2 Planning table within the Enterprise Resource Planning (ERP) system R/3 by SAP

(Author's own photograph)

user-friendly-ness. The handiness is the other side of the tabular structure. It turns every clear and unequivocal order into a tentative and temporary order since it can be rearranged without too much hassle.

I come to this abstract and generalized description by looking at different real-world planning tables as well as collecting images, reports, literature, and anecdotes.

However, my exploration of planning tables as a media technology of organization started with an impressive specimen of a planning table that I encountered while doing field research at company N., a midsized metal working business in the south of Germany.[2] I will focus on this case so as to flesh out the organizational powers of planning tables. And in this narrative, too, I will first focus on the *table* aspect of the planning table—its capacity to remove ambiguity and to create tabular clarity—and then on the *planning* aspect—its capacity to hold different drafts of how to intervene into a certain context. After that, I will attend to resistances as well as attachments to the planning table. They, once again, help to pin down the features that actually constitute a planning table.

PLANNING TABLE AT COMPANY N.

The 'Plantafel'[3] has existed at company N. 'ever since', so since its foundation in the 1950s. Nobody remembers that it has ever not been there. It consists of four pegboards mounted to the wall in the spacious office of the company's department with the curious, but locally unquestioned name 'labour preparation' (*Arbeitsvorbereitung*). There is ample space for people from different departments to gather in front of the board pointing out and paraphrasing the allocations it displays.[4] This is the planning table's main function: presenting the assignment of certain orders to a date and to an assembly line of production (Figures 29.3 and 29.4).

The planning table has seven columns corresponding to the company's seven production lines. The table's 60 rows correspond to 60 calendar days. The rows consist of rails that are scaled about each other. They hold hand-lettered cardboard cards in different colours. Pink cards represent confirmed orders. Blue cards represent reservations, hence not yet confirmed orders. Green cards stand for free capacities and weekends. White cards point to initial productions demanding special attention. There can be

---

[2] I spent thirty days over a period of three years there (2012–15), doing interviews, hanging out, noting observations, taking pictures, gathering other materials, and trying to understand what was going on. My focus was on the practices and especially the means of coordination. What are the tools used in order to reach coordinated action and how do they, in turn, feed back on the situation? I asked my interviewees about their job activity and how they know what to do and when to hand over, etc. I inquired after the specific tools that are used so as to coordinate divided labour and how they have changed during the last decades and especially in the course of the company's transition to a new enterprise software by SAP in 2013 (Conrad, 2017).

[3] Quotation marks that are not followed by a reference indicate terms originating from the field.

[4] The department 'labour preparation' could be characterized as a 'centre of coordination', a term Lucy Suchman uses to label a certain 'family of settings'. These are settings 'dedicated to the ongoing management of distributed activities in which one set of participants is charged with the timely provision of services to another' (Suchman, 1993: 114).

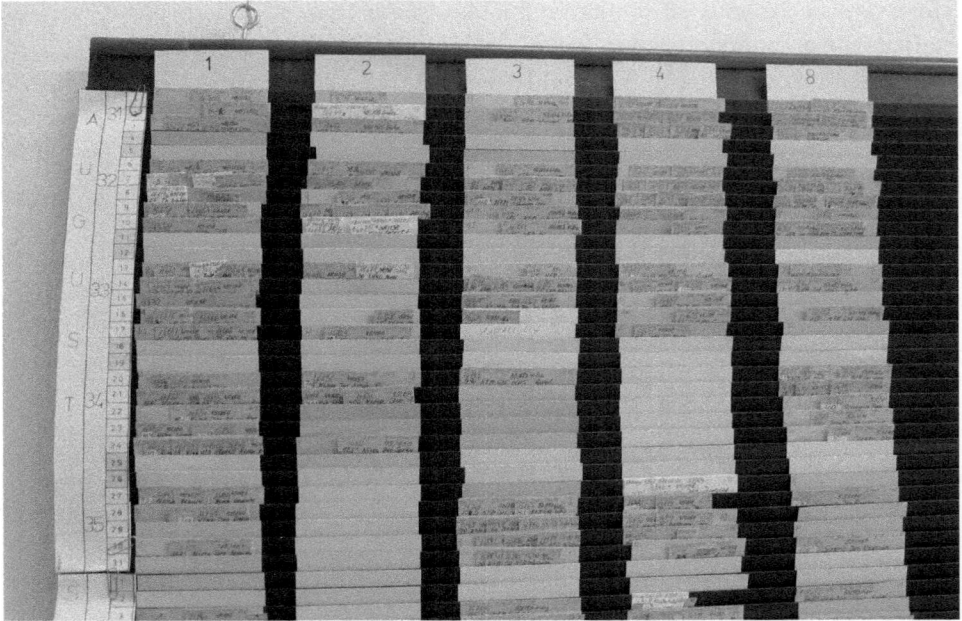

FIGURE 29.3 One of four pegboards making up the planning table. The columns correspond to production lines, the rows correspond to the days

(Author's own photograph)

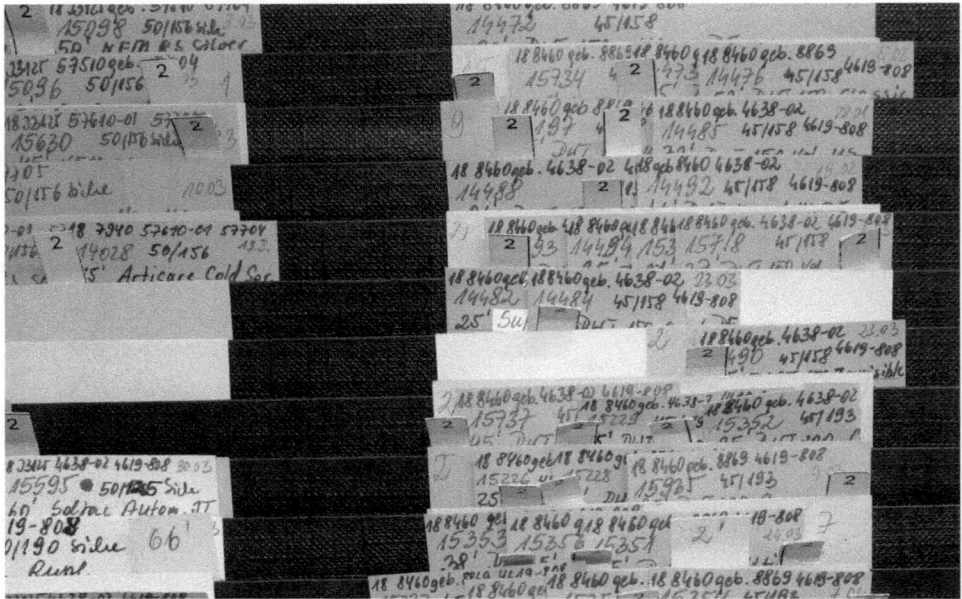

FIGURE 29.4 Planning table close-up. Depending on the colour of the cards, they stand for confirmed orders or free capacities. The small stripes indicate time needed to reconfigure the machines between two orders

(Author's own photograph)

orange arrow-shaped cards pointing out special attention in general. There are also green and orange stripes tucked between the pink cards. Numbered either 2 or 4 they stand for the amount of time (2 or 4 hours) needed to reconfigure the machines so that they can run the upcoming order.

Tabular Structure

In his critique of tables as a means to generate anthropological knowledge about non-literate societies, Jack Goody wonders what a table actually is. He puts down the basics: 'It consists essentially of a matrix of columns and rows, or of what can be regarded from another angle as one or more vertical lists' (Goody, 1977: 75). A one-dimensional table is a list. Hence, a two-dimensional table combines two lists. In the case of company N.'s planning table, these are a list of days and a list of assembly lines. Crossing them at their respective zero point creates a gridded surface. The lists turn into the grid's metadata: they describe whatever data enters into the grid. The emerging gridded surface is initially empty. This is what Bernhard Siegert stresses as the 'ontological effect of the grid' as a modern and Western cultural technique: the grid 'presupposes the ability to write absence, that is, to deal equally efficiently with both occupied and empty spaces' (Siegert, 2015: 97). Hence, in the beginning there are just addresses (such as assembly line 1, 6 August) leading to empty cells. The empty cells allow to accommodate a third entity that is defined by its relation to the x and y coordinates, hence by its place within the grid. In company N.'s planning table, the third entity are existing orders, almost existing orders (reservations), and non-existing orders (free capacities).

   According to Markus Krajewski, the table is a formation and by putting data into this formation (formatting it), in-formation emerges (Krajewski, 2007: 39). Getting data in shape and combining it with other pieces of data creates its informational value. In the case of the planning table, too, orders need a certain format so as to fit into the planning table: they have to take on the shape of a cardboard card with numbers on it, such as production number, article number, amount, customer, and so on (Figure 29.4). It is J., one of two employees making up 'labour preparation', who formats orders in this way, hence writes the numbers in a certain order on a pink card. Tucking the card into the planning table puts it in the diagrammatic relation of rows and columns (Krajewksi, 2007: 307). It turns into an order that is attributed to a date and a production line. Further, the order enters into a relation with all other orders in the board. It is now possible to assess the position of a single order with regard to its surrounding area as well as the situation as a whole. The planning table assembles all orders on a flat, homogeneously gridded surface that allows zooming in and zooming out. Looking at the planning table is not always directed at 'the individual case', but quite often it 'can be better described as a scanning of a surface, a gaze which glides over the rows and columns, in search of table-specific enunciations' (Gregory, 2013: 314). In the case of company N.'s planning table, those table-specific enunciations derive from densifications of either pink or green cards. Areas of pink inform about high machine utilization and hence 'full order books' whereas areas of green inform about low machine utilization and hence low return on investment.

Handiness

The planning table is a table, but a particular one. Its particularity lies in the movability of the entries—the movability of the coloured cards. Researchers studying tables—such as Jack Goody and Stephan Gregory—often insist on the table's relation to fixity and clarity. For Goody, the table is a tool to eliminate ambivalence and vagueness. It 'freeze[s] a contextual statement into a system of permanent opposition' (Goody, 1977: 72–3). Gregory also underlines this effect: aiming at creating 'a static, quasi-frozen world', the table extinguishes 'the temporal dimension' (Gregory, 2013: 320). 'Unmistakably, the construction of the tabular order is bound to an act of standstill' (Gregory, 2013: 319). However, the material support of a table is decisive for its capacity of fixing and freezing: 'Of course, a table realized in a computer program is not the same as a table drawn on a chalkboard or printed in a book' (Gregory, 2013: 323). Due to its material composition, the planning table at company N. demonstrates a different relation to the 'act of stand-still'. As said before, the planning table is basically a pegboard with horizontal rails scaled about each other. Its entries—the coloured cardboard cards—are merely tucked into the rows. Of course, the planning table only permits unequivocal entries and, in this sense, it is a typical table. But these unequivocal entries are not permanent. Quite the contrary, they can be reshuffled without much difficulty.

This is exactly what happens in the process of planning at company N. Planning means attributing incoming orders to a date and to an assembly line. Attributing orders to a date and to an assembly line—deciding on when and where an order will be produced—means turning towards the planning table and searching for a place where to tuck the new card. This is not a simple or single task, but rather a process of continually re-tucking cards. J. talks of this procedure as going 'from rough to fine planning'. The primary criterion guiding the planning is 'causing the least expense possible whilst complying to the date of delivery'. All other criteria operationalize this goal or are subordinate to it. For instance, the time needed to convert the machines should be as short as possible. Therefore J. forms blocks out of those orders that demand the same or a very similar adjustment of machines. However, large orders should go to the fastest production lines: 'Processing time is faster and hence profit is higher there', says J. Other orders have to run on specific production lines because technically they cannot run on others. Further, J. considers whether feedstock is missing or whether machine parts are broken. Also, customers might increase or decrease the number of items they have ordered. In the process of planning, J. juggles all these factors in finding the right date and line of production. And juggling, again, translates into constantly re-tucking cards. The planning table captures the momentary situation as well as her current draft of how to attend to it 'causing the least expense possible whilst complying to the date of delivery'. Like a buffer memory, the planning board retains this draft until it is time to reshuffle: a new order has to fit in, an order has been cancelled, an order has been increased, a machine is down, etc.

Representing, Simulating, Operating

Its light and mobile make-up turns the planning table into a tool for drafting and simulating different set-ups of producing in the most efficient, but still timely way. This process, however, sooner or later has to come to an end. Approximately two weeks in advance, the process of planning—the continuous re-tucking of cards—is eventually settled. Then, the workers in the production hall know that the allocations shown in the planning table are binding: order X will be produced on date Y and on assembly line Z. They have to comply with what the planning table tells them to be done on their respective assembly line. Hence, the 'symbolic operations' are now having 'effects in the real' (Siegert, 2015: 98). This is—according to Bernhard Siegert—the crucial capacity of tables or *grids*, his preferred term. The grid 'effectively merges representation and operation' (Siegert, 2015: 98).

Company N.'s planning table shows this merging of representation and operation very well. On the one hand, it is designed to represent the production facilities, its seven production lines, confirmed orders, reservations and free capacities. It translates the three-dimensional space of the production hall and the activities going on there onto the flat and gridded surface that is manageable from the point of human capacities of perception and cognition. In Siegert's words, it 'submits the representation of objects to a theory of subjective vision' (Siegert, 2015: 98). Or in Latour's words, it translates 'an undifferentiated continuum' into 'a one- or two-meter-square flat surface that a researcher with a pen in hand could carefully inspect' (Latour, 1999: 53). The planning table grants an overview by representing a vast and jumbled situation on a reduced and ordered scale. But on the other hand, it also allows users to manipulate the world it represents. The planning table is a tool for creating an arrangement that can then be applied to the world. It is a means 'for acting upon things, for manipulating them' (Gregory, 2013: 321). J. uses the planning table to continually draft different plans of how the workers in the production hall should act. It is much easier to manipulate the tokens or the signifiers of the world than their signified, the world itself. But as said before, the table's allocations are binding. Everybody and everything has to comply to its arrangement. In the end, the workers translate J.'s plan into action.

This is, then, how the planning table moves from representation to simulation and eventually to action. This is how it is related to governance (Siegert, 2015: 97) and to the generation of organizational powers. Handling a situation via handling a ready-to-hand version of it could be considered as a constitutive practice of modern business. Siegert traces it back to the emergence of double-entry bookkeeping in the sixteenth century. Business people begin to cope with the large space by coping with arithmetic operations on paper. They stop accompanying the goods and start doing business from the office (Siegert, 2003: 43). 'Office' is also the term he attributes to the practices of writing that are relevant here: they are not about building sentences that rely on a consecutive utterance, but about retaining and administering data across a two-dimensional surface of inscription (Siegert, 2003: 33). He calls this the diagrammatic option of writing that is more related to processing and computing than to speaking or narrating.

The kind of governance afforded by tools like the planning table can be said to work in two directions. Firstly, it aims at representing the situation, at creating correspondence with it. Secondly, it aims at intervening in the situation, at designing it. Thus, the arrangement created within a planning table is 'at one and the same time plan, register, and cadaster' (Siegert, 2015: 107). It documents and takes stock of a situation, but it also imagines and models the situation. It does all of this on the same surface and within the same frame. Here, the planning table's materiality comes into play again. The interesting thing that differentiates a (well-functioning) planning table from just any table is its light, mobile, and user-friendly architecture. This way, planning tables allow for constant updates with regard to a constantly changing environment (representation). They also allow the drafting of different plans of how to attend to it (simulation). And lastly, they initiate the translation of the plan into the world (action).

Resistance and Attachment

It does not always work in as smooth and unobstructed way as the text has suggested so far. There are some resistances to the functioning of the planning table, but also—interestingly—some strong attachments to it. First of all, there is the story of the planning table's authority being undermined: once, a worker secretly re-tucked some of the pink cards, therefore changing the sequence of orders to be produced on his line. He did not want to work on the order the planning table had assigned him to because it would have meant tedious cleaning of some parts of the machine.[5] This story shows the planning table's power at the same time as its fragility. It testifies to its monopoly to lay down the production schedule. Removing a card corresponds to cancelling an order, placing a card corresponds to initiating its production. But despite this crucial role, there is no material barrier regulating access to it. It consists of cardboard cards that can be repositioned easily in an unobserved moment. In this story, the planning table makes an appearance in itself. It presents itself as a potential troublemaker, or at least as a potential collaborator in making trouble.

However, this is a real exception. Most of the time, the planning table at company N. is a completely ready-to-hand tool. It governs in the background without its workings being subject of discussion or even noticed. Due to its existence at company N. 'ever

---

[5] The story goes that this act of rebellion was discovered quite quickly. The planning table interconnects different tasks and different departments. If one of them does not comply, it will soon have an impact on the others. The worker remained anonymous though.

since', it is entirely attuned to and mingled with the context. This means that the processes of planning and coordinating on the one hand, and the capacities of the planning table on the other, are thoroughly intertwined as a result of 'reciprocal tuning' (Pickering, 1995: 20). There are ample experiences and skills (often tacit and visceral) related to the planning table—knowing how to use it, how to read it, how to tweak it—and they are passed on from old to new employees. Reversely, its features and capacities are accepted as the fixed framework of what can and cannot be done at company N.

In the rare cases where the planning table turns into a subject of discussion, it often elicits praise and even affection. F. is an especially good example for this. He joined company N. in 2011 in order to support the introduction of a new firm-wide software system by SAP. Obviously, this is a big investment for company N. accompanied by many concerns and headaches. One of F.'s tasks is to get to know the existing processes, especially those of production planning, and help to reconcile them with the structure of SAP. During the process of transition, though, he turns into an admirer of the planning table: 'For me, this is the thing. There is no APS system [Advanced Planning and Scheduling] in the world able to replace this. Because here you really have everything at a glance. You immediately see how and what and where.' At the same time, there is something embarrassing about the planning table. The managing director—also the principal pusher of the decision to introduce SAP—calls it 'antiquated' and a 'thorn in the side'. Other employees share this perception and evoke a future of networked touch-screens where there is no more 'carrying back and forth pieces of paper'. At company N., there is a strong belief in technology and progress. The planning table seems to stand for backwardness. Therefore, the admiration of the planning table and its capacities of producing overview and coordination is mingled with doubts and embarrassment. There should be an updated version of it corresponding to the digital age.

With the new SAP system eventually up and running the planning table is supposed to disappear. SAP's module 'production planning' comprises an electronic planning table (Figure 29.2) which is expected to become the only tool for assigning an order to a day and a line of production. And this is also what has happened since then—though not really. Despite F. and J. 'railing now and again' about the SAP system, they eventually 'got to know it'. By now, they have become attuned to the electronic planning table's capacities, features, and obstinacies. They know how to handle it. This includes, however, making use of 'safety measures' and a back-up system. For one thing, their knowledge and experience serve as such a safety measure. J. says: 'I would say, knowledge is key. If you have been working here for 20 years, you immediately see if there is something wrong.' The system's planning module is riddled with mistakes, such as wrong connections (i.e., of a material to an article). Also, it automatically assigns orders to dates and lines of production, even if those are actually not available. Further, the electronic planning table is not helpful when dealing with variances and deviations. Even though it allows entering 'additional texts' that point out peculiarities or uncertainties, those texts are not really visible. 'You don't see them, but you have to search for them', J. says. Therefore, another safety measure is the planning table that is still hanging on the wall in the department for 'labour preparation'. J. and F. just continued equipping and using it—below the radar in a way. 'As long as we have this tool, this is how it is', J. says. The analogue planning table still serves as a display of the attribution of orders. It has remained a meeting point for people from different departments gathering in front of the planning table in order to get an overview and talk it through. Also, J. continues to use it to mark deviations: 'I'm putting a card into it saying "Attention customer is coming" or "Attention print template might change" '. This is something that she notices and that actively reminds her. Here is, then, the second kind of resistance, this time related to the computer-based variant of the planning table, but also a case of attachment. Instead of replacing the planning table, SAP's electronic planning table now coexists with it. The two planning tables have come to supplement each other.

OUTLOOK

Planning tables can be resisted and when this resistance becomes regular, they will fail to work, hence to plan and to coordinate. But looking for the reasons of their failure reveals the features actually constituting a planning table. Firstly, a planning table can fail to sufficiently represent a situation. This is the case with company N.'s electronic version of the planning table. It rests on an erroneous database; it does not take into account whether there are free capacities on the production lines it picks automatically; it does not represent deviations from the standard procedure in a flashy, noticeable way; and lastly, while it is able to represent the situation as whole, it does so not in a way that allows groups of people to perceive it collectively. The screen is made to be scanned by one pair of eyes only. Planning tables can also fail at allowing for the quick and easy drafting of plans. They can be disproportionately unhandy. According to J. and F., planning with SAP demands more 'handling' since there are 'more levels'. They see this complexity and granularity as disproportionate to the company's rather simple structure. However, a couple of years ago the company has become part of a group consisting of four companies. Introducing SAP is based on the idea of networking those locally dispersed sites. F. says: 'Of course, we need this in order to network the facilities. We cannot take a picture of the planning table and send it over every day.' Hence, the idea of working towards a distributed structure justifies the hard to handle tool. But if there is no such justification, a disproportionately unhandy digital planning tool might not withstand the criticism. Sywottek (2016) tells the story of a Munich based agency for digital services that has been running an encompassing enterprise software. But in the area of resource and personnel planning, they soon give up. They 'invested more effort into handling the tool than into actual planning', one of the employees says. Also, it lacked possibilities to accommodate 'the dynamics of day-to-day business'. They were just not envisaged by the system. And again, the size of the screen was a problem: 'You just cannot display a whole year on the monitor'. The software solution to resource planning was

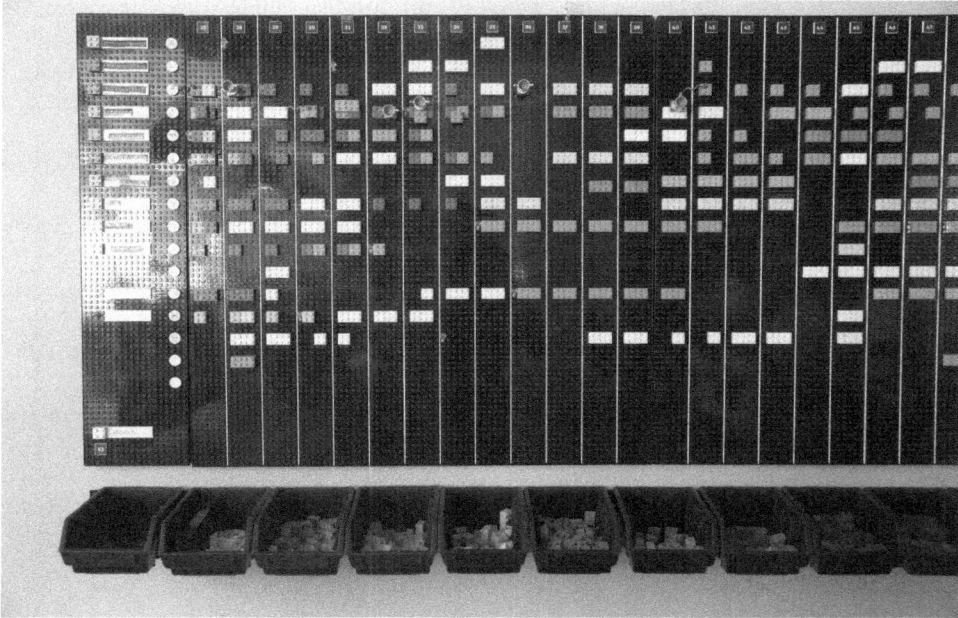

FIGURE 29.5 Planning table made of Lego

(Photograph by Sigrid Reinrichs, published in Christian Sywottek,
'Später Sieg der Zettelwirtschaft', brand eins 06 (2016))

replaced by a mechanical one. Now, a large 'Lego Wall' covers one of the walls of their open-plan office (Figure 29.5).

Planning tables are—like many other tools of organizing—subject to changes in technologies available for 'drawing things together' (Latour, 1986). Their size, shape, and grip are diverse and will certainly continue to be so. For the great majority of organizations and businesses today, networked computers relying on a shared database are the undisputed standard everything else has to comply to. However, when transferring the principles of a heretofore paper-based tool to digital technology, something gets lost. Deducing from the present text, this seems to be foremost tangibility and collective overview. But something is won, too. When there is the need to interlink locally dispersed activities, there is probably not much leeway for non-computerized solutions. Yet, depending on the context of use there can be occasions for local and idiosyncratic fixtures and for revelling in their features and capacities.

REFERENCES

Conrad, Lisa. 2017. *Organisation im soziotechnischen Gemenge*. Bielefeld: Transcript.

Goody, Jack. 1977. *The Domestication of the Savage Mind*. Cambridge: Cambridge University Press.

Gregory, Stephan. 2013. 'The Tabulation of England: How the Social World was Brought in Rows and Columns'. *Distinktion: Scandinavian Journal of Social Theory*, 14(3), 305–25.

Krajewski, M. 2007. 'In Formation: Aufstieg und Fall der Tabelle als Paradigma der Datenverarbeitung'. *Nach Feierabend: Züricher Jahrbuch für Wissensgeschichte*, 3, 37–57.

Latour, Bruno. 1986. 'Visualization and Cognition: Drawing Things Together'. In H. Kuklick (ed.), *Knowledge and Society: Studies in the Sociology of Culture Past and Present*. Greenwich, CT: Jai Press, 1–40.

Latour, Bruno. 1999. *Pandora's Hope: Essays on the Reality of Science*. Cambridge, MA: Harvard University Press.

Pickering, Andrew. 1995. *The Mangle of Practice: Time, Agency, and Science*. Chicago: University of Chicago Press.

Schmandt-Besserat, Denise. 1992. *Before Writing, Vol. 1: From Counting to Cuneiform*. Austin: University of Texas Press.

Siegert, Bernhard. 2003. *Passage des Digitalen: Zeichenpraktiken der neuzeitlichen Wissenschaften*. Berlin: Brinkmann & Bose.

Siegert, Bernhard. 2015. *Cultural Techniques: Grids, Filters, Doors, and Other Articulations of the Real*. New York: Fordham University Press.

Suchman, Lucy A. 1993. 'Technologies of Accountability: Of Lizards and Aeroplanes'. In Graham Button (ed.), *Technology in Working Order: Studies of Work, Interaction, and Technology*. London and New York: Routledge, 113–26.

Sywottek, Christian. 2016. 'Später Sieg der Zettelwirtschaft'. *brand eins*, 06.

CHAPTER 30

··

PREZI

··

ANNIKA SKOGLUND

PREZI, a cloud-based zooming user interface presentation tool developed with software programming languages such as Flash, iOS, Javascript, Typescript, and C, has since the early twenty-first century been shaped by an intense concern with how human beings communicate, with the wish to facilitate the sharing of ideas worldwide. Presentation software has mainly been developed for the educative or business sectors, to offer tools that better convey messages (Levasseur and Sawyer, 2006), and the very first vision of the Hungarian company behind the technology was to create a more aesthetic and pedagogic presentation tool; a market dominated by Microsoft's PowerPoint, Keynote, and Google Slides. The co-founders and employees of the company, also known as Prezi, have since then spoken and programmed the technology into being in three overlapping ways: transmission, collaboration, and augmentation. Based on participant observations since 2008, this chapter will trace how these three discursive themes have acted as an organizing force in the development of the technology, from Prezi Classic to Prezi Next, and finally Prezi AR (Augmented Reality).

The three themes that have shaped Prezi, and by extension the company itself and its users, follow well-known conceptual shifts and increased ontological fluidity in both media and organization studies (e.g., see introduction in Packer and Crofts Wiley, 2012). The first theme remains in the old idea that communication should be optimized into an effective transmission of information via a medium—Prezi developed in a transformation of face-to-screen situations. The second theme grapples with the idea that communication should rather be conceived as human relations of collaborative interaction, with more emphasis on storytelling and meaning due to the complex temporality and non-linearity of human encounters and thinking (e.g., see Knorr-Cetina, 2009)—Prezi developed in the death of face-to-face meetings. And the third theme revolves around a belief in socio-material global interconnectivity, increased post-human embodiment and overall cybernetic embeddedness in systems. An embed-dedness technically supported by increased digitalization that is aimed at enhancing human senses and bodily affectivity that intensify attention assemblages and finally update communication to an augmentation of reality (e.g., see Wise, 2012)—Prezi

FIGURE 30.1 Canvas to zoom in and out on, including the presenters themselves, which exemplifies a self-conception of connective bodies-on-screen

(*Source*: Prezi)

developed as a sense-to-sense prosthesis. This chapter traces the 'productiveness' in the shifts between these three themes, moving from transmission through collaboration towards the mobilization of technologies for the more-than-human (Massumi, 2002), and argues that there is a correlating critical position of productiveness present in the alternative entrepreneurship, i.e., more-than-economic organization, that Prezi seeks as a company (Figure 30.1).

Presentation as Transmission

Ádám Somlai-Fischer, principal artist and co-founder of Prezi, meets me in his Budapest office, a spacious tower room in the Merkur Palace, a building constructed in 1903 to host Budapest's first telephone exchange. Notes and sketches are scattered over whiteboards that run along two of the walls. A large cluttered table surrounded by fifteen chairs of different types occupies the middle of the room. Beanbags, a couple of colourful medicine balls, and a comfy armchair, have been left in a corner to keep, it seems, the creative conversation going on between them. 'It is a little bit messy', Adam understates, quickly adding that messy is exactly how he likes it to be (Somlai-Fischer, 2017a). Principal artist or not, this office called 'the Lab' looks much like the rest of the office space, where approximately 220 Prezi employees enjoy an intellectual 'free space'. Prezi's assumption is that the organizational culture is inseparable from the technology, and both are under continuous development.

The grounding idea of the three founders is that communication is at the core of society, an understanding they suggest permeates both Prezi as an organization, and its technology. Along the lines of current branding trends, Péter Halácsy or 'HP' (former Chief Technology Officer, and co-founder) states Prezi is 'a very purposeful organization', one that wishes to 'change society by changing how people interact with others, and how people communicate ideas' (Halácsy, 2017). The increasing demand to orally present ideas and put oneself on stage, or a screen, is thus mixed with a general turn to technologies for communication. The will to self-promote is merged with a will to connect. Continuously linking humans to the technology is therefore fundamental for Prezi, which 'must be a humanistic organization'. HP explains that 'you can write a code here and you can write a code in another company, but this code [Prezi's] is different because it's meant for something else, so the purpose of the technology is different, and we talk a lot about this purpose' (Halácsy, 2017). The presentation tool has been designed with communication in mind, where the social mission is the basis for the technical development that already began in the late 1990s.

With a background in architecture and well known within the field of new media art, Ádám talks about being influenced by theories on space that arose at the end of the twentieth century. He was particularly interested in how to communicate his own ideas better, believing that PowerPoint was aesthetically inadequate. In 2000, he made his first Prezi presentation by coding in Flash, inspired by how architects visualize space iteratively to simulate constructions. 'For them, an artistic idea is furthered by how you make one visualization after another. For me, it was the same, I wanted to continuously code new presentations, to use visualization as a way to further ideas' (Somlai-Fischer, 2017a). Coding the presentation not only became a way to transform the ideas into something tangible, taking the conceptual to the material, but also to experience the conceptual materially. The first Prezi presentation code was a technology, or tool, with which to think. Dressed in a dark pink T-shirt with 'unreal' written on it, Ádám nevertheless emphasizes that communication is about bringing the 'idea world' into the real world, mainly by means of interactive people who wish to co-produce effectively.

Since Ádám's initial steps of individual coding had to be repeated before every new presentation, Péter Halácsy proposed building a first prototype and interface directly usable by anyone without coding skills. This first version, launched in 2009, used an open canvas on which users could place zoomable visuals and text. The zooming presentation tool also provided more fluidity and a possibility for flexible, less static and linear, storytelling. The presenter could take the audience from the particular to the general and back to facilitate a braiding of perspectives. The purpose was to enable the user to create a clearer route for uncharted ideas, where the Prezi presentation would serve as a first exploration of a potential way forward. The assumption was that a more spatial presentation could produce a stronger response and be more memorable for the audience. With the social aim to spread ideas, users who enjoyed a free subscription of Prezi also had to share their presentations online, which has resulted in over 325 million public Prezis. Hence, the presentation tool seeks to internalize and externalize memories at the same time. Memories are to be shared and co-edited.

The presentation tool was further developed with the wish to keep the audience attentive by imbibing meaning, rather than information only. Prezi thus designed the first version with the aim to help humans to better remember what had been said. However, ensuring that an idea is received with greater impact is demanding, especially since the technology requires the users to be attentive to how their own designs may affect the audience in unintended ways. In comparison to PowerPoint's linear structure, the design of Prezis can be loosely compared to how engineers design rollercoasters. For rollercoaster engineers, it is crucial to know how to balance between twists and turns, velocity and g-power, to attain the required thrill levels. Overly twisting and zooming Prezis could in the worst case make live audiences dizzy and unwell, and in mild forms distract the audience from the message. Hence, Prezi developers had to think about the aesthetic balance between digital moves and visual tricks that disrupt the static, and spatial structures that enforce the static. The dilemma is to merge flexibility with rigidity and thus ground the fun functions on a balanced platform. One way to address the users' needs and lack of knowledge on how to balance visual effects was to provide templates. That is, prepared backgrounds for those who are less inclined to creatively craft presentations on a blank canvas without a given structure and design. The design is furthermore suggested to align with a long history of storytelling among the human species. Crafting a technology for storytelling was not only important for live performances and a listening audience, including face-to-face meetings, but for the increasing popularity of face-to-screen communication.

PRESENTATION AS COLLABORATION

Talking to the employees in the Budapest office requires familiarization with software terminology such as 'change of engine' and 'platform'. Péter Halácsy jokes about the number of technical phases Prezi has gone through since 2008, guessing that it must be around 2,400. The presentation tool is ceaselessly changing shape through constant updates, necessary due to a powerful surrounding technical development, and specifically the increased popularity of real-time collaboration. Prezi was for example forced to change its technology completely, with a new product launch in 2016, due to browsers beginning to abandon Flash. After three years of hard work, the presentation tool was rebuilt with the help of HTML5 and WebGL, structures that afforded greater collaborative potential. Peter Arvai, the CEO and co-founder, considers this to be a completely new product that is the only full-lifecycle presentation platform on the market that allows users to create, present, and analyse their presentations. Designed for customers in a wide range of industries, including education and non-profit, Prezi Next scales from the individual user up to large business teams. Hence, in addition to transmission, which remains a grounding aspect of communication, comes increased awareness of possibilities for the collaborative generation of meaning (Figure 30.2).

```cpp
code.cpp

1   #include "Ball.h"
2   #include "Globals.h"
3
4   Ball::Ball(Ball* parent_) {
5       parent = parent_;
6   }
7   void Ball::initFromParent() {
8       if (parent == NULL) return;
9       d = parent->d / 2.8;
10      r = parent->d * 0.52 + d * 0.52;
11      theta = atan( 1.32 * d / r ) * index;
12      theta -= 1.2 * atan( 1.32 * d / r );
13  //if (parent->theta != 0) theta += parent->theta;
14      pos = ofVec3f(parent->pos.x - cos(theta) * r, parent->pos.y - sin(theta) * r);
15      rotation = theta / 2;
16  }
17  void Ball::draw() {
18      if (skinRatio == -1) skinRatio = Globals::contentImageOpen[myID].getWidth()
19                                     / Globals::contentImageOpen[myID].getHeight();
20      bool containsSelectedChild = false;
21      for (Ball *c : children) {
22          if (c->myID == Globals::selectedBall) {
23              containsSelectedChild = true;
24          }
25      }
26      if (parent != NULL) fadeMeOut = (myID != Globals::selectedBall && (parent -
27                                     > myID != Globals::selectedBall) && !containsSelectedChild);
28  //fadeMeOut = false; // turn of level of detail zoom reveal for leap paying
29      alpha -= (alpha - (fadeMeOut ? 0 : 255) ) / 22;
30      ofPushMatrix();
31      ofPushStyle();
32      ofTranslate(pos.x, pos.y);
33  // fit skin to object
34      float skinW = d * 1.22;
35      float skinH = skinW / skinRatio;
36  //ofTranslate(- d * 0.04, 0);
37      float pushUntil = Globals::mPrezi->screenToWorld( 300 );
38      float pushFrom = Globals::mPrezi->screenToWorld( 100 );
39  //ofTranslate(-smoothRollover * 1, smoothRollover * 1 );
40      if ( myID != 0 && Globals::ballUnderMouse == myID &&
41              Globals::mouseInactivityCounter < Globals::MOUSE_INACTIVE )
42      {
43          smoothRollover += ( Globals::mPrezi->screenToWorld( 100 ) -
44                          smoothRollover ) / 6;
45      }
46      else
47      {
48          smoothRollover -= (smoothRollover ) / 16;
49      }
50      if ( myID != 0 && Globals::ballUnderFingerId == myID &&
51              Globals::enableHoverEffect )
52      {
53          float s = 1.8;
54          float z = Globals::fingerZ >= Globals::pressedAtZ ? 0 :
55                  Globals::fingerZ - Globals::pressedAtZ;
56          if ( z < -20 )
57          {
58              z = -20;
59          }
60          if ( myID != Globals::selectedBall || myID == Globals::ballUnderFingerId
61              )
62          {
63              if ( Globals::ballPressed )
64              {
65                  smoothRollover += ( ofMap( z, -20, 0, pushUntil, pushFrom ) -
66                                  smoothRollover ) / 4;
67              }
68              else
69              {
70                  smoothRollover += ( pushFrom - smoothRollover ) / 4;
71              }
72          }
73          else
```

FIGURE 30.2 Program code in C++, used for Prezi Augmented Reality

(*Source*: sent by Ádám Somlai-Fischer, 2018. Image: Prezi)

This shift to collaboration was accentuated as Prezi turned to a new user, business professionals, who use presentation tools to collaborate and make effective decisions. In addition to Prezi's mission of a global sharing of ideas, the facilitation of decision-making thus became a core concern. In comparison to the first Prezi, now called Prezi Classic, which introduced movement and liquid effects of transmitting imagery, the new product, Prezi Next, objectifies the content to a higher degree. This turn back to an emphasis on successful communication via a presentation tool that actually facilitates a clarity of the transmission, is however suggested to encourage the creation of narrative structures and thereby engagement. The question is how this generally affects creativity, for example of pretend stereotypical customers, such as 'Ian the accountant', 'Sarah the creative consultant', and 'Mitch the investment banker'. The presentation tool is offered to business professionals as a way to inspire with 'imagery' that can 'shape your business content, messaging, and approach'. The assumption behind this is that 'visuals and imagery matter so much when it comes to customer engagement and retention'. Visuals are said to grant impact and get the audience 'to take action', not least because they make the message of the narrative memorable (Prezi, 2017a). With an accentuated use of visuals for business users, Prezi Next is still there to enrich transmission, but now with a more explicit intent of creatively creating collaborative communicative structures such as storylines.

PRESENTATION AS AUGMENTATION

In 2017, Prezi launched a first showcase of Prezi AR via a remote TED talk (Sapolsky, 2017). Prezi AR seeks to 'build experiences that include people' instead of replacing them, so taking the user's transmission and collaboration into technologically enriched forms of 'immersive storytelling' (Prezi, 2017b). Thus far, however, it is a technology that is only used internally by Prezi employees, or by external professional presenters who work closely with the company. The users are to apply their body differently within the presentation, and be able to physically engage with a topic through gesture control, perhaps of increased importance in future remote conferencing and meetings and webinars. This means the user will need to think about how their own body or face fits into the presentation, and understand how they interact with the visuals. The idea is the presenter should become embedded in, and extended by, the presentation tool, while the audience is injected with sense-mixing. However, extensive future technical development is needed before presentation as augmentation results in audience immersion by an intensified merger of the senses.

At a conference presentation by Ádám titled 'Augmented Ideas: Taking Back Augmented Reality from the Terminator' (Somlai-Fischer, 2017b), he starts by pointing to how people thoughtlessly move their hands when they present. They use 'gestures to augment their body language', he says, and it would be 'great to see the ideas around the speaker' as they gesticulate. An augmentation of our storytelling can be very powerful, he continues, adding that there is a challenge in front of us. Either we will create a future where service providers prosper from the augmentation, as in the creation of 'hyper reality', or we can seek to take control of how we wish ourselves to be augmented and immersed in the digital. Hyper reality, as problematized in a dystopic film by Keiichi Matsuda, 'has nothing to do with the people who you encounter', Ádám argues. In comparison, Ádám turns to art, and a film by Luis Buñuel titled *That Obscure Object of Desire*. In this film, two female actors play the same character, but they are still very different, which creates confusing shifts in temperament and looks. A similar play on the disjunction between what we feel and how we come across can be found in Japanese animation art, anime. The point Ádám makes is that our facial muscles show less than we actually feel, something that the exaggerated face expressions of the anime characters work against. There is thus a discrepancy between our inner life and what we express bodily and facially, where a presentation as augmentation aims to 'create a new language that people can express themselves with'. The mission for Prezi is to be at the forefront of creating this language 'before the technology arrives', so that you yourself are in control of what someone else is to see in you (Somlai-Fischer, 2017b). Conclusively, a potential cultural shift is envisioned before a material shift, to prioritize how personal sensibilities are to be digitally targeted.

Digital Aesthetics and Affective Digitalization

The early digital aesthetics of Prezi questioned the foundations of linear communication and objectifying vision characterized by the likes of PowerPoint, with the aim to reach meaningful storytelling and a material appearance of ideas, set in comparison to some existing reality. Psychologists have also concluded that whilst PowerPoint's linear slide format is simpler to process for the human, and so may result in more focused attention, Prezi's visualizations and animations support narrative storytelling and thereby facilitate conceptual understanding (Moulton et al., 2017). The overlap between architecture and communication that inspired the turn to 'ideas' and stories in Prezi's digital aesthetics is echoed in early conceptualizations of cyberspace (Bell, 2007), when the computer was treated as a 'life form that [could] understand people better than people [could] understand themselves' (Lanier, 2010: 29), an example being 'neural architecture' that has been claimed to accomplish 'cyber-consciousness' (The LifeNaut Project, 2015).The notion of cyberspace has in addition encouraged the human, as well as humanoid machines, to live out their stories by applying mythical symbols in a Popperian 'mind-space' (Bell, 2007: 18). Suggesting that if architecture is about transcending and stretching ideas about space, so too is digitized connective communication, with its webs of interrelations and re-imagination of a body/mind unity (Wegenstein, 2010). Tellingly, optimistic and idealistic philosophies have also proven to come to fruition when digitalized, i.e., when an abstract piece of code encounters users who create a 'non-abstract continuation of reality' (Lanier, 2010: 163). And in the case of Prezi, the optimistic philosophy starts with an abstract mind-space, processed via codes and bites, ending in a non-abstract presentation.

This positive outlook of digital aesthetics and its effects on the human has nevertheless been broadly debated in media studies and beyond. Some argue that digital designs thus far only can manage to establish a mind-space for stationary bodies in the wake of a Neoplatonic techno-metaphysics that thrives on 'telepresence' (Hillis, 2012: 256). That is, communication technology that synthesizes the ideal and disembodied on the screen, with the flesh that is only possible to digitally mobilize behind the screen (Hillis, 2012: 256). Others, however, criticize a historically strong biopolitical agenda, i.e., a life optimizing trend prevalent in cybernetic thinking about information, circulation, and connectedness (Thacker, 2010). Scrutinizing the turn to technical stimulations of the human mind, there is also an important aesthetic tension between space and structures in digital designs. Just as built structures may limit architecture, computerized structures may limit digital aesthetics.

The philosophical questions engaged in by the Prezi founders also follow from a problematization of human agency and autonomy in this tension between space and structures. At the same time as Prezi seeks to become a humanitarian purposeful company, it has been suggested that we live in an increasingly cybernetically interconnected world, bereft of 'the human' as we knew it. This post-human trend had been problematized already during the digital aesthetic boom in the 1990s (Parisi and Terranova, 2001), when Prezi was in its infancy, and it is observable that the three co-founders take an active stance to show how they wish to shape our socio-material future. An overall critique of cybernetic totalism has thus directed the development of the presentation tool, with a clear emphasis on storytelling and an active co-producing user. Similarly, critics of cybernetics have turned away from structures and 'networks' to leave behind 'information', and the correlating alienation of experience (Lanier, 2010: 25). In the case of Prezi, the user and audience are opened up to flows of energy, which goes beyond or adds to flows of information (cf. Ticineto Clough, 2004). This design attempts to reach beyond a dialectical approach to position the audience directly within the idea world by means of a new language; a language that is invented to function in the dissolved distinctions between the technology, the presenter, and the audience. By facilitating '[h]igher levels of description', the idea is to make the 'world more creative, expressive, empathic, and interesting'—to enhance it instead of escape it (Lanier, 2010: 28–34). The three versions of the presentation tool, Classic, Next, and AR, also seek this sort of productiveness in a balancing act between spatial thinking and structural ordering in the death of face-to-face communication and birth of sense-to-sense interaction. This tension between digital space, dependent on human coding that can offer a fluidity of potential moves that stimulate uncharted creativity, and digital structures, dependent on human coding of predetermined moves that stimulate charted creativity, can promote either habitual subjectification or critical thinking (Parisi and Terranova, 2001). Hence, different presentation tools open up for users to have a more or less active or passive relation to technology, and address the subjective capacities of the presenter and the audience differently.

The debate about cybernetic totalism has thus transformed and moved critical thinking to the more positively technically immersed human that is furthered with the architectural genesis of AR, as observable in the development of Prezi. It is no longer enough to position critical thinking as the adoption of an external position by which to scrutinize identity politics and volunteer labour on the Internet from the outside. Instead, critical thinking now affirms embeddedness and emergence, and needs to be positioned within the very systems it seeks to scrutinize through 'productivism' (Massumi, 2002: 12):

> The world is in a condition of constant qualitative growth. Some kind of constructivism is required to account for the processual continuity across categorical divides and for the reality of the qualitative growth, or ontogenesis: the fact that with every move, with every change, there is something new to the world, an added reality. The world is self-augmenting.

Massumi's extension of constructivism into productivism and repositioning of critical thinking echoes early cybernetic assembly, and the calls for emergence, interconnectedness, and co-creation. His critical position of productiveness also resonates with how the likes of Prezi attempt to augment reality. In the beginning, AR was conceived as being obtainable with the help of a 'virtual world that enriches, rather than replaces, the real world' (Feiner et al., 1993: 54) with the aim of narrowing in on certain information to secure the production of attention (Wise, 2012). By immersing the user in some sort of digitally construed system that creates a specific, but still flexible, genre of behaviours, AR aimed to bridge the false divide between 'the virtual' and 'the real' (cf. Bell, 2007), whereby AR technology attempted to upgrade the human technically by amplifying the senses and making them more sensitive to that which otherwise would have been invisible, inaudible, intangible, and broadly unexperienced. In this sense Prezi AR embraces a productiveness to bridge the disjunction between the inner life and our inability to express this inner life fully. The coding seeks to better align how we humans really feel with how others interpret what we feel, feelings that are, inevitably, touched by the code. Limited to the so-called 'actual environment', AR is nevertheless dependent on a production of digital objects that 'appear to human users as colourful and visible beings' (Hui, 2012: 387), where the key capacity of these digital objects is that they can transform matter continuously by data processing. The Prezi user is there to speak a new language and produce the data by thinking with the help of the presentation tool, whereby the tool transforms this data to serve the sense-to-sense communication. If the user's thinking is stretched by the technology, so is the capacity of the technology to reach the senses of the users and the audience. It exemplifies an inherently optimistic digital aesthetic approach that assumes that what is thought and sensed can be shared with the technology and stimulated by the human and machine interaction. The digital aesthetics is turned into a co-creational affective digitalization.

Affect and productiveness via co-creation have not only been situated to happen between humans and machines, for example by 'engagement systems' supported by connected platforms (Ramaswamy and Ozcan, 2014: 32), but are also manifest between machines, as in the Turing game (Kittler, 1997), or in between platforms and algorithms (Pasquale, 2015), and it is with affective digitalization that the modification of co-creational space is supposed to augment, rather than alienate, experience. Somehow then, the individual subject is not supposed to disappear when embedded in the environment, but a more productive socio-material co-creation should undergird the sought after sense-to-sense interaction, emphasizing that flow and movement can be obtained in a directed stimulation of the human senses.

> Each sense has its field of emergence. Each such field is in a unique relation of noncoincidence with what emerges from it. In conceptualizing the senses, the way in which the emerged diverges from its emergence needs to be positively described. The virtual is too broad a concept for conditions of emergence.
>
> (Massumi, 2002: 154)

The virtual, understood as a unified connectivity between body, mind, and technology that creates affective shifts in between these (Ticineto Clough, 2004: 8), is, with AR, disrupted by the body. The body must inevitably lend the sensory to the technology, rather than the other way around. If we are to believe AR's possibilities then, the body vanishes in digital space only to dynamically reappear with senses that are algorithmically conditioned. Since digital objects are computational objects, this sensibility is generated under calculation (Hui, 2012), where control of how sense-mixing augments reality is difficult. We do not yet know how, but AR requires some-thing other than contingent chaos and detachment in emergence. By creating digital objects that resemble veridical hallucinations that coincide with actual events, Prezi AR may thus seek to deliver hallucinatory effects without deception. This does, however, require a recovery of the scattered parts and de-subjectification that the unruly product-iveness and embeddedness potentially creates. To fulfil the conventional idea that communication of a message should be optimized, presentation as augmentation will, hypothetically, form an enhanced totality of the mixed senses, perhaps with an ability to technically zoom into and out of experiences.

In contrast to post-humanly dissolving the boundaries between mind/body and subject/object (Knorr-Cetina, 2009), Prezi AR addresses the potential loss of autonomy and the tension between the subjective and objective in their targeting of sensibilities (Hansen, 2012). The technology fits in a timely way in the aftermath of the debased human in its attempt to function as a human-centred prosthetic for thinking, as the first prototype suggests with its intensification and strengthening of bodily experiences of concepts (the mind). The healing of de-subjectification and unifying of sense-mixing is, in other words, following the lentitude of the body. Similarly to AR technology, the pres-entation tool must balance between the natural rhythm of the body and the coded rhythm of the digital technology. It is an algorithmic condition that needs to feed the body with the right amount of emergence to accomplish pleasurable sense-mixing, whereby the presentation tool functions as a medium for the enhancement of embodied experience by how it stitches together human and technically infused senses. Consequently, future experiences of the more-than-human condition could potentially be what a presentation tool such as Prezi aims to deliver, perhaps by making 'thought bend back to participate in its own emergence from sensation', as Massumi suggests (2002: 136). In other words, Prezi AR exemplifies how a critical position of productiveness digitally morphs into an optimistic embracement of affective shifts in between humans and technology, to fortify experience and create more-than-economic value.

REFERENCES

Bell, David. 2007. *Cyberculture Theorists: Manuel Castells and Donna Haraway*. London: Routledge.

Feiner, S., B. MacIntyre, and D. Seligmann. 1993. 'Augmented Reality: Reflections on Its Contribution to Knowledge Formation'. *Communications of the ACM*, 36(7), 52–62.

Halácsy, Péter. 2017. Interview (9 July), Prezi Budapest office.

Hansen, Mark B. N. 2012. 'Ubiquitous Sensibility'. In Jeremy Packer and Stephen B. Crofts Wiley (eds), *Communication Matters: Materialist Approaches to Media, Mobility and Networks*. New York: Routledge, 56–65.

Hillis, K. 2012. 'Virtual Mobility, the Sign/Body of Pure Information'. In Jeremy Packer and Stephen B. Crofts Wiley (eds), *Communication Matters: Materialist Approaches to Media, Mobility and Networks*. New York: Routledge, 256–64.

Hui, Yuk. 2012. 'What is a Digital Object?' *Metaphilosophy*, 43(4), 380–95.

Kittler, Friedrich A. 1997. *Literature, Media, Information Systems*. New York: Routledge.

Knorr-Cetina, Karin. 2009. 'The Synthetic Situation: Interactionism for a Global World'. *Symbolic Interaction*, 32(1), 61–87.

Lanier, Jaron. 2010. *You Are Not a Gadget*. New York: Alfred A. Knopf.

Levasseur, David G. and J. Kanan Sawyer. 2006. 'Pedagogy Meets PowerPoint: A Research Review of the Effects of Computer-Generated Slides in the Classroom'. *Review of Communication*, 6(1–2), 101–23.

Massumi, Brian. 2002. *Parables for the Virtual: Movement, Affect, Sensation*. Durham, NC: Duke University Press.

Moulton, Samuel T., Selen Türkay, and Stephen M. Kosslyn. 2017. 'Does a Presentation's Medium Affect Its Message? PowerPoint, Prezi, and Oral Presentations'. *PLoS ONE* 12(7).

Packer, Jeremy and Stephen B. Crofts Wiley (eds). 2012. *Communication Matters. Materialist Approaches to Media, Mobility and Networks*. New York: Routledge.

Parisi, Luciana and Tiziana Terranova. 2001. 'A Matter of Affect: Digital Images and the Cybernetic Re-Wiring of Vision'. *Parallax*, 7(4), 122–7.

Pasquale, Frank. 2015. *The Black Box Society: The Secret Algorithms that Control Money and Information*. Cambridge, MA: Harvard University Press.

Prezi. 2017a. 'Prezi + Augmented Reality: A Peek into the Future of Presenting'. *Prezi* blog. https://blog.prezi.com/prezi-augmented-reality-preview/ [Accessed 10 September 2017].

Prezi. 2017b. 'On-Demand Webinar: Images that Influence—Getty Images 2017 Visual Trends'. *Prezi* blog. https://blog.prezi.com/2017-visual-trends/ [Accessed 15 October 2017].

Ramaswamy, Venkat and Kerimcan Ozcan. 2014. *The Co-Creation Paradigm*. Stanford, CA: Stanford Business Books.

Sapolsky, Robert. 2017. 'TED Talk: The Biology of Our Best and Worst Selves'. https://www.ted.com/talks/robert_sapolsky_the_biology_of_our_best_and_worst_selves [Accessed 15 October 2017].

Somlai-Fischer, Ádám. 2017a. Interview (2 September). Prezi Budapest office.

Somlai-Fischer, Ádám. 2017b. 'Augmented Ideas: Taking Back Augmented Reality from the Terminator'. *Youtube*, Augmented World Expo. https://www.youtube.com/watch?v=-a7qcrRtNRI [Accessed 15 January 2018].

Thacker, Eugene. 2010. 'Biomedia'. In W. J. T. Mitchell and Mark B. N. Hansen (eds), *Critical Terms for Media Studies*. Chicago and London: University of Chicago Press, 117–30.

The LifeNaut Project. 2015. 'Bina 48 Meets Bina Rothblatt—Part One'. https://www.youtube.com/watch?v=KYshJRYCArE [Accessed 15 October 2017].

Ticineto Clough, Patricia. 2004. 'Future Matters: Technoscience, Global Politics, and Cultural Criticism'. *Social Text*, 22(3), 1–23.

Wegenstein, Bernadette. 2010. 'Body'. In W. J. T. Mitchell and Mark B. N. Hansen (eds), *Critical Terms for Media Studies*. Chicago and London: University of Chicago Press, 19–34.

Wise, J. Macgregor. 2012. 'Attention and Assemblage in the Clickable World'. In Jeremy Packer and Stephen B. Crofts Wiley (eds), *Communication Matters: Materialist Approaches to Media, Mobility and Networks*. New York: Routledge, 159–72.

CHAPTER 31

···

PRICE BOOK

···

DAMIAN O'DOHERTY

DESPITE rumours of its demise the old-style hardcopy print bound 'book' continues to act on organization in potent ways. This chapter takes one particular book, from the genre called 'price books', namely *SPON's Architects' and Builders' Price Book*, first published in 1873 and currently in its 143rd edition, edited by AECOM, published 2018 by Taylor and Francis. In publication terms alone, the book would appear to be a remarkable success, outliving even the most popular and enduring textbooks in Media Theory and Organization Studies. Its powers of organization are unrivalled in that respect; and yet Organization Studies especially is not good at addressing books, nor the print cultures associated with their production and circulation. No doubt, this is because, at least in part, modern research and scholarly practice is inscribed into the modern culture of the book in which scholars and their disciplines are subject positions within the operations of the book. So close to the book, they are at the same time very far from it. It is this reflexive aporia that helps make the book so elusive, strange or ghost-like.

If the era of the book is coming to an end (McLuhan, 1964) we must find ways of treating this elusiveness and find a vantage point or rather 'mode of transport' to move through the practices of book use. Hence this chapter takes on the role of 'bookie', which may serve as a 'conceptual personae' for followers of Deleuze and Guattari (1994), helping to bring together things normally severed and kept apart—distributed into the classic dualisms, for example, of structure/agent. Adopting the personae of the bookie helps follow the practices involved in the work of SPON's and show that gambling and speculation is central to the ongoing production of organization. Here, 'the market' is neither structure nor agent, but occupies an apparently bizarre ontology, in part made to appear and disappear by a book, which itself flickers in a half-life of presence and absence. Occupying a difficult and unreliable reality, the book constitutes contingencies and volatilities in the production and reproduction of organization whose market-based interplay has largely been missed in Organization and Media Studies.

OPENING THE BOOK

It was a minor, almost accidental inscription made into a notebook that helps introduce and distil the outline of this thing: SPON's price book. I had been working with Martin Garnish on what was called 'the restroom project', a £1million refurbishment of airport staff restroom facilities. At that time he and I were going through what seemed to me an impossibly long and intricately detailed document listing materials and prices to form what is a called a set of contractor's proposals—or CPs. In recent times the idea is that tenders go out to 'market' where competition can ensure best value and efficiency. Martin was pouring over each densely typed page of the tender bids during which he finds a whole series of costs that can be stripped out; 'We don't need stainless steel splash-back all around the worktops', he says, 'but only at certain areas'. In doing this stripping out, Martin is reliant on a 'good quantity surveyor'—the client QS—who has a technical training in the pricing and measuring of work. In team meetings with the client QS there is discussion about the type of material used for the flooring. We are told vinyl tiles might be cheaper before discovering that all flooring needs a plywood base otherwise there are likely to be problems of uneven floors and getting the tiles to stick. We have to tread carefully in stripping out these costs. Not only are there complex technical specifications that might be compromised by seeking cheaper cost solutions, but Gavin, the owner of the preferred tender bid, also has a reputation of taking things personally, which might cause relationship problems. Martin wants to keep the contractor in good spirits because projects depend on trust and close working relationships; there are also future projects on which Martin is going to need a good and trusted contractor.

With a reputation for high cost capital projects, the current senior project management team at the airport were under pressure to embrace a more 'commercial mind-set' and Martin had to find ways of bringing estimated costs down because the budget for the restroom project was far less than the contractor's proposals. As we proceed laboriously through line upon line of detailed proposals—that included things like 'angle section bearers', '108 × 125 mm half-round gutters; on brackets; screwed to timber', and 'ventilating two piece adjustable capping to ridge'—Martin shows me 'SPON's'. SPON's looks like a thick oversized red house brick, or even a breeze block. It is almost of biblical proportion. On Amazon websites current hardback editions retails around £150. 'With tender prices rising slightly, looking at price lists is not enough', the site reads, '—you need SPON'S ARCHITECTS' AND BUILDERS' PRICE BOOK 2018 to get a competitive edge'.

Over 950 pages, SPON's takes two hands to lift and open. In fact we need the support of a desk so that we could consult its contents. It is only a glimpse I have of Martin lifting this red volume and carrying it over to a desk, but it quickly took form as a memory that stayed with me bothering my research over the next two-and-a-half years (Strathern, 1999: 9). At some point I wondered if it was precisely in these apparent inconveniences that clues might be found to aspects of organization not normally considered in the management of budgets and projects. Might the price book form part of an elaborate collective ritual of 'window dressing', or stage-managed performance, an object to which deference and obligation was duly observed? However, users did not seem preoccupied with representational or symbolic features of SPON's; instead they were engaging with it practically and, to all intents and purposes, obsessively.

In working through the contractor's proposals we consult page 373 (Figure 31.1) listing prices and materials classified under 'R Disposal Systems', which includes items such as rainwater pipework, gutters, and drains, and what appears to be an endless list of their sub and sub-sub (and even in places sub-sub-sub!) components measured to the exacting details of millimetres and two decimal points for the labour hours estimated for their on-site assembly. It's a bewildering and dizzying read. Page after page of seemingly infinite detail presented in monotonous lists typeset in 6-point font or smaller, descending into a seeming abyss marked by endless indents and tabulations. Now, contractors' proposals respond to the publication of what is known as employers' requirements (ERs), which suggests this is the *second* time that this page of SPON's has appeared in the work of our project manager. Presumably this page of SPON's was used in building up what was needed by the client and to price the project so that the client team could evaluate the rigour and price efficiencies achieved by the various contractors vying for the work. However, as we delve deeper into the mysteries of this book, the reliability of SPON's for the purposes of establishing a project budget becomes increasingly uncertain. Let's go back a little and work out what is happening at an earlier stage in the project where the price of half-round gutters becomes an object of concern.

The Quantity Surveyor

In order to secure the budget for the restroom project, the project manager needs some idea of the materials and labour prices ahead of 'going to market' with a tender. To do this, he relies on the expertise of a project client team, which includes the QS and his SPON's. So, assuming the client *knows* they will need '108 × 125 mm half-round gutters; on brackets; screwed to timber', the QS can go to SPON's and get an accurate price. For the gutters, SPON's offers prices, set out in tabular form (Figure 31.2).

The book tells you what is called 'the prime cost' (PC) of the unit, which is £9.67. You need 108, so you might think that would sum to a total cost of £1044.36. You may not have considered delivery though. Imagine you have bought the items; however, even if you have bought them, are they going to be delivered? And if so, who is going to deliver them? Delivery might be difficult and expensive. As it happens, the PC includes an allowance for a delivery cost. So, you've not been fired yet as a QS or project manager. However, £1044.36 is *not the total cost to the project*. The total cost is in fact the 'total rate' (Figure 31.2, final, right hand column) times the metres (m) required. However, can we rely on this as a *true total cost*? The guttering is no use to you unless it has been *installed*. The building is what you are buying—or a kitchen in this case. Guttering, in and of itself, is not really what you are after, but you cannot have your building without it. The full price is actually calculated by adding the costs of the material and the costs of labour

Prices for Measured Works – Major Works   373

R DISPOSAL SYSTEMS

Item	PC £	Labour hours	Labour £	Material £	Unit	Total rate £
Aluminium gutters and fittings; BS 2997; polyester powder coated finish						
100 mm half round gutters: on brackets; screwed to timber	8.22	0.32	5.78	11.70	m	17.48
Extra for						
stop end	2.23	0.71	2.71	4.74	nr	7.45
running outlet	4.94	0.31	5.60	5.41	nr	11.00
stop end outlet	4.39	0.15	2.71	6.38	nr	9.09
angle	4.57	0.31	5.60	4.20	nr	9.79
113 mm half round gutters; on brackets; screwed to timber	8.61	0.32	5.78	12.17	m	17.94
Extra for						
stop end	2.34	0.15	2.71	4.88	nr	7.59
running outlet	5.38	0.31	5.60	5.87	nr	11.46
stop end outlet	5.04	0.15	2.71	7.06	nr	9.11
angle	5.14	0.31	5.60	4.73	nr	10.32
125 mm half round gutters; on brackets; screwed to timber	9.67	0.37	6.68	14.60	m	21.27
Extra for						
stop end	2.85	0.17	3.07	6.57	nr	9.64
running outlet	5.83	0.32	5.78	6.26	nr	12.04
stop end outlet	5.35	0.17	3.07	8.45	nr	11.51
angle	5.71	0.32	5.78	6.24	nr	12.01
100 mm ogee gutters; on brackets; screwed to timber	10.26	0.34	6.14	14.97	m	21.10
Extra for						
stop end	2.35	0.16	2.89	3.12	nr	6.00
running outlet	5.79	0.32	5.78	5.72	nr	11.50
stop end outlet	4.49	0.16	2.89	7.14	nr	10.03
angle	4.88	0.32	5.78	3.71	nr	9.49
112 mm ogee gutters; on brackets; screwed to timber	11.41	0.39	7.04	16.47	m	23.51
Extra for						
stop end	2.52	0.16	2.89	3.30	nr	6.19
running outlet	5.86	0.32	5.78	5.70	nr	11.47
stop end outlet	5.03	0.16	2.89	7.82	nr	10.70
angle	5.82	0.32	5.78	4.53	nr	10.31
125 mm ogee gutters; on brackets; screwed to timber	12.60	0.39	7.04	18.11	m	25.15
Extra for						
stop end	2.75	0.18	3.25	3.57	nr	6.82
running outlet	6.41	0.34	6.14	6.20	nr	12.34
stop end outlet	5.71	0.18	3.25	8.72	nr	11.97
angle	6.79	0.34	6.14	5.38	nr	11.52
Cast iron pipes and fittings; EN 1462; ears cast on; joints						
65 mm pipes; primed; nailed to masony	18.32	0.48	7.90	21.05	m	28.95
Extra for						
shoe	16.07	0.30	4.94	16.39	nr	21.33
bend	9.84	0.53	8.72	9.52	nr	18.24
single branch	19.34	0.67	11.03	19.34	nr	30.36
offset 225 mm projection	17.54	0.53	8.72	16.76	nr	25.48
offset 305 mm projection	20.54	0.53	8.72	19.66	nr	28.38
connection to clay pipes; cement and sand (1:2) joint	-	0.14	2.30	0.11	nr	2.41

FIGURE 31.1 Page 373 from the 2007 edition of *SPON's Architects' and Builders' Price Book*

Item	PC £	Labour Hours	Labour £	Material £	Unit	Total Rate £
125 mm half round gutters; on brackets; screwed to timber	9.67	0.37	6.68	14.60	m	21.27
Extra for						
Stop end	2.85	0.17	3.07	6.57	nr	9.64
Running outlet	5.83	0.32	5.78	6.26	nr	12.04
Stop end outlet	5.35	0.17	3.07	8.45	nr	11.51
angle	5.71	0.32	5.78	6.24	nr	12.01

FIGURE 31.2 Detail from the 2007 edition of *SPON's Architects' and Builders' Price Book*, p. 373

together. Material costs include allowances for 'delivery, waste, sundry materials and an allowance (currently 7½%) for overhead charges and profit for the unit concerned' (Langdon, 2007: 86). Hence the full material costs of a metre of guttering are more than the PC, and in this case £14.60 per metre (so long as you are building outside inner London). In addition, to fit a metre of guttering is going to need a fitter or builder, which is calculated to take 0.37 of an hour. Labour rates are estimated from 'typical gang costs' for the trade concerned, and then divided by the number of primary working operatives, with additional allowances for supervision costs.

So all these published costs are based on calculations of sub-components and assumptions that seem guesswork, and presumably by definition, out of date as it takes time to gather, calculate, and publish these costs, by which time, prices will have moved on. However, we seem to be at least moving towards a price that can be used for estimating the costs of our refurbishment project. Further potential confusion emerges when we consider what '108' listed in the Employer's Requirements refers? Is it metres or units? And do you want 'stop ends' and 'running outlets'? Be careful. If not, the pointing on your brickwork might get ruined by rainfall—which is a major consideration for projects in Manchester.

Under pressure to achieve cost-efficient and economically viable projects, our project manager is in trouble here. We started out with a contractor's proposal, but were not sure whether we were comparing like with like when we sought to refer these back to the ERs. Using SPON's opened up the prospect of an infinite regress of price components, built on various assumptions and attempts at normalizing particularity and contingency. It takes a great deal of skilled reading to recover this from SPONs, and yet as we shall now see, the very *raison d'être* and practice of SPON's is one that promises quick and efficient project costing.

THE MARKET AND THE TAUTOLOGIES OF SPON'S

In reaching for SPON's Martin is going to get as close to 'the market' as he will do during any other time in the life of this project (Callon et al., 2007). However, recall the estimates of prices for labour and materials in the contractors' proposals also make use of

SPON's. The significance of this becomes clearer if we elaborate a little further the respective membership of the contractors and client design teams. Martin is working with a complex team of many members, with different skills and trades. His team also has internal 'stakeholders', senior management and group executives who variously 'sponsor' and act as 'client'. However, at the design stage his immediate team is a little simpler, composed of an architect, a quantity surveyor, and a 'Mechanical and Electrical' (M&E) consultant. SPON's has been used by all three to establish a budget and write out a set of employer's requirements. On receiving this document, the contractor similarly draws on an architect, a quantity surveyor, and an M&E consultant, to help make sense of the requirements, and to price them accordingly. Hence, we can see how the two sides, contractor and employer, mirror and replicate one another, with all being paid for their time and skill. And right in the middle is SPON's—which tells you the price of a job! Surely, one might ask, the book does the job, on its own? However, it does so only as *media*—and it is a form of media that will prove increasingly complex and paradoxical as our analysis proceeds.

As a senior QS explains to me, there are at least four prices concealed in SPON's— wholesale prices, retail prices, preferential customer account discount, and the price calculated on what is called a relative annualized turnover adjustment between a buyer (on account) and a seller, which depends on the profit and loss of their respective annual accounts. SPON's prices are also derived and calibrated on the assumption of a £3 million project that takes place in the outer London area. All kinds of adjustments, therefore, have to be made to arrive at an accurate estimate for any particular job. The fact that Martin's kitchens will be built 'airside' in the airport, for example, means restrictions— safety procedures, driving permits, liability insurance—and therefore costs. So, as far as the QS is concerned, far more 'accurate' price estimates can be achieved by telephoning local suppliers, and particularly those who have been used before on airport projects. Indeed, it is not unusual for a QS to ring a contractor with whom they are on *friendly* terms (hence, vulnerable to accusations of nepotism, which is precisely what the efficiencies of markets promise to eradicate). Our QS may also have worked at one time for a contractor and as with all professions personal relations (carefully managed) can be drawn upon for information and favours.

In simple terms, everyone is spending all their time phoning around everyone else trying to get a price! And, at the same time, they are all charging each other for their respective work. In simple terms, the client QS might get the same information from the same supplier that the contracting QS will use! And, yet, it's all in SPON's at the same time—which also employs a workforce busy soliciting for prices. At its limit a bookie might easily imagine a situation in which the client QS rings a 'friend' to ask the price of 1 08 units of '125mm half-round gutters; on brackets; screwed to timber'. On the basis of these prices the QS compiles the ERs, which are then circulated to initiate the tender process. It is possible the 'friend' works for a contractor who decides they're going to bid for the work, at which point the friend becomes 'a contractor', but who presumably will plug the same price into the CPs that the client QS listed in the ERs. Marginal gains might be sought by contractors eager to win (or in industry speak 'buy') work and cut costs below what might be considered necessary to turn a profit. However, the client QS is under professional obligation to alert their client to such practices, and to alert the contractor if they have missed a critical item from their list of quantities. Moreover, if one contractor has found a cheaper solution to the one that has been asked for in the tender pack then the QS is also duty bound to ask the other contractors who have submitted their tenders to re-price by using this item. One might wonder where room might be found for the free play of the market.

THEATRICAL INTRIGUE

Whilst all this is happening, the SPON's book continues to sit quietly in the background. We might say it lurks, like a spectral presence, the items and their prices 'disconcerting' commodity-like things that—according to Marx—one has to approach with 'metaphysical subtlety' and 'theological niceties'. We thought we were getting close to the market in reaching out for SPON's, but in so doing we appear to be participating in some masque ball or a strange séance-like set of practices where tables turn and characters shift identities (cf. Derrida, 1994). Nor can we simply attribute the dance to human agency, as if SPON's contributed nothing. More than a mere collection of prices or a representation of 'the market', SPON's in fact helps *realize* and *activate* prices—at the very least as comparator or 'back-stop'. It is therefore not representational, but constitutive, or 'performative' (MacKenzie, 2006; Callon et al., 2007). This means that human agency risks being diminished to the status of relay, or amanuensis, as everyone might simply lift the price from SPON's. But the end point entailed by this logic seems endlessly deferred, in part because the book remains complexly woven into supplementary practices of 'market making'. If SPON's disguises a whole series of prices, and is both used and not-used by actors looking to buy or sell work, the fact that it is also, in a sense, shared, lends the book an even greater degree of fantastical powers, mischief-making and ontological instability. At this stage we might also be prompted to ask: who are the performers, or bookmakers of SPON's?

BEHIND THE STAGE . . .

A clue is provided by Lesley, one of the QSs at the airport, who one day shows me an email from someone chasing her for price information. The email comes from a cost researcher and it has been sent to a long list of recipients, collectively titled 'Data Contacts'. The tone of the mail suggests that price information is difficult to retrieve: in part cajoling and at times more threatening, marked by occasional but very prominent electronic carbon copies sent to senior members of the organization. In her reply Lesley provides a price of '£261,789 for 148t hot rolled and 23t cold rolled (priming only)', but in her mail she also notes that she may be able to provide more information because on this occasion 'we are lucky to have received well detailed tender returns for this project, including quotes for alternative specifications'. However, most astonishing is the discovery that the cost researcher works for the same QS firm (Davis Langdon (DL)), who hold the contract with Taylor & Francis to publish SPON's! Apparently, it is considered prestigious, even an honour, to work on SPON's—'DL are seen as knowing what they are doing by leading the QS world by generating this information', I am told by another. This obviously raises the question: who prices the SPON's book? How much work goes into SPON's? How much does SPON's cost to produce? Is it profit-making or loss-making? Who pays for the time of the QSs busily collecting data or responding to emails for price information? Does this somehow get incorporated into the costs carried by our restroom refurbishment project? Could it add to the price of an item requested in the ERs, in part because more time is being spent providing or searching for prices? The very success of the book would then seem to simultaneously undermine the value of its presence.

In pursuit of answers to these questions, and in an effort to try and study how the book is practically made, I travel down to London to the headquarters of Davis Langdon, located in High Holborn. Here I am introduced to Roger, an associate cost researcher who works full time on SPON's. He tells me there are over 10,000 items listed in SPON's, but also that the prices listed in the book are aggregates and build-ups, abstractions in many ways, that must be tempered by things like season, time of the year, the type and size of the building, etc. Moreover, 'behind' those published prices is a long list of *sub-prices* out of which the aggregate and build-up is formed. That tiny item the 'stop end' we were introduced to earlier, due to form part of the kitchen refurbishment project, which is the fixing at the end of a run of guttering, and listed as an extra in SPON's at a PC of £2.85, is a little misleading. The full price of the stop-end, Roger tells me, includes things like allowances for fascia brackets, 0.17 of an hourly rate for a carpenter or joinery gang, and perhaps remarkably 0.0065 of a tube of silicant sealant (presumably the estimated amount needed to affix the stop end to the guttering) (Figure 31.3).

Behind this screen though lie others, including Excel spreadsheets, which are themselves composite assemblies made from a trail of phone calls, correspondences, emails and attachments, like the one exchanged with Lesley earlier. There appears to be an endless decomposition of products and prices such that it becomes difficult to know what is what, and what its cost is. The spreadsheets of course do not record the wavering and prevarications included in the emails, the tussle of motivation and resistance amongst the central team and the data collectors, and all the provisos that caveat an estimate such as the observation that the figures come from 'well detailed tender returns' or not so well-detailed returns. Suppliers can also be 'lazy' and simply consult SPON's from last year and add 5 per cent, which according to Roger undermines the seriousness and rigour required to achieve accurate prices. This attachment to accurate prices is remarkable, however, in part because as we have seen, Roger simultaneously questions the accuracy of whatever is printed in SPON's. Contractors will often submit prices to SPON's with half an eye on pecuniary gain; they are in effect chancing their arm, trying to get a price circulated and normalized that they know is too high. 'The book reflects what we see of the market not what it wants you to see', Roger explains, as small adjustments are made

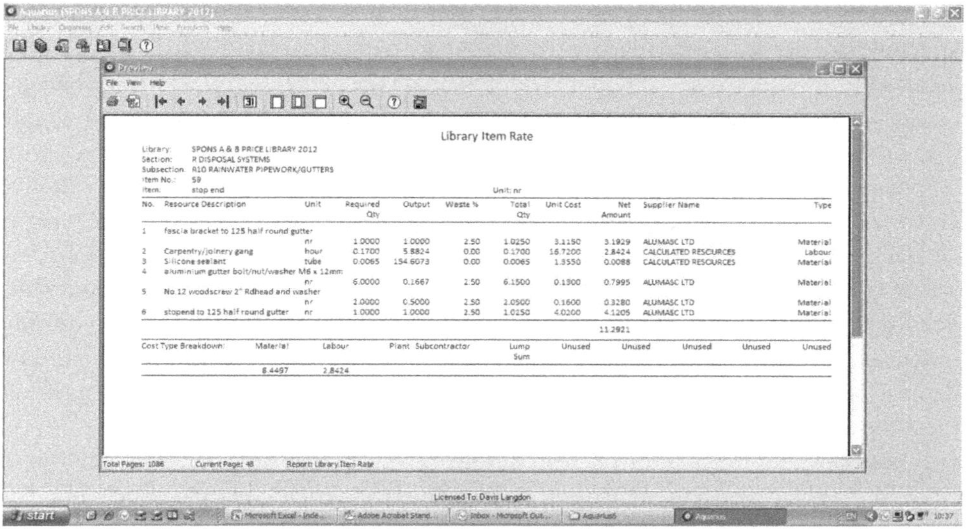

FIGURE 31.3   *SPON's* library item 'Rate: Stop End'

to mitigate the worst excesses of the braggarts and chancers, but in effect registering a price that has no referent in 'the market'.

As we progressively follow more closely the practical work of collecting prices the line between collection of a pre-existing reality and its collective negotiation, or settlement, blurs, then disappears (cf. Latour and Woolgar, 1979; Çalişkan, 2010). If the use and circulation of the book in constructing prices erodes and transgresses the identities and boundaries of those ostensibly occupying positions of demand and supply, a similar loss of identity and boundaries seems to afflict the contents. Moreover, in this composition the book loops back into precisely the practices of those very same characters who are simultaneously making use of the book, adding further complexity and paradox to practices that are already so complex and shrouded in gesture, feint, and ruse that it is often difficult to know what is going on. Those making use of the book are in effect enrolled into the process of making or writing, and hence they are both readers and writers at the same time, which also serves to *un-make* the book, at least as a stable and bounded media object.

ORGANIZING THE LOGIC OF THE MARKET: IN SEARCH OF AN ELECTRIC WACKER?

The book is also itself a commodity, with its own market, and one facing competition from other price lists such as those produced by Wessex, Griffiths, and Laxtons. How does SPON's keep its market leading edge? In part, the answer lies in looking for things like an 'electric wacker'. There are rumours that such an item has come into existence and confirmation of its existence by the discovery of a price might help SPON's steal a lead on Griffiths or Wessex and thereby differentiate the SPON's product. As the current edition of SPON's sells at £150, members of the cost team believe you cannot simply make the same as the one the year before and Roger is also performance managed with targets and incentives, which means that there is always pressure to add new items. However, for Roger an *electric* wacker did not make sense. As he explained to me, a regular wacker is a square hod-like tool which is used by road tarmac operatives to compress newly laid tarmac, but the logic of an *electrically assisted* wacker seemed impossible. For Roger one cannot *hold* a wacker, which must be held above the ground requiring 'wacking', *and at the same move* enable the hod to move down towards the ground by depressing the trigger of an electric switch. How can something move, and not move, at the same time?

Eventually after many hours of search Roger gives up. The only prices he could find were for regular wackers, appearing to confirm his conviction that much like the infamous left-handed screwdriver (or Lewis Carroll's 'Snark') no such thing as an electric wacker could possibly exist. However, if he had changed his search for an *electric wacker plate* or an *electric compactor plate wacker*, reality might have been more accommodating, or generous, and confirmed his request for a price. In the meantime, and if SPON's was a pure vanity project, the costs of searching for this Snark are presumably being carried by wages and project budgets that would have to be recovered from somewhere else in the business. To take a wager on this we might say that QS hourly rates for Davis Langdon would be marginally higher, a reflection of a proliferating series of meta-analyses and abstractions that parasitically create more markets whilst undermining the integrity of those from which they build. No concept for this exists in Organization Studies and it seems unlikely that the form of writing required to weave its understanding could be contained within the books of our discipline...

CONCLUSIONS

Much like the hard-to-see frame that surrounds and lifts the wacker plate above the ground, those stuck in the 6-point font of the textual world of SPONs might not appreciate the groundless and suspended quality of their price book. Yet, just as we are leaving the office, we step towards a large plate glass window that looks out high over High Holborn. On the other side of the road there is a busy building site. Looking down we see what appears to be a complex orchestration of bright yellow-vested builders with hard hats moving ant-like in and out of clusters. These clusters form only for a brief moment, and then quickly disperse before new clusters form. As we watch there is that slight delay and de-synchronization of sound and vision making the heavy steel of the wrecking ball appear to silently strike a wall moments before a large boom and crash sounds out. As we watch this balletic performance I ask Roger whether the economy is 'picking up'. He looks at me; then speaks a few words. 'Mmm. These things *are* in the air. You have to ... feel these things, and you *can* kind of feel it. Yes, in the air. Something's happening, yes'. Suspended in the air, we might say, much like SPON's ... or an electric wacker.

In this suspension SPON's is a book that proliferates prices and with the keen eye of a bookie we have encountered a set of practices associated with it that are productive of a series of inversions and *reductio ad absurdum* threatening aporias and reflexivities that stretch our thinking of organization to the limit. SPON's is a book without origin or ground, undermining neat divides between demand and supply and the idea that price is a spontaneous and natural market mechanism that clears any discrepancies between demand and supply. We are also uncertain as to who or what deserves the status of agent, or who or what is acted-upon. Solving these problems practically demands remarkable skills and dexterity of organization that seem to emerge collectively in the give and take and the push and pull of a range of diverse and heterogeneous materials. In preparing our own book of ethnography (and this book here) we have perhaps had to take a gamble. If the 'bookie' is a person of undecidable reputation—both a mythic *fons et origo* of the market *and* a disreputable figure that undermines the legitimacy and pretension of 'the economy'—then we might have to bear with this undecidability to gain a temporary vantage point from which to see the more obscure forces of organization that are at work shaping our reading, writing, and thinking.

REFERENCES

Çalişkan, Koray. 2010. *Market Threads: How Cotton Farmers and Traders Create a Global Commodity*. Princeton University Press.

Callon, Michel, Yuval Millo, and Fabian Muniesa (eds). 2007. *Market Devices*. Oxford: Blackwell.

Deleuze, Gilles and Félix Guattari. 1994. *What is Philosophy?*, trans. G. Burchell and H. Tomlinson. London and New York: Verso.

Derrida, Jacques. 1994. *Specters of Marx*, trans. P. Kamuf. New York and London: Routledge.

Langdon, Davis. 2007. *SPON's Architects' and Builders' Price Book 2007*, 132nd edn. Abingdon: Taylor & Francis.

Latour, Bruno and Steve Woolgar. 1979. *Laboratory Life: The Social Construction of Scientific Facts*. Beverly Hills, CA: Sage.

MacKenzie, Donald. 2006. *An Engine, Not a Camera: How Financial Models Shape Markets*. Cambridge, MA: MIT Press.

McLuhan, Marshall. 1964. *Understanding Media: The Extensions of Man*. New York: McGraw-Hill.

Strathern, Marilyn. 1999. *Property, Substance and Effect: Anthropological Essays on Persons and Things*. London: Athlone.

CHAPTER 32

....

PUSH BUTTON

....

L. ROMAN DUFFNER

PUSH buttons, keys, and switches are attached to a wide spectrum of functional devices varying from simple ones, like the doorbell or the light switch, to very complex digitized systems. The pushing action has become a basic and enduring gesture of interacting with technical apparatus: we can find pushable items on almost every technical device whose inner condition and output can be effected. Although the proliferation of such objects is strongly related to electrification—J. H. Holms presented the first electrical switch in 1884 (Dummer, 1983)—the push-function is not exclusively used in electrical devices. Mechanical mechanisms like the start/stop of the stopwatch, the shutter of the camera, the flush of the toilet, the keys of the typewriter or the piano also act on the pushing input. The push-function has become so practical and effective that it also found its way into digital and virtual space, where it characterizes working with computers, terminals, and mobile phones. In computer programs, apps, and websites, manifold virtual buttons graphically imitate the special transformation of the physical push.

With a simple push a lot can happen. A single person is able to activate or stop a powerful technical apparatus, including weaponry of mass destruction. More prosaically, online, we can sign contracts (like terms of conditions) with a click, vote, express our position about published content, and become ever more implicated in providing our own consumer services (think of the now redundant roles of elevator or bell boy [Bernard, 2014], service checkout operators, bank tellers, all superseded by push buttons). In fact, push buttons are a crucial mediating device in many social interactions. The success of this technique finds us almost continuously surrounded by input-devices in both the private and the public sphere, so much so that the philosopher Hans Blumenberg (2010: 210) argues the (modern) world 'is increasingly characterized by triggering functions', leading to the rise of a 'push button culture'.

Despite the fact they colonize our lifeworld and represent an integral part of organizational and communicative processes, push buttons, keys, and switches have received only little attention in organizational studies. In the research on 'scopic media' and 'synthetic situation' (Knorr Cetina, 2003; Knorr Cetina and Preda, 2007), or in the study of 'planned and situated action' (Suchman, 1987) for example, the researchers mention, in passing, the use of input-devices but do not consider the physical activities in greater detail. There are a few exceptions. The QWERTY keyboard (David, 1985; Kittler, 1986; Reinstaller and Hölzl, 2009) and the synthesizer keyboard (Pinch, 2008), for example, draw some attention to actual buttons and keys to then examine the interlocking of technological developments, practical experiences, and the social context of the keyboard usage. Such studies reveal the impact of socio-technical assemblages on establishing and institutionalizing, for example, the key arrangement of the QWERTY keyboard. By focusing on the remarkable (and arbitrary) structure of key arrangements, these studies raise the question of underlying social processes that lead to the successful standardization of tools. Yet the trivial objects—like push buttons—are almost incidental in these studies. The functionality of the technique and the distribution of action among different actors are not in focus, since these authors are not engaged with the 'inter-' or 'intra-actional' (Barad, 2007) dimension but rather focus on the social and structural consequences of the fully composed machine: the push button is only there in passing.

It seems that the more apparently trivial the one-to-one-relationship of input and output is, the more it is taken for granted. The underlying mechanisms appear to be self-explanatory, permanent, predictable, and thoroughly technical. Even though the entangled objects are involved in everyday activities, and thus a fundamental part of our human interactions, we struggle to give them attention. Because of this trivialization of the functional properties, which typically constitute constructed machines (von Foerster, 1972), the not-so-trivial agency of the object as well as its co-constructing activities stays hidden.

Exceptions are found outside of the study of organization. Beside Blumenberg's (2010) brief theoretical thoughts on buttons, detailed consideration of the push button can be found in socio-historical studies which discuss the control mechanism of the elevator (Bernard, 2014), or the rise of the panic button to alarm people (Plotnick, 2015). In media studies, researchers address how media technologies afford a certain tactility of the media and deal with, for instance, gestures of touching and pushing as crucial practices of media use (Flusser, 1985; Heilmann, 2011; Pias, 2011). In reflecting the (single) input-item, they show the distinctive characteristics of the push button in simplifying and homogenizing activities, enabling control and remote action, allowing fast activation of complex processes, tempting users into manipulation, and being used as a metaphor for power and efficiency.

In this chapter, the focus on the push button, key, or switch is a focus on the mediating activity (Latour, 2005) of the pushable input-item, and on situations in which more than a single button or key is involved (e.g., the creative production of electronic music). Thus, a closer look will be taken at the practice of pushing, the abilities of both user and input-items, the transparency of the mechanism's functionality, and the place of action. Through this focus, a number of binary patterns become visible that demonstrate what buttons do and how they are entangled in everyday practice. Input-entities do not just enable remote action at the push of a button, but rather co-create human activities and mechanical, electronic, or digital processes, and they unify opposite or contradicting properties and conditions. These are relevant for the organization of space and practice, as they provide possibilities of control, and allow participation in what become pre-organized socio-technical processes.

THE PRACTICE OF PUSHING:
ACTING VS. ACTIVATING

Using a pushable input-item can be described succinctly: the pressing of a switch (= input) causes a technical reaction (= output). Although the different practices of 'pushing' closely resemble each other, the mediating and transforming process at the interface differs considerably, thus affecting not only the practice but also the position, and in the case of creative practices, the appreciation of the user.

Keys and push buttons, which apply a press key mechanism, link the human input-force immediately to an output-event of the machine. The machine's activity only lasts as long as the button or key is being pressed. Applied to doorbells or electric strikes as well as to the push-to-talk function of walkie-talkies, or the dead-man switch in trains, the direct link between human activity and technical output integrates the user physically and actively in the performance, providing a high degree of control over the accomplished technical action. The use of such input-items also characterizes music performances: the keyboards of pianos and synthesizers work on this principle. Here the artistic activity reveals how the key-pressing practice affects the entire performative role and position of the player. As the technically produced sound refers directly to the activity of the musician, the player appears as the creator, and the sound as an expression of skilfulness and individual feelings. The player is not only present in the technical process, but bears responsibility for what is happening. The involved buttons or keys and the assembled technology are easily ignored or assumed as intermediary, even though as mediating input-items they transform the induced pressure into an electric signal, or into the acceleration of a hammer.

This becomes visible when the input-item perpetuates a short human activity, such as a DJ operating a controller, where the user turns more into a passive participant alienated from the technical performance (Blumenberg, 2010; Bernard, 2014). Switches like push button switches and rocker switches enable the temporal extension of the human activity. They are the most common form of controllers that operate accordingly to the binary on/off principle. Optically and technically, many types of push button switches hardly differ from press keys, like the power-on button or the caps-lock key on the computer. However, they transform the physical push of the user into an enduring—sometimes reversible—technical change of the inner setting of the machine (Figure 32.1).

Though it requires a user to start the activity of the machine, it is the switched-on button that ensures the ongoing creation of the output, independently from the user. As Bernard (2014) points out, with the use of push button switches the user does not

FIGURE 32.1 Acting during an activated function. The user halts the activated operation of the elevator by actively pushing the door-opener button

(Photo by the author)

actively produce, but only activates a technical process. The delegation of the task to the switch frees the user from a tying and therefore restricting interaction (otherwise he/she would have to press the button the whole time).

In the case of playing music, this enables the musician to dedicate themself to other activities. However, in doing so, the player is separated from the occurring sound-event, reducing her or his direct contribution to the process of playing music. Therefore the audience may question his or her capability, skills, and performance: pushing the play button is considered as a trivial and less skilful act. Moreover the player—the DJ— disappears and turns into an intermediary (he or she makes no difference to the practice), while the machine starts to reproduce something already created.

DE-SKILLING AND EMPOWERING

To push or to press and hold a button or a key means to draw on, mobilize, and bring in a persistent technical process. Despite the difference of the gestures, all these movable input-items affect the abilities of the users in two ways, since their use entails simultaneously de-skilling and empowering characteristics. They are de-skilling insofar as a user does not need specific (technical) abilities or physical power to initiate the technical process. Even children and animals are able to handle them intentionally. Additionally, the possibilities of actions are often limited to the default setting, like a singular agreement (push on the play/start button) or a dual choice (on/off). Blumenberg (2010: 210) emphasizes that push buttons homogenize the input, making 'the human action increasingly unspecific' because the technical output is always the same. As Bernard (2014: 84) shows, in the 1920s the industry expected that push-controlled machines would work more efficiently, as the minimal and unspecific human contribution seemed to compensate for the unreliability of unskilled workers (see Bernard, this volume).

However, the pushable input-item also empowers users (Plotnick, 2015: 50). Together they can initiate or manipulate a machine's activity, whereby the user decides when and how to participate. Hence, the users still have the decisive power at least about their involvement in, and contribution to, a specific setting (Flusser, 1985), although the possibility of intervention seems minimal and limited. It is the sense of decisiveness rendered by an immediately available technical effect and the simple way to activate it that makes the usage of the push button an often-quoted metaphor for power and control (Bernard, 2014). The trivial mechanism stands for the prompt and unstoppable execution of a decision: the push button metaphor becomes a call for action, no matter whether more generally, with the Chemical Brothers (2005) exhorting the world to push the button and Sun Ra (1983) observing how, when the button is pushed 'your ass gotta go', or more specifically, such as to start a love relation (Sugababes and Austin, 2005).

COMBINING IMMOVABILITY AND MOVABILITY

With regard to the basic abilities of buttons and keys, pushable input-items have a hybrid character, as they combine immovability and stability with movability and thus (easy) changeability. Both properties are crucial to the practice, as they determine the agency of the object and thus influence the human activities.

The mechanical and technical principles require a fixation in a device or a room. When pushing a button, for example, a counterforce that stops the movement is required. Buttons and switches are therefore anchored in a specific place, causing a spatial arrangement. Once installed, it is almost impossible to change their position. Apart from this, the input-items always perform the same mechanism. Therefore, they reliably create a durable and expectable setting, and support the user in the practice as well as in the spatial coordination. This reduces the situational uncertainty as it is not necessary to monitor or observe the arrangement; a stable work environment is thus created.

Furthermore, the (assumed) stable setting facilitates orientation, and understanding, allowing users to repeat similar actions and foster the development of routinized practices

(Reckwitz, 2002) that refer to previous experiences, and relate to the expectation that the setting remains (almost) the same. Turning the light on in a familiar room does not require a deliberate action, as we move our hand to the switch in a habitual, purposeful, and routinely embodied way. The action appears smooth and controlled. This also applies for input-devices assembling several buttons and switches into one spot, like the computer or piano keyboard. The stable set-up and the spatial proximity enable the development of a highly routinized usage like touch-typing (Reinstaller and Hölzl, 2009), or the blind playing of the piano.

The movability, in contrast, ensures communication with a technical device (Plotnick 2015), and serves as the mechanism for transforming the human activity. It keeps the inner status of the apparatus easy and almost immediately changeable. With the push of a button, the user can activate the predefined effect built into the device at any time (Blumenberg, 2010: 210). The 'always-ready-to-use' operation of the apparatus accelerates the action, allowing us to promptly intervene in ongoing processes. We are able to call an operating elevator to our floor, to activate the fire alarm, to stop an entire production machinery in case of an emergency, or in the case of electronic music to actively participate in (and create) the ongoing acoustic event.

The property of and need for ubiquitous changeability entails a susceptibility to mistakes and misuse. Once pushed, the input-item causes an immediate and (initially) unstoppable reaction. In a musical performance the audience immediately hears an accidently-touched or wrongly-pressed key, perhaps forcing the musician to pretend that nothing amiss has happened. Therefore the fast transformation may cause an unintended and (positively or negatively) surprising outcome and in the case of creative work it keeps the practice open for serendipity. In contrast, the abuse or accidental activation of a fire alarm or an emergency button affects organizational processes, as they immediately stop the production process or mobilize an emergency team. Since a false activation causes a waste of effort, various measures help avoiding misuse such as educating the users, protecting the alarm button with a glass cover (Bernard, 2014), or sanctioning the abuse with a fine.

SIMPLIFICATION AND COMPLEXITY

When using the input-function, the user faces contradictory properties since buttons and keys on the one hand simplify the human–machine interaction and on the other hand increase the situational complexity. Beyond easy usability, the binary function (on/off) successfully trivializes the operation. It is quite simple and 'intuitive' (Pias, 2011) for the user to link a distinct machine's outcome to the induced action. The assignment of one specific activity to one specific input-item reduces complexity, as the user is able to learn and subsequently anticipate what will happen, when he/she uses the button. This also means that the input-item constrains the possible scope of action to a distinct event.

However, the limitation to the easy-to-understand, one-to-one relationship causes an increase in the number of the input-items if more than one function is to be triggered. The piano keyboard, for example, is made up of 88 keys, each assigned to a pitch. Although the functionality of the keys remains trivial the number of the keys induces complexity as it becomes more difficult to keep track of the assembled input-items. Arranging the functions in distinct patterns like the cross-design of the directional pad on the Game Boy (Figure 32.2) and using colour codes like the black and white keys on the keyboard may improve orientation and understanding.

A further increase in complexity follows from the need for a compact size, which requires a reduction in the number of input-items—as we see with synthesizers and computer keyboards (Figure 32.3). To maintain a high number of feasible functions by a restricted number of keys, two or more functions must be allocated to a single button. In using a control (or shift) button, the user can quickly switch to the secondary function of the button. The relation of simplicity and complexity on the one hand ensures an easy use and allows us to develop fast movements. On the other hand, to act skilfully

FIGURE 32.2 Arrangement of buttons on the Nintendo Game Boy

(*Source* : https://commons.wikimedia.org/wiki/File:Game-Boy-FR.jpg?uselang=en)

FIGURE 32.3 A compact digital synthesizer (three octaves) with red lightning 'performance buttons' and green lightning 'control knob circle', showing the actual setting of the instrument

(Photo by the author)

and purposefully requires an engagement with the equipment and the development of embodied, habitual knowledge.

In addition, there is a high but rather unremarkable complexity in the limited communication abilities of the input-items, causing complication in the practice. We are used to assuming that the object's reaction to a force is a machine's communicative act—such as the 'back-bounce' of a button. However, all the information we get from the bouncing button is not about the machine's activity, but only about the mechanism of the input-items, which show in most cases the same reaction when switched off. Therefore, the input-device acts as a nested trivial machine and accomplishes two different activities. On the one hand the device reacts and gives feedback on the induced force and shows a certain response. On the other hand, the machine may also react, whereby the first reaction does not necessarily lead to the second one.

This nested function is the only but crucial possibility of communication that a simple button or switch provides. If the user is able to observe the outcome and to keep the overview over the number of input-items, this information may be enough. If this is not possible, or the triggered function is delayed, the ability to act towards a purpose becomes restricted or error-prone. A doorbell panel without accurate imprints, for example, makes it almost impossible for an inexperienced visitor to ring the right bell (Figure 32.4). In addition, if the user cannot see or hear the effect of a pushed button, he or she may repeatedly activate the input-item, resulting in an annoying use of the doorbell, or in the case of print orders at the computer, in tons of wasted paper. To act on a controller item requires not only the ability to activate, but also the opportunity to oversee the situation, and to keep track of the technical activity.

Adding different additional objects diminishes the limited communication ability of the input-item and may increase situational control. Signifying objects, such as letters, numbers, abbreviations, or other symbolic information printed on or next to the buttons

FIGURE 32.4 The doorbell panel shows challenges for users like an illogical arrangement of buttons (Top 12) or missing imprints and coping strategies such as DIY-corrections of incorrect imprints by pasting over the wrong ones

(Author's own photograph)

or switches, enable the input-entities to reveal their function. Although such signifiers help in the human–machine interaction, they do not ensure understanding, and therefore may only partially reduce difficulties. The firmly attached signifier can even be a source of misunderstanding and failure. If the linked function changes, like the virtual key assignment, the symbol gives incorrect information. Additionally, the user has to know the meaning of the printed words, abbreviations, or symbols to relate a specific function to them.

The limited communication ability of the input-entity has led to expand the input-items with LEDs or light signals. Similar to the imprints, the LEDs are situated beneath the buttons illuminating them or are located next to them. The light signal primarily helps to increase control over the human–machine interaction, as it informs the user about the activity of a button or switch. If a control light lights up, we assume that the button or switch next to it is turned on.

Apart from the visual information, other forms of signals, like beeping or buzzing tones, or the vibration of a mobile device, confirm the successful input. Due to these signals, the input-entity comprises a control item detached from the machine's output increasing the remote control of machine activities in situations which are more difficult to grasp. The flashing or shining light, the beeping and vibrating point out that the input is captured and transmitted. Therefore it enables us to respond more independently to the related technical effect. The extension of the input-item with specific information entities, like imprints or light signals, facilitates the difficult information exchange between the human and the machine.

LOCALITY AND DISTANCE

Taking into account the place of activity, the translation at the input-device links the proximity and locality of the human–machine interaction to a distant and often unreachable sphere of the machine's mechanical, digital, and virtual processes (Flusser, 1985: 24). Indeed, the act of mediation enables the user to influence physically remote and inaccessible places, like computer programs or the sphere of a sound. Nonetheless, the activity itself is associated with a certain location in which the human body is present. Even if more and more activities are transferred to the virtual places of digital programs, at least a minimum of practice remains, like the push of a (virtual) button. This means that mechanically or digitally mediated activities and interactions are always locally bound and thus shape a local practice.

This becomes visible in the use of laptops and digital audio workstations (DAW) in live music performances. Although DAW programs are essential to the performance today, many musicians engage little or infrequently with the laptop. This is because the usage of the point-and-click system of the mouse requires a 'melting' with the machine (Knorr Cetina and Preda, 2007), restricting the human activity to single clicks one after another. As the screen-based work constrains the performance, most

FIGURE 32.5 Novation Launchpad DAW-controller: the back-lit red and yellow buttons indicate assigned function and the bright glowing buttons show activated function

(Author's own photograph)

musicians use specific input-interfaces, such as a midi-controller, which enable them to change the DAW settings with the push of a button, and without looking at the screen. The input-item embodies the effect or function the user can trigger directly in his or her physical world, making the performance a more local, intuitive, and interactive event (Figure 32.5).

However, the user is always tied to the places where the input-items are located. By mobilizing them, the human actor becomes spatially flexible, whereby remote controls for televisions, doors, cars, and even musical instruments separate the input and output operation entirely. In doing so, the input-device merges even more with the user, so they become a constant companion in daily life. The pendant alarm for elderly people for example is a mobile emergency button which is worn around the neck allowing the user to call for help in the case of a fall or other emergency situations. Pushable input-items provide the co-presence of remote effects and imply a (minimal) local activity that is necessary to contribute to distant events.

In conclusion, the input-item is a member of two spheres of action (Geser, 2002). On the one hand, it belongs to the material/human world consisting of a touchable surface, which immediately reacts to forces, and allows the human actors to interact with a technical system. On the other hand, it belongs to the distant world of the machine with its programmed and fixed automatism, and thus it is part of the processes executed by the machine. The input-items keep remote spaces, dimensions, and mechanisms of actions not only accessible but also (partly) influenceable, so that human actors might participate actively in social and organizational processes. Push buttons and keys can open organizational processes to non-members permitting the consumer to execute tasks. Despite their trivial mechanism, they are not always self-explaining and can cause difficulties. A closer look at the involvement of trivial objects in practice reveals their strong entanglement in collective agency and its co-construction of the (material) phenomena. Although the mundane use of the pushable input-item is a small detail of socio-material performance, its consideration shows the variety of conditions that can become meaningful in the interaction, affecting media use and organizational practice.

REFERENCES

Barad, Karen. 2007. *Meeting the Universe Halfway: Quantum Physics and the Entanglement of Matter and Meaning*. Durham, NC: Duke University Press.

Bernard, Andreas. 2014. *Lifted: A Cultural History of the Elevator*. New York: New York University Press.

Blumenberg, Hans. 2010. 'Lebenswelt und Technisierung unter Aspekten der Phänomenologie'. In M. Sommer (ed.), *Theorie der Lebenswelt*. Berlin: Suhrkamp, 181–224.

Chemical Brothers. 2005. 'Galvanize'. On *Push the Button* [CD]. London: Virgin.

David, Paul. 1985. 'Clio and the Economics of QWERTY'. *The American Economic Review*, 75(2), 332–7.

Dummer, G. W. A. 1983. *Electronic Inventions and Discoveries. Electronics from its Earliest Beginnings to the Present Day*. Oxford: Pergamon Press.

Flusser, Vilém. 1985. *Into the Universe of Technical Images*. Minneapolis: University of Minnesota Press.

Geser, Hans. 2002. *Towards a (Meta-)Sociology of the Digital Sphere*. http://socio.ch/intcom/t_hgeser13.htm [Accessed 5 July 2018].

Heilmann, T. A. 2011. 'Buttons and Fingers: Our "Digital Condition"'. Paper presented at Media in Transition 7 'Unstable Platforms: the Promise and Perils of Transition'. Massachusetts Institute of Technology, Cambridge, MA. http://www.tillheilmann.info/mit7.php [Accessed 22 July 2018].

Kittler, Friedrich A. 1986. *Gramophone, Film, Typewriter*. Stanford, CA: Stanford University Press.

Knorr Cetina, Karin. 2003. 'From Pipes to Scopes: The Flow Architecture of Financial Markets'. *Distinktion: Scandinavian Journal of Social Theory*, 4(2), 7–23.

Knorr Cetina, Karin and Alex Preda. 2007. 'The Temporalization of Financial Markets: From Network to Flow'. *Theory, Culture & Society*, 24(7–8), 116–38.

Latour, Bruno. 2005. *Reassembling the Social: An Introduction to Actor-Network-Theory*. Oxford: Oxford University Press.

Pias, C. 2011. 'The Legibility of Movement'. In A. Zinsmeister (ed.), *Figure of Motion/Gestalt der Bewegung*. Berlin: Jovis, 130–59.

Pinch, Trevor. 2008. 'Technology and Institutions: Living in a Material World'. *Theory and Society*, 37(5), 461–83.

Plotnick, Rachel. 2015. 'Panic Button: Thinking Historically about Danger, Interface, and Control-at-a-Distance'. In Robert C. MacDougall (ed.), *Communication and Control: Tools, Systems and New Dimensions*. Lanham, MD: Lexington Books, 45–58.

Reckwitz, Andreas. 2002. 'Toward a Theory of Social Practices: A Development in Culturalist Theorizing'. *European Journal of Social Theory*, 5(2), 243–63.

Reinstaller, Andreas and Werner Hölzl. 2009. 'Big Causes and Small Events: QWERTY and the Mechanization of Office Work'. *Industrial and Corporate Change*, 18(5), 999–1031.

Suchman, Lucy. 1987. *Plans and Situated Actions: The Problem of Human–Machine Communication*. Cambridge: Cambridge University Press.

Sugababes and D. Austin. 2005. 'Push the Button' [recorded by Sugababes]. On *Push the Button* [Single]. London: Island Records.

Sun Ra. 1983. 'Nuclear War'. On *A Fireside Chat With Lucifer* [LP]. Chicago: Saturn Research.

von Foerster, Heinz. 1972. 'Perception of the Future and the Future of Perception'. *Instructional Science*, 1(1), 31–43.

CHAPTER 33

....

PUSSYHAT

....

SINE NØRHOLM JUST

On 8 November 2016, the day Donald Trump was elected president, a woman from Hawaii posted a note in a Facebook forum, suggesting a march on Washington in response to the election. On 21 January 2017, the day after Trump's inauguration, almost half a million people participated in the Women's March on Washington (Figure 33.1), with millions more joining across the United States—and worldwide. In total, an estimated five million people took to the streets, and many of them wore pink knitted hats with characteristic pointy ears, so-called 'pussyhats'. In just under three months what began as a social media post had arguably turned into a social movement, and the pussyhat had become one of the central organizing objects of this process.

The Women's March on Washington was a spectacular sea of people dotted in pink, and the iconic pussyhat was worn by many protesters around the world as well. However, in Copenhagen, where I joined the march, it was not nearly as ubiquitous as in the march on Washington. Thus, it was only as images of the march and commentary on the hat began circulating in both traditional news media and on social media platforms that I became fully aware of the central role the hat had played in the organization of the march. Intrigued by its apparent affective force, I decided to follow the pussyhat's pattern, literally and figuratively. Thus, I have knitted a pussyhat at the same time as I have explored the socio-cultural significance of knitting and knitwear, using both my practical experience with the specific knitted object and my encounter with the broader social trajectory of knitting as inroads to explaining the organizational force of the pussyhat in relation to the Women's March.

This chapter retraces the organizational force of the pussyhat. How did this object interrelate with other media and technologies in the making of the Women's March? And how does it continue to do so as the Women's March transforms itself from a

FIGURE 33.1 Women's March on Washington, 2017
(*Source*: https://commons.wikimedia.org/wiki/File:Women%27s_March_2017-01_(12).jpg)

singular event to a social movement? In answering these questions, I will focus on the material quality of the pussyhat as a *knitted* object.[1]

KNITTING AS A SOCIAL TECHNOLOGY

Both knitted objects and the process of knitting are steeped in culture (Turney, 2009). First, knitted objects, e.g., sweaters, hats and scarves, are markers of social identity. Second, the activity of knitting offers occasions for social gathering and communion (Prigoda and McKenzie, 2007). Third, both the process and objects of knitting have been reappropriated for various purposes, e.g. social criticism and resistance, instigating a veritable 'knitting revolution' (Groeneveld, 2010). Fourth, the varying meanings of knitting are intimately connected with developments in and reconfigurations of the broader socio-cultural landscape of media and technologies. Most notably, the recent

---

[1] Some pussyhats are crocheted and others, indeed, sewn from woven materials, but I focus on knitting as the most common practice and take it to be (at least partially) representative of the pussyhat's other modalities.

repurposing of knitting as a tool for social change would have been impossible without the advent of online communication, broadly speaking, and social media, more specifically (Pentney, 2008).

As an object, the pussyhat clearly draws its meaning potential from the third cultural dimension of knitting/knitwear, and as the centre of a social activity its organizing force is greatly enhanced by digital technology. Thus, these two features will form the core of my analytical engagement with the pussyhat, but I will briefly introduce all four.

Knitted Objects as Markers of Social Identity

Today, knitwear is often associated with tradition; hand-crafted objects made from patterns and using techniques that pertain to particular places and peoples. Such associations, however, are examples of the manufacture of culture: although knitting does have a long and intricate, albeit fragmented history (Wills, 2007: 6–7), most of the patterns and designs that we now think of as cultural icons are of fairly recent invention— they are prime exemplars of 'the invention of tradition' (Hobsbawm, 2012).

An extreme case in point is the Icelandic sweater, which dates back to the mid-twentieth century and was explicitly invented as a marker of social identity (Helgadottir, 2011). While popular from the outset, the sweater's status as a national icon only became dyed-in-the-wool in the aftermath of the financial crisis (Helgadottir, 2011). Simultaneously, the sweater's international repute has grown greatly as a side-effect of the rise of 'Nordic noir', with Sarah Lund, the be-sweatered protagonist of the hugely popular drama *The Killing* (Stougaard-Nielsen, 2016), as a main ambassador. Strictly speaking, Lund's sweater is not even Icelandic, but a mix of various patterns with similar histories. Even so, it draws on and contributes to a broader narrative of 'nostalgia for times and practices (knitting, family, warmth, neighbourliness, truth) which remain in the past, nearly lost and in need of revival' (Turney, 2014: 29). Within this narrative, the Icelandic sweater (in a generalized Nordic sense) exists outside chronological time and, instead, is central to the mythology of life in a cold climate.

While some 'traditional' knitted objects can be traced further back than the Icelandic sweater, they do tend to be traceable. That is, rather than being ancient pieces of handicraft whose origins are lost in the fog of history, iconic knitted objects tend to be patterned socio-cultural fabrications. The Aran sweater is another case in point as is the British gansey (Gordon, 2010; Carden, 2017). In each instance, narratives of belonging are knitted into these objects with every stitch.

Knitting as a Social Activity

Just as the process of establishing associations between knitted objects and social identities is, in many cases, of quite recent design, so the currently dominant connotations of knitting as a social activity were not established as far back in time as one might think. This is, of course, not to say that knitting and knitwear were ever free of socio-cultural significance, but rather that the prevailing interpretation of them is a projection of present needs and desires onto the past: 'As a craft, and specifically an ordinary or domestic craft, knitting communicates a tradition that is not only gendered but is also vernacular, providing evidence of peoples and places that appear distant to the industrial or post-industrial urban contemporary world' (Turney, 2009: 45).

Today, then, the social activity of knitting is often perceived as traditional and traditionally feminine, as well as folksy. This perception, however, has emerged and gained force in step with the industrialization of knitting. That is, only when knitting was standardized and automated, did hand knitting move out of the sphere of practical necessity and into that of cultural symbolism. As the knitting industry was established around factories and/or distribution centres (Porac et al., 1989; Lazerson, 1993), hand knitting was put out of the practical loop—and became reconfigured as a harmless, but trivial way of keeping women occupied (McDonald, 1988).

When identified as a feminine and domestic activity (Turney, 2009: 11), knitting enters into a 'violent hierarchy' (Derrida, 1981: 41) with masculinity and the public sphere and, hence, is barred from the privileged domains of modernity (see Zundel, this volume). While such positioning may serve to marginalize knitting and knitters, it may also prove useful to those who want to provide an alternative to the currently dominant social order—whether in the form of a nostalgic return to/construction of tradition, as discussed in the previous section, or as a means of advocating social change, as will be discussed below.

Never a talented knitter, I had not laid hand on yarn and needles for more than twenty-five years when I decided to embark on the pussyhat project (Figure 33.2). Fortunately, there are plenty of patterns of the hat out there, complete with YouTube video instructions, and I thought knitting would prove to be like biking once I got the needles going. This assumption proved partially right, but only after I had overcome an initial obstacle; the current American way of casting on and looping the stitches turned out to be rather different from the one I learned in Denmark as a child. Faced with this fact, and unable to decipher the American codes no matter how pedagogically delivered, I had no other

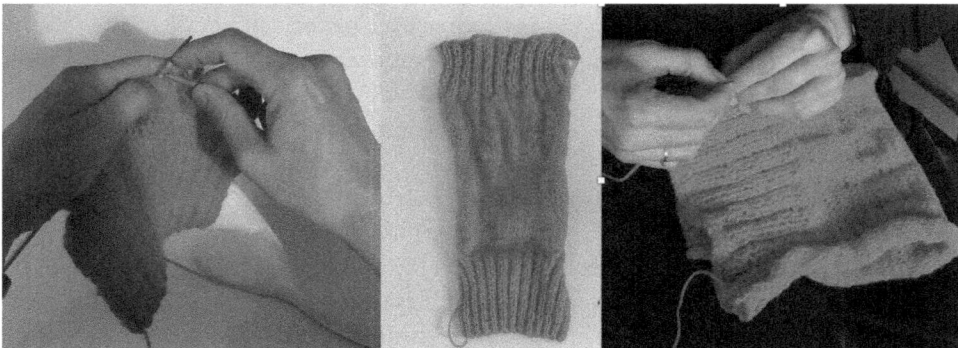

FIGURE 33.2 Knitting

(Photo by author)

option than to ask my mother for help. An enthusiastic knitter herself, Mom was more than happy to help me get my knitting in gear. Even if she was a bit puzzled by the concept of a pussyhat (she had not heard of it before I made my request), she was thrilled that I was taking up knitting again. And, as it turned out, my hands did remember the old loops even as my mind was still in some sort of denial of the memory. With some hesitance, then, I settled into the process of knitting the pussyhat.

Knitting and Knitted Objects as Technologies of Change

Feminists have variously viewed knitting as an oppressive technology aimed at keeping women out of the loops of economic and political power, an expression of genuine feminine identity and sociality that offers an alternative to and reprieve from the bustling world, and as a subversive tool for criticizing and altering social norms and structures (Turney, 2009: 9–11). The first of these positions lashes out at the social activity of knitting, as established above, while the second seeks to recuperate and revalue this activity. The third, which will be unfolded in this section, both sees positive value in knitting and criticizes those who espouse its 'traditional' feminine values (Dirix, 2014; Kelly, 2014). Here, knitting becomes a technology of change by engaging creatively with and offering playful alternatives to dominant interpretations of the objects and practices of knitting rather than being either directly opposed to or engaged in positive revaluations of these (Groeneveld, 2010; Myzelev, 2009).

The use of knitting and knitted objects as technologies of change has thrived alongside a popular cultural revival of knitting that goes beyond the nostalgic turn to an imagined past to also posit knitting as a hip or trendy thing to do as, for instance, evidenced by the phenomenon of 'celebrity knitting' (Parkins, 2004). Thus, knitting is becoming a multifaceted practice as it is both reinterpreted within mainstream social contexts and reappropriated by individuals and organizations who seek to challenge social norms. Examples of the latter include explicitly third-wave feminist groups and projects that seek to unravel and reshape social norms by using dominant markers, e.g., knitting, in creative and productive ways: 'Reclamation in this arena does not simply recreate these traditional art forms but rather uses historically undervalued means of artistic expression to discuss very contemporary issues in fresh new ways' (Chansky, 2010: 682).

Similarly, men have taken up knitting as a means of challenging gender stereotypes (Kelly, 2014: 135). While this can be a personal project for the individual male knitter, it can also be an explicitly activist public practice as is the case of the 'hombres tejedores' of Chile—a group of men who take to knitting in public places so as to advocate a more inclusive and tolerant society (Ventas, 2016). In the same vein, some groups that organize knit-ins or define themselves as knitting circles have moved beyond playfully messing about with the social norms and codes of knitting in order to reclaim it as a more radical, perhaps outright revolutionary, technology of social change. In this latter regard, 'where knitting succeeds is in crossing boundaries of age, gender, ethnicity, class and politics. Its fluidity and its community nature, its encouragement of interpersonal connection and conversation are all the basis of a quiet, slow, revolutionary movement' (Robertson, 2007: 220).

While some focus on a—more or less radical—reinterpretation of the social activity of knitting, others use knitted objects to make political statements. Most conspicuously, this latter practice takes the form of yarn bombing; i.e., knitted interventions in public places, e.g. covering trees, buildings, bridges, and so on in knitwear. Yarn bombing is sometimes explicitly political, e.g. forming part of protests and other interventions, but more often than not this use of knitwear is deviously playful and conspicuously open-ended; aimed, first and foremost, at re-enchanting the mundane and only secondarily at promoting one specific message or another (Goggin, 2015).

Other activist uses of knitted objects include the making of blankets, scarves, and other items to mark particular occasions, forward specific messages, and/or as gifts for those in need of clothing and comfort (Newmeyer, 2008; O'Donald et al., 2010). Further, some activists seek to combine the processes and products of knitting for maximum impact (Robertson, 2007; Stops, 2014). Whether focused on the knitted object or including the process of knitting as well, such activist initiatives offer more specific and specifically readable messages than the more spectacular, but also less easily decipherable displays of yarn bombing (but see Baldini and Pietrucci, 2016 for an example of how yarn bombing can also be used for specific activist purposes). What unites all these recent activities and objects, however, is how digital technologies of communication are used to bring knitters together and to circulate knitting in and to broader publics.

Interweaving Knitting and Other Social Technologies

Just as the industrial revolution relegated hand knitting to the realm of domestic leisure (Wills, 2007: 9), so the digital revolution is instrumental to the rediscovery and repurposing of knitting and knitwear. Web 2.0 technologies enhance the visibility of activities that were previously relegated to the private realm, thereby promoting and reshaping the sociality and practice of knitting and other material crafts (Orton-Johnson, 2014). For instance, knitting circles need no longer meet in specific physical locations, but can be entirely virtual places of gathering. And even when there is an offline dimension to the new knitting communities, social media facilitate the organization of these—as is particularly important when moving beyond small and close-knit groups. New communication technologies, then, have enabled a reinvigoration of knitting communities, but they have also patterned the many other ways in which knitters are now putting their craft to activist use (Pentney, 2008).

Further, new media technologies not only help organize knitters—whether for politically activist or purely social purposes. Knitted objects are also receiving increased attention as they circulate online. While new technological affordances have facilitated a renaissance of knitting that is 'truly unprecedented and remarkably different' (Wills, 2007: 5), knitting (and other traditional crafts) has also proven to be truly versatile and to interconnect remarkably well with broader social trends and transformations. Thus, knitting has become an integral part of the wild and messy weave of emergent forms of political activism, characterized by playful repurposing, intertextual references, memetic circulation, and performed 'by any media necessary' (Jenkins et al., 2016).

THE ORGANIZING FORCE
OF THE PUSSYHAT

The Pussyhat Project was launched in late November 2016 with the express purpose of creating a 'sea of pink' at the Women's March. The founders of the project present the hat as 'a symbol of support for women's rights and resistance' (Pussyhat Project, n.d.). The symbolism combines three elements—the colour pink, the term 'pussy', and the craft of knitting—each of which is reclaimed and repurposed in order to endow the acts of knitting and wearing the hat with power. First, 'pink is considered a very female colour representing caring, compassion, and love—all qualities that have been derided as weak, but which are actually STRONG' (Pussyhat Project, 2016, capitals in original). Second, the satirical reference to Donald Trump's infamous comment about being able to 'grab them by the pussy', aims to 'reclaim the term [pussy] as a means of empowerment' (Pussyhat Project, 2016). Third, the project celebrates 'women's crafts'; and, more particularly, seeks to recast knitting circles as 'powerful gatherings of women, a safe space to talk, a place where women support women' (Pussyhat Project, 2016). The project, then, is explicitly feminine, but also explicitly feminist in its strategy of turning dominant, stereotypical, and/or denigratory understandings of 'the feminine' upside down so as to empower women. Further, the project relies on the affordances of social media to establish communities and circulate messages. More specifically, a community of knitters and wearers of the hat was created; here, the former provided the latter with the material means of representing them at the march. Of course, some knitted *and* wore the hat, but many knitters who could not themselves attend the march ensured that non-knitters had hats to wear—and the Pussyhat Project provided the impetus and infrastructure for this exchange (Pussyhat Project, n.d.).

Thus, the Pussyhat Project draws directly on a Web 2.0-facilitated recasting of knitting as a feminist project, as presented above, in order to establish both the making and wearing of the hat as powerful acts of resistance. In the context of the Women's March this proved to be an effective organizing strategy as the sheer number of pussyhats worn at the march indicates, and as is evident from the amount of attention given to the hat both before and after the march. As knitters around the world joined forces to ensure that everyone who wanted to wear a pussyhat at the Women's March might do so, the project established itself as 'a global group of loose-knit activists' (BBC, 2016). And it has subsequently sought to apply the affective force of the moment to the construction of a more stable organizational base for the women's rights movement.

While knitting the pussyhat, two strands of thought got entangled in my mind. One was: 'I really don't like knitting'. The other: 'How strange to be a part of this'. Of course, my engagement with the Pussyhat Project was a reconstruction, removed in space and time

FIGURE 33.3 Pussyhat

(Photo by author)

from the Women's March and, hence, what I felt in the process of knitting may not reflect the feelings of any other knitters of the object. Yet I did, even as I sat alone with my knitting, feel the pull of community, the desire to take part. I wondered what those thousands of women felt as they prepared for the march, knitting hats for themselves and for others, putting them on and taking to the streets. A great sense of solidarity must have swelled and surged through the crowds that day. As, indeed, it did in Copenhagen with not nearly as many pussyhats in the mix. Here, those of us not wearing pussyhats smiled appreciatively at those who did. My retrospective knitting, however, gave more pause for reflection and concern than opportunity to relive those exuberant moments. The making of the hat was accompanied by readings on not only the Women's March, but also on activist knitting, and in the process, I become ever more convinced of the activist potential of crafting, but also more and more aware of the limitations of the pussyhat for organizing a feminist movement. As I put the finishing stitches to my pussyhat (Figure 33.3), I knew I would not continue the intimate association with this object that making it had been—I will neither wear it nor give it away.

From March to Movement?

There is widespread recognition of the role of the pussyhat in organizing the Women's March; indeed, it became a central symbol or 'totem' around which the march gathered momentum (Davis, 2017). Further, the hat may be part of the continued organization of a social movement. Indeed, the 8 February 2016 issue of *Time Magazine* suggests as much; here, the cover story that ran under the headline of 'The Resistance Rises', illustrated with a picture of a single pink pussyhat casting a long shadow on a blank white background, queried the very issue of 'how a march becomes a movement'.

The organizers of the Pussyhat Project also espouse the continued role of their invention. Claiming that the hat has 'grown into a symbol of support and solidarity for women's rights around the globe' (Pussyhat Project, n.d.), they encourage people to keep knitting and wearing it—both as acts of everyday resistance and for special occasions like International Women's Day. In this respect, social media continue to be used as a means of not only joining knitters and wearers together, but also—and increasingly—of spreading the image and, hence, the message of the pussyhat.

This very message, however, is not uncontested. To the contrary, the hat has received severe criticism—and the harshest critique has been raised from within the movement that the Pussyhat Project has helped organize. Here, the main issue is the interpellation of the hat; identifying women with their genitalia may have been an effective ploy in the specific context of Trump resistance, but as an organizing force it interpellates a certain subject position from which some people are excluded (not all women have pussies, not all pussies are pink) and into which other people are included against their will (not all people with pussies are women). In short, the pussyhat has been called out as a symbol of transphobia and racism (Kozol, 2017; Rachel, 2017). And even if one does not want to take this argument all the way, the essentialism of the Pussyhat Project's notions of femininity and solidarity may still be questioned (Livingstone, 2017).

In sum, the organizing force of the pussyhat springs from the tensions between the object itself and the movement it purports to represent. Being an affective force, the power of the pussyhat is neither fully articulated nor articulable, but grips everyone it touches (Pedwell, 2014; Pullen et al., 2017). As such, the 'sea of pink' at the Women's March was truly spectacular; moving participants and onlookers alike. But the pussyhat may turn out to have been more of the moment than the movement. Insofar as its meaning stabilizes, it also loses affective momentum. Thus, the hat now more clearly demarcates a particular position within the broader movement—it is becoming an object that draws boundaries rather than a boundary object (Star, 2010). One indication of the unravelling of the organizing force of the pussyhat is that while Women's March organizers and activists were coming together under the label of 'intersectional feminism' in preparation for the US midterm elections in 2018 (Women's Convention, n.d.), pussyhats were being solicited for art projects and put in museums (Pussyhat Project n.d.; Brooks, 2017; Jones, 2017).

The relation between the pussyhat and the Women's March, then, may be said to have moved from an unspecified affective flow to a more specific emotional bond. Whereas the hat was definitely central to the initial success of the march, it is now a divisive sign that some women may identify with, but which others feel alienated by. Going forward, advocates of intersectional feminism may seek to detach the pussyhat from the broader social movement they aim to build.

Nevertheless, the story of the pussyhat remains interesting for what it may teach us of the use of symbolism within feminist organizing and social movements and as an illustration of the interlinkages between 'old' and 'new' technologies and social formations.

The pussyhat connects back to identities and activities traditionally associated with knitting and women's crafts, in general, but also links up with current affairs and new media. It is, in this sense, one specific element in a dynamic socio-technical assemblage, and we may only begin to understand the configuration of the whole by studying the movements and relations of each part (Pierides and Woodman, 2012; Müller, 2015).

Thus, one methodological implication of seeking to understand the organizing force of the pussyhat is that we should pay attention to how craft and crafting matter for organizations—we should turn to objects in use (Gajjala, 2013). A further implication, and one more closely related to the specific materiality of the hat, is that we should focus more on the making of objects (Jungnickel and Hjorth, 2014). Finally, the interrelations of traditional crafts and new technologies provide alternative avenues for research (McLean et al., 2017). In the case of the pussyhat, this object came to signify and organize in various ways because of how it related to other objects, technologies, and forces, but in following these trajectories we should not lose sight of its specifics as a hand-knitted object; one that demands time and effort on the part of its maker—and one that invites alternative research methodologies (Rippin, 2017).

In this chapter, I have juxtaposed the practical process of knitting the hat with the intellectual endeavour of understanding knitting as a social technology, thereby seeking to feel and grasp the affective force of the pussyhat for organizing the moment and the movement of the Women's March. Other—and more radical—methodologies are surely needed if we are to engage more deeply with organizing 'in the making'.

References

Baldini, Andrea and Pamela Pietrucci. 2016. 'Knitting a Community Back Together: Post-Disaster Public Art as Citizenship Engagement'. In Laura Iannelli and Pierluigi Musarò (eds), *Performative Citizenship, Public Art, Urban Design, and Political Participation*. London: Mimesis International, 115–32.

BBC. 2016. '"Pussyhat" Knitters Join Long Tradition of Crafty Activism'. *BBC*, 18 January. https://www.bbc.com/news/world-us-canada-38666373.

Brooks, Katherine. 2017. 'How Pussyhats are Making Their Way into Museums around the World'. *Huffington Post*, 18 April. http://www.huffingtonpost.com/entry/how-pussy-hats-are-making-their-way-into-museums-around-the-world_us_58c6dd50e4b081a56dee37f3.

Carden, Siún. 2017. 'Cable Crossings: The Aran Jumper as Myth and Merchandise'. *Costume*, 48(2), 260–75.

Chansky, Ricia A. 2010. 'A Stitch in Time: Third-Wave Feminist Reclamation of Needled Imagery'. *Journal of Popular Culture*, 43(4), 681–700.

Davis, Ben. 2017. 'Those Pink Hats at the Women's March Can Teach Us Something about Political Art'. *Artnet News*, 25 January. https://news.artnet.com/art-world/pussyhats-womens-march-art-829571.

Derrida, Jacques. 1981. *Positions*. Chicago: University of Chicago Press.

Dirix, Emmanuelle. 2014. 'Stitched Up: Representations of Contemporary Vintage Style Mania and the Dark Side of the Popular Knitting Revival'. *Textile: The Journal of Cloth and Culture*, 12(1), 86–99.

Gajjala, Radhika. 2013. 'Use/Use Less: Affect, Labor and Non/Materiality'. *Normorepotlucks*. http://nomorepotlucks.org/site/useuse-less-affect-labor-and-nonmateriality-radhika-gajjala/.

Goggin, Maureen D. 2015. 'Joie de fabriquer: The Rhetoricity of Yarn Bombing'. *Peitho*, 17(2), 145–71.

Gordon, Jennifer. 2010. 'Maritime Influences on Traditional Knitwear Design: The Case of the Fisherman's Gansey: An Object Study'. *Textile History*, 41(1), 99–108.

Groeneveld, Elizabeth. 2010. ' "Join the Knitting Revolution": Third-Wave Feminist Magazines and the Politics of Domesticity'. *Canadian Review of American Studies*, 40(2), 259–77.

Helgadottir, Gudrun. 2011. 'Nation in a Sheep's Coat: The Icelandic Sweater'. *FORMakademisk*, 4(2), 59–68.

Hobsbawm, Eric. 2012. 'Introduction: Inventing Traditions'. In Eric Hobsbawm and Terence Ranger (eds), *The Invention of Tradition*. Cambridge: Cambridge University Press, 1–14.

Jenkins, Henry, Sangita Shresthova, Liana Gamber-Thompson, Neta Kligler-Vilenchik, and Arely Zimmerman. 2016. *By Any Media Necessary*. New York: New York University Press.

Jones, Jo. 2017. ' "Pussyhat" Acquired for Rapid Response Collection'. Victoria and Albert Museum,8March.http://www.vam.ac.uk/blog/network/pussyhat-acquired-for-rapid-response-collection.

Jungnickel, Katrina and Larissa Hjorth. 2014. 'Methodological Entanglements in the Field: Methods, Transitions, and Transmissions'. *Visual Studies*, 29(2), 136–45.

Kelly, Maura. 2014. 'Knitting as a Feminist Project?' *Women's Studies International Forum*, 44, 133–44.

Kozol, Wendy. 2017. 'White Privilege and the Pussyhat'. *Reading the Pictures*, 2 March. http://www.readingthepictures.org/2017/03/feminism-race-pussy-hat/.

Lazerson, Mark. 1993. 'Future Alternatives of Work Reflected in the Past: Putting-Out Production in Modena'. In Richard Swedberg (ed.), *Explorations in Economic Sociology*. New York: Russell Sage Foundation, 403–28.

Livingstone, Josephine. 2017. 'The Problem with "Pussy" '. *New Republic*, 24 January. https://newrepublic.com/article/140063/problem-pussy.

McDonald, Anne L. 1988. *No Idle Hands: The Social History of American Knitting*. New York: Ballantine Books.

McLean, Alex, Ellen Harlizius-Klück, and Janis Jefferies. 2017. 'Introduction: Weaving Codes, Coding Weaves'. *TEXTILE: Cloth and Culture*, 15(2), 118–23.

Müller, Martin. 2015. 'Assemblages and actor-Networks: Rethinking Socio-Material Power, Politics and Space'. *Geography Compass*, 9(1), 27–41.

Myzelev, Alla. 2009. 'Whip Your Hobby Into Shape: Knitting, Feminism and the Construction of Gender'. *TEXTILE: Cloth and Culture*, 7(2), 148–63.

Newmeyer, Trent S. 2008. 'Knit One, Stitch Two, Protest Three! Examining the Historical and Contemporary Politics of Crafting'. *Leisure/Loisir*, 32(2), 437–60.

O'Donald, Sarah, Nikki Hatza, and Stephanie Springgay. 2010. 'The Knitivism Club. Feminist Pedagogies of Touch'. In Jennifer A. Sandlin, Brian D. Schultz and Jake Burdick (eds), *Handbook of Public Pedagogy: Education and Learning beyond Schooling*. New York: Routledge, 327–32.

Orton-Johnson, Kate. 2014. 'Knit, Purl and Upload: New Technologies, Digital Mediations and the Experience of Leisure'. *Leisure Studies*, 33(3), 305–21.

Parkins, Wendy. 2004. 'Celebrity Knitting and the Temporality of Postmodernity'. *Fashion Theory*, 8(4), 425–41.

Pedwell, Carolyn. 2014. *Affective Relations: The Transnational Politics of Empathy*. New York: Palgrave Macmillan.

Pentney, Beth A. 2008. 'Feminism, Activism and Knitting: Are the Fibre Arts a Viable Mode for Feminist Political Action?' *Third Space: A Journal of Feminist Theory & Culture*, 8(1). http://journals.sfu.ca/thirdspace/index.php/journal/article/view/pentney/210.

Pierides, Dean and Dan Woodman. 2012. 'Object-Oriented Sociology and Organizing in the Face of Emergency: Bruno Latour, Graham Harman and the Material Turn'. *British Journal of Sociology*, 63(4), 662–79.

Porac, Joseph F., Howard Thomas, and Charles Baden-Fuller. 1989. 'Competitive Groups as Cognitive Communities: The Case of Scottish Knitwear Manufacturers'. *Journal of Management Studies*, 26(4), 397–416.

Prigoda, Elean and Pamela J. McKenzie. 2007. 'Purls of Wisdom: A Collectivist Study of Human Information Behaviour in a Public Library Knitting Group'. *Journal of Documentation*, 63(1), 90–114.

Pullen, Alison, Carl Rhodes, and Torkild Thanem. 2017. 'Affective Politics in Gendered Organizations: Affirmative Notes on Becoming-Woman'. *Organization*, 24(1), 105–23.

Pussyhat Project. N.d. 'Pussyhat Project'. https://www.pussyhatproject.com.

Pussyhat Project. 2016. 'Pussyhat Project Knit Pattern'. https://drive.google.com/file/d/0BwBjtQGbV7gEZU1TdUd2b1JIZGM/view.

Rachel. 2017. 'Are Pussy Hats Inherently Transphopic?' *Transphilosopher*, 26 January. https://transphilosopher.com/2017/01/26/are-pussy-hats-inherently-transphobic/.

Rippin, Ann. 2017. 'Writing with Eve: Queering Paper'. In Alison Pullen, Nancy Harding and Mary Phillips (eds), *Feminists and Queer Theorists Debate the Future of Critical Management Studies*. Bingley: Emerald Publishing, 171–94.

Robertson, Kirsty. 2007. 'The Revolution Will Wear a Knitted Sweater: Knitting and Global Justice Activism'. In Stevphen Shukaitis, David Graeber, and Erika Biddle (eds), *Constituent Imagination: Militant Investigations//Collective Theorization*. Oakland: AK Press, 209–22.

Star, Susan L. 2010. 'This is Not a Boundary Object: Reflections on the Origin of a Concept'. *Science, Technology & Human Values*, 35(5), 601–17.

Stops, Liz. 2014. 'Les Tricoteuses: The Plain and Purl of Solidarity and Protest'. In Anne Brennan and Patsy Hely (eds), *Craft, Material, Memory*. Canberra: Australian National University Press, 7–28.

Stougaard-Nielsen, Jakob. 2016. 'Nordic Noir in the UK: The Allure of Accessible Difference'. *Journal of Aesthetics and Culture*, 8(1). http://dx.doi.org/10.3402/jac.v8.32704.

Turney, Joanne. 2009. *The Culture of Knitting*. London: Bloomsbury.

Turney, Joanne. 2014. 'A Sweater to Die For: Fair Isle and Fair Play in The Killing'. *TEXTILE: Cloth and Culture*, 12(1), 18–33.

Ventas, Leire. 2016. 'Hombres tejedores: El grupe chileno que desafía prejuicios con agujas e hilo'. *BBC Mundo*, 13 December. http://www.bbc.com/mundo/noticias-37884575.

Wills, Kerry. 2007. *The Close-Knit Circle: American Knitters Today*. Westport, CT: Praeger.

Women's Convention. N.d. http://www.womensconvention.com.

CHAPTER 34

RAILWAY TRACKS

CHRISTIAN DE COCK

INTRODUCTION

IN choosing railway tracks as my particular form of organizational-technological mediation I aim to bring together a personal and perhaps somewhat tenuous collection of memories and reflections. The personal memory concerns, for example, the railway tracks just outside the house where I have lived for over seven years and am now leaving behind (Figure 34.1). I see these tracks from my bedroom window every day as I wake and go to sleep and can even glimpse them from my study while typing these very words. Whilst these tracks are a part of a very concrete technology—they carry thousands of commuters into London every weekday—they also hold a significant affective charge.[1] The apparent permanence of these tracks speaks of familiarity and routine which engages and enrols me/us in ways which are both conscious and unconscious; it speaks of 'the very relationship between human beings and time, which leaves its traces on things' (Orlando, 2006: 3). Railway tracks as a technology offer a fusion of the imaginary and the real par excellence.

In developing my argument, I will follow W. G. Sebald in seeing the key to our contemporary world not in the surface phenomena of our immediate present but in technologies developed in the nineteenth century, whose effects continue to be felt (Long, 2007). The anti-hero Austerlitz suggests, in Sebald's (2002: 9) eponymous novel, when reflecting on the construction of the imposing Antwerp central station, that the nineteenth century is a time 'now so long ago although it determines our lives to this day'. Maybe railway tracks these days stand for that clash of temporalities in a modernity that is somehow out of sync with itself now that technological progress is itself ageing? We live in the ruins of the nineteenth century which we constantly try to incorporate in

---

[1] This can include boredom (Johnsen, 2016). As McGuinness (2017: 46) suggests, whilst for some they 'represent escape, travel and bohemia', to others they connote 'drudgery, offices, the rut of life and a particular sort of existential stasis that we only notice, paradoxically, because we're moving'.

FIGURE 34.1 Railway tracks

(Author's own photograph)

the present (Boym, 2007; De Cock and O'Doherty, 2017). One small manifestation of this influence on our current modes of organizing is the way in which payment companies like Visa and MasterCard refer to themselves as 'rails' for transiting value. The payment industry's use of the term 'rails' dates back to when telegraph lines ran alongside railroad tracks during the period of westward expansion (Tooker and Maurer, 2016).

I will be further guided by Walter Benjamin's notion of 'the physiognomy of the thingworld' (Marx et al., 2015: 57) which suggests a relationship to objects, 'which does not emphasize their functional, utilitarian value—that is their usefulness—but studies and loves them as the scene, the stage of their fate'. For Benjamin things are like fossils in which a constellation of social tensions and forces are petrified. Things also produce to a large extent their own subjects and it is in 'those minute points of contact between new things and old habits' that we learn to 'include in our sense of history the power of things themselves to impress and shape and evoke a response within consciousness' (Trachtenberg, 1986: xv).

The nineteenth-century railways greatly enhanced the ability of governments to project power, to the point where one can state that 'the creation of the modern state would have been impossible' (Sheehan, 2017: 21) without trains and tracks. If we fast-forward to the immediate present we can still see this dynamic very much at work.

When in 2017 the 34 wagon *East Wind* train, connecting several Chinese cities by direct rail to Europe, rumbled into London it was very much portrayed as a powerful symbol of China's *One Belt, One Road* initiative, opening up the old Silk Road routes. And yet, here again the we can still detect the presence of the nineteenth century with the persistence of old frictions: differing rail gauges in countries along the route meant a single locomotive and set of wagons could not travel the whole route and had to be changed four times (Hillmann, 2017). Perhaps the major 'new' development here was the actual organizational choreography—between railway operators, terminal owners, and national authorities—required to overcome a specific set of technological barriers.[2] The fashionable reclaiming of the Silk Road has to be located somewhere halfway between the real and the imaginary.

OF TRACKS AND TIME

For Wagner (2012) two themes dominate our self-understanding as modern human beings: that we are self-determining subjects and that we can exercise this autonomy to choose reason as the basis for the pursuit of domination over nature. The railroad as technology is an expression of these themes, whilst simultaneously putting them in a dialectical tension. It is beyond dispute that this nineteenth-century technology has had a central role in the making of industrial capitalism through the transformation of ways of production and consumption and the organization of work in emerging modern societies (Schivelbusch, 1986). The network of rail tracks which emerged in the mid-nineteenth century (by 1880 there were more than 100,000 miles of track in Europe) represented the visible presence of modern technology *as such*, radically foregrounding the role of machinery and technology within everyday life (Trachtenberg, 1986). It was perhaps the first time in human history that required the negotiation of such complex relations between technology and society, bringing together 'machines, workers, structures, finance, route ways and various means of recording and decision-making assembled into functional networks of communication' (Revill, 2012: 36). Railway tracks projected a vivid and dramatic sign of modernity, a promise of imminent Utopia. Yet, from its very beginnings the railroad was never free of an undercurrent of disquiet and foreboding (Schivelbusch, 1986). It provoked two intense contradictory reactions: the

---

[2] Seeking to burnish its international image in the wake of Brexit, the UK was of course keen to tap into this imaginary globalism. As one British minister boasted: 'This new rail link with China is another boost for global Britain' (Hillmann, 2017). Yet, as ever the story of technology and organization is never quite so simple. As Hillman (2017) elaborates: 'Chinese trains could run on UK railway tracks since both use standard gauge track. But they are separated by broader gauge that runs across former Soviet states. As a result, cargo was transferred to new locomotives at Dostyk, Kazakhstan and Brest, Belarus…The journey was like an Olympic torch relay, with four subcontractors responsible for different parts of the trip. In addition to transfers between different gauges, containers were loaded on to special platforms at Duisburg approved for the Channel tunnel.'

enthusiastic acceptance and celebration of a progressive triumph, and on the other side the troubled denunciation of its destructive effects on the social landscape and individual sensibility (Ceserani, 1999). Some of the initial responses in the mid-nineteenth century praising the railroads as an emblem of America's greatness and an expression of the idea that history is a record of steady, continuous, cumulative material progress, capture the first affective reaction well. Its spokesmen were the dominant economic and political elites of the day who fashioned a hyperbolic rhetoric of the technological sublime. Leo Marx (2000: 197) quotes a luminary called George Ripley from his 1846 speech:

> The age that is to witness a rail road between the Atlantic and Pacific... will also behold a social organization, productive of moral and spiritual results, whose sublime and beneficent character will eclipse even the glory of those colossal achievements which send messengers of fire over the mountain tops, and connect ocean with ocean by iron and granite bands.

Yet, the speed of the new way of travelling and all the tracks that started cutting across the plains and mountains of various continents also created the contrapuntal trope of the 'interrupted idyll' (Marx, 2000: 27) (Figure 34.2). The straight lines of the railway tracks, representing the visible presence of modern technology, brought to the emerging physical and social landscapes of modernity an element of forced acceleration, dislocation, and foreboding (Ceserani, 1999). Railroad travel produced novel experiences and required new modes of perception of the landscape which was now no longer perceptible as a totality but became reduced to a series of isolated fragments

FIGURE 34.2 Railway tracks from a train

(Author's own photograph)

(Long, 2007). Ruskin complained that 'all travelling becomes dull in exact proportion to its rapidity' (quoted in Schivelbusch, 1986: 58), and Flaubert caustically suggested that, 'the railway would merely permit more people to move about, meet and be stupid together' (quoted in Barnes, 2009: 108). Flaubert wrote to a friend in 1864: 'I get so bored on the train that I am about to howl with tedium after five minutes of it'. He was reported to stay up all night before a rail journey in order to be able to sleep through it and not experience it at all (Schivelbusch, 1986: 58). Ruskin and Flaubert are witnesses to a particular condition of modernity in which the sense of subjective autonomy and authenticity is jeopardized by the technological innovations of modernity symbolized by trains and tracks (Long, 2007).

Whatever the initial responses, it is clear that the railways played a significant part in engineering changes in people's experience of time and space by which the dynamics of modernity were encountered on a day-to-day basis. As such railway tracks stand for the simultaneously creative and destructive energy which Karl Marx made central to the making of modernity (Berman, 1983). In dialectical opposition to Ripley's rhetoric of the technological sublime, we can quote Walter Benjamin (1974: 1232) from his preparatory notes to *On the Concept of History*, which do not actually appear in the final versions of the document.

> Marx said that revolutions are the locomotive of world history. But perhaps things are very different. It may be that revolutions are the act by which the human race travelling in the train applies the emergency brake.

The image seems to suggest that if history simply follows the progressive course mapped out by the steel structure of the rails we would be heading straight for a plunge into the abyss (Löwy, 2016). Without attention to the destructive side of the dialectic of modernity, the rails of history simply accelerate us to disaster. Yet, the image of tracks simple stretching into a distant future 'leaves revolution as a receding moment—the station we never quite arrive in' (Noys, 2013).

It is undeniable that railways, both as constructive and disruptive influence, deeply affected (and continue to affect) the collective sensibility and imagination of those living in modern societies. It should not be a surprise then that they have been aesthetically represented by various artists who tried to articulate the ambivalence of this technology; we only have to think of Turner's famous *Rain, Wind and Speed*, Duchamp's cubist painting *Jeune homme triste dans un train*, and the impressionist canvases of Parisian stations. More recently, the eponymous centre piece of the exhibition by conceptual artists Ilya and Emilia Kabakov at the Tate Modern (18 October 2017–18 January 2018) called *Not Everyone Will Be Taken into the Future*, featured a life-size train that has almost left a dimly lit room dissected by rail tracks. These tracks are carrying the selected few into the future whilst discarded canvases on the abandoned platform remind us of all those left behind and relegated to obscurity. The melancholic setting furthermore makes us wonder what that future to which the tracks point actually holds. Joseph Beuys also reflects on this hold of rail tracks on our imaginary of the future in a work called

A monument to the Future (1976; 2. Fassung) which consists of a pair of old rails spread out on the floor.[3] Michaels and Berger (2011), in their poetic book *Railtracks*, talk of tracks as a symbol of progress that now also evokes nostalgia for a bygone era. In their short conversation pieces that usually span a handful of pages, the rail system becomes a living and breathing body of technology and memories as the narrators interweave personal histories and historical reflections with references to stations, tracks, rail yards, and timetables. They gently make us aware that the continual production of the new under the capitalist mode of organization has as its concomitant the acceleration of obsolescence.

The range of cultural materials on railway tracks is huge and their proliferation is closely related to the technologically mediated society that railways helped to produce (Revill, 2012). Yet, the most sensitive and telling responses to the complex societal and organizational effects of trains and tracks are necessarily rather singular (Marx, 2000). The artists I have chosen to introduce in the remainder of this chapter try to engage the collective, subjective, and visceral aspects of railway tracks. Their work can thus be read as a space where a variety of tensions and forces converge and become visible, allowing for a range of aesthetic judgements.

The Eternal Way and Auschwitz

> Almost always, at the beginning of the memory sequence, stands the train which marked the departure towards the unknown not only for chronological reasons but also for the gratuitous cruelty with which those (otherwise innocuous) convoys of ordinary freight cars were employed for extraordinary purposes.   (Levi, 1989: 85)

Anselm Kiefer's work evokes a complex and contradictory set of resonances about the hold that railway tracks have on our imagination. At the time of Kiefer's 2017 summer exhibition in New York, a journalist accompanied him on a walk from his hotel to the gallery in Chelsea. Suddenly Kiefer

> shook his head at a stretch of preserved rail track running along the path. 'It's the domestication of the wonderful symbol of the endless way', he said. 'And of Auschwitz...'   (Parker, 2017)

When collecting materials for this research project I was amazed to find how many rail track paintings Kiefer had produced: *Lot's Wife*, *Abendland [The Occident]*, *Abendland [Twilight of the West]*, *Iron Path*... All of these are clearly inspired by the ominous image by Stanislaw Mucha of a depopulated Auschwitz-Birkenau, with tracks

[3] On display in Staatliche Museen zu Berlin, Nationalgalerie, Sammlung Marx.

pointing the viewer's eye towards the entrance gate while in the foreground belongings lie strewn across the rails in the snow. Beyond this gate lies a reality only those who experienced it can know; a horror we can never truly appreciate second-hand.[4]

Kiefer's *Iron Path* shows a bleak and blasted landscape in which a railway track leads from the foreground to a junction at which the tracks split and recede sharply into the distance. In *Lot's Wife* and *Twilight of the West*, diverging tracks set in an equally forbidding landscape also take us to the horizon, seemingly evoking Benjamin's reflection of how the human race seems to be rushing to its final vanishing point. These bifurcating tracks also can be considered a conduit for the concealed and distanced implications of human action (Latour, 1993). On top of the tracks in the foreground of *Lot's Wife*, a metal heating coil has been attached to the painting; in *Iron Path* cast-iron replicas of climbing shoes are attached. All of the paintings incorporate lead and some canvases almost seem to drown in the substance. These are examples of Kiefer's transvaluation of industrial items into objects of darkly poetic significance. The works remind us of the central role the railway played in genocide, with death camps located at convenient places on the network, thus allowing Adolf Eichmann to claim at his trial that he was merely a 'transport officer' (Revill, 2012). By evoking the technology that enabled the Final Solution, Kiefer implies that technology can never be simply neutral or value free. Indeed, the trauma of deportations as a sensory and embodied history of train experiences radicalizes both nineteenth- and twenty-first century responses to transit by rail technology (Gigliotti, 2009). And yet, the elements making up Kiefer's paintings are also caught in a rich network of contradictions (Biro, 2013). The blasted earth connotes a catastrophe, yet may also be perceived as a mythical primordial chaos. These are landscapes that are simultaneously ancient and modern. The lead which is such a feature of these paintings is simultaneously a poisonous element and a means of protection against radiation. The *Abendland* paintings have impressions of manhole covers stamped into the lead representing the sun, suggesting perhaps an escape from the desolation. Kiefer's choice of materials—lead, gold, silver, iron, salt—are full of alchemical symbolism which point to the possibility, however tentatively, of some kind of redemption.

Kiefer's 'railway track' paintings all represent landscapes that seem somehow stranded between possibility and catastrophe, making them into what Smithson (1996) called dialectical landscapes. For Kiefer ruination is not simply the end of something but always suggests a beginning: 'I like things that are ruined. That's a starting point for me' (quoted in Gayford, 2014). Precisely because of the rubble and destruction, we have the possibility of remaking the world differently (Gordillo, 2014).

---

[4] Images for all these can be found on the following webpages:

Stanislaw Mucha: https://commons.wikimedia.org/wiki/File:Bundesarchiv_B_285_Bild-04413,_KZ_Auschwitz,_Einfahrt.jpg

Lot's Wife: http://www.clevelandart.org/art/1990.8

Iron Path: http://www.christies.com/lotfinder/Lot/anselm-kiefer-b-1945-eisen-steig-5621943-details.aspx

Abendland [The Occident]: http://www.tamuseum.org.il/collection-work/3419

Abendland [Twilight of the West]: https://artsearch.nga.gov.au/Detail.cfm?IRN=14804&PICTAUS=True-

SEFT OR THE TRANSIENCE OF TRACKS

When Michaels and Berger (2011: 22) 'imagined all the towns brought into boom by the laying of the rails, all those that gradually vanished because the rails passed them by', they could have been writing about the project of artists Ivan Puig and Andres Padilla Domene. As much of the railway that connected Mexico City to the Atlantic Ocean was abandoned in the mid-1990s, communities were stranded and tracks were left to decay. Puig and Domene set off along these abandoned railway tracks in a vehicle they called SEFT: *Sonda de Exploración Ferroviaria Tripulada* (or Manned Railway Exploration Probe). I encountered SEFT during an exhibition at Finsbury Park (London) in 2014, and this discovery no doubt was also a significant influence on my choice of technological object for this book project. SEFT, pictured in Figure 34.3, is a kind of retro-futuristic 1960s vision of small spaceship.

Describing their time in SEFT as 'retracing the routes that the train used to pass while spreading the promise of the future', Puig and Domene (2014) found ghost towns and isolated communities where the railway had left a clear mark. In comparing the promises of progress to the current status of these towns, they engaged people in a questioning of what happened to them and learnt about their yearning for community

FIGURE 34.3 SEFT in Finsbury Park Puig and Domene

(Author's own photograph)

as their experience of place, temporality, and mobility was unravelling. We witness with them how technology created an artificial environment which people had become used to as second nature; but when this technological base collapsed the world that was built around it collapsed with it, including a whole web of perceptual and behavioural forms (Schivelbusch, 1986).

As an icon of the modern world, the railway's importance rested very much on its collectivity (Revill, 2012); and it is this notion of collectivity which now seems strangely out of time.[5] During his fieldwork in north-west Argentina Gordillo (2014) mapped the initial destructive impact of the railroads in the early twentieth century and then the dislocation produced by the privatization of those same railroads in the 1990s, in a process that mirrored the one described by Puig and Domene. He detected a nostalgia for a national collectivity that the railways once embodied when they were publicly owned and managed; but beyond that he observed a sense of disbelief among people that the railroads proved to be so transient as now abandoned rail tracks were until recently 'solid, visible objects of a progress that seemed to have arrived for good...reminders of that which they thought was a present of modernity and spatial inclusion, but that abruptly became a fractured past' (Gordillo, 2014: 174).

The abandoned railway tracks, like SEFT navigating them, suggest a dystopian future, where a technology which was once the quintessential marker of modernity has become obsolete; not because it was no longer functioning but because of a range of ideological and economic reasons. Within two decades, the railway tracks that once interconnected whole regions 'had become relics from another era, ruins of progress leading nowhere' (Gordillo, 2014: 173). Not unlike Kiefer's paintings, Puig and Domene (2014) present us with images of an eerie landscape which offers us 'memory traces of an abandoned set of futures' (Smithson, 1996: xxi). Their eloquent descriptions of loss make us look at modern history, 'not solely searching for newness and technological progress but for unrealized possibilities, unpredictable turns and crossroads; the unrealized dreams of the past and visions of the future that have become obsolete' (Boym, 2007: 10).

This dialectic of old and new is probably the defining characteristic of the railway track in our age. At the present time bullet trains and their special tracks are being introduced or expanded at rapid pace throughout the world (for example, in 2007 there was no high-speed rail in China; by 2015 the country had 12,000 miles of such tracks which is more than Europe). From China to India, from Korea to Malaysia, tens of billions of dollars are being spent on such projects. And yet, these high-tech projects can be found to coexist with ageing carriages and crumbling tracks once one looks beyond the high-profile, high-speed connections between metropolises. As for 'my' little bit of track: we were informed by Network Rail in 2017 that the railway crossing visible in my photograph—a public footpath which has been in operation for

---

[5] Régis Debray's (2017) reflection on the ideological dimension of technology provides an interesting twist to this: 'The material and mental toolkit of our time was invented in America, and it is the vehicle for a way of being, of living, of imagining and feeling. For example, each mode of transport is a world-view. *The train is collectivist and social-democratic*...The plane is globalist...The car is neoliberal and individualist' (emphasis added).

150 years, is used by about 200 people on a daily basis, and never has witnessed an accident in living memory—is earmarked for closure. This will shave between ten and twenty seconds off the journey time from London (a seventy-minute commute) as it will allow trains to accelerate faster out of the station. It will also allow the company to save on the cost of maintaining the crossing. Technology, as ever, is dialectically entwined with the political and ideological games of the age, both at the macro and the micro level.

References

Barnes, Julian. 2009. *Flaubert's Parrot*. London: Vintage.

Benjamin, Walter. 1974. *Gesammelte Schriften*, vol. I. 3, ed. R. Tiedemann and H. Schweppenhäuser. Frankfurt am Main: Suhrkamp.

Berman, Marshall. 1983. *All That Is Solid Melts Into Air: The Experience of Modernity*. London: Verso.

Biro, Matthew. 2013. *Anselm Kiefer*. London: Phaidon.

Boym, Svetlana. 2007. 'Nostalgia and its Discontents'. *The Hedgehog Review* (Summer), 7–18.

Ceserani, Remo. 1999. 'The Impact of the Train on Modern Literary Imagination'. *Stanford Humanities Review*, 7(1). https://web.stanford.edu/group/SHR/7-1/html/ceserani.html

De Cock, Christian and Damian O'Doherty. 2017. 'Ruin and Organization Studies'. *Organization Studies*, 37(1), 129–50.

Debray, Régis. 2017. 'Macron, or the Coronation of America: A Conversation with Régis Debray'. https://www.versobooks.com/blogs/3246-macron-or-the-coronation-of-america-a-conversation-with-regis-debray.

Gayford, Martin. 2014. '"I like vanished things": Anselm Kiefer on Art, Alchemy and his Childhood'. *The Spectator*. https://www.spectator.co.uk/2014/09/meet-everyones-favourite-post-catastrophic-romantic-anselm-kiefer/

Gigliotti, Simone. 2009. *The Train Journey: Transit, Captivity, and Witnessing in the Holocaust*. New York: Berghahn.

Gordillo, Gastón R. 2014. *Rubble: The Afterlife of Destruction*. Durham, NC: Duke University Press.

Hillmann, J. E. 2017. 'Trains from China Laden with Hype and Subsidies'. *The Financial Times*, 26 July. https://www.ft.com/content/dd6196f8-715e-11e7-aca6-c6bd07df1a3c.

Johnsen, Rasmus 2016. 'Boredom and Organization Studies'. *Organization Studies*, 37(10), 1403–15.

Latour, Bruno. 1993. *We Have Never Been Modern*. Cambridge, MA: Harvard University Press.

Levi, Primo. 1989. *The Drowned and the Saved*. London: Abacus.

Long, Jonathan J. 2007. *W. G. Sebald: Image, Archive, Modernity*. Edinburgh: Edinburgh University Press.

Löwy, Michael. 2016. *Fire Alarm: Reading Walter Benjamin's 'On the Concept of History'*. London: Verso.

McGuinness, Patrick. 2017. 'Diary: Railway Poetry'. *London Review of Books*, 39(21): 46–7.

Marx, Leo. 2000. *The Machine in the Garden*. Oxford: Oxford University Press.

Marx, Ursula, Gudrun Schwarz, Michael Schwarz, and Erdmut Wizisla (eds). 2015. *Walter Benjamin's Archive: Images, Texts, Signs*, trans. E. Leslie. London: Verso.

Michaels, Anne and John Berger. 2011. *Railtracks*. London: Go Together Press.

Noys, Benjamin 2013. 'Emergency Brake'. *No Useless Leniency*, 3 March. http://leniency.blogspot.com/2013/03/emergency-brake.html.

Orlando, Francesco. 2006. *Obsolete Objects in the Literary Imagination*, trans. G. Pihas, D. Seidel, and A. Grego. New Haven, CT: Yale University Press.

Parker, Ian. 2017. 'Anselm Kiefer's Beautiful Ruins'. *The New Yorker*, 3 July.

Puig, Ivan and Andres P. Domene. 2014. 'Crossing Mexico in a Home-Made Spacecraft'. https://www.bbc.co.uk/news/av/magazine-27929846/crossing-mexico-in-a-home-made-spacecraft.

Revill, George. 2012. *Railway*, London: Reaktion Books.

Schivelbusch, Wolfgang. 1986. *The Railway Journey: The Industrialization of Time and Space in the 19th Century*. Leamington Spa: Berg.

Sebald, W. G. 2002. *Austerlitz*. London: Penguin.

Sheehan, James. 2017. 'Echoes from the Far Side'. *London Review of Books*, 39(20), 21–2.

Smithson, Robert. 1996. *Robert Smithson: The Collected Writings*. Berkeley: University of California Press.

Tooker, Lauren and Bill Maurer. 2016. 'The Pragmatics of Payment: Adventures in First-Person Economy with Bill Maurer'. *Journal of Cultural Economy*, 9(3), 337–45.

Trachtenberg, Alan. 1986. 'Foreword'. In Wolfgang Schivelbusch, *The Railway Journey: The Industrialization of Time and Space in the 19th Century*. Leamington Spa: Berg.

Wagner, Peter. 2012. *Modernity: Understanding the Present*. Cambridge: Polity Press.

....

CHAPTER 35

REAL TIME BIDDING
SYSTEM

....

THEODORE VURDUBAKIS

In early 2017, Euro-American publics were suitably shocked to be told they had unwittingly been subsidizing those 'sworn to destroy' their way of life. 'Some of the world's biggest brands' including, *inter alia*, car-makers such as Jaguar and Mercedes-Benz, supermarkets such as Waitrose, banks, travel agencies, charities, and universities were apparently funding Islamic extremists, white supremacists, pornographers, and child abusers 'by advertising on their websites' (Mostrous, 2017: 1; Bridge and Mostrous, 2017). As 'sites/sights of organization' (O'Doherty et al., 2013), the display of advertisements for consumer goods and services framed by demands for jihad, or of appeals to charity set against a backdrop of calls to racial violence, appear surreal, almost Dada in their absurdity.

Foucault (1973: xv), among others, has drawn attention to the methodological significance of seemingly bizarre moments and paradoxical juxtapositions as affording, perhaps peripheral glimpses of the workings of much broader processes of social organization. What is being glimpsed in this instance, it is suggested, is the 'parasitical' (Serres, 2007) machineries and machinations of the contemporary 'attention economy'. The workings of this economy, of which banner ads, pop-ups etc., are the most visible products, are perhaps most tellingly organized by devices such as Real-Time (or 'programmatic') Bidding systems (RTB). RTB (on which the blame for the above 'transgressions' was eventually laid by the media) enables advertisers in the form of automated agents (bots), to select and target Internet or social media users in 'real time' and through multiple third-party websites. Such devices then allow every online *viewing* of an advertisement (impression) to be evaluated in terms of its commercial potential, sold and bought, within milliseconds (Wang et al., 2017; Papadopoulos et al., 2017). Thus, as soon as a 'surfer' clicks on a site (e.g., a YouTube video) their 'profile' (IP address, geolocation, browsing history, search keywords, time of day, etc.) are passed on to an auction site (such as Google's DoubleClick RTB Ad Exchange or Yahoo!'s Right Media). Immediately, an auction takes place whereby (automated) agents 'bid' to place their advert in front of that particular user. For example, an agent representing a resort chain will typically place a higher bid to advertise to a visitor with a particular profile who has been browsing travel-related sites. A proportion of the successful bid will then be paid to the owner of the website where the ad was displayed. Based on users' browsing history, further suggestions of content (e.g., 'similar' YouTube clips) will be made. Increasingly complex machine learning algorithms seek to establish (or rather construct) the relations of 'relevance' that underpin such 'user recommendations' (e.g., Alaimo and Kallinikos, and Chun this volume): the longer visitors remain at the site the more adverts can be 'impressed' upon them. 'How often', wonders one commentator, 'have you sat down with a plan say...to buy one thing online, only to find yourself, hours later, wondering what happened?' (Wu, 2016: 344).

It is worth noting at this point, that industry 'blacklists' are supposed to police the commercial logic of 'Real-Time Buying' by preventing the subsidizing of offensive material. We are thus assured that 'armies' of Google and Facebook staff (BBC, 2017) 'combining human judgment with powerful machine learning' (Wojcicki, 2017) are constantly scanning media platforms to identify and remove unsavoury content.[1] Nonetheless, as media researchers have recently shown, it is still possible to use sites such as Facebook in order to pitch to audiences who identify with topics such as 'why Jews ruin the world' and 'how to burn Jews' (Angwin et al., 2017).

It is a commonplace to say we live in an 'information age' (Castells, 1996) characterized by an abundance, an 'overload' even, of readily and freely available information (Gantz et al., 2008). As information becomes abundant, however, it is *attention* that becomes scarce—and thus the proper object of economic reasoning and calculation (Goldhaber, 1997; Davenport and Beck, 2001; Lanham, 2007). Nicholas Carr (2008, 2010) for instance, has made the case that the flood of information, and the omnipresence of information technologies, has brought in their wake a kind of attention deficit, a deterioration of the ability to focus on any particular task at hand. Compared to the days when *print* media were the main vehicles of information, contemporary expectations of quick and constant access have, he argues, eroded subjects' attention spans and left them unable to, for instance, read texts 'in depth' (Carr, 2008). According to those who claim to measure such things (e.g., the National Centre for Biotechnology Information at the US National Library of Medicine), the average attention span has now shrank to 8 seconds, apparently one second shorter than that of a goldfish.[2]

The increasing inability to pay attention means that attention has increasingly to be paid for. The by now long established practices of advertising (McFall, 2004) appear, in retrospect, to have pre-figured the nostrums of this new 'attention economy' (Davenport and Beck, 2001). This is hardly surprising. Unlike other forms of 'information', advertisements never had a *prima facie* legitimate claim to the subject's attention and as a result, one way or another, had to pay their way. Thus, in the days when

[1] Recent targets of this zeal have included Facebook images of the neolithic Willendorf Venus ('pornography'; Chigne, 2018) and the US Declaration of Independence (as 'hate speech'; BBC, 2018c).
[2] E.g. https://www.statisticbrain.com/attention-span-statistics/.

the 'information marketplace' had been dominated by a limited number of media channels—such as newspapers, radio and television—'publishers' would attempt to sell the attention of audiences with particular characteristics ('demographics') to advertisers. News, knowledge, and entertainment were the 'bait' that had been laid out for these audiences, in the spaces between advertisements. This long-established model was in turn challenged by the rise of the Internet and of *user* generated' content. The press, we are told, has suddenly become 'free' in the sense that anyone can (in principle) have her/his own (Rosen, 2011): 'Everyone is a media outlet' (Shirky, 2009: 55). The new environment, argues Webster (2008: 23) 'can be thought of as a virtual marketplace in which the purveyors of content compete with one another for the attention of the public'. In the new era of '*platform* capitalism' (Langley and Leyshon, 2016; Srnicek, 2016; see also Ridgway this volume), increasingly monopolistic 'platforms', such as Google or Facebook, have set themselves up as *the* virtual market*places* where attention is to be traded. The inauguration of the major ad exchanges from 2007 onwards, notes Google (2011: 3–5) 'brought more liquidity to the marketplace for online inventory', which enabled advertisers, businesses, and individual users 'to transact in online display... Now with large pools of liquidity... and a robust ecosystem of buyers capable of accessing it, the market was ripe for innovation. RTB was the missing piece.'

Cometh the hour, cometh the artefact. RTB is thus called upon to complete the advertising *system* and to real-ize the market for attention. The new economic sociology (Callon and Muniesa, 2005; Callon et al., 2007; MacKenzie, 2008) has highlighted the role of such 'market devices' in summoning markets into being (McFall, 2009). RTB, as we have seen, enacts the commercial reallocation of attention as an auction. As Google's (2011) white paper notes, it constitutes a crucial node in the system of attention capture and resale, where the devices of 'behavioural' forms of digital advertising endeavour to harness the agency of the 'user' in the organization of the targeting process. These devices range from 'permanent' cookies (which are not deleted when the browser is closed), to 'flash cookies' (which cannot be deleted by browsers), to 'deep packet inspection' (DPI) tools (which collect data at ISP level) (OFT, 2010). Even when search histories are being deleted, the users' 'clickstreams' can still be retained. Actions such as closing an advert, or selecting a menu option such as 'This advert isn't relevant', are routinely harnessed as the (negative) feedback necessary to further train the targeting algorithm. The subject is meant to be nudged, so to speak, little by little, along consumption pathways pre-dictated by the algorithm. In a series of patent applications for instance, Facebook has sought to develop devices for analysing data derived from its users' text communications and status updates which would enable it to infer those users' 'personality characteristics' thus increasing the precision of its targeting processes (Nowak and Eccles, 2016). The digital technologies which now mediate social life, argues Lash (2007: 60) increasingly seek to organize 'from the inside: there is self-organization... now the brain... is immanent in the system itself'.

The histories and functions of contemporary media technologies, their voracious appetite for personal data, and their complex roles as both products and producers of the 'organized worlds' that they set in motion, are analysed in a number of contributions to this volume (see also Beer, 2013). Devices such as RTB can therefore be seen as the other side of this particular equation (Kaplan, 2014). By allowing measurable monetary value to be ascribed to the outputs of such technologies, RTB helps the advertising *system* cohere. What remains to be explored then, is the specific role of this system in organizing and disorganizing the ever-expanding 'attention economy'. Its influence is of course most evident in the transformation of the Internet *away* from the participative gift economy, subversive of commercial interests, not so long ago celebrated by the libertarian 'digerati' (Barlow, 1996; Rheingold, 2000; Benkler, 2006; Jenkins et al., 2006). Now, according to former Facebook data team leader Jeff Hammerbacher, '[t]he best minds of my generation are thinking about how to make people click ads' (Vance, 2011). An infrequently asked question therefore is what accounts for the advertising system's conquest of the Web? Put another way, how did advertising come to appear as *the* most obvious way to extract monetary value out of data?

'If we were sensibly materialist', argued Raymond Williams (1980) in his discussion of the 'magic system' of advertising, 'we should find most advertising to be of an insane irrelevance'. For nearly a century, social science and public discourse alike 'have been preoccupied with proving whether or not advertising does influence an otherwise autonomous subject' (Slater, 1989: 118). For (neoclassical) economics on the one hand, the prevalence of advertising hints of the possibility of irrational forces at work in the operations of a rational economy. In order then to safeguard the autonomy of the 'rational choices' that ground it, economics had to ascribe to advertising the strictly 'supplementary' (in the Derridean [1976] sense of the term) role of 'conveying information' (regarding the availability, prices, and characteristics of goods and services) to otherwise autonomous utility-maximizing consumers (Marshall, 1919; Stigler, 1961; Nelson, 1974). For Marxist-influenced social science on the other hand, advertising, with its claimed ability to illicitly substitute 'sign value' for 'use value', provides the hidden persuader, the generator of false needs required in order to smooth out capitalism's endemic crises of overproduction (Adorno and Horkheimer, 1944; Marcuse, 1964; Packard, 1977). In what Lash (2007) calls our 'post-hegemonic age', where the routines of everyday life are increasingly organized and colonized 'from within' (Lash, 2007: 59) by the powers of the algorithm, Williams' (1980) question of how the advertising system's magic actually works, appears to have a clear and unambiguous answer.

Much has accordingly been made of the growing sophistication of digital advertising, and of its claimed power to stoke and reshape the consuming desires of the subject (Arvidsson, 2005). The much-vaunted ability of the current generation of tracking and targeting algorithms to match advertisers with *individual* Internet users is typically contrasted (by advocates and critics alike) with the 'wastefulness' of the traditional (mass) advertising on billboards, newspapers, and television. The increasing deployment of algorithmic personalization technologies, notes Eli Pariser (2011), facilitates the emergence of 'filter bubbles', intellectual environments which select and deflect information sources, giving their inhabitants a distorted picture of both the online and the offline worlds. Within such filter bubbles, technological lore has it, precisely targeted adverts will be both relevant and timely and no longer experienced as unwelcome diversions. As Cathy O'Neil (2016: 69) (critically) sums up this line of argument, until now 'most people objected to advertisements because they were irrelevant to them. In the future they... [will no longer] be'. In line with this kind of narrative, Cambridge Analytica, a (now defunct) British data mining and analytics *cum* political consulting firm, has been credited (if that is the right word) with using Facebook data to surreptitiously influence voters in a number of elections across the world including the 2016 US Presidential election and the UK 2016 EU referendum (BBC, 2018a).

It is evident that whatever we might call the 'digital advertising system' at the very least *aspires* to the invidious post-hegemonic powers described by Lash (2007: 59) and others. Wendy Chun (2006: 9) has cautioned, however, against the tendency to, as she put it, accept 'propaganda as technological reality, and [to routinely conflate] possibility with probability'. Indeed, there is, so far, little evidence that user tracking and targeting has increased effectiveness (Blake et al., 2015; Hoffman, 2017) and, on the whole, users' experience of digital adverts remains akin to that of an infestation (Serres, 2010): pop ups and banner ads obscure a site's content; interstitials appear to delay the loading of webpages. Ads open behind the main window or masquerade as search results or as content native to the site ('advertorials'). 'Videos [have] a way of popping up and starting to play unbidden...the stop button...[is] the tiniest of all, and often oddly located. And something of a ruse as well; if you missed hitting it directly, yet another website will open with yet more ads' (Wu, 2016: 324). In addition, it is becoming clear that the layers of complexity which behavioural advertising is introducing to the code of any website is increasing loading times—on average by 'five seconds or more' in 2015—and often causing the system 'to slow or freeze... sometimes preventing the page from loading altogether' (Wu, 2016: 324). It is therefore hardly surprising that digital advertising has long been shadowed by practices and technologies of ad blocking. As far back as the 1990s, users of Prodigy (an early online service provider) would often paste a strip of paper at the bottom 1/5th of their screen, the space where adverts appeared at the time (see Introna, 2014). More recently the ever-expanding usage of ad blockers—which prevent ads from loading and also block tracking information—is said to threaten the 'free' nature of the Web itself prompting moral denunciations ('using an ad blocker is stealing'),[3] accusations of blackmail (payments to get on ad blocker 'whitelists'), and technological countermeasures. Furthermore, because of various 'viewability' problems (such as the ad not loading on time) and, more importantly, the burgeoning business of ad fraud (in which adverts are 'shown' to bots) only a relatively small percentage of targeted ads may be actually shown to humans (Hoffman, 2017). But of course, there are signs that digital ad platforms like Google are increasingly using advertising precisely for, what we might call, its nuisance value rather than in the hope of enticing consumers to the product being promoted. As Lyor Cohen, YouTube's current (2018) Global Head of Music, set out the company's 'frustrate and seduce' strategy, YouTube intends to

---

[3] E.g., ad blockers 'are no different from a lock-picking kit for burglars or a lead-lined bag for shoplifters... Every time you block an ad, what you're really blocking is food from entering a child's mouth' (Piltch, 2015).

'frustrate music listeners by playing more adverts' in order to 'seduce' them into paying for its new subscription service (BBC, 2018b).[4] Which gives a rather different inflection to the notion of 'targeted advertising' and appears to also 'frustrate' the market logic of RTB.

DISCUSSION

The picture of the 'marketplace of attention' (Webster, 2008) that emerges from even such a cursory sketch, is less that of an 'algorithmic configuration that organizes the encounter of calculative agencies' (Callon and Muniesa, 2005: 1242) and more akin to that of an ecosystem in the process of transition (Parikka, 2010). As already mentioned or alluded to, Serres' (2007) concept of the 'parasite'—from the Greek para ($\pi\alpha\rho\dot{\alpha}$), 'beside' + *sitos* ($\sigma\tilde{\iota}\tau o\varsigma$) 'wheat' and *siteúō* ($\sigma\iota\tau\varepsilon\dot{\upsilon}\omega$) feed/fatten (Chambers, 2003)—can provide us with a suitable point of entry. As is well known (Brown, 2002, 2004; Pasquinelli, 2008; Brown and Stenner, 2009) Serres invokes three different, but in practice closely interrelated uses of the term: the biological parasite, as an organism that lives off another organism; the social parasite, as a free loader or uninvited quest; and the communicative parasite, as the static noise that interferes with and distorts communication. The 'logic' of parasitism is that of taking without giving: in Serres' (2007: 80) paraphrasing of Marx, it is that of 'abuse value' (see also Brown, 2002). For Serres, parasitism is a typical, if typically unacknowledged, feature of all forms of social, economic, and technological mediation. By means of their ongoing interferences and interceptions, parasites introduce complexity into the systems that they have come to inhabit: 'The bit of noise, the small random element, transforms one system or one order into another' (Serres, 2007: 21).

Viewed in this light, RTB and the machinations of digital advertising can be seen as ideal typical manifestations of parasites jockeying for position. Adverts appear uninvited in online communications, searches, and social interactions in order to siphon away and divert attention. They come in between and frustrate the desire for speed, immediacy, and direct access. The ostensibly 'minimal actions', 'noise', and 'small fluctuations', that they introduce, have set in train systemic changes in their hosts. Were we to follow Serres in his predilection for metaphors and examples that disrespect long-established science–fable distinctions, then one way to describe the apparent direction of the changes catalysed by the advertising parasite, would be as akin to the actions of *Cymothoa exigua*. *Cymothoa exigua*, better known as the 'tongue-eating louse' parasitizes on fish, typically entering though the gills and attaching itself to the tongue (Brusca and Gilligan, 1983). Once in position, the parasite drains the tongue of blood causing it to atrophy. The parasite's body remains attached to the muscles of the tongue-stub and begins to function as a substitute tongue. The fish now depends for its survival

---

[4] 'You're not going to be happy after you are jamming Stairway to Heaven and you get an ad' (Lyor Cohen quoted by BBC, 2018b).

on the parasite as much as the parasite depends on the fish. Perhaps, we could see in *Cymothoa exigua* a parable of the ongoing commercial appropriation of the 'voices' of the Internet by commercial interests (Post Brothers and Fitzpatrick, 2011). This appropriation, in turn, creates various forms of mutual parasitism: since advertising parasitizes on 'free' content and 'free' content parasitizes on advertising neither is said to be able to survive without the other.

Parasitical inhabitations, Serres (2007) notes, commonly take the form of chains where what is a parasite in one relationship frequently reappears as a host in another. As we have seen, in digital advertising, as elsewhere, entities are always in transition from host to parasite and vice versa. We might even speak of a constant struggle among parasites and would-be parasites (platforms, devices, fraudsters, bots, extremists, advertisers, etc.) for the best positions from which to intercept and divert the greatest number of circuits: money, attention, etc. (see Roque, 2010: 35). In this environment the various techno-logical fixes deployed, rather than restoring identity and means–end rationality, further add to the complexity and the proliferation of 'host–parasite' and 'parasite–host' transformations.

REFERENCES

Adorno, Theodor W. and Max Horkheimer. 1944. *The Dialectic of Enlightenment*. Stanford: CA: Stanford University Press.

Angwin, Julia, Madeleine Varner, and Ariana Tobin. 2017. 'Facebook Enabled Advertisers to Reach "Jew Haters" '. https://www.propublica.org/article/facebook-enabled-advertisers-to-reach-jew-haters.

Arvidsson, Adam. 2005. *Brands: Meaning and Value in Media Culture*, London, Routledge.

Barlow. John Perry. 1996. 'A Declaration of the Independence of Cyberspace'. https://www.eff.org/cyberspace-independence.

BBC. 2017. 'Google pledges 10,000 staff to tackle extremist content', 5 December. http://www.bbc.co.uk/news/technology-42232482.

BBC. 2018a. 'Cambridge Analytica: Facebook boss summoned over data claims', 20 March. http://www.bbc.co.uk/news/uk-43474760.

BBC. 2018b. 'YouTube wants to "frustrate" users with ads so they pay for music', 22 March. http://www.bbc.co.uk/news/newsbeat-43,496,603.

BBC. 2018c. 'Facebook finds Independence document "racist" ', 5 July. https://www.bbc.com/news/technology-44722728.

Beer, David. 2013. *Popular Culture and New Media*. Basingstoke: Palgrave Macmillan.

Benkler, Yochai. 2006. *The Wealth of Networks*. New Haven: Yale University Press.

Blake, Thomas, Chris Nosko, and Steven Tadelis. 2015. 'Consumer Heterogeneity and Paid Search Effectiveness'. *Econometrica*, 83(1), 155–74.

Bridge, Mark and Alexi Mostrous. 2017. 'Child Abuse on Youtube'. *The Times*, 18 November. https://www.thetimes.co.uk/article/child-abuse-on-youtube-q3x9zfkch.

Brown, Steve. 2002. 'Michel Serres: Science, Translation, and the Logic of the Parasite'. *Theory, Culture & Society*, 19, 1–27.

Brown, Steve. 2004. 'Parasite Logic'. *Journal of Organizational Change Management*, 17(4), 383–95.

Brown, Steve and Paul Stenner. 2009. *Psychology Without Foundations*. London: Sage.

Brusca, Richard and Matthew Gilligan. 1983. 'Tongue Replacement in a Marine Fish (Lutjanus guttatus) by a Parasitic Isopod'. *Copeia*, 3(3), 813–16.

Callon, Michel, Yuval Millo, and Fabian Muniesa (eds). 2007. *Market Devices*. Oxford: Blackwell.

Callon, Michel and Fabian Muniesa. 2005. 'Economic Markets as Calculative Collective Device'. *Organization Studies*, 26(8), 1229–50.

Carr, Nicholas. 2008. 'Is Google Making Us Stupid?' *The Atlantic*. http://www.theatlantic.com/magazine/archive/2008/07/is-google-making-us-stupid/6868/.

Carr, Nicholas. 2010. *The Shallows: How the Internet is Changing the Way We Think, Read and Remember*. London: Atlantic Books.

Castells, Manuel. 1996. *The Information Age: Economy Society and Culture, Vol. I*. Oxford: Blackwell.

Chambers. 2003. *Etymological Dictionary of the English Language*. London.

Chigne, Jean-Pierre. 2018. 'Facebook Censors 30,000-Year-Old Statue Venus of Willendorf Calling It Pornographic'. *TechTimes*, 28 February. http://www.techtimes.com/articles/222074/20180228/facebook-censors-30-000-year-old-statue-venus-willendorf-calling.htm.

Chun, Wendy H. K. 2006. *Control and Freedom: Power and Paranoia in the Age of Fiber Optics*. Cambridge, MA: MIT Press.

Davenport, Thomas H. and John C. Beck. 2001. *The Attention Economy*. Cambridge, MA: Harvard Business Review Press.

Derrida, Jacques. 1976. *Of Grammatology*. Baltimore, MD: Johns Hopkins University Press.

Foucault, Michel. 1973. *The Order of Things: An Archaeology of the Human Sciences*. New York: Vintage.

Gantz, John, Christopher Chute, Alex Manfrediz, Stephen Minton, David Reinsel, Wolfgang Schlichting, and Anna Toncheva. 2008. *The Diverse and Exploding Digital Universe*. IDC White Paper.

Goldhaber, Michael. 1997. 'The Attention Economy and the Net'. *First Monday*, 2(4). http://journals.uic.edu/ojs/index.php/fm/article/view/519/440.

Google. 2011. *The Arrival of Real-Time Bidding and What it Means for Media Buyers*. Google White Paper. https://static.googleusercontent.com/media/www.google.com/en//doubleclick/pdfs/Google-White-Paper-The-Arrival-of-Real-Time-Bidding-July-2011.pdf.

Hoffman, Bob. 2017. *BadMen: How Advertising went from a Minor Annoyance to a Major Menace*. San Francisco, CA: Type A. Group LLC.

Introna, Lucas D. 2014. 'The Ontological Choreography of the Impressionable Subject in Online Display Advertising'. Paper presented at the Algorithmic Cultures Workshop, Konstanz University, 23–25 June.

Jenkins, Henry, Katie Clinton, Ravi Purushotma, Alice J. Robison, and Margaret Weigel. 2006. *Confronting the Challenges of Participatory Culture*. Chicago, IL: The MacArthur Foundation. https://www.curriculum.org/secretariat/files/Sept30TLConfronting.pdf.

Kaplan, Frederic. 2014. 'Linguistic Capitalism and Algorithmic Mediation'. *Representations*, 127(1), 57–63.

Langley, Paul and Andrew Leyshon. 2016. 'Platform Capitalism: The Intermediation and Capitalization of Digital Economic Circulation'. *Finance and Society*, 3(1), 11–31.

Lanham, Richard A. 2007. *The Economics of Attention*. Chicago, IL: University of Chicago Press.

Lash, Scott. 2007. 'Power after Hegemony: Cultural Studies in Mutation'. *Theory, Culture & Society* 24(3), 55–78.

McFall, Liz. 2004. *Advertising: A Cultural Economy*. London: Sage.

McFall, Liz. 2009. 'Devices and Desires: How Useful Is the "New" New Economic Sociology for Understanding?' *Sociology Compass*, 3(2), 267–82.

MacKenzie, Donald. 2008. *Material Markets: How Economic Agents Are Constructed*. Oxford: Oxford University Press.

Marcuse, Herbert. 1964. *One-Dimensional Man: Studies in the Ideology of Advanced Industrial Society*. Boston, MA: Beacon Press.

Marshall, Alfred. 1919. *Industry and Trade*, London, MacMillan.

Mostrous, Alexi. 2017. 'Big brands fund terror'. *The Times*, 9 February, 1 and 6.

Nelson, Philip. 1974. 'Advertising as Information'. *Journal of Political Economy*, 82, 729–54.

Nowak, Michael and Dean Eccles. 2016. 'Determining User Personality Characteristics from Social Networking System Communications and Characteristics'. US Patent Application number 15/173,009. Patent Grant number: 9740752.

O'Doherty, Damian, Christian De Cock, Alf Rehn, and Karen Ashcraft. 2013. 'New Sites/Sights: Exploring the White Spaces of Organization'. *Organization Studies*, 34(10), 1427–44.

O'Neil, Cathy. 2016. *Weapons of Math Destruction: How Big Data Increases Inequality and Threatens Democracy*. London: Penguin.

Office of Fair Trading. 2010. *Online Targeting of Advertising and Prices: A Market Study*, London: OFT.

Packard, Vance. 1977. *The Hidden Persuaders*. Harmondsworth: Penguin.

Papadopoulos, Panayotis, Nicolas Kourtelis, Pablo Rodriguez, and Nikolaos Laoutaris. 2017. 'If you are not paying for it, you are the product: How much do advertisers pay to reach you?' Paper presented at the IMC '17, London. https://conferences.sigcomm.org/imc/2017/papers/imc17-final193.pdf.

Parikka, Jussi. 2010. *Insect Media: An Archaeology of Animals and Technology*. Minneapolis, University of Minnesota Press.

Pariser, Eli. 2011. *The Filter Bubble: How the New Personalized Web Is Changing What We Read and How We Think*. London, Penguin.

Pasquinelli, Matteo. 2008. *Animal Spirits: A Bestiary of the Commons*. Rotterdam, NAi Publishers.

Piltch, Avram. 2015. 'Why Using an Ad Blocker Is Stealing', 22 May. https://www.tomsguide.com/us/ad-blocking-is-stealing,news-20962.html.

Post Brothers and Chris Fitzpatrick. 2011. 'A Productive Irritant: Parasitical Inhabitations in Contemporary Art'. *Filllip 15*, Fall.

Rheingold, Howard. 2000. *The Virtual Community*. Cambridge, MA: MIT Press.

Roque, Ricardo. 2010. *Headhunting and Colonialism: Anthropology and the Circulation of Human Skulls in the Portuguese Empire*. Basingstoke: Palgrave Macmillan.

Rosen, Jeffrey. 2011. 'The Great Horizontal: 8 Key Ideas'. Keynote presentation #media140, Barcelona, 13 April.

Serres, Michel. 2007. *The Parasite*. Minneapolis: University of Minnesota Press.

Serres, Michel. 2010. *Malfeasance: Appropriation Through Pollution?* Stanford, CA: Stanford University Press.

Shirky, Clay. 2009. *Here Comes Everybody*. London: Penguin.

Slater, Don. 1989. 'Corridors of Power'. In J. Gubrium and D. Silverman (eds), *The Politics of Field Research*. London: Sage, 113–31.

Srnicek, Nick. 2016. *Platform Capitalism*. Cambridge: Polity Press.

Stigler, George J. 1961. 'The Economics of Information'. *Journal of Political Economy*, 69, 213–25.

Vance, Ashlee. 2011. 'This Tech Bubble Is Different'. *Bloomberg Businessweek*, 14 April. https://www.bloomberg.com/news/articles/2011-04-14/this-tech-bubble-is-different.

Wang, Jun, Zhang Weinan, and Shuai Yuan. 2017. *Display Advertising with Real-Time Bidding (RTB) and Behavioural Targeting*. Boston: Now Publishers.

Webster, James. 2008. 'Structuring a Marketplace of Attention'. In J. Turow and L. Tsui (eds), *The Hyperlinked Society*. Ann Arbor: University of Michigan Press, 23–38.

Williams, Raymond. 1980. 'Advertising: The Magic System'. In *Problems in Materialism and Culture*. London: Verso, 170–95.

Wojcicki, Susan. 2017. 'YouTube is taking on extremists by combining human judgment with powerful machine learning'. *The Daily Telegraph*, 5 December. https://www.telegraph.co.uk/news/2017/12/05/youtube-taking-extremists-combining-human-judgment-powerful/?li_source=LI&li_medium=li-recommendation-widget.

Wu, Tim. 2016. *The Attention Merchants*. London: Atlantic Books.

....

CHAPTER 36

RECOMMENDER SYSTEM

....

CRISTINA ALAIMO AND JANNIS KALLINIKOS

PERSONALIZATION increasingly mediates the experience of users on the Web. Online platforms and organizations use personalization services to retain users, achieve longer user or customer engagement, and, ultimately, higher profits. Cast in this light, personalization is a ubiquitous modality by means of which organizations seek to structure interaction with their users. Amazon, for instance, mediates the buying experience of its customers through computational systems that advance recommendations concerning relevant products to buy upon nearly every transaction. Similarly, Spotify uses the listening habits of its users to recommend tunes which they may find relevant to listen to. In a rather different context, Facebook modulates its news feed to the interests of individual users by mapping each user's ongoing interaction with his/her network of other users, and Google famously personalizes its search engine results, gathering, in turn, relevant information on the search habits of users.

Collaborative filtering recommender systems are one amongst a complex and differentiated landscape of technologies of personalization. Such systems are used alone or, increasingly, in combination with other systems to develop and implement so-called hybrid recommenders (for a useful taxonomy see Burke and Ramezani, 2011). The form of user mediation the various recommender systems offer may slightly vary in procedures and types of data required but their logic and operations remain largely similar (Adomavicius and Tuzhilin, 2005). All these systems work with a steady collection of carefully structured data produced by myriad individual transactions or interactions and the continuous feedback from users. This engineering of experience through which individual actions are first produced, then tracked, inferred, and transferred over to users presupposes a set of standardized and automated procedures which—paradoxical as it may sound—become the backbone of personalization. User interaction has to be shaped along actions standardized enough to leave a computable data footprint underlain by carefully crafted user models that make individual users comparable or commensurable with one another.

Famously conveyed by Amazon's statement 'Customers who bought this item also bought...', collaborative filtering was first implemented in 1993 as a Usenet news recommender system called GroupLens. The system recorded user ratings of news articles, stored them in a user profile database and once it had enough data was able to relate ratings to recommend other articles that users might find relevant (Konstan et al., 1997). A second experiment was done with the site MovieLens. There, users were asked to rate movies on a scale of 5 stars and paired into groups of similar users, that is, users who had rated movies similarly. Soon after that, MIT built Ringo, a music recommender system (Shardanand and Maes, 1995).

Riedl and Konstan are credited with having built one of the first collaborative filtering recommender systems. Collaborative filtering, as the name suggests, uses information from groups (hence, the collaborative dimension) to filter relevant items to individuals. Quite aptly described as 'any mechanism whereby members of a community collaborate to identify what is good and what is bad' (Riedl and Konstan, 1999: 330–1), collaborative filtering embeds also elements of information retrieval and filtering, and a large list of automated and often unsupervised computational operations. The 'collaborative' element resonates with the so-called 'wisdom of the crowd', the expectation that information produced by masses of people is somehow more precise or relevant than information produced by experts (Surowiecki, 2005). Recommender systems take these insights along a distinctive route that reflects partly the quest for personalization and partly the technological rendition of online experience we outlined above.

Recommender systems are largely automated systems that derive from the application of AI methods to information filtering and techniques of data representation and inference that have their roots in the expert systems of the 1980s (Jannach et al., 2010). Collaborative filtering gathers information on user interactions, although users seldom interact with each other in these settings, they only interact with the system to provide relevant information in the form of buying transactions and ratings. Users, therefore, do not collaborate as the name collaborative filtering might suggest. Nowhere is a community or group to be found. Those systems compute affinities between users by gathering the independent preferences of atomized individuals. Collaboration is euphemistically deployed to refer to these statistically mediated comparisons of user ratings. In fact, users are most of the time unaware of contributing to the development and working of recommender systems and, when they happen to be, they are seldom fully aware of the ways in which their contribution occurs. Advances in AI and machine learning amplify such trends. As these systems grow in complexity and automation the risk of overemphasizing implicit assumptions on the basis of which the system operates also increases. Take the two-stage approach to personalization of the deep neural network system for YouTube recommendation illustrated by Covington et al. (2016). Here the system does not need to rely on user generated ranking anymore but has itself learned how to rank. The system is made by two sub-systems: one for the generation of videos to be recommended and another one for ranking those videos. The first network uses collaborative filtering as it takes a user's YouTube activity history and filters a smaller sample of items to be recommended. The second system, called the ranking network, automatically assigns a score to each video by gathering large volumes of data both on videos and users (see Figure 36.1). It is not by accident that YouTube has been at

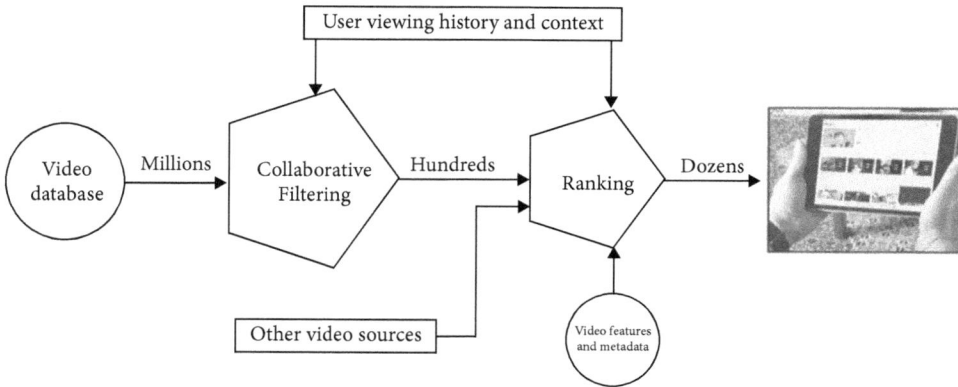

FIGURE 36.1 The deep neural network recommender system for YouTube. The figure illustrates the system architecture with its recommendation flow and relative data sources. From left to right: the video database; the first sub-system of collaborative filtering; the second sub-system of ranking; and the recommendations to users

(*Source*: Adapted from Covington et al., 2016)

the centre of media scrutiny recently as its modus operandi seeking prolonged user engagement led the system to recommend mostly crude or offensive content.[1]

The basic assumptions of collaborative filtering are that (i) the history of individual preferences together with (ii) the history of the preferences of similar individuals are better *predictors* than experts or the derivation of user preferences from market segments to which users are assumed to belong in standard marketing practices. In other words, the system assumes that if users shared similar preferences in the past they will also share similar preferences in the future. This broadly means that suggestions are tightly coupled with past preferences and the preferences of similar others as mapped and recorded by the system.

As a general rule, in all recommender systems the kind of information selected and the ways it is structured into databases and user profiles are directly connected with the core technology of the system. In collaborative filtering, user profiles are modelled as a list containing the history and quality of their ratings (Ricci et al., 2011); other systems may have different user models and thus require different kinds of information (e.g., data on items or a mix between data on items and user behaviour). Information can be gathered by explicit actions such as buying, rating, liking, watching, listening, etc. but it can also be collected by using implicit behaviour such as browsing, searching, or time spent on web pages, etc. The denominations of explicit and implicit refer to the indicator of consumer preferences which are explicit when the indicators stem from actions that can be straightforwardly linked to preferences; when the system does not have

---

[1] See for instance: https://www.theguardian.com/technology/2018/feb/02/how-youtubes-algorithm-distorts-truth; https://www.theguardian.com/technology/2018/feb/02/youtube-algorithm-election-clinton-trump-guillaume-chaslot.

indicators of preferences, or data are sparse, it infers them by interpreting as preferences any implicit actions. Other systems may use data derived from items or from items and users and gather additional data from third-party organizations. It is relevant to note that even when the system uses item data, it does not simply record information but rather carefully designs specific data formats so as to fit the core technology of the recommender system. A good example of the increased sophistication in information design is given by the different ways music personalization techniques work.[2]

In collaborative filtering, once the system is up and running with a rating database and active user profiles, it needs to cluster users or items into groups of similar users or items. To do so, collaborative filtering recommender systems use two different approaches: user-based or item-based algorithms. The first is the simplest and relies on the idea that given a rating database the distance between users is determined by each individual user's ratings of the same item. The second instead computes the distance between items on the basis of how closely users who have rated these items agree (Figure 36.2).

With a user-based algorithm, users who rate items similarly form a neighbourhood of users. With an item-based algorithm, items that are rated similarly form a neighbourhood of items. The grouping of users or items (sometimes called nearest neighbours or peer users) is how the system establishes segments, that is, groups of similar users. Differently from traditional marketing, in recommender systems this happens automatically and under computational rules. There are no real similarities between user-user or item-item; users or items are grouped together and deemed similar on the basis of the affinities emerging from rating patterns provided by large amounts of data. The 'segments' so created by the automated recommender system become groups of

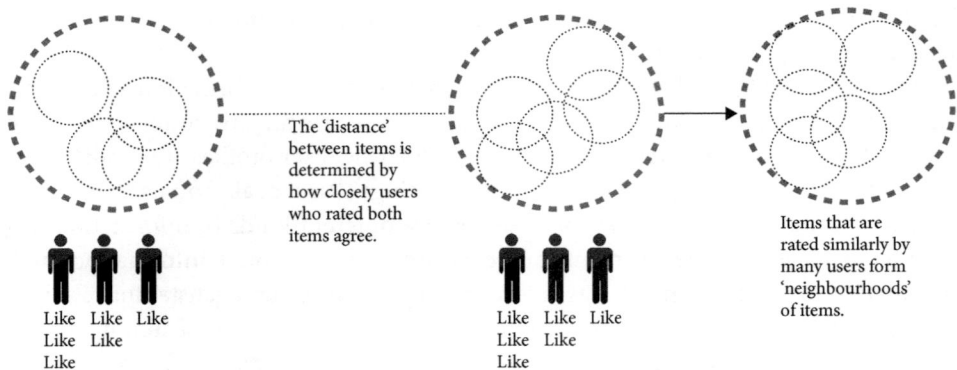

The 'distance' between items is determined by how closely users who rated both items agree.

Items that are rated similarly by many users form 'neighbourhoods' of items.

Like Like Like
Like Like
Like

Like Like Like
Like Like
Like

FIGURE 36.2 Illustrative example of item-item personalization algorithm

[2] Pandora's music genome project and The Echo Nest, the music intelligence platform empowering Spotify recommender engine, are two hybrid recommenders that approach the same problem (music personalization) in an entirely different way. See https://www.pandora.com/about/mgp and http://the.echonest.com/.

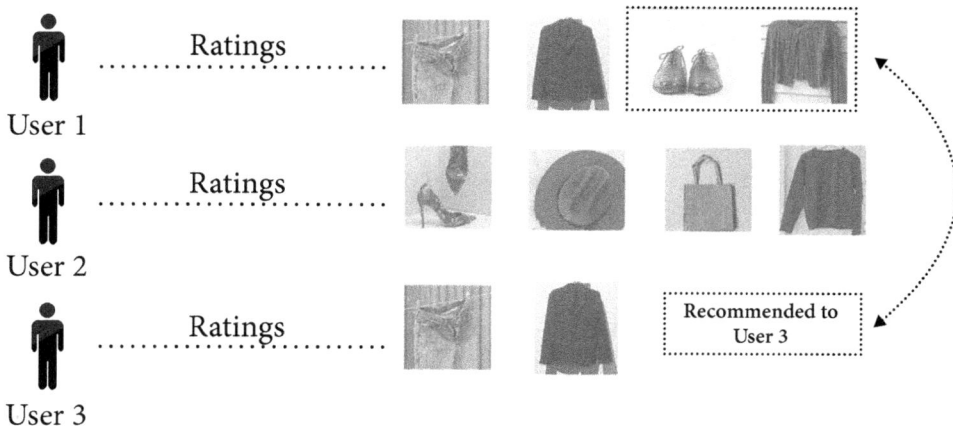

FIGURE 36.3 Illustrative example of the prediction based on nearest neighbours: 'for every item y that user x has not seen, a prediction is computed on the basis of the rating of the item y made by nearest neighbours'

(*Source*: Images reproduced from Unsplash and Flickr/Jeff M for Short, m01229, and Robert Sheie)

predictors: for every item y that user x has not seen, a prediction is computed on the basis of the rating of the item y made by nearest neighbours (Figure 36.3). The majority of recommender systems use variants of a weighted, k-nearest-neighbour prediction algorithm (roughly synthesized in 'how much a target user u will like a target item i by first selecting a neighbourhood of other users with tastes most similar to that of u') (see Konstan and Riedl, 2012a, 2012b).

To automatically select neighbours, however, the system needs first to compute a measure of similarity between users or items. The computation of similarity is a fundamental problem for the success of the prediction. That is, the measure by which two or more users or items are deemed similar by the system and grouped together needs to be as accurate as possible. There is no consensus yet as to which of the different measures applied to the construction of the similarity function works best (Alaimo, 2013, 2014). A common approach remains the Pearson's correlation coefficient (another is the vector cosine similarity measure). 'It computes the missing rating of user u according to the average value of ratings made by its neighbours weighted by each of their degree of similarity with the user u' (Ricci et al., 2011) (Figure 36.4).

The coefficient is successful in factoring ratings so as to make users commensurable. If a user rates movies only with 4 and 5 stars and another user instead deploys the whole scale from 1 to 5, the ratings of the two users need to be adjusted in order to allow comparison. However, the coefficient cannot solve sparsity problems, it cannot factor out the difference between the ratings of two users if the former has rated 500 movies and latter only 5. Also, the coefficient cannot make any difference between ratings that concern popular movies (most people like *Star Wars* anyway) and ratings that concern less popular movies (which are arguably more indicative of similarity of taste). Two other problems concern size and time. Size has always been a problem for the successful

$$S_{i,j} = \frac{\sum_{u \in U} (R_{u,i} - \bar{R}_i)(R_{u,j} - \bar{R}_j)}{\sqrt{\sum_{u \in U} (R_{u,i} - \bar{R}_i)^2} \sqrt{\sum_{u \in U} (R_{u,j} - \bar{R}_j)^2}}$$

FIGURE 36.4 Pearson's correlation coefficient. It is used to compute 'the missing rating of user u according to the average value of ratings made by its neighbours weighted by each of their degree of similarity with the user u'

computation of predictions. If the group of similar users or items is too small, it becomes impossible to compute good predictions but if the size of the group is too large this may be because the threshold of similarity is too low and predictions will be inaccurate. Time instead refers to the complexity of re-computing the whole model once a user adds a new rating. For this last reason, an item-based collaborative filtering approach is considered a better one. In this approach, the item-item algorithm calculates the distance between two items according to how much users agree. It is important to bear in mind that here similarity is computed between ratings; it is the pattern of user rating behaviour overtime that is analysed and computed to make or more items nearest neighbours. Therefore, in this case the distance between items is pre-computable as it is dependent on thousands of available ratings thus remaining relatively stable over time.

CUSTOMIZATION AND PERSONALIZATION

The ideas presented above indicate that personalization and recommender systems are closely associated with the online, data-intensive environments in which most organizations currently operate. However, personalization is more than a technical response to the ubiquity of data that characterize our age. It is above all an organizational practice that seeks to modulate a space of interaction between organizations and users or customers in an economic, cultural, and social context that is increasingly marked by the fragmentation of consumer needs and the individualization of consumption. Placing personalization within the larger historical context of customization gives a better appreciation of the origin of this organizational practice and the ways it has evolved with the use of digital technologies.

Customization has historically emerged as the organizational antidote to the typified consumer experience characteristic of mass production and the long-driven standardization of products or services which meticulous specialization and far-driven economies of scale have brought about (Chandler, 1977; Lampel and Mintzberg, 1996). Despite significant variation which the model of mass production has been subject to over the last few decades (Pine, 1993), a great deal of products and services are still produced under conditions that recount the exigencies of low unit cost, achieved through specialization, economies of scale, and standardization. Seen in this light, customization has been a response to this one-size-fits-all consumer experience associated with standardized products and services. It is an organizational practice that seeks to alleviate some of the negative implications of a long-driven standardization and expand the possibilities of consumer choice.

In reality, customization operates by designing a space of interaction with consumers whereby the latter are claimed to have ample freedom for exercising their choice. For standard market segmentation techniques, individuals are just singular instances of wider market segments associated with class, demographic, educational, or income attributes. Under these conditions an individual cannot but be one among a large group of similar others with whom she/he shares a predictable set of needs. Customization assumes that such a rather wholesale method for dissecting taste distribution among large populations is no longer well attuned to the faster production cycles of modern organizations nor to the context of free-will individuals that seem to characterize hyper-modern societies. As a response to these changes, marketing segmentation techniques have been redefined to conceive individuals as active consumers driven by desires mostly associated to lifestyle and other expressive attributes of contemporary ways of living.

It is true that such a shift in marketing practices seldom moves far beyond the market segmentation techniques with which standardized products and services have always been closely associated. In fact, such consumer space of choice is more fictional than real. In a great deal of cases, the exercise of personal choice assumed by customization is linked to products and services that are mass-produced and then mass-customized. The adaptation to individual customers is still mediated by segmentation techniques, certainly more finely differentiated, that place individuals into smaller target groups (Zuboff and Maxmin, 2003; Zwick and Denegri Knott, 2009). Yet, targeting individual consumer desires instead of product needs signalled a shift of capital importance for the development of the modern individual consumer which we find closely relates to personalization. In this regard, customization and personalization can be linked to a broader culture of individualism and commodious consumerism characteristic of post- or late-modern societies which they variously reinforce (Bauman, 2000; Lipovetsky, 2005). The current fragmentation of consumer tastes is thus a broader societal phenomenon (Beck, 1992; Anderson, 2006) which has been certainly escalated by the attempt of organizations to grapple with social changes (Zwick and Cayla, 2011). By redefining the space of consumption as a fictional space of limitless consumer choice, organizations have reinforced the increasing fragmentation of taste promoting a standardized notion of hyper-individuality.

FROM INDIVIDUALS TO DATA

Personalization as an organizational practice and recommender systems as a primary technology of personalization have transformed the ways data and data-based systems mediate human experience of consumption and the space of consumer choice

(Kallinikos, 1992, 2007; Manovich, 2001). This was noted as far back as 1999 when, in *The New Yorker*, Malcom Gladwell (1999) wrote:

> The really transformative potential of collaborative filtering, however, has to do with the way taste products—books, plays, movies, and the rest—can be marketed. Marketers now play an elaborate game of stereotyping. They create fixed sets of groups—middle-class-suburban, young-urban-professional, inner-city-working-class, rural-religious, and so on—and then find out enough about us to fit us into one of those groups. The collaborative-filtering process, on the other hand, starts with *who we are*, then derives our cultural 'neighborhood' from those *facts*. And these groups aren't permanent. They change as we change...   (Italics added)

If we follow this account, personalization seems to go much further than customization. Against the 'lazy, prejudice philosophy' of demographic profiling, as Riedl and Konstan (1999: 113) call traditional marketing segmentation, recommender systems claim to offer an unbiased account of individual user needs and desires. Recommended systems aspire to champion the resurgence of 'who we are' out of the coarse taxonomies of traditional marketing but also out of the late techniques of target groups and lifestyles characteristic of customized marketing. Once again, personalization operates by re-engineering the space of consumption. The space of limitless consumer choice is given an interesting tweak by being transformed into a data field that is assigned the status of facts. Yet as we have seen, recommender systems engineer the space of user interaction to fine-tune it to practices of data gathering, user profiling and computation. Data are fashioned as facts by black-boxing the numerous technological operations sustaining them which are far removed from user interface and thus from user awareness. As distinct from past practices of customization, interaction is now staged between an individual user and a digital system (not an organization or other users) and shaped by iterative feedback loops between (i) programmed behaviour and an initial dataset, (ii) suggestions, and (iii) user reactions and feedback by which the system learns and readjusts its outputs.

Cast in the digital medium, the space of consumer choice articulated by the organizational practice of personalization transforms traditional assumptions concerning individuals. On most counts, personalization is the process of *inference* of what individuals and small taste groups are or would like to be based on the clustering of data that are supposed to stand for their actual preferences. There are of course other cultural, social, or personal references to which the identity of individuals and groups is related. Yet, in the context of personalized recommendations there is often no reference to personal attributes or actual buying behaviour but only ratings and other clicking-related behaviours (Alaimo and Kallinikos, 2017).

A critical look at these practices suggests that the practice of personalization is reductive in ways that undermine any genuine concern for persons as unique cultural individuals. Contrary to what Malcom Gladwell seems to suggest in the quote above, personalization—mediated in recommender systems—is not concerned with persons but data. It compiles profiles of individuals or taste groups out of digital marks (data) and what these are engineered to represent. Disturbing as it may seem, users online are seldom individuals in the literal sense of the world. They are not real-world persons who exercise choice as the outcome of their unique make-up of life experiences. They are rather data aggregates which are put together on the basis of specific computational models (or profiles) of users imposed by the core technology of the system (i.e., users are lists of such engineered operations such as rating, listening, liking, etc.). This, in turn, conditions what kind of behaviour the system needs to design so as to have the data required to perform adequately. As distinct from the fictional lifestyles of late-consumerism and the rigid stereotypes of customization, individuals online are just another digital item, constantly re-modulated, repositioned, and updated every time a new click, like, or rating is produced. Personalization systems help online consumers in solving a problem that technology itself has caused. Overabundance of choices linked to endless data-fields is a daunting scenario for users who therefore need to be aided by constant personalized suggestions. In this respect personalization has many positive effects for organizations as it is effectively correlated with higher user engagement and higher margins. Personalization may also have some positive effects for users, as far as it helps users discover new items, learn something new about themselves, and over-come the anxiety of making decisions (Anderson, 2006). Yet, there is high price to pay for these functional gains. As shown in this chapter, whatever the implications of personalization, these need to be appreciated within a larger time purview and the ways recommender systems encode and engineer user experience.

CONCLUDING REMARKS

In this chapter we have reviewed the organizational practice of personalization by deconstructing the ways collaborative filtering recommender systems work (see also Chun, this volume). Although different recommender systems may be based on varying computational paradigms (Adomavicius and Tuzhilin, 2005; Jannach et al., 2010; Burke and Ramezani, 2011; Ricci et al., 2011) and may rely on several different data principles and algorithms, all of them work by following a similar logic: (i) they need data on users and user behaviour (some gather data on products as well); (ii) they need to construct and update a user model or profile whereby they gather user preferences and past behaviour; (iii) they offer automated personalized suggestions; (iv) they require continuous feedback from users to adapt and learn.

We have placed the emergence of personalization within the broader historical pro-cess of customization and the quest of producing goods and services that are supposed to address the distinctive needs of individuals and small groups. We have drawn atten-tion to the mediating properties of personalization systems and the current practice of data clustering and data-based techniques that deeply impregnate personalization processes. Although the accuracy of data-based techniques makes personalized services look like an empirically grounded mediation through which users can discover their own, allegedly true, needs and predispositions, it is important to realize that there are no genuine individuals in these systems, at least not in the sense we understand the term in real-life contexts (Elmer, 2004; Hildebrandt and Rouvroy, 2011; Alaimo and Kallinikos, 2016, 2017).

The datification of user experience which underlies personalization has a number of consequences. Online involvement is heavily shaped by first translating individuals into user profiles or computable models that render them steadily knowable entities. Any recommender system actively fashions what an individual user or consumer is by creating models of users which are highly dependent on the core technology-in-use. Such models constitute the backbone for designing a set of clearly defined online interactions that enable the computability of user preferences. In this respect, the individuality that personalization constructs is no more than a changeable data profile. User models or profiles are assembled out of strategies of attributing preferences to individuals through a complex journey of technologizing experience whereby standardized expressions of individual behaviour—clicking, liking, or rating—are interpreted as expressions of taste and assessed by comparison to a network of standardized behavioural expressions of others. Online individual consumers are automatically fashioned by computational models as digital objects that are always updatable and constantly in the making.

References

Adomavicius, Gediminas and Alexander Tuzhilin. 2005. 'Toward the Next Generation of Recommender Systems: A Survey of the State-of-the-Art and Possible Extensions'. *IEEE Transactions on Knowledge and Data Engineering*, 17(6), 734–49.

Alaimo, Cristina. 2013. 'Technology of Consumption on Social Shopping Platforms: Deconstructing Similarity'. Presentation at EGOS: Montreal, Canada.

Alaimo, Cristina. 2014. 'Computational Consumption: Social Media and the Construction of Digital Consumers'. PhD dissertation, London School of Economics.

Alaimo, Cristina and Jannis Kallinikos. 2016. 'Encoding the Everyday: The Infrastructural Apparatus of Social Data'. In C. Sugimoto, H. Ekbia, and M. Mattioli (eds), *Big Data is Not a Monolith: Policies, Practices, and Problems*. Cambridge, MA: MIT Press, 77–90.

Alaimo, Cristina and Jannis Kallinikos. 2017. 'Computing the Everyday: Social Media as Data Platforms'. *The Information Society*, 33(4), 175–91.

Anderson, Chris. 2006. *The Long Tail: Why the Future of Business is Selling Less of More*. New York: Hachette Books.

Bauman, Zygmunt. 2000. *Liquid Modernity*. New York: Wiley.

Beck, Ulrich. 1992. *Risk Society: Towards a New Modernity*. London: Sage.

Burke, Robin and Maryam Ramezani. 2011. 'Matching Recommendation Technologies and Domains'. In Francesco Ricci, Lior Rokach, Bracha Shapira, and Paul B. Kantor (eds), *Recommender Systems Handbook*. New York: Springer, 367–86.

Chandler Alfred D., Jr. 1977. *The Visible Hand: The Managerial Revolution in American Business*. Cambridge, MA: Harvard University Press.

Covington, P., J. Adams, and E. Sargin. 2016. 'Deep Neural Networks for YouTube Recommendations'. In *Proceedings of the 10th ACM Conference on Recommender Systems*. New York: ACM, 191–8.

Elmer, Greg. 2004. *Profiling Machines: Mapping the Personal Information Economy*. Cambridge, MA: MIT Press.

Gladwell, Malcolm. 1999. 'The science of the sleeper'. *The New Yorker*, 4 October. https://www.newyorker.com/magazine/1999/10/04/the-science-of-the-sleeper.

Hildebrandt, Mireille and Antoinette Rouvroy (eds). 2011. *The Philosophy of Law Meets the Philosophy of Technology*. London: Routledge.

Jannach, Dietmar, Markus Zanker, Alexander Felfernig, and Gerhard Friedrich. 2010. *Recommender Systems: An Introduction*. New York: Cambridge University Press.

Kallinikos, Jannis. 1992. 'The Significations of Machines'. *Scandinavian Journal of Management*, 8(2), 113–32.

Kallinikos, Jannis. 2007. *The Consequences of Information: Institutional Implications of Technological Change*. Cheltenham: Edward Elgar Publishing.

Konstan, Joseph A., Bradley N. Miller, David Maltz, Jonathan L. Herlocker, Lee R. Gordon, and John Riedl. 1997. 'GroupLens: Applying Collaborative Filtering to Usenet News'. *Communications of the ACM*, 40(3), 77–87.

Konstan, Joseph A. and John Riedl. 2012a. 'Recommender Systems: From Algorithms to User Experience'. *User Modeling and User-Adapted Interaction*, 22(1–2), 101–23.

Konstan, Joseph A. and John Riedl. 2012b. 'Recommended for You'. *IEEE Spectrum*, 49(10), 54–61.

Lampel, Joseph and Henry Mintzberg. 1996. 'Customizing Customization'. *Sloan Management Review*, 38(1), 21–30.

Lipovetsky, Gilles. 2005. *Hypermodern Times*. Cambridge: Polity Press.

Manovich, Lev. 2001. *The Language of New Media*. Cambridge, MA: MIT Press.

Pine, B. Joseph. 1993. *Mass Customization: The New Frontier in Business Competition*. Cambridge, MA: Harvard Business School Press.

Ricci, Francesco, Lior Rokach, and Bracha Shapira. 2011. 'Introduction to Recommender Systems Handbook'. In Francesco Ricci, Lior Rokach, Bracha Shapira, and Paul B. Kantor (eds), *Recommender Systems Handbook*. New York: Springer, 1–35.

Riedl, John and Joseph A. Konstan. 1999. *Word of Mouse: The Marketing Power of Collaborative Filtering*. New York: Hachette.

Shardanand, Upendra and Pattie Maes. 1995. 'Social Information Filtering: Algorithms for Automating "Word of Mouth"'. In *Proceedings of the SIGCHI Conference on Human Factors in Computing Systems*. New York: ACM Press/Addison-Wesley, 210–17.

Surowiecki, James. 2005. *The Wisdom of Crowds*. New York: Anchor Books.

Zuboff, Shoshana and James Maxmin. 2003. *The Support Economy: How Corporations Fail Individuals and the Next Episode of Capitalism*. London: Allen Lane.

Zwick, Detlev and Julien Cayla (eds). 2011. *Inside Marketing: Practices, Ideologies, Devices*. Oxford: Oxford University Press.

Zwick, Detlev and Janice Denegri Knott. 2009. 'Manufacturing Customers: The Database as New Means of Production'. *Journal of Consumer Culture*, 9(2), 221–47.

CHAPTER 37

....

SEARCH ENGINE

....

RENÉE RIDGWAY

ORGANIZING THE WORLD'S INFORMATION

IN previous centuries analogue querying by monarchies, private entities, or state institutions accounted for the collation of subject and citizen data and could be considered the 'pre-history of search engines' (Tantner, 2014: 123). King Philip II's 'elaciones topográficas', Louis XIV's administrator Jean-Baptiste Colbert's enquêtes, and the harvesting of information by the Habsburg dynasty exemplify the compiling of data by monarchies (Tantner, 2014). Analogous to modern-day search engines, the 'bureau d'adresse' (Figure 37.1), 'Universal Register Office', 'Intelligence Office', and 'Fragamt' in the seventeenth to nineteenth centuries in various European cities used human 'crawlers' (maids, servants, and journeymen) to first gather information and then share it with interested parties (Tantner, 2014). These 'offices' created storage technologies of public and secret registers of personal data along with publishing commercial listings, bringing to mind contemporary search engine 'platforms' that index, filter, and structure information through ads.[1] In the nineteenth century, media measured through statistics became the means by which the state communicated its activities to the public sphere,[2] yet eventually private entities and social scientists also started to produce research based on data collection.[3]

---

[1] The process of indexing, and the search methods that algorithms apply (e.g., linear, binary, prefix trees, AVL trees, and distributed hash tables) determine how URLs can be found and how to discover or find them. Also called webcrawlers, spiders are programs that retrieve web pages and crawl the World Wide Web that are then read by an indexer, another program, which subsequently builds an index of all of the gathered documents. These indexers discover new documents and as they are archived, decide what to index and how items should be structured and organized through parsing.

[2] The term statistics (*Statistik*) was defined in 1749 by the German political scientist Gottfried Achenwall (1719–72) 'as the science dealing with data about the condition of a state or community' (Stalder, 2010).

[3] 'Academic social scientists began to analyse data for their own purposes, often entirely unconnected to government policy goals. By the late 19th century, reformers such as Charles Booth in London and W. E. B. Du Bois in Philadelphia were conducting their own surveys to understand urban poverty' (Davies, 2017).

FIGURE 37.1 'Bureau d'adresse pour les curieux', Théophraste Renaudot

(*Source*: Bibliothèque nationale de France, département Estampes et photographie)

In the first half of the twentieth century, speculative essays that attempted to organize the world's information appeared in newspapers and publications, such as 'World Brain' (Wells, 1937) and 'As We May Think' (Bush, 1945), with the imagined *memex* machine.[4] Mostly utopian in spirit and visionary in nature, the *Mundaneum* (1910) created by Paul Otlet and Henri LaFontaine, was a physically assembled corpus comprising diverse data: writings, newspaper clippings, photos, and drawings. The general public could submit queries by post and human assistants searched for answers via an 'index card' catalogue (see Figures 37.2 and 37.3). Influenced by Bush, Eugene Garfield's *Scientific Citation Index* (SCI) (1964; see Garfield, 2007), organized references hierarchically,

FIGURE 37.2 Employees of the Mundaneum working on the Universal Bibliographic Repertory, circa 1900.

(Collection Mundaneum, Mons (Belgium))

[4] A complex figure, Vannevar Bush was the director of the Office of Scientific Research and Development for the US Government when he wrote the influential article As We May Think. Published in *The Atlantic* in 1945, Bush maps out his vision of the 'memex' machine that foregrounds associative indexing by creating trails or linkages similar to hypertexting on the World Wide Web. In 2015 the military industrial complex (DARPA) decided to honour Bush by developing a search engine dubbed *Memex*, which scrapes content not indexed by commercial engines along with indexing sites on the so-called Dark Net, in order to find criminals.

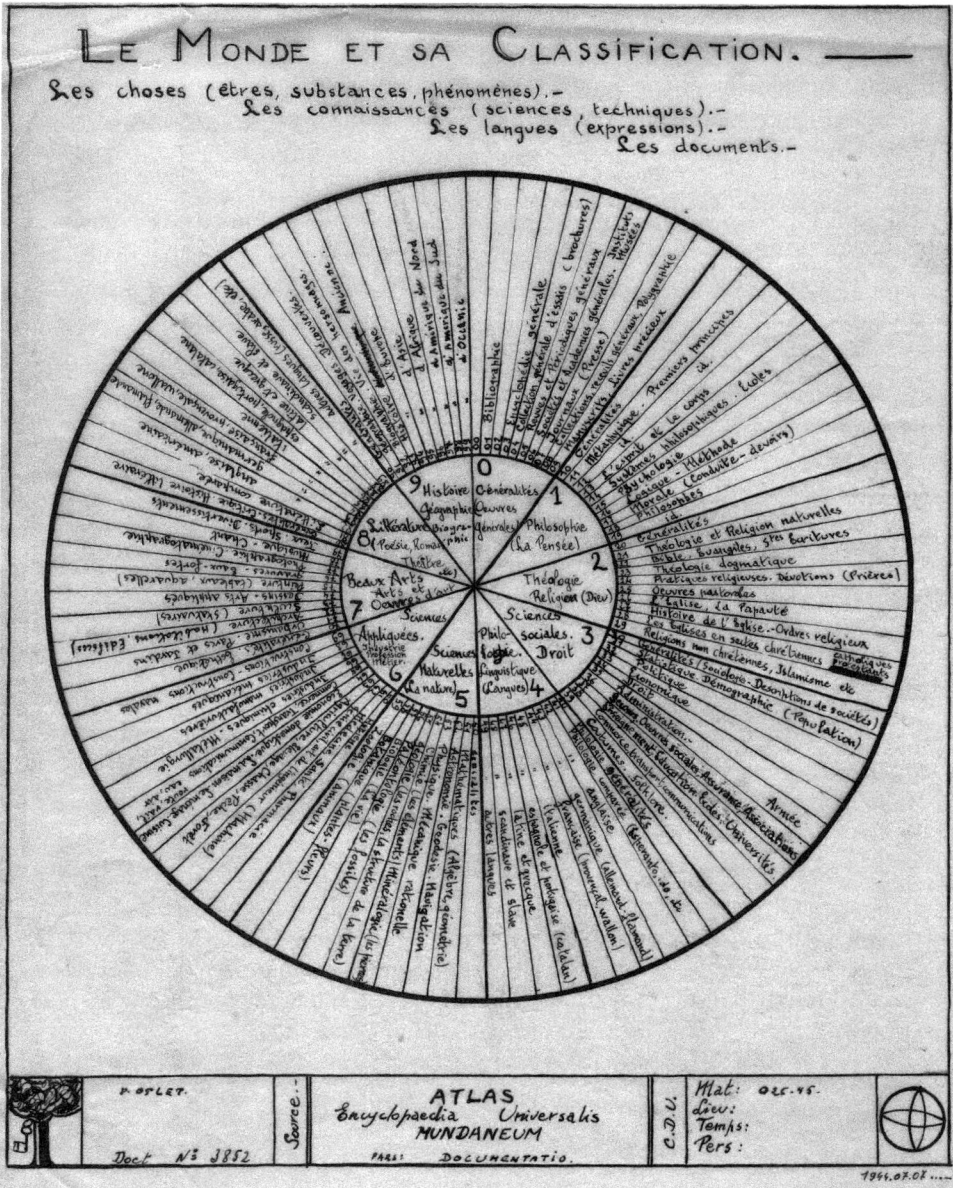

FIGURE 37.3 'Le Monde et sa Classification', Atlas, Encyclopaedia Universalis,
(Mundaneum by Paul Otlet. Collection Mundaneum, Mons (Belgium))

becoming the status quo for authorship and along with it, authority in academia.[5] Inspired by the law index, *Shephard's Citations* (1873), SCI is based on the number of citations attributed to an article and measures their so-called 'impact factor'. This is key to understanding contemporary 'hyperlink culture', along with *hypersearch* (Marchiori, 1997), which introduced link analysis and influenced the future of search engine development, specifically Google's PageRank.

The end of the twentieth century ushered in new technologies and media enabling data to be gathered and aggregated into computational forms or 'relational databases', which take on new data, or make possible their deletion. However, data are never just 'given'; in the electronic era they contain bits of annotations, characters, texts, and numbers that are 'machine readable'. The bit (a combination of the term binary and digit) of data is the fundamental particle that comprises information (Gleick, 2011: 10), which has been given form, or processed, making it 'human readable'. 'The future is in data' (Rosenberg, 2013: 28) yet 'data has no truth' because 'it may be that the data we collect and transmit has no relation to truth or reality whatsoever beyond the reality that data helps us to construct' (Rosenberg, 2013: 37).[6] In this way, 'data are effectively made independent of their organization, and users who perform logical operations on the data are thus "protected" from having to know how the data have been organized' (Gitelman and Jackson, 2013: 9).

As the World Wide Web grew so did the amount of data, accruing into millions of documents that needed to be organized in order to be retrieved. Concomitantly, the question of how to navigate such a space in order to find what users were seeking was answered by search engines.[7] When Introna and Nissenbaum wrote their seminal text at the dawn of the millennium, 'Shaping the Web: Why the Politics of Search Engines Matters', the Web was a growing space of HTML webpages on the Internet that were not necessarily interconnected. For some 'enthusiasts' the Internet was a:

> new medium, a democratizing force that will give voice to diverse social, economic, and cultural groups, [and] to members of society not frequently heard in the public sphere. It will empower the traditionally disempowered, giving them access both to typically unreachable nodes of power and to previously inaccessible troves of information.   (Introna and Nissenbaum, 2000: 177)

---

[5] SCI does not measure content but links or references and their reputation: the probability effect in which the more a paper or article is cited the more it will be cited afterwards, enabled a significantly greater likelihood of references being made to works that were already popular.

[6] The etymology of word 'data' is actually the plural of the Latin word datum, meaning 'given'. Daniel Rosenberg provides a fascinating account of the shift from Latin datum to data and its application to the English language the past centuries, using data to analyse corpuses of textual data by drawing on quantitative digital humanities measures such as Google's Ngram viewer, either online through search terms or literature publications.

[7] In 1995, AltaVista, the first full-text search engine appeared. Built on automated information gathering and indexing, it established the now standard interface paradigm of a simple search box where users enter a query and receive a ranked list of search engine result pages (SERPs) containing uniform resource locators (URLs).

In the 1990s early net programmers and users with their 'bulletin board' postings, chat rooms, and networks envisioned a 'digital democracy'; there was also the belief that searching the Web wasn't only about information retrieval but knowledge exploration. The concept of serendipity, or the discovery of web pages that occurred through surfing—clicking on hyperlinks—was the modus operandi for many users. However, by the mid-2000s the political discourse was already censored as it emerged, with information being filtered through 'Googlearchy' (Hindman, 2009).[8] Thus 'deliberative democracy' was prohibited by the infrastructure itself, 'the social, economic, political and even cognitive processes that enable it' (Hindman, 2009: 130).

During the past twenty-five years, search engines have increasingly become a powerful force that not only facilitate accessibility to users and give direction in regard to navigation, but also organize online activities. Two aspects determine how users search: one is the user or seeker (demand) and the other, sites that wish to be discovered (supply). According to the economic sociologist David Stark, 'search is the watchword of the information age' (Stark, 2009: 1). Not knowing what you are looking for, yet being able to recognize it when found, is what search engines can do (Harrington, 2010).[9] With only a few keywords, databases open up and reveal answers to queries, hence 'search engines power the information economy' (Harrington, 2010). The question of why people search, their motivations for entering this search term and not another, and whether the results are satisfying all play a role (Halavais, 2009: 32). Thus search is not merely an abstract logic but a daily practice that helps organize, sort, and filter information users seek as well as rank the results of their queries.

Since the early 1990s search engines have played a determining role in 'orientating online traffic, distributing content and constructing knowledge' (Van Couvering, 2008) with commercial interests woven into the very fibre of the modern media networks through legislation, market mechanisms, and the like (Introna and Nissenbaum, 2000: 169; McChesney, 1996). Google's PageRank emerged from the competition not only as an algorithm for sorting and filtering information on the Web but also a dominant paradigm that established the new social, cultural, and political logics of search-based information societies—a phenomenon that Siva Vaidhyanathan characterized as the 'Googlization of everything' *(and why we should worry)* (2011: 20).[10] The implications of

---

[8] The websites that are the most heavily linked 'rule'.

[9] David Stark discusses his book, *A Sense of Dissonance* with researcher Brooke Harrington, where he also cites John Dewey's 'open ended inquiry' and focuses on identifying the problem instead of problem solving (Harrington, 2010).

[10] In 1998, a relatively late entry arrived into this burgeoning field, a small company called Google, comprising two Stanford students, Sergey Brin and Larry Page. The concept of relevancy that Brin and Page (1999) took into consideration when designing PageRank was based on the discovery that '[p]eople are still only willing to look at the first few tens of results' (Brin and Page, 1999: 108). This statement is crucial because it takes the position of a user who has to be able to search through large indexes of information. It is this aspect that comprised the efficiency of navigability and the quality of search results as deemed relevant by and Brin and Page. Google's PageRank had its basis in the SCI that has now been grafted as a conceptual paradigm for the way searchers find information and how that information is prioritized and organized. Hyperlinks became recognized as more than just connections

this political hegemony in regard to questions of identity, free speech, mobilization, and relevant search results, however, should not be underestimated.

At the turn of the past century, apprehension about corporate control was already expressed through a concern with 'the evident tendency of many of the leading search engines to give prominence to popular, wealthy, and powerful sites at the expense of others' (Introna and Nissenbaum, 2000: 181). With hyperlinks continuously being added ad infinitum, the bias in search engine results simultaneously became more noticeable—'bias that invites users to click on links to large websites, commercial websites, websites based in certain countries, and websites written in certain languages' (Van Couvering, 2010: 3). In other words, surfing the net through fortuitous clicks on hyperlinks that take users to other unknown spaces and places, outside of known comfort zones, disappeared. 'Cyberspace' (Gibson, 1984), as the Internet was affectionately termed back in the 1990s, morphed into 'Cybercapitalism' (DeLillo, 2003), where the returned search results are shaped by a highly intricate series of communication networks and commercial platforms (ads, personalization, and surveillance mechanisms). The user is now directed, manipulated, and organized by the invisible infrastructures that corporations have implemented in the design of the World Wide Web.

As of writing (September 2018) Google is the most used search engine worldwide, with around 5.8 billion requests per day.[11] Instead of drawing on acquired memory of knowledge, querying Google 'the oracle' for directions, names, articles, or locations has become commonplace with users, along with accepting answers as 'truths'. The performative action of entering a keyword in the search box has become 'habitual' (Chun, 2016) and has been grafted as a conceptual paradigm for the way users find information. Search engines nowadays are part of a larger 'media ecosystem' comprising various actors that attempt to control the results of what the user queries and receives as returns, yet they are not neutral. By promoting not only algorithmic authority but authorship as 'socio-epistemological machines' that source certain topics or interests (Rogers, 2013), search engines prioritize information for us(ers).

PLATFORM CAPITALISM

Google's original mission statement (1998) 'to organise the world's information and make it universally accessible and useful' has become instead a platform for mission control. For as much as information is desired by the user, Google orchestrates results in

---

and are considered votes, with high link popularity leading to an improved ranking. The scheme assigns two scores for each page: its authority, which estimates the value of the content of the page, and its hub value, which estimates the value of its links to other pages (this also refers to backlink value). Moreover, the PageRank algorithm considered the frequency and location of keywords within a web page and how long the web page has existed.

[11] Other commercial search engines such as Yandex, Yahoo, Bing, and Baidu also organize information for users.

turn—by (re)organizing its large accumulation of user data based on criteria that only it is able to provide. However, as Geoffrey Bowker points out, data are never entirely 'raw' but rather already 'cooked' (Gitelman and Jackson, 2013: 3) during their collection, storage, and transmission. As the search histories of users are captured, patterns appear that leave traces in the corpus of 'big data' and this 'vast archive is Google Inc.'s key monetizable resource, as its contents are sold to advertisers to generate the bulk of the company's revenues' (Jarrett, 2014: 17).

> The basic exchange [infra]structure of Web search consists of *users* querying the *engine* to find information made available by *content providers* competing for attention; *advertisers* hoping to grow their visitor numbers or sales finance the system.
> (Rieder and Sire, 2013: 3 emphasis in original)

Virtually all search engines today have a dual purpose: they provide search results to users and users to advertisers. Over the course of the past twenty years, Google transformed itself into an advertising company; first by producing not search results but rather audiences as its primary commodity. Later on the 'business focus shifted from the need to attract more users to the need to monetize what the viewers see' (Pasquale, 2015: 98). Advertisers wish to place ads on Google because it has a monopoly on the search engine market, with advertising still its prime source of revenue (around 87 per cent).[12] With a treasure trove of user data and online sealed auction bids for keywords from advertisers (see Vurdubakis, this volume), Google is able to organize the results through AdWords, where advertisers pay when users click on their paid ads.[13]

> The subsequent development of pay-per-click advertising is central to the current industry structure, an oligarchy of vertically integrated companies managing networks of syndicated advertising and traffic distribution.   (Van Couvering, 2010: 3)

Progressively, AdWords determine the monetary value of keywords, which, much like their discernment, provides a certain semantic governmentality. Keywords reflect users' thoughts and their monetization organizes the returned search results, thereby adding another dimension into the discursive patterning of human–machine interaction.[14] This helped restructure advertising according to the logic of semantic capitalism, 'in

---

[12] In 2014, 91 per cent of its revenue came from advertising. In 2016, Google held a 71.41 per cent market share of global desktop search and 91.61 per cent of the global mobile and tablet web search (Bilić, 2018).

[13] AdWords is an online advertising system that enables competition between bidders based on keywords, or search terms to display certain web pages and advertisers pay when users click on the ads. AdWords are a crucial part of Google's business model and comprise the lion's share (2016: 87 per cent) of their revenue.

[14] Until there becomes a more efficient and relevant means for the development of search strategies, whether through only hyperlinking within social media or through the potential of semantic search, keywords are the dominant methodology of querying (and researching) online.

which any word of any language has its price, fluctuating according to the laws of the market' (Bruno, 2006 cited by Feuz et al., 2011). Just as market devices assign prices to goods, the function of Google rankings assigns visibility, creating a 'market of relevance' in response to specific queries (Madsen, 2012: 12). Both types of actors, human and machinic, 'play a role in organizing this "market of relevance" that draws boundaries around a given issue', enabling it to be quantified and thereby monetized (Madsen, 2012: 11). Nowadays algorithms ostensibly know what users want before they even type them, as with Google's 'autocomplete' (Figure 37.4), revealing the user not as a user but as a content provider. It is self-evident that Google does not display the complete sum of possible words and, most critically, controls the process of deciding which words remain secret—the individual context for each user is prioritized—with users only noticing the top results and rarely looking beyond the first page (Groys, 2012). 'In some sense, Google has extended capitalism to language, transforming linguistic capital into money' (Kaplan, 2014: 58).

As of July 2018, there are globally more than 1.3 billion websites connected to the Internet and Google.com is the 'world's most popular website' (Zuboff, 2015: 77). Google Search facilitates the connectivity between users and advertisers as it concurrently usurps the middleperson. Search engines such as Google are platforms that regulate the flow of capital and obtain profit through an advertising business model, 'which extract information on users, undertake a labour of analysis, and then use the products of that process to sell ad space' (Srnicek, 2016: 49). In this way search engine platforms such as Google 'intervene' (Gillespie, 2014), serving up ads that influence the user's experience and diverting their path to information. This type of 'platform capitalism' is disrupting entrenched business models by highlighting as well as concealing, by downplaying the labour of users and the horizontality of services. Both search results and the corresponding advertisements shown are now optimized according to their potential market value based on pre-emptively calculated individual 'user relevancy' (Feuz et al., 2011). Nonetheless, many users still assume that 'Google is a search engine, rather than a multimillion dollar cooperation making large profits from devising personalized advertising schemes' (Stalder and Mayer, 2009: 99).

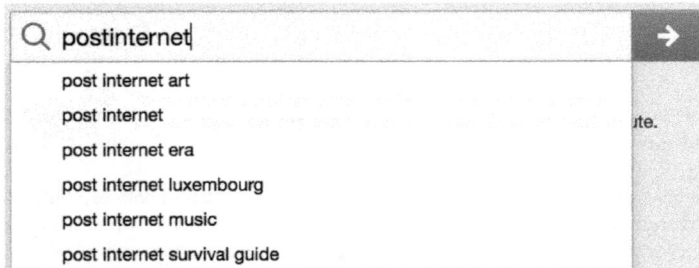

FIGURE 37.4 Screenshot of Google 'autocomplete' with keyword 'postinternet'

PERSONALIZATION

In order to connect users to advertisers, Google utilizes the IP address (Internet Protocol) for communication, a numeric label assigned to any type of device that is connected to a network. Similar to the 'pre-history of search engines', where the 'office' collected residential addresses of citizens, in contemporary digital society, the IP address serves as a 'signal', or location marker and determines the device and identity of the user online. IP addresses also facilitate data transfer, with search engines creating a database on each user containing their respective search histories and adopting this means to target users. Moreover, Google is a 'database of intentions'—'a massive clickstream database of desires, needs, wants, and preferences that can be discovered, subpoenaed, archived, tracked, and exploited for all sorts of ends' (Battelle, 2006: 6). One of these 'ends' is personalization where, since 4 December 2009, Google Search logs users' previous search queries, even if one is not signed into a Google account. The more data accrued (past search queries, locative data, and other hidden criteria) the more relevant the personalization. With this type of customization algorithms gather, extract, filter, and monitor online behaviour, offering suggestions for subsequent search requests. In exchange for their data, users receive free search results and ostensibly 'tailored' advertising, turning themselves into commodities for advertisers. Most users searching every day allow this personalization to occur without deleting 'cookies' or installing adblocking plug-ins that would inhibit it (see, e.g., Ad Block Plus, n.d.). This personalization is then a currency, with data correlated through algorithmic technologies and acquired by marketers, or third parties (Ridgway, 2014).

The pre-eminent search paradigm of the present, personalization is an opaque algorithmic process customized for each user, what Eli Pariser termed the 'filter bubble' (Pariser, 2011), which also limits knowledge by controlling what is visible, often leading to a 'distortion effect'.

> Like a lens, the filter bubble invisibly transforms the world we experience by controlling what we see and don't see. It interferes with the interplay between our mental processes and our external environment.   (Pariser, 2011: 82–3)

With this algorithmic organization, certain information on the World Wide Web is kept invisible and obscured, thus users are deterred from learning about things they do not already know. 'We are led—by algorithms and our own preference for the like-minded—into "filter bubbles", where we find only the news we expect and the political perspectives we already hold dear' (Gillespie, 2014: 88). In this way personalization has legitimized an online public sphere that is manipulated by algorithms.

The orchestration of these 'socio-technical arrangements' informs the logics of informational and affective organization and comprises the 'second index', which is fed back to users as search results, or recommendations, becoming 'device cultures' (Rogers, 2013: 161 cited in Weltevrede et al., 2014).

> Sites hope to anticipate the user at the moment the algorithm is called on, which requires knowledge of that user gleaned at that instant, knowledge of that user already gathered, and knowledge of users estimated to be statistically and demographically like them (Beer 2009)—drawing together what Stalder and Mayer (2009) call the second index.   (Gillespie, 2014: 173)

As more information is gathered on users and the data accrue in Google's proprietary, second index, seemingly a more complete profile is composed of the individual user. Yet Google's personalization assigns users to predefined criteria—categories that are constructed not through individuation but 'collaborative filtering'—a technique used in recommendation systems (see both Chun, and Alaimo and Kallinikos, this volume), which sorts through the data for patterns. Advertisers target not only individual users based on the user's consumption behaviour and search histories, but other users' search histories that are similar.

ORGANIZATION OF THE SELF

During the past eighteen years Google has constantly updated and actuated its proprietary algorithm PageRank, which reportedly has now more than 200 'signals' in its 'recipe' (Sullivan, 2010).[15] Although the exact workings of the black box remain unknown, one aspect that is known is RankBrain, a machine-learning artificial intelligence system that ostensibly interprets what people are searching for, even though they may have not entered the exact keywords. As of June 2016 RankBrain is being implemented for every Google Search query and the SEO (Search Engine Optimization) industry speculates it's summarizing the page's content (Sullivan, 2016). Rumour has it that the algorithm is adapting, or 'learning' from people's mistakes and its surroundings by applying 'deep neural networks' that are modelled after the human brain. By combining hardware and software in an attempt to copy the human web of neurons, RankBrain is fed vast amounts of data to train the deep-learning neural networks splitting computing tasks across machines and now there is enough computational power at Google's data centres to handle much more data.[16] Previously, humans—programmers—wrote the code and then tweaked the results; now with RankBrain the models are machine-readable and therefore less human-readable.

[15] Google usually describes that it has around 200 major ranking signals, yet there have been discussions of 1,000 or even 10,000 sub-signals (Sullivan, 2010).
[16] 'Training the many layers of virtual neurons in the experiment took 16,000 computer processors—the kind of computing infrastructure that Google has developed for its search engine and other services. At least 80 percent of the recent advances in AI can be attributed to the availability of more computer power, reckons Dileep George, cofounder of the machine-learning startup Vicarious' (Hof, 2012).

These digital technologies 'fundamentally alter the ways we collect, circulate, and make sense of information' (Flyverbom et al., 2016: 99). Machine-learning technology disrupts human ontologies and taxonomies of keywords that previously structured queries and the search results obtained. '[H]uman bodies as cyborgs—as human machine systems—are in turn systematically combined into modes of "cyberorganization"' (Parker, 2000: 73), which controls the flow of information back to users. With cyberorganization 'as [a] continually shifting set of relationships' (Parker, 2000: 81), the subject becomes the site of data collection as well as being constantly evaluated by algorithms. Moreover,

> there is a case to be made that the working logics of these algorithms not only shape user practices, but also lead users to internalize their norms and priorities.
>
> (Gillespie, 2014: 187)

Nowadays with the user's IP address fully recognized, along with captured search histories, user data shape users reciprocally. 'Algorithms are made and remade in every instance of their use because every click, every query, changes the tool incrementally' (Gillespie, 2014: 173). Users reinvent themselves in all kinds of ways because 'technology participates in what people become' (Goriunova, 2015). Through searching, users' thoughts and values are transferred into predictions, which then produce changes in deciding which keywords will be entered back into the recursive loop. To what extent do humans adapt to algorithms in this machine learning process and how much do algorithms affect human learning?

> As these algorithms nestle into people's daily lives and mundane information practices, users shape and rearticulate the algorithms they encounter; and algorithms impinge on how people seek information, how they perceive and think about the contours of knowledge, and how they understand themselves in and through public discourse.   (Gillespie, 2014: 183).

In this way these interfaces organize the self (selves) through continuous human–computer interaction(s). The lesson to be learned from the 'big data revolution' discussion incited by Mayer-Schöneberger and Cukier (2013) is that 'it would instigate no less than a change in human beings' (Beyes, 2020: 2). With users constantly consulting the machine they become more like them, and '[t]hrough habits we become our machines' (Chun, 2016: 4). This 'self-organization' has precipitated direct interference by algorithms not only with personalized search results that are obtained by the user but also with the machine-learning RankBrain algorithm that could eventually informate (Zuboff, 1981) as well as automate, producing the information it organizes, making users thereby redundant.[17]

---

[17] Google and big data break with the past, 'its populations are no longer necessary as the source of customers' employees. Advertisers are its customers along with other intermediaries who purchase its data analyses' (Zuboff, 2015: 80).

Users search habitually, producing vast amounts of data, which they 'voluntarily' provide. With the extraction activities of user data by Google, Shoshana Zuboff proposes that it is the logic of accumulation that comprises 'surveillance capitalism', 'of which "big data" is both a condition and an expression' (Zuboff, 2015: 77). This logic of accumulation of data decides what is left out, what is accumulated and how it is organized, producing 'its own social relations and with that its conceptions and uses of authority and power' (Zuboff, 2015: 77).[18] Although the black box curation of search algorithms remains non-transparent, what has become clear is that corporations (GAFAM)[19] gather this user data and according to technology cognoscenti Evgeny Morozov, the great secret of Silicon Valley has been revealed to the public: the data that users supply to digital platforms has a greater economic value than the value of the platforms' services (Morozov, 2017). With the 'reorganization' of Google in 2015, it was actually (re)branded as a research corporation, Alphabet. In 2016, Alphabet still earned most of its revenue from advertising (US$90.272 billion, 88 per cent of total revenue) and is therefore not a 'search engine'—Google Search is just its most profitable service. Whatever the name, its mission statement 'organizing the world's information and making it universally accessible and useful' no longer only applies to users. Gathering as much user data as possible (Morozov, 2017), Alphabet presently focuses on lucrative and creative applications with its treasure chest of collected data, extracted and refined through AI (artificial intelligence).

ALTERNATIVES

Instead of supplying data to corporations, users could hide it, control it, or even delete it and therefore need not give it away in exchange for free service.[20] The creation of new organizational methods could challenge corporate online search practices. 'Vaidhyanathan (2011) imagined a "human knowledge project" to approach the "task

---

[18] With Google's accumulation of user data, a new asset class is created, what Zuboff deems *surveillance assets*. Data are extracted from users today just as the earth's natural resources (minerals and elements) were extracted in colonial times and now have become 'big business' along with 'rare earth elements'. Karl Polanyi wrote over seventy years ago about the 'three fictions' of market economies where life, nature, and exchange become transformed into a 'commodity fiction', which can be bought and sold. This fiction 'disregarded the fact that leaving the fate of soil and people to the market would be tantamount to annihilating them' (Polanyi cited by Zuboff, 2015: 85). Zuboff's argument is that it is the 'fourth fictional commodity'—the logic of accumulation—in an era of surveillance capitalism that marks the market economics of the twenty-first century. 'Reality' is now transformed and reborn as data through processes of commodification and monetization and marketed as 'behaviour' (Zuboff, 2015).

[19] Google, Apple, Facebook, Amazon, Microsoft.

[20] Well-designed browser extensions such as *Ad Nauseum* 'obfuscate browsing data and protect users from surveillance and tracking by advertising networks' (Nissenbaum et al., 2014). Working in conjunction with Ad Block Plus (https://adblockplus.org), an open source plug-in that removes ads whilst browsing, this intervention clicks and likes *all* ads, concomitantly visualizing the ads over time. By 'clicking ads so you don't have to', it addresses the lack of standards for tracking, privacy issues, user profiling, and 'excessive universal surveillance' (Nissenbaum et al., 2014).

of organizing the world's information and making it universally accessible in" a non-corporate way' (Mager 2012: 783). Non-proprietary indices would be accessible to a variety of search engines and enable counter strategies of exploring new methods of finding information through human directed filtering or human sharing through networks. According to Dirk Lewandowski (2014) such indices could facilitate competition on the search engine market and allow for many smaller search projects to be realized.

Drawing on these non-proprietary indices, other means of finding specific information on the World Wide Web fall under the heading of 'vertical search'.[21] YaCy is a decentralized, peer-to-peer (p2p) privacy search engine that offers an alternative to the tracking, censorship, and targeted ads of mainstream commercial search (Figure 37.5).[22] With its 'expert' crawler that follows links to any depth, the user decides which sites and pages get indexed and can limit the scope of the search for a specific topic. Imparting control of indexing and searching to the user, YaCy emphasizes unique or special collections, which are dependent on the p2p indices of the

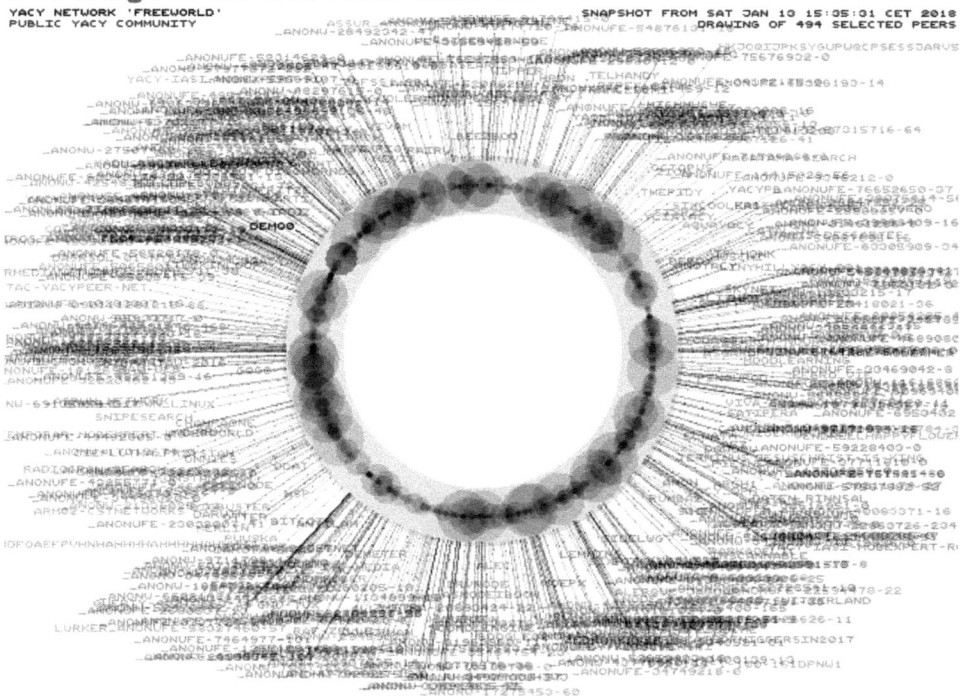

FIGURE 37.5 Live image of the 'freeworld' YaCy network, screenshot

[21] Also referred to as topical, or specialized, vertical search targets a special area of interest, one area of knowledge or topic and does not index the entire Web, but instead uses a focused web crawler.

[22] YaCy can be downloaded for free and once installed on a user's computer the *freeworld* is the network of the peers who take part in this distributed network. https://yacy.net/en/index.html.

collective. Additionally, it is possible to query with plug-ins that prevent tracking,[23] or to use Tor (The Onion Router), a p2p browser that enables privacy online and facilitates anonymous communication by not divulging the IP address of the user (Figure 37.6).[24]

Besides governmental actors in the security industries, activists, dissidents, and whistleblowers using Tor, there are those who wish to search regions of the Web that have not yet been indexed by Google to form the surface web.[25] These users embody agency, evincing a hacker freedom with the goal of being off the radar, able to control their data and deciding, just as interfaces do, what to show and what not to. The habit of being online and searching could instead become a new form of (in)visibility management. Similar to the proprietary corporate search algorithms of Google, the evaluative criteria and code of which are concealed from the user, the user instead can find ways to

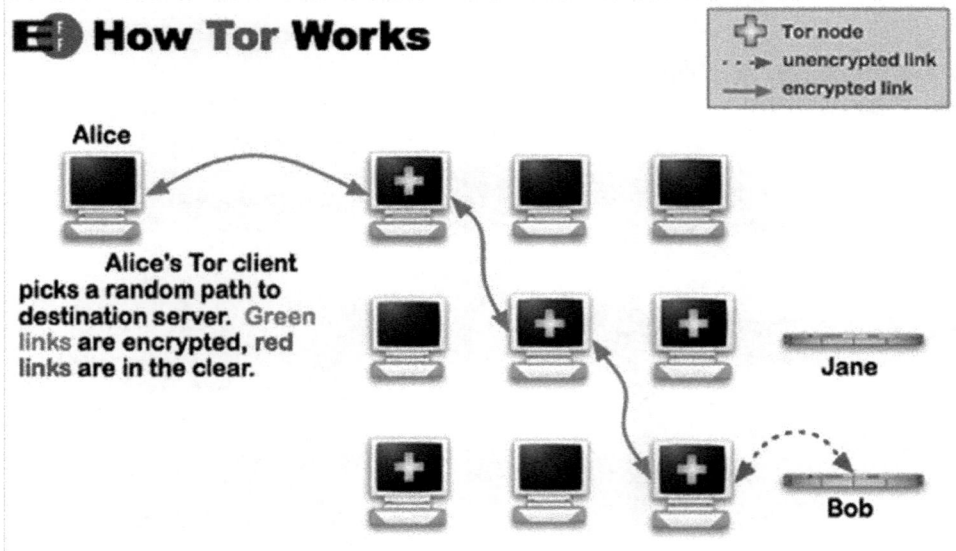

FIGURE 37.6 'How Tor Works'. Diagram originally contributed by Ludovic Rembert via Privacy Canada for the Electronic Frontier Foundation (EFF), December edition 2011

[23] 'TrackMeNot is a lightweight browser extension that helps protect web searchers from surveillance and data-profiling by search engines. It does so not by means of concealment or encryption (i.e. covering one's tracks), but instead by the opposite strategy: noise and obfuscation. With TrackMeNot, actual web searches, lost in a cloud of false leads, are essentially hidden in plain view' (Howe and Nissenbaum, 2006).

[24] The Tor p2p network is a mesh of proxy servers in which data are bounced through three relays (entry, middle, exit), adding a layer of encryption at every node while decrypting the data at every 'hop' and forwarding it to the next onion router (see The Tor Project, 2017).

[25] 'Tor today is an influential anticensorship technology that allows people in oppressive regimes to access information without the fear of being blocked, tracked or monitored. The importance and success of Tor is evident from recent global uprisings where the usage of Tor spiked as people used it as a revolutionary force to help them fight their social and political realities' (AlSabah et al., 2012: 1).

obfuscate their online presence.[26] Hidden from the very algorithms that are designed to be obscure and that obscure, the user could become much more like the algorithms, stealthy and arcane, shrouded in the (onion) layers of the Tor Browser instead of remaining inside the filter bubble of Google Search.

REFERENCES

Ad Block Plus. N.d. https://adblockplus.org.

AlSabah, Mashael, Kevin Bauer, and Ian Goldberg. 2012. 'Enhancing Tor's Performance Using Real-Time Traffic Classification'. Presentation at CCS '12, Raleigh, North Carolina, USA, 16–18 October.

Battelle, John. 2006. *The Search: How Google and Its Rivals Rewrote the Rules of Business and Transformed Our Culture.* New York: Penguin.

Beer, David. 2009. 'Power through the Algorithm? Participatory Web Cultures and the Technological Unconscious'. *New Media & Society* 11(6), 985–1002.

Beyes, Timon. 2020. 'Organization'. In N. B. Thylstrup et al. (eds), *Uncertain Archives.* Cambridge, MA: MIT Press (forthcoming).

Bilić, Paško. 2018. 'A Critique of the Political Economy of Algorithms: Brief History of Google's Technological Rationality'. *WIAS*, 16, 315–31. Westminster Institute for Advanced Studies (WIAS).

Brin, Sergey and Lawrence Page. 1999. 'The Anatomy of a Large-Scale Hypertextual Web Search Engine'. http://infolab.stanford.edu/~backrub/google.html.

Bruno, Christophe. 2006. 'Adwords'. http://www.iterature.com/adwords/.

Brunton, Finn and Helen Nissenbaum. 2015. *Obfuscation: A User's Guide for Privacy and Protest.* Cambridge, MA: MIT Press.

Bush, Vannevar. 1945. 'As We May Think'. *The Atlantic.* https://www.theatlantic.com/magazine/archive/1945/07/as-we-may-think/303881.

Chun, Wendy H. K. 2016. *Habitual New Media: Updating to Remain the Same.* Cambridge, MA: MIT Press.

Davies, William. 2017. 'How Statistics Lost their Power—and Why We Should Fear What Comes Next'. *The Guardian*, 19 January. https://www.theguardian.com/politics/2017/jan/19/crisis-of-statistics-big-data-democracy.

DeLillo, Don. 2003. *Cosmopolis.* New York: Scribner.

Feuz, Martin, Matthew Fuller, and Felix Stalder. 2011. 'Personal Web Searching in the age of Semantic Capitalism: Diagnosing the Mechanics of Personalisation'. *First Monday, Peer-Reviewed Journal on the Internet*, 16(2–7). http://firstmonday.org/article/view/3344/2766.

Flyverbom, Mikkel, Paul M. Leonardi, Cynthia Stohl, and Michael Stohl. 2016. 'The Management of Visibilities in the Digital Age'. *International Journal of Communication*, 10, 98–109.

Garfield, Eugene. 2007. 'The Evolution of the Science Citation Index'. *International Microbiology*, 10, 65–69.

Gibson, William. 1984. *Neuromancer.* New York: Ace.

---

[26] *Obfuscation: A User's Guide for Privacy and Protest*, by Finn Brunton and Helen Nissenbaum (2015) offers a comprehensive overview of tactics and why obfuscation is necessary in an era of online tracking and digital surveillance.

Gillespie, Tarleton. 2014. 'The Relevance of Algorithms'. In Tarleton Gillespie, Pablo Boczkowski, and Kirsten Foot (eds), *Media Technologies: Essays on Communication, Materiality, and Society*. Cambridge, MA: MIT Press, 167–95.

Gitelman, Lisa and Virginia Jackson. 2013. 'Introduction'. In Lisa Gitelman (ed.), *Raw Data Is an Oxymoron*. Cambridge, MA: MIT Press, 1–14.

Gleick, James. 2011. *The Information: A History, A Theory, A Flood*. New York: Pantheon.

Goriunova, Olga. 2015. 'Digital Subject'. https://www.youtube.com/watch?v=yAIheBLmG6M.

Groys, Boris. 2012. *Google, Words beyond Grammar*. 100 Notes, 100 Thoughts: Documenta Series 046. Berlin: Hatje Cantz.

Halavais, Alexander. 2009. *Search Engine Society*. Cambridge: Polity Press.

Harrington, Brooke. 2010. 'Economic Sociology: The Sense of Dissonance – An Interview with David Stark'. *The Society Pages*. https://thesocietypages.org/economicsociology/2010/04/14/the-sense-of-dissonance-an-interview-with-david-stark/.

Hindman, Matthew. 2009. *The Myth of Digital Democracy*. Princeton, NJ: Princeton Press.

Hof, Robert D. 2012. 'Deep Learning'. *Technology Review*. https://www.technologyreview.com/s/513696/deep-learning/.

Howe, Daniel and Helen Nissenbaum. 2006. *Track Me Not*. https://cs.nyu.edu/trackmenot/.

Introna, Lucas D. and Helen Nissenbaum. 2000. 'Shaping the Web: Why the Politics of Search Engines Matters'. *The Information Society*, 16(3), 169–85.

Jarrett, Kylie. 2014. 'A Database of Intention? Why We Need an Independent Index of the Web'. In René König and Miriam Rasch (eds), *Society of the Query Reader #9: Reflections on Web Search*. Amsterdam: Institute of Network Cultures, 16–29.

Kaplan, Frederic. 2014. 'Linguistic Capitalism and Algorithmic Mediation'. *Representations*, 127(1), 57–63.

Lewandowski, Dirk. 2014. 'Why We Need an Independent Index of the Web'. In René König and Miriam Rasch (eds), *Society of the Query Reader #9: Reflections on Web Search*. Amsterdam: Institute of Network Cultures, 49–59.

McChesney, Robert W. 1996. 'The Internet and U.S. Communication Policy-Making in Historical and Critical Perspective'. *Journal of Computer-Mediated Communication*, 1(4), 98–124.

Madsen, Anders K. 2012. 'Web-Visions as Controversy-Lenses'. *Interdisciplinary Science Reviews*, 37(1), 51–68.

Mager, Astrid. 2012. 'Algorithmic Ideology'. *Information, Communication & Society*, 15(5), 769–87.

Marchiori, Massimo. 1997. 'The Quest for Correct Information on the Web: Hyper Search Engines'. *Proceedings of the Sixth International World Wide Web Conference (WWW6)*. https://www.w3.org/People/Massimo/papers/WWW6/.

Mayer-Schönberger, Viktor and Kenneth Cukier. 2013. *Big Data: A Revolution That Will Transform How We Live, Work, and Think*. London: John Murray.

Morozov, Evgeny. 2017. 'Big Tech een digital wereldrijk wankelt'. *NRC*. https://www.nrc.nl/nieuws/2017/09/08/big-tech-een-digitaal-wereldrijk-wankelt-12,901,285-a1572789.

Nissenbaum, Helen, Daniel Howe, and Mushon Zer-Aviv. 2014. *Ad Nauseum*. http://dhowe.github.io/AdNauseam.

Pariser, Eli. 2011. *The Filter Bubble*. New York: Penguin Books.

Parker, Martin. 2000. 'Manufacturing Bodies: Flesh, Organization, Cyborgs'. In John Hassard, Ruth Holliday, and Hugh Willmott (eds), *Body and Organization*. London: Sage, 71–86.

Pasquale, Frank. 2015. *The Black Box Society: The Secret Algorithms that Control Money and Information*. Cambridge, MA: Harvard University Press.

Ridgway, Renée. 2014. 'Personalisation as Currency'. *APRJA* (*A Peer-Reviewed Journal About*). http://www.aprja.net/?p=2531.

Rieder, Bernhard and Guillaume Sire. 2013. 'Conflicts of Interest and Incentives to Bias: A Microeconomic Critique of Google's Tangled Position on the Web'. *New Media & Society*, online first, 1–17.

Rogers, Richard. 2013. *Digital Methods*. Cambridge, MA: MIT Press.

Rosenberg, Daniel. 2013. 'Data before the Fact'. In Lisa Gitelman (ed.), *Raw Data Is an Oxymoron*. Cambridge, MA: MIT Press, 15–40.

Shepard, Frank. 1873. *Shepard's Citations*. New York: Frank Shepard Company.

Srnicek, Nick. 2016. *Platform Capitalism*. Cambridge: Polity Press.

Stalder, Felix. 2010. 'Autonomy and Control in the Era of Post-Privacy'. In *Open. Cahier on Art and the Public Domain # 19*: Beyond Privacy: New Notions of the Private and Public Domains.

Stalder, Felix and Christine Mayer. 2009. 'The Second Index: Search Engines, Personalization and Surveillance'. In Konrad Becker and Felix Stalder (eds), *Deep Search: The Politics of Search beyond Google*. Innsbruck: Studien Verlag, 98–116.

Stark, David. 2009. *The Sense of Dissonance*. Princeton, NJ: Princeton University Press.

Sullivan, Danny. 2010. 'Dear Bing, We Have 10,000 Ranking Signals to your 1000. Love Google'. *Search Engine Land*, 11 November. http://searchengineland.com/bing-10,000-ranking-signals-google-55,473.

Sullivan, Danny. 2016. 'All about the Google RankBrain Algorithm'. *Search Engine Land*, 23 June. http://searchengineland.com/faq-all-about-the-new-google-rankbrain-algorithm-234,440.

Tantner, Anton. 2014. 'Before Google: A Pre-History of Search Engines in Analogue Times'. In René König and Miriam Rasch (eds), *Society of the Query Reader #9: Reflections on Web Search*. Amsterdam: Institute of Network Cultures, 121–36.

The Tor Project. 2017. https://www.torproject.org/.

Vaidhyanathan, Siva. 2011. *Googlization of Everything (And Why We Should Worry)*. Oakland, CA: University of California Press.

Van Couvering, Elizabeth. 2008. 'The History of the Internet Search Engine: Navigational Media and the Traffic Commodity'. In A. Spink and M. Zimmer (eds), *Web Search: Information Science and Knowledge Management*, vol. 14. Berlin: Springer, 177–206.

Van Couvering, Elizabeth, 2010. 'Search Engine Bias: The Structuration of Traffic on the World-Wide Web'. PhD dissertation, London School of Economics and Political Science (LSE).

Wells, H. G. 1937. 'World Brain: The Idea of a Permanent World Encyclopaedia'. *Encyclopédie Française*, August. https://sherlock.ischool.berkeley.edu/wells/world_brain.html.

Weltevrede, Esther, Anne Helmond, and Carolin Gerlitz. 2014. 'The Politics of Real-Time: A Device Perspective on Social Media Platforms and Search Engines'. *Theory, Culture & Society*, 31(6), 125–50.

Zuboff, Shoshanah. 1981. 'Psychological and Organizational Implications of Computer-Mediated Work'. MIT Working Paper, Center for Information Systems Research.

Zuboff, Shoshanah. 2015. 'Big Other: Surveillance Capitalism and the Prospects of an Information Civilization'. *Journal of Information Technology*, 30, 75–89.

CHAPTER 38

SMARTPHONE

JENNIFER WHYTE

THE SMARTPHONE AS ORGANIZATIONAL FORCE

A smartphone is a 'mobile phone that performs many of the functions of a computer, typically having a touchscreen interface, Internet access, and an operating system capable of running downloaded apps' (Oxford Living Dictionaries, 2017). It is a personal device owned by one user. Such devices have become widely used: while there are many different brands and generations of smartphones, a familiar image of one is shown in Figure 38.1. It is a multipurpose physical device of internal complexity with processors, sensors, GPS, camera, microphone, speaker, and display (Brynjolfsson and McAfee, 2014). Users treat it as solid (with no internal circuitry or components), or as a display window through which to access other worlds. In choosing a device, many seek something small enough to fit in a hand or pocket, but yet also large enough to provide a screen that is legible and comfortable to use. In their various offerings to the market, technology providers explore the limits of the device dimensions and functionality while also aiming to make it thinner.

The development and rapid diffusion of smartphones followed organizational uses of personal digital assistants (PDAs), which became used by business professionals in the late 1990s (Palm PDA was introduced in 1996, and Blackberry in 1999), and the increasingly pervasive use of mobile phones in both business and personal lives for voice calls and for text messaging. Early smartphones became available in the late 2000s, with the first generation of iPhone released in 2007, and android devices reaching the market in 2008–9. The transformative effect was realized, in part, through the iPhone App Store, opened in July 2008, and Android Market following a few months later.

As an object, the smartphone has substantial organizational force. Through it, people distribute their organizing activity across locations, juggled between other home and office activities—such as the shopping trip, commute, business meeting, and socializing—with partial attention. It alters the bodily work involved in organizing. Rather than the

FIGURE 38.1 A smartphone 'selfie'

(Photo by the author)

use of fingertips on a keyboard, use of the smartphone emphasizes the thumb (where the fingers are used, as in Figure 38.1, to cradle the device). The artefact itself dissolves from view (Middleton et al., 2014) as focus is given to the services and remote virtual interactions that it enables. The locus of organizing becomes distributed. The data involved also become held in different locations, remote from the activities, and potentially in servers in other jurisdictions across the globe. The temporal pace of work changes, with new forms of organizing in the moment rather than pre-planned activity; and new forms of asynchronous as well as synchronous interaction through social media as well as person to person communication.

THE SMARTPHONE'S ORGANIZATIONAL EFFECTS

There are hence a number of organizational effects. One is to blur the boundary between work activities and home activities (Derks and Bakker, 2014; Derks et al., 2014a, 2014b, 2015, 2016). The smartphone has taken 'the office' into the commuter train and into the home, even, as previous authors note, as far as the toilet (Cecchinato et al., 2014) and bedroom (Perlow, 2012, Lanaj et al., 2014). Because of its multifunctional nature, using the smartphone may transform the home, replacing the music collection, book collection, city map and tourist guides, phonebook, photo albums, video, camera, wallet, cookery books, and radio and consigning physical analogues to the attic, basement, cupboard, or recycling bin. It also takes 'the home' into 'the office', as both managers and employees use their phones to organize the evening meal, shop, check on their children, change the home heating, pay bills, buy tickets, and communicate with or update friends. It increases the potential for real-time communication with peers, within and outside working hours, and changes the language, introducing neologisms, e.g., 'selfie' and 'app', and adding new graphical symbols to alphabets, e.g., '☺' to the correspondence involved in both business and leisure. While early phone users might have a corporate phone, or juggle two—one for work and one for home—the single privately paid-for smartphone symbolizes increasing norms of integration between home and work.

Another organizational effect is to involve different constituent actors and forms of power that bypass traditional organizational gatekeepers. The smartphone hence changes which activities become visible or hidden in organizational settings. Isolated experiences in one organization may become contextualized and connected by the smartphone user, for example through the #metoo movement that called out sexual harassment, and led to a reconsideration of the acceptance of behaviours that had become institutionalized and unseen in organizational practices across industries. What might seem like a private and personal interaction can have substantive organizational effects, coordinating social movements and formal organizations, and enabling the subversion or consolidation of power. While there may be organizational guidelines for smartphone use, as a personal device that is often paid for privately it is not fully under the control of the formal work organization. President Trump's late night tweets, for example, are an expression of the potentially far-reaching effects. Such smartphone-enabled communication may operate outside of formal organizational structures and outside of the organizational checks, balances, and controls used by external communications teams.

As an interface to a technological ecosystem, the smartphone changes accepted norms of practice across devices, with the organizational effect of raising new ways to access services and new forms of exclusion. People in organizations now access the Internet, social media, and an ecosystem of apps through a range of computing devices (including tablets, watches, and wearables). Neff and Nafus note that 'Most connectable wearable devices come with the presumption that their owners will already have both a smartphone and a computer but this is not always the case for the elderly, poor and sick' (Neff and Nafus, 2016: 127).

Hence the smartphone is a small device that impacts large-scale organizing. As businesses and customers are increasingly connecting through the smartphone, the coordination mechanisms of bureaucracy, market, and trust (Alvesson and Thompson, 2006) are altered. In the 'gig' economy, a computing device that is near-to-hand becomes essential, and it is the smartphone that becomes the taskmaster. Examples are the work of Airbnb apartment owners, where customer response times matter and owners are rated online; and the work of Uber drivers, where the smartphone is used in monitoring their availability and location, making it available to potential customers, and is also employed as the 'satnav' showing the route to their destination. The smartphone is a convenient interface in the new forms of work, across the formal and informal economy, including the new forms of work variously described as 'microwork', 'crowdsourcing', or 'the sharing economy'. Yet, while organizational activities are changed by smartphone use, this use and the technologies that support it are less regulated than traditional mechanisms, with for example the terms and conditions on smartphone apps usually unread by individual users, and, in the work environment, also by the legal representatives of their organizations.

A Technology of Power, Protest, and Surveillance

The smartphone is relevant to organizational scholars as it becomes a site of interaction between the individual and organizing. Use of the smartphone becomes integral to the lives of organizational members, raising questions of power. It can be seen as a locus for self-expression and self-broadcast, thus invoked in the creation of social movements:

> The omnipresence of smartphones and social media such as YouTube, Twitter and Facebook (with WhatsApp) allow (sic) to post telling images, videos, emails or documents in real time; finding a broad audience, quickly, in conjunction with the use of classic mass media; and triggering spontaneous forms of protest without any major organizational effort.   (Dolata, 2017: 13; Dolata and Schrape, 2018: 40)

Scholars have examined how smartphones have been used in street protest (Neumayer and Stald, 2014) and in the 'Arab Spring' (Steinert-Threlkeld, 2017). Yet Han (2017) argues that such limitless freedom and communication is an illusion, and that such technologies are primarily used for monitoring and surveillance, turning the citizen into an onlooker, spectator, and consumer of events (see Figures 38.2 and 38.3, for the social use of smartphones as spectators). Through surveillance, the citizen's use of the smartphone can reinforce and amplify existing power structures, as patterns of use give information to those in power to intervene and influence the smartphone at a 'pre-reflexive level'. Han (2017) argues that transparency as a 'dispositive' requires both conformity and 'exteriorization':

> Every dispositive—every technology or technique of domination—brings forth characteristic devotional objects that are employed in order to subjugate. Such objects *materialize* and stabilize dominion. Devotion and related words mean 'submission' and 'obedience'. Smartphones represent *digital* devotion—indeed, they are the *devotional objects of the Digital*, period.   (Han, 2017: 12)

FIGURE 38.2 Smartphones being used to photograph the 'Trump Baby' in London protest
(Photo by the author)

FIGURE 38.3 Smartphones used to show devotion at an event
(Photo: 'Rumšiškės Culture Center, Rumšiškės, Lithuania' by Kipras Štreimikis on Unsplash)

Thus for Han, the smartphone as a material artefact becomes understood as a devotional object, used to subjugate its user. It is compared to the rosary, a near-to-hand artefact that is used for self-monitoring and surveillance (see Bunz, and Gregg and Kneese, this volume). Han argues that the effects of the smartphone are ones of submission and obedience in neoliberalist societies:

> Confession obtained by force has been replaced by voluntary disclosure. Smartphones have been substituted for torture chambers. Big Brother now wears a friendly face. His friendliness is what makes surveillance so effective.
>
> (Han, 2017: 38–9)

While in totalitarian regimes, the use of such technologies for control may be explicit, here it is delegated to the user through self-surveillance. Neff and Nafus (2016) observe extensive self-tracking (see, for example, Figure 38.4). Actuators in the smartphone enable a user to monitor their steps, activity levels, sleep, and orientation and Neff and Nafus (2016) argue that the associated data are fundamentally social, with the potential for social and organization effects. The smartphone becomes a pivotal device in organizing the self in late capitalism, with far-reaching effects for self-monitoring and surveillance, both legally and illegally, by corporate organizations and states. Through

FIGURE 38.4 Runner with smartphone used as a tracking device
(Photo: unnamed, Toronto, Canada—by Filip Mroz on Unsplash)

people's interactions with the phone, social and cultural norms are also internalized, with, for example, beauty apps leading to a 'forensic scrutiny' of the female body through smartphones (Elias and Gill, 2017).

The smartphone's agency in terms of power and surveillance has had an influence on scholarly work on organization studies, though it has been relatively under-researched within this field. Smartphone users in an engineering firm studied by Symon and Pritchard (2015) were found to use their phones to perform identities, such as being 'contactable and responsive', 'involved and committed', and 'in-demand and authoritative', Their research also examined the role of the smartphone camera in the organizational work, articulating how the smartphone does not simply enhance a sense of proximity as the new evidence that could be collected from the field and shared back to office locations, as it provides only partial representations, where for example visual and verbal evidence may disagree (Pritchard and Symon, 2014). As the use of the smartphone extends and amplifies social and organizational impacts observed with earlier devices, this earlier work is a useful theoretical starting point (e.g., Jarvenpaa and Lang, 2005), with its insights into how devices can make interactions visible or invisible, for example, mediating in the lives of veiled women in Saudi Arabia (Lobo and Elaluf-Calderwood, 2012). Rather than arranging in advance the location or time of a meeting, smartphone users may make arrangements or coordinate details on the day, a few hours or minutes in advance. In work on smartphone use in consulting project work, Azad et al. (2016) argue that such projects have a logic of 'clockwork coordination', with front-stage practices having an affinity with synchronous use of the smartphone and back-stage practices having an affinity with asynchronous uses. As we fit more into the day through multi-tasking with devices that do not require full attention (and thus time is not consumed exclusively by a single activity), Mullan and Wajcman (2019) suggest the smartphone may change our experience and perception of time itself.

WELL-BEING AND INDIVIDUAL PERFORMANCE

A relatively mature part of the extant literature addresses the implications of smartphone use for the individual smartphone user, and their interactions across work and home. Dery et al. (2014) argue that, for many smartphone users, it is no longer possible or desirable to disconnect from work and advocate a reframing of work activities as part of a connective flow. Karlson et al. (2009) describe patterns of use in which the smartphone is checked on when waking up:

> Although we might expect the bleeding of work into personal time to cause stress, the continuous access to work email instead seemed to provide peace of mind (e.g., P21: "I'll wake up in the middle of the night, check [work] email, if it's something I can answer right then, I'll go ahead and do it. It's more convenient to go ahead and do it, and then I can forget about it."   (Karlson et al., 2009: 404)

However, many authors raise concerns about the long-term implications of this connectivity on physical and mental well-being of individuals. Lack of sleep has a significant detrimental effect on an individual's performance and health (Walker, 2017). Lanaj et al. (2014) find the impact on sleep leads to diminished daily engagement in work in late-night smartphone users. Derks and Bakker (2014) find greater evidence of burnout in intensive smartphone users who lack engagement in recovery activities (that foster psychological detachment and relaxation), advocating organizational policies on out-of-hours smartphone use. Perlow (2012) provides techniques for limiting use; while Yun et al. (2012) also associate smartphone use with work overload and job stress. Piszczek (2017) distinguishes between individuals who prefer role integration and hence experience control of the boundary between their family and work through their use of a smartphone, and those that prefer separation and hence experience loss of control, arguing that individuals in this latter category can more easily experience emotional exhaustion in these conditions.

Organizational implications are also suggested by the growing psychology literature on problematic smartphone use and addiction; and the relationship between smartphone use and cognition. In general, problematic smartphone use is understood to both drive, and be driven by, mental disorders (Elhai et al. 2017). Elhai et al. (2016) find 'fear of missing out' to be strongly correlated with intensive smartphone use (see Figure 38.5, for the commuters using smartphones). There is a growing interest in the impact of smartphones on cognition, though here (as in the above) results are also currently contradictory and inconclusive (Wilmer et al., 2017). Ward et al. (2017) find a cognition capacity reduction in users when their smartphone is present. Such work leads to new questions about how to ensure attention within organizations across a range of forms of knowledge work. Personal uses of smartphones at work are positively associated with end-of-workday affective well-being (Kim and Park, 2017), but to avoid distractions and ensure attention to the task-in-hand (and hygienic conditions), guidelines have been developed for the use of smartphones in the organizational context of healthcare (Gill et al., 2012). The workplace becomes one of a number of loci for organizing that vies for the employee's divided attention.

The Smartphone's Hidden Practices of Organizing

There are relevant insights for organization scholars in the reviewed body of work in this chapter, which draws on organization studies and related fields of psychology, human factors, organizational behaviour, information systems, political science, and sociology. However, there is substantial work for organizational scholars to do to unpack the implications of the smartphone for organizations and organizing, to address issues such as work–life balance, self-monitoring, trust, bureaucracy, and markets, to contribute to wider debates on the digital workforce (Colbert et al., 2016) and the relationships

FIGURE 38.5 Commuters using their smartphones

(Photo: unnamed, Naha, Japan—by Jens Johnsson on Unsplash)

between organizations and media (Roulet and Clemente, 2018), and to understand the wide set of organizing activities that are enabled by the smartphone across the formal and informal economy, including the new forms of work variously described as 'microwork', 'crowdsourcing', or 'the sharing economy'. This chapter concludes by considering some related organizing practices that have become hidden and some methodological issues of organizational research where organizational members use smartphones.

Although the smartphone is associated with transparency, many aspects of the artefact and its use are hidden. We know little about the organizing practices through which raw materials become smartphone artefacts. Organization scholars could examine a range of organizational settings related to the production of the physical artefact of the smartphone, perhaps exposing the hidden forms of organizing that popular media suggests may include modern slavery, child labour, corruption, and war (Hindess, 2018). They could examine the organizational settings involved in disposal of smartphones and the network of related technologies involving humans and other species, where the popular media suggests these involve toxic working conditions (Holgate, 2017). How do organizations account for these and how are these deemed safe? These may seem remote from the smartphone in the corporate office, or the pocket, but they are constitutive practices of organizing on which the organizational force and effects of the smartphone are built. Likewise, we know little about the organizing practices that underpin the distributed nature of the virtual interactions through the smartphone, where these include the physical servers that support the digital information and the activities of groups that may, unknown to the smartphone user, have access to and use the data. The smartphone raises new security concerns as it is not apparent to users how the connectivity they value can also be used to monitor their location; potentially access their passwords, camera, and microphone; or access information related to their work. For example, what are the organizational implications when smartphones become a vehicle for 'fake news' undermining trust in institutions, where new external groups are becoming more sophisticated at targeting users.

The physical nature of smartphone use, and the smartphone itself, has organizational consequences which raise new methodological demands for research in organizations. It becomes harder to grasp, through ethnographic and practice-based studies, where the action in organizations is taking place. It may not be apparent what meeting participants are doing on a smartphone (which may or may not be work related), and key decision makers may not be in the room. The distributed nature of work requires distributed forms of ethnography (Marcus, 1995). We might have to accept that the researcher may not be at the location of decision making and trace the actions out across time and space (Whyte et al., 2016). Erickson (2018: 301) argues that 'what counts for professional and organizational praxis today cannot be understood without the collection and analysis of trace data'. Thus, to understand organizing practices requires a more explicit consideration of the power and effects of the smartphone and other digital devices and new questions arise about how organization and management can be accomplished.

REFERENCES

Alvesson, Mats and Paul Thompson. 2006. 'Post-Bureaucracy?' In Stephen Ackroyd, Rosemary Batt, Paul Thompson and Pamela S. Tolbert (eds), *The Oxford Handbook of Work and Organization*. Oxford: Oxford University Press, 485–507.

Azad, Bijan, Randa Salamoun, Anita Greenhill, and Trevor Wood-Harper. 2016. 'Performing Projects with Constant Connectivity: Interplay of Consulting Project Work Practices and Smartphone Affordances'. *New Technology, Work and Employment*, 31(1), 4–25.

Brynjolfsson, Erik and Andrew McAfee. 2014. *The Second Machine Age: Work, Progress, and Prosperity in a Time of Brilliant Technologies*. New York: W. W. Norton.

Cecchinato, Marta, Anna L. Cox, and Jon Bird. 2014. '"I check my emails on the toilet": Email Practices and Work–Home Boundary Management'. Presentation at the ACM Conference on Human Factors in Computing Systems (CHI).

Colbert, Amy, Nick Yee, and Gerard George. 2016. 'The Digital Workforce and the Workplace of the Future'. *Academy of Management Journal*, 59(3), 731–9.

Derks, Daantje and Arnold B. Bakker. 2014. 'Smartphone Use, Work–Home Interference, and Burnout: A Diary Study on the Role of Recovery'. *Applied Psychology: An International Review*, 63(3), 411–40.

Derks, Daantje, Arnold B. Bakker, Pascale Peters, and Pauline van Wingerden. 2016. 'Work-Related Smartphone Use, Work–Family Conflict and family Role Performance: The Role of Segmentation Preference'. *Human Relations*, 69(5), 1045–68.

Derks, Daantje, Desiree Duin, Maria Tims, and Arnold B. Bakker. 2015. 'Smartphone Use and Work–Home Interference: The Moderating Role of Social Norms and Employee Work Engagement'. *Journal of Occupational & Organizational Psychology*, 88(1), 155–77.

Derks, Daantje, Lieke L. ten Brummelhuis, Dino Zecic, and Arnold B. Bakker. 2014a. 'Switching On and Off…: Does Smartphone Use Obstruct the Possibility to Engage in Recovery Activities?' *European Journal of Work & Organizational Psychology*, 23(1), 80–90.

Derks, Daantje, Heleen van Mierlo and Elisabeth B. Schmitz. 2014b. 'A Diary Study on Work-Related Smartphone Use, Psychological Detachment and Exhaustion: Examining the Role of the Perceived Segmentation Norm'. *Journal of Occupational Health Psychology*, 19(1), 74–84.

Dery, Kristine, Darl Kolb, and Judith MacCormick. 2014. 'Working with Connective Flow: How Smartphone Use Is Evolving in Practice'. *European Journal of Information Systems*, 23(5), 558–70.

Dolata, Ulrich. 2017. 'Social Movements and the Internet: The Sociotechnical Constitution of Collective Action'. Research Contributions to Organizational Sociology and Innovation Studies SOI Discussion Paper 2017–02.

Dolata, Ulrich and Jan-Felix Schrape. 2018. *Collectivity and Power on the Internet: A Sociological Perspective*. Cham: Springer.

Elhai, Jon D., Robert D. Dvorak, Jason C. Levine, and Brian J. Hall. 2017. 'Problematic Smartphone Use: A Conceptual Overview and Systematic Review of Relations with Anxiety and Depression Psychopathology'. *Journal of Affective Disorders*, 207, 251–9.

Elhai, Jon D., Jason C. Levine, Robert D. Dvorak, and Brian J. Hall. 2016. 'Fear of Missing Out, Need for Touch, Anxiety and Depression Are Related to Problematic Smartphone Use'. *Computers in Human Behavior*, 63, 509–16.

Elias, Ana S. and Rosalind Gill. 2017. 'Beauty Surveillance: The Digital Self-Monitoring Cultures of Neoliberalism'. *European Journal of Cultural Studies*, 21(1), 59–77.

Erickson, Ingrid. 2018. 'Working, Being and Researching in Place: A Mixed Methodological Approach for Understanding Digital Experiences'. In Raza Mir and Sanjay Jain (eds), *The Routledge Companion to Qualitative Research in Organization Studies*. New York: Routledge, 291–305.

Gill, Preetinder S., Ashwini Kamath and Tejkaran S. Gill. 2012. 'Distraction: An Assessment of Smartphone Usage in Health Care Work Settings'. *Risk Management and Healthcare Policy*, 5, 105–14.

Han, Byung-Chul. 2017. *Psychopolitics: Neoliberalism and New Technologies of Power*. New York: Verso.

Hindess, Kathryn. 2018. 'Not So Smart: Inside the Hazardous World of Making Smartphones'. *The Ecologist*, 2 March. https://theecologist.org/2018/mar/02/not-so-smart-inside-hazardous-world-making-smartphones.

Holgate, Peter. 2017. 'The Model for Recycling Our Old Smartphones is Actually Causing Massive Pollution: Millions of New iPhones Will be Sold this Month. What Really Happens to the Millions that Get Thrown Out?' *Recode*. https://www.recode.net/2017/11/8/16621512/where-does-my-smartphone-iphone-8-x-go-recycling-afterlife-toxic-waste-environment.

Jarvenpaa, Sirkka L. and Karl R. Lang. 2005. 'Managing the Paradoxes of Mobile Technology'. *Information Systems Management*, 22(4), 7–23.

Karlson, Amy K., Brian R. Meyers, Andy Jacobs, Paul Johns, and Shaun K. Kane. 2009. 'Working Overtime: Patterns of Smartphone and PC Usage in the Day of an Information Worker'. In H. Tokuda, M. Beigl, A. Friday, A. J. B. Brush, and Y. Tobe (eds), *Pervasive Computing: Lecture Notes in Computer Science*, vol. 5538. Berlin: Springer, 398–405.

Kim, Sooyeol and YoungAh Park. 2017. 'A Daily Investigation of Smartphone Use and Affective Well-Being at Work'. *Academy of Management Annual Meeting Proceedings*, 2017, 1–1.

Lanaj, Klodiana, Russell E. Johnson, and Christopher M. Barnes. 2014. 'Beginning the Workday yet Already Depleted? Consequences of Late-Night Smartphone Use and Sleep'. *Organizational Behavior & Human Decision Processes*, 124(1), 11–23.

Lobo, Sunila and Silvia Elaluf-Calderwood. 2012. 'The BlackBerry Veil: Mobile Use and Privacy Practices by Young Female Saudis'. *Journal of Islamic Marketing*, 3, 190–206.

Marcus, George A. 1995. 'Ethnography in/of the World System: The Emergence of Multi-Sited Ethnography'. *Annual Review of Anthropology*, 24, 95–117.

Middleton, Catherine, Rens Scheepers, and Virpi K. Tuunainen. 2014. 'When Mobile is the Norm: Researching Mobile Information Systems and Mobility as Post-Adoption Phenomena'. *European Journal of Information Systems*, 23(5), 503–12.

Mullan, Killian and Judy Wajcman. 2019. 'Have Mobile Devices Changed Working Patterns in the 21st Century? A Time-Diary Analysis of Work Extension in the UK'. *Work, Employment and Society*, 33(1), 3–20.

Neff, Gina and Dawn Nafus. 2016. *Self-Tracking*. Cambridge, MA: MIT Press.

Neumayer, Christina and Gitte Stald. 2014. 'The Mobile Phone in Street Protest: Texting, Tweeting, Tracking, and Tracing'. *Mobile Media & Communication*, 2(2), 117–33.

Oxford Living Dictionaries. 2017. https://en.oxforddictionaries.com/definition/smartphone [Accessed 6 September 2017].

Perlow, Leslie A. 2012. *Sleeping with Your Smartphone: How to Break the 24/7 Habit and Change the Way You Work*. Boston, MA: Harvard Business Review Press.

Piszczek, Matthew M. 2017. 'Boundary Control and Controlled Boundaries: Organizational Expectations for Technology Use at the Work–Family Interface'. *Journal of Organizational Behavior*, 38(4), 592–611.

Pritchard, Katrina and Gillian Symon. 2014. 'Picture Perfect? Exploring the Use of Smartphone Photography in a Distributed Work Practice'. *Management Learning*, 45(5), 561–76.

Roulet, Thomas J. and Marco Clemente. 2018. 'Let's Open the Media's Black Box: The Media as a Set of Heterogeneous Actors and Not Only as a Homogenous Ensemble'. *Academy of Management Review*, 43(2), 327–9.

Steinert-Threlkeld, Zachary C. 2017. 'Spontaneous Collective Action: Peripheral Mobilization during the Arab Spring'. *American Political Science Review*, 111(2), 379–403.

Symon, Gillian and Katrina Pritchard. 2015. 'Performing the Responsive and Committed Employee through the Sociomaterial Mangle of Connection'. *Organization Studies*, 36(2), 241–63.

Walker, Matthew. 2017. *Why We Sleep: The New Science of Sleep and Dreams*. New York: Scribner.

Ward, Adrian F., Kristen Duke, Ayelet Gneezy, and Maarten W. Bos. 2017. 'Brain Drain: The Mere Presence of One's Own Smartphone Reduces Available Cognitive Capacity'. *Journal of the Association for Consumer Research*, 2(2), 140–54.

Whyte, Jennifer, Kjell Tryggestad, and Alice Comi. 2016. 'Visualizing Practices in Project-Based Design: Tracing Connections through Cascades of Visual Representations'. *Engineering Project Organization Journal*, 6(2–4), 115–28.

Wilmer, Henry H., Lauren E. Sherman, and Jason M. Chein. 2017. 'Smartphones and Cognition: A Review of Research Exploring the Links between Mobile Technology Habits and Cognitive Functioning'. *Frontiers in Psychology*, 8, 605.

Yun, Haejung, William J. Kettinger, and Choong C. Lee. 2012. 'A New Open Door: The Smartphone's Impact on Work-to-Life Conflict, Stress, and Resistance'. *International Journal of Electronic Commerce*, 16(4), 121–52.

CHAPTER 39

··

SUIT

··

BARBARA VINKEN

In the bourgeois era, the suit is the icon of menswear, and its triumph is global. The suit is worn by men around the world. Anne Hollander has rightly called it THE garment of modernity (Hollander, 1994: 113).

'Suit' comes from French 'suite' and refers to a set of garments made of the same cloth and colour, worn together: jacket, trousers, and eventually a waistcoat. It is a formalized garment, standardly worn with a collared shirt and a tie. The jacket can be single breasted—with usually three buttons—or double breasted with two columns of four to six buttons. Each cuff has three to four buttons. It can come unvented, single vented, or double vented. Its lapels come notched, peaked, or in a style called shawl, usually reserved for the dinner jacket. The trousers, flared, bell-buttoned, wide legged or slim can since Edward VII be turned up at the bottom. They eventually have a break and usually two pleats. The classical suit is a highly constructed garment: between the outer fabric and the inner lining of the jacket, there is a sturdy interfacing, called canvas. The suit is all about cut, the high art of bespoke tailoring. Its acid test is the wrinkle. Savile Row has become shorthand for this unwrinkled perfection of the trade: the body has to be smoothly followed in its movements. During the second half of the twentieth and the beginning of the twenty-first century, Italian tailoring became key for unconstructed, slim fitting unvented jackets with shorter, slim leg trousers.

Since the 1980s, the suit has started to lose its universality both in business and civil life. From universal business attire, 'suits' has become a shorthand term for middle and upper corporate management. The technical, the creative, and the academic world has widely abandoned the suit—or turned it upside down. Counter cultures and fashion designers have deconstructed and perverted its message.

The origin of the suit is to be found in British country wear and the clothes of the French third estate. If Fashion begins in the middle of the thirteenth century with men leaving behind their long folds to show lots of leg, the fashion of Modernity begins roughly around the French Revolution, with men concealing their legs in trousers—pantalons. While dress had, until 1789, largely separated society's 'estates'—the nobility from the clergy and the tiers état, and all the three from the peasants—fashion after the

Revolution ceased to divide classes as openly, and instead divided the sexes (Outram, 1989: 156). Since then, the sexes have a different relation to fashion (Vinken, 2005). The spectacular modernization of male clothing consists basically in renouncing all ostentatious display. Access to power, authority, and wealth depends on the de-sexualization of the body. The suit is the means of that. The psychoanalyst John Flügel has wittily paralleled the Great French Revolution with what he called the Great Male Renunciation (Flügel, 1930: 111). Menswear does not carry the stigma of the fashionable. The modern opposition of male and female consists in unmarked vs. marked sexuality. This opposition defines post-revolutionary, modern fashion (Vinken, 2013: 36). Aristocratic display of the body and its erotic play of possibilities became, after the Revolution, the privilege—or burden—of women.

During the Great French Revolution, two parties opposed each other, named after their leg wear: the aristocratic 'culottes' stood against the 'sans-culottes'. The sans-culottes did not go naked, but appeared in a garment that carried the ridicule in its name: the 'pantalon' derives from a figure of the *commedia dell'arte*. Pantalone is a vain, old, intriguing, lusting, and avaricious man, the very opposite of the elegant, generous, amorous courtier, who shows off his legs in tight tricot stockings, fitting like a glove. As a fashion fop Pantalone tries to imitate the courtier, but gets it all wrong: his trousers are not tied under or above the knee, but fall straight down to his shoes. These trousers, concealing the leg, became the basics of modern menswear.

Menswear in the bourgeois era was constituted in a deliberate contrast to the aristocratic fashion that came to its last flourish during the ancien régime. The story of the suit and thus the story of modern male-hood starts with the renunciation to show off—last but not least to show off the body (Kraß, 2006).[1] The clothes worn by the aristocracy displayed masculinity. By adorning a wildly enhanced sex, they highlighted the potency of the male (Wolter, 1988). By fitting like a glove, they showcased a body that can run, ride, fight, chase, and dance—an elegantly disciplined body. The readiness for a phallically connoted violence remains readable in this well-trained body.

The bourgeois of modernity is eminently civil. He does not display a beautiful, capable body, but intellectual capacities. He renounces ostentatious adornment. The bourgeois man does not wear feathers and laces, diamonds, rubies, emeralds, and sapphires; he does not peacock around in the brilliant colours of the rainbow; he does not display powdered lion locks; he does not enter a room in a cloud of perfume.

The bourgeois suit is the negation of the aristocratic attire: less became more (Loos, 1987). The suit manages the paradoxical speech act to cross itself out, to not attract any attention to itself. The suit comes in plain, muted colours: navy, night blue, light grey, anthracite. A lighter grey, brown, beige, and stone are possible, but bottle green remains border line. Black has largely become the colour of mourning and of the evening. The material of the suit has to be of solid colour, lacklustre and monochrome—usually wool—and not patterned or shining. No ornaments. Some traditional, almost

---

[1] In disagreement with this thesis is the strictly empirically oriented British research that does not see any dialectical dynamics here (Breward, 1999).

invisible patterns are allowed—pin stripes, plaids, and checks. Splashes of brighter colour are reserved for the shirt, neck tie, or handkerchief.

The bourgeois is suspicious of everything ornamental and shining that distracts from inner values. Little by little, menswear gets rid of anything surprising, of fantastic frills that single the man out. The waistcoat, made from gaily coloured and patterned damask or brocade, from plaid velvet or embroidered satin, was a remnant of the glorious fashion of bygone days. Around 1835 already, it was supposed to be too showy and was reduced to the tie. The very desire for elegance became inelegant. The bourgeois citizen does not get tired of demonstrating that he does not have to distinguish himself through his clothes. The suit had become second nature to men, who covered up his first nature, the individuality of his flesh (Gautier, 1858). The enforcement of this radical puritanism was no easy game since it is not only less amusing, but surely more difficult to be dressed correctly than to be dressed elegantly. To be dressed correctly means now above all not to be dressed ostentatiously. The art of the artless is difficult to master. Distinction, in men's fashion, is insider knowledge; you have to be able to discern the smallest differences. In male fashion, less is more and form follows function.

The suit is only barely and intermittently subject to the fashion cycle. While the female silhouette has changed radically over the last two hundred years, the male silhouette has evinced an astonishing, one might even say a *classical* constancy. The classic men's suit is worn equally in public space, in the City, and in the private workplace, the office. The only alternative was, at least in rural areas, 'traditional costume'. Even 'evening dress' for men is laid out clearly as the choice between tuxedo or tails, white tie or black tie (Figure 39.1).

Those who put their body on display are now, with the emerging figure of the 'homosexual' (Foucault, 1976), easily tainted as fops or ladies' men. The bourgeois does not need ostentatious display, which seen as effeminate. He does not have to present, he simply *is*. Real men, sober and anti-theatrical, put their effort in the effortless. To dress as a man is an art one has to master. The question that accompanies menswear since, roughly, the Revolution, is not that of elegance or beauty, but to be dressed correctly. Distinction consists in not attracting the eye. The suit should underline the individuality, but sublate the sexual body.

The suit is an incarnation of civic, Stoic–Christian virtues the citizen should embody: *continentia, modestia, abstinentia.* The unadorned soberness, the disciplined austerity, the appearance of only the 'personality' in its unvarnished truth replicates the bourgeois ethics. The *constantia* of the person, who does not float with the tide, is highlighted by the *constantia* of the suit that varies only minimally. The suit constitutes the bourgeois man as an authentic being in opposition to the aristocrat who is cast by bourgeois discourse as somebody indulging in mere appearances, in empty but dangerous frivolities. The suit, to put it with Roland Barthes, does not connote arbitrarily changing artificiality—i.e., fashion—but its strict functionality without any ado (Barthes, 1976).

The corporate identity of the body politic and other corporations can only occur if each man's unique, eccentric body becomes invisible. The suit fits loosely and does not cling to the body. Its muted colours are a rebuke to vivid aristocratic hues, and its

FIGURE 39.1 Philip William May, 1864–1903, British, 'Jack' Millage, undated, black wash and graphite with pen and black ink on very thick, smooth, cream wove paper

(*Source*: Yale Center for British Art, Gift of Andrew Wilton)

material—wool, linen, cotton—negates the gleaming silks, the delicate lace, and the ostentatious furs of the nobility. The modern, classic suit is not fitted closely to the body, but idealized on the V form of statuary, with narrow hips and broad shoulders. Buttocks and genitals are covered by the suit jacket. There is no gaping of fabric between bare skin and garment, and with the exception of hands and face, all skin is covered. 'Love handles' and other imperfections of the human form are smoothed out, carnality de-materialized. No thighs swell under tightly clinging fabric. The ideal suit follows the movements of the body without losing its defining, idealizing function. Variation is quite minimal. The classic suit is largely external to the fashion cycle. Men are not 'fashion-conscious', or, in the more recent idiom: 'trendy'. Once the Republic had been declared and the New Era had begun, there was no need for constant change.

In 1878, Theodor Friedrich Vischer highlighted the radical change in male fashion at the beginning of the nineteenth century, the change from an aristocratic order of estates to a bourgeois order of classes:

> The male dress shouldn't say anything for itself; rather the man should bring to the fore his allure, his face, his words and deeds, his personality... Our grandfathers

thought it the most natural thing in the word that one person would distinguish himself by wearing a red frock lined with a golden ribbon and blue stockings, and the other by wearing a green one lined with silver, and peach-coloured stockings. We, blasé against such bathos, have finished off with all that. We just have a tired smile if somebody wants to distinguish himself through anything else but himself.... Although this disillusioned soberness of the male dress is not even half a century old, we could still say that it incarnates the very character of modern fashion, once it has become what by its very nature it was meant to be.

(Vischer, 1986: 63)

Under the header 'fashion and modernity', Friedrich Nietzsche promoted the suit as the very emblem of the fashion of the Modern: the suit fashions Modernity. In the collection of aphorisms *Human, All Too Human—A Book for Free Spirits* Nietzsche managed this improbable collusion of fashion and suit by turning upside down what till then was understood as fashion. According to Nietzsche, the suit expresses the virtues of a modern, industrialized, enlightened Europe. Fashion—and by fashion Nietzsche means the suit—does not divide and separate, it rather equalizes and unifies. Fashion's true character is not seasonal change, but constancy: 'On the whole, therefore, it is not change that will characterize fashion and the modern, for change is a sign of backwardness and that the men and women of Europe are still immature: what will characterize it is repudiation of national, class and individual vanity' (Nietzsche, 1986: 364). While women and some idle young men, namely dandies, still strived to distinguish themselves by means of their clothes, European men had in their vast majority already reached Modernity: they were unified by means of their clothes, the suit. The fickle change in fashion—and change is, after all, what commonly defines fashion—will come to an end, if dandies and women finally grow up to become mature Europeans. To put it differently: it is the indifference towards all things commonly thought of as fashionable that distinguishes. The suit performs the paradoxical speech act to express through dress the indifference towards dress. The mature European, obviously an intellectual, shows in the way he dresses that he is 'industrious and has little time for dressing and self-adornment, likewise that he finds that everything costly and luxurious in material and design accords ill with the work he has to do; finally, that through his costume he indicates the more learned and intellectual callings as those to which as a European he stands closest or would like to stand closest' (Nietzsche, 1986: 363).

The suit, the only dress that corresponds to the norms of the Modern, is defined by not being stigmatized as fashionable. By making himself almost invisible, it stresses the personality only. It precludes the vain desire to distinguish oneself through lavish display, which is now associated with all things effeminate, aristocratic, decadent. The suit sublimates the sexualized flesh into character and personality, and constitutes a collective of equals. This suspension (*Aufhebung*) of the flesh into a body politic or a corporate identity is the very condition of access to the public sphere of a democratic republic.

But from the very beginning, the suit did not only have fans. One of its most prominent adversaries was the philosopher Georg Wilhelm Friedrich Hegel. Graceless, the suit is a straitjacket for Hegel. He described the limbs as 'stretched out sacks with stiff folds'. The suit was for Hegel:

> something produced for an external purpose, something cut, sewn together here, folded over there, elsewhere fixed, and, in short, purely unfree forms, with folds and surfaces positioned here and there by seams, buttons, and button-holes. In other words, such clothing is in fact just a covering and a veil which throughout lacks any form of its own but, in the organic formation of the limbs which it follows in general, precisely conceals what is visibly beautiful, namely their living swelling and curving, and substitutes for them the visible appearance of a material mechanically fashioned. This is what is entirely inartistic about modern clothes.
>
> (Hegel, 1975: 746)

Torn, the suit disfigures the natural beauty of our limbs. The dress does not follow the body: the body has to follow the dress. Men, forced to adapt their movements to the stiffly uncomfortable garment, are turned into mechanical puppets by wearing the suit. The sociologist Edmond Goblot also deplored the gracelessness of the suit: a woman who likes the suit is, according to Goblot, ready to give up on male beauty for the advantage of class and standing that the suit signals (Goblot, 2010: 41). She is into the man of means.

For fashion scholar Anne Hollander, the suit is, on the contrary, not only the most modern, but at the same time the sexiest fashion item ever. It turns every man into an antique hero, into a tiger, ready to jump; the suit teases out the lurking, velvety elegant eros. But such panegyrics are an exception.

By not singling the man out, the suit has as important a political function as the ostentatiously adorned clothes of the aristocracy—although an altogether different politics is at stake. The suit levels, equalizes, neutralizes, in short, de-sensualizes. The dark, simple suit of the post-revolutionary bourgeoisie creates the body politic, creates corporations of all sorts, establishes corporate, civil identity. The suit is the uniform of Republican democracies and their only legitimate expression. The suit transcends the way of all flesh into the constancy of the organized institution. It is the suit that sublates the particular body into a collective one. Day in, day out, bourgeois men stage the spectacle of the unspectacular. With lots of rhetorical effort, whose main task resists its invisibility, men display authentic a-rhetoricity. If the Republic is often represented through female allegories—Britannia, Marianne, Bavaria—the dominant institutions and organizations are male collectives. All the corporations of the modern states, their body politics, their administrations, their courts, their armies, their universities, their guilds and professions, were exclusively and are predominantly male. The modern states thus translate the theory of the two bodies of the king from the single person of the king to their collective institutions (Kantorowicz, 1998). The bourgeois male secularized body unites also two bodies in one. It is the private body of an individual as well as part of a body

FIGURE 39.2 Council of Europe 2016

(*Source*: https://www.consilium.europa.eu/en/media-galleries/european-council/
meetings/2016-06-28-29-euco/ © European Union, 2017)

corporate (Weinelt, 2016). He is more than and not only the concrete person that holds
an office. The organization outlives the person who holds the office. The second body is
not situated in a transcendent sphere as with the king's divine right, but guarantees the
constancy of a body corporate within history beyond the individual who holds the
office. The suit thus performs the permanence of the republican, democratic institutions
beyond the individual. Officials (*Funktionsträger*) do not have to be dressed beautifully,
but correctly for the office they hold. To fit in, not to stand out, is the goal. The suit has
therefore been called a civil uniform: the Washington uniform of the politicians, the
Frankfurt uniform of the bankers. To perform corporate identity, to perform the body
politic, the garment you are wearing should not point to your flesh, and thus to your
death, but make everybody look alike with only slight variations (Figure 39.2). The suit
should not meet the eye and attract attention to itself: in order to be effective, it has to be
overlooked. The suit is the ideal garment to perform this speech act. Such is its organizational power. It articulates through the clothes that clothes, mere superficialities, are not
what you are about. It takes a lot of artful techniques and know-how to make the suit
say that.

From the very beginning, this speech act that informs modern male-hood was contested—by appropriation and deconstruction of the suit. First came the dandies, who
were defined by Carlyle as 'clothes wearing men' (Carlyle, 1869: 253; Garelick, 1998).
Furthermore, Paul de Saint Victor called the dandy a 'black prince of elegance'[2] and
thereby expressed Baudelaire's dandyism best. Nothing could be more spectacular than
the way they looked. The idea that they could not care less about how they dressed would
not have occurred to anybody, their understatement was too ostentatious. The long line

[2] Paul de Saint-Victor, following an article in *La Presse*, 21 August 1859.

of the movements that countered the ideology of the suit cannot be traced here in its entirety, so we have to stick to a few examples.

The Zoots Suits just after the war, worn first by black men, overstressed the understatement of the suit and thereby undid the very speech act the suit was meant to perform. The Mods, of course, who were mostly blue collars wearing the suit, fitted it to the body so that there was clearly something *very* wrong with their way of wearing it. David Bowie wore the suit as a woman would her feathered cocktail dress and also undid the suit's speech act. The Gentlemen of Bacongo, finally, the latest incarnation of the dandy, appropriated the suit of the colonizers and turned it into the most spectacular garment ever (Gandoulou, 1989; Tamagni, 2009; Loreck, 2011).[3] By overstressing any single rule of the suit, they stood out.

Finally, the designers have deconstructed the suit by bringing back the singular, vulnerable, erotic, beautiful body that does not merge into a body politic. It was Pierre Cardin who started it all with his pencil suits that fitted the body like a dancer's ballet outfit. Armani took out the canvas in his unconstructed jackets that followed the body closely and looked more like a pullover than like the properly constructed suit jacket. Helmut Lang continued the stressing of the erotic, individual, and mortal body within the suit, undoing its required sublation. It was Hedi Slimane for Dior, who introduced techniques of female haute couture to make the suit fit like a glove. Gucci featured fabrics that reminded us of the gobelins of the ancien régime: English roses exploded in all their gorgeousness. They certainly had nothing in common with the classically 'cool wool' materials. The play between skin and fabric, traditionally the hallmark of female fashion, invaded the male suit. Finally, Vuitton for the 2018 summer collection brought all the ostentatious glamour of the ancien régime back to the suit version 'Versailles'. Might the bourgeois, republican speech act of the suit and its organizational force simply be undone by these counter-suits? They single the men out in their strength, but also in their mortal vulnerability, and turn them into heroic, effeminate birds of paradise. These bodies cannot be reintegrated into the republican body politic.

REFERENCES

Barthes, Roland. 1976. *Système de la Mode*. Paris: Éditions du Seuil.
Breward, Christopher. 1999. *The Hidden Consumer: Masculinities, Fashion and City Life 1860–1914*. Manchester: Manchester University Press.
Carlyle, Thomas. 1869. *Sartor Resartus: The Life and Opinions of Herr Teufelsdröckh*. London: Chapman & Hall.
Flügel, John Carl. 1930. *The Psychology of Clothes*. London: Hogarth Press.
Foucault, Michel. 1976. *Histoire de la sexualité I: La volonté de savoir*. Paris: Gallimard.
Gandoulou, Justin-Daniel. 1989. *Dandies à Bacongo. Le culte de l'élégance dans la société contemporaine congolaise*. Paris: L'Harmattan.
Garelick, Rhonda. 1998. *Rising Star: Dandyism, Gender, and Performance in the Fin de Siècle*. Princeton, NJ: Princeton University Press.

[3] For a fictional treatment of this phenomenon see Mabanckou (2010).

Gautier, Théophile. 1858. *De la Mode*. Paris: Poulet-Malassis et De Broise.

Goblot, Edmond. 2010. 'La barrière et le niveau. Étude sociologique sur la bourgeoisie française moderne' [1925]. In *Penser la mode*, ed. Frédéric Godart. Paris: Éditions du regard.

Hegel, Georg Wilhelm Friedrich. 1975. *Aesthetics: Lectures on Fine Arts*, trans. T. M. Knox. London: Oxford University Press.

Hollander, Anne. 1994. *Sex and Suits: The Evolution of Modern Dress*. New York: Alfred Knopf.

Kantorowicz, Ernst H. 1998. *Götter in Uniform—Studien zur Entwicklung des abendländischen Königtums*, ed. Eckhart Grünewald and Ulrich Raulff. Stuttgart: Klett-Cotta.

Kraß, Andreas. 2006. 'Das Geschlecht der Mode'. In Gertrud Lehnert (ed.), *Die Kunst der Mode*. Oldenburg: Deutscher Buchverlag, 26–51.

Loos, Adolf. 1987. *Spoken into the Void: Collected Essays, 1897–1900*. Cambridge, MA: MIT Press.

Loreck, Hanne. 2011. 'La Sape. Eine Fallstudie zu Mode und Sichtbarkeit im postkolonialen Kontext'. In Katharina Knüttel and Martin Seeliger (eds), *Intersektionalität und Kulturindustrie. Zum Verhältnis sozialer Kategorien und kultureller Repräsentation*. Bielefeld: Transcript, 259–82.

Mabanckou, Alain. 2010. *Black Bazar*. München: Liebeskind.

Nietzsche, Friedrich. 1986. *Human, All Too Human*, trans. R. J. Hollingdale. Cambridge: Cambridge University Press.

Outram, Dorinda. 1989. *The Body and the French Revolution: Sex, Class and Political Culture*. New Haven, CT: Yale University Press.

Tamagni, Daniele. 2009. *Gentlemen of Bacongo*. London: Trolley Books.

Vinken, Barbara. 2005. *Fashion—Zeitgeist: Trends and Cycles in the Fashion System*. Oxford: Berg.

Vinken, Barbara. 2013. *Angezogen. Das Geheimnis der Mode*. Stuttgart: Klett-Cotta.

Vischer, Friedrich Theodor. 1986. 'Mode und Zynismus.' In Silvia Bovenschen (ed.), *Die Listen der Mode*. Frankfurt am Main: Suhrkamp, 33–79.

Weinelt, Nora. 2016. *Minimale Männlichkeit. Figurationen und Refigurationen des Anzugs*. Berlin: Neofelis Verlag.

Wolter, Gundula. 1988. *Die Verpackung des männlichen Geschlechtes. Eine illustrierte Kulturgeschichte der Hose*. Marburg: Jonas.

CHAPTER 40

....

TELEGRAPH

....

MIKKEL FLYVERBOM AND ANDERS KOED MADSEN

INTRODUCTION

ON 1 May 1844, the Whig Party assembled in Baltimore to nominate its presidential candidate at their national convention. The announcement of the winning nominee was eagerly awaited by political pundits and bureaucrats in Washington. However, this year the choice of Henry Clay as presidential candidate did not travel by train or horse to reach the political establishment. Rather, the newly established electrical telegraph wires connecting Baltimore to the Capital transmitted the news with a previously unseen speed. With this demonstration of the capabilities of electrical communication, a new pace of information flows had entered American politics (Carpenter, 1892).

The telegraph, like other earlier media technologies, still deserves our attention. This magic invention was a key moment in the history of technological innovation—the telegraph effectively detached transportation and communication and made it possible for messages to fly through air and wires and across space and time. Like contemporary technological developments, this spurred reflections about the relationship between innovation and social order, and communication and power, concerns that continue to be played out today as we inquire into Internet and social media organizations and their broader societal role and impact.

A SHORT, MATERIAL HISTORY

Talking about *the* telegraph in the singular is in many ways misleading because, in the 19th century, different types of telegraphs—or technologies for writing at a distance—were competing to be the preferred standard (see Figures 40.1 and 40.2). The first telegraph was optical—it circulated information through visual signals between towers

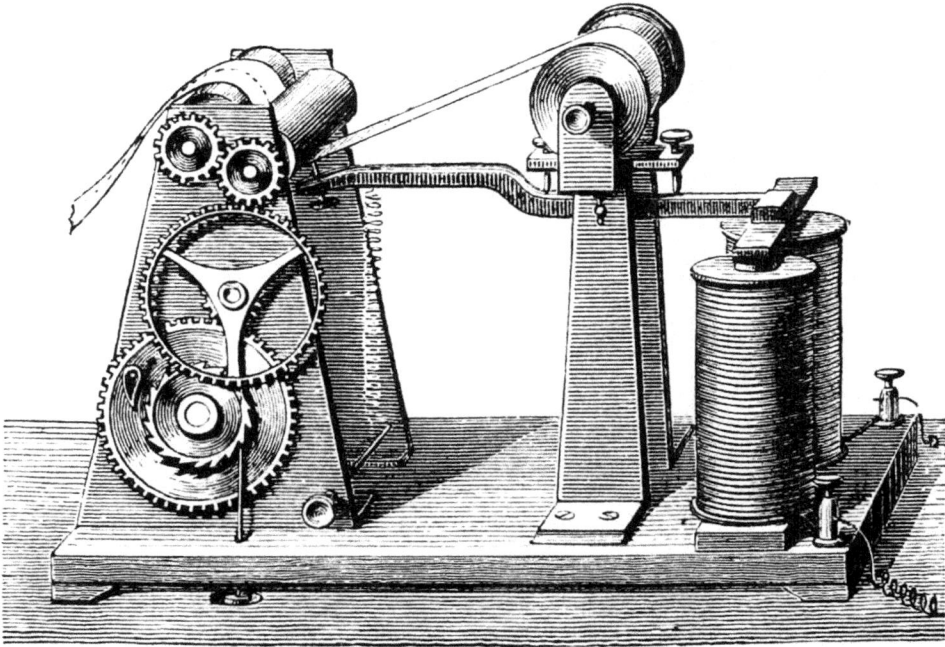

FIGURE 40.1 This version of the receiving end of Morse's electrical telegraph consisted of an electro-magnet, a movable lever, and pencil that wrote code on a paper roll

(*Source*: https://commons.wikimedia.org/wiki/File:PSM_V03_D423_Morse_telegraph.jpg)

on hills—and was invented in France in the 1790s as part of Napoleon's military complex and the restructuring of France after the revolution. The optical telegraph could carry signals between towers if they were less than twenty miles apart and the number of towers, along with weather conditions, set limits to the distance that messages could travel.

Optical telegraphs were put to use in both Great Britain and United States, but each of these countries eventually embarked on a different technological trajectory by using electricity to circulate information. The *electric* version of the telegraph was first prototyped in the 1820s, turned into a workable technology in the 1830s and ultimately patented by Samuel Morse in 1840. A crucial demonstration of this version of the telegraph appeared in May 1844 when Samuel Morse sent the famous phrase, 'What hath God wrought?' through electrical signals from Baltimore to Washington on a wire funded by Congress.

The electrical telegraph soon became the preferred standard over the optical telegraph, also in France. Compared to its optical counterpart it ensured fast circulation of, for example, stock prices even in bad weather conditions. Also, whereas the optical telegraph communicated at phrase-level, the so-called Morse code worked on letter-level (John, 2010). This meant communication was not restricted to phrases pre-defined by 'men in power'. Any combination of words could be circulated, and the Morse code is

FIGURE 40.2 Contrary to the electrical version of the telegraph, the optical variant worked by circulating visual signals through the air between towers on hills

(*Source*: ITU Pictures/flickr.com)

an obvious precursor to the binary language used in digital computers and networks of our age (Balbi and John, 2015). Furthermore, the electrical telegraph made the circulation of information more visible because its receivers could leave permanent records of the signals going through it. It was also a technology enabling more secure flows of information because communication through wires was harder to intercept and hack than visual signals.

At the birth of the electrical telegraph, it was an open question as to who was to control this new system. But during the nineteenth century it was effectively turned into a private enterprise. In Great Britain, venture capitalists quickly bought huge stakes in the invention, which became a crucial component in the unification and spread of the British Empire. By 1851 it was set up as a public service, including a money order system (Thussu, 2000). In the United States, privatization was more ambivalent. Morse called for federal control of the technology and the Congress-funded demonstration line was originally transferred to the Post Office Department as a fee-based service. However, the telegraph was not profitable and federal interest in controlling the system faded. In 1845, Congress made the decision not to buy Morse's patents and subsequently leased the demonstration line to private investors.

During the 1850s more than fifty telegraph companies existed in the USA, which resulted in a chaotic industry often troubled by poorly constructed lines. In 1852, the USA had more than 23,000 miles of wires owned by a multitude of investors (Czitrom, 1982). However, in the 1860s, Western Union emerged as *the* most powerful telegraph company. As a consequence, the number of standards and different types of telegraphs were reduced, thus facilitating more efficient transmission (Scholte, 2005). The position of Western Union was partly ensured by the US civil war (1861–9) where fresh information was in high demand. In 1861, Western Union opened the first transcontinental telegraph line, in 1865 a telegraph line to Russia, in 1866 a stock-ticker to transmit stock prices, and in 1871 a money transfer service built on top of the telegraph. In 1881, Jay Gould—a prominent railroad investor—acquired major parts of the company.

In a global perspective, the telegraph network also expanded during the latter part of the nineteenth century. The first underwater cable linking Britain and France was established in 1851, and the first reliable transatlantic cable connecting Britain and the USA was put in operation in 1866. Five years later, Australia, Japan, China, and Europe were connected by telegraph cables (Scholte, 2005). However, using this global information network was often quite expensive: a message between Australia and Europe took days to arrive and cost £10 for 20 words, five times the average weekly wage at the time (Scholte, 2005). Telegraphy was possible, but it was not a technology for the average citizen. In the United States, the telegraph remained a specialist medium used by merchants and journalists and was not turned into a mass service until the 1910s.

In sum, it seems fair to say that the material development of the telegraph afforded new forms of social ordering. Its invention opened questions about how to organize societies, publics, and markets in the face of technological transformations.

A CHICAGO SCHOOL READING
OF THE TELEGRAPH

One way to learn about the way such connections between media, technology, and social order were formed, is to study how social theorists of the time thought about this relation. This approach starts from the assumption that we cannot separate ways of thinking about the telegraph from the modes of social ordering afforded by this technological innovation. Such insights are well-developed in scholarly work in Science and Technology Studies and related approaches focusing on socio-material assemblages (Latour, 2005; Hackett et al., 2007) and socio-technical imaginaries (Jasanoff and Kim, 2015). Along these lines, we set out to provide a material and ideational history of the telegraph and to articulate how the social ordering afforded by the telegraph was also connected to the way it was theorized, discussed, and imagined.

Reading accounts from a time where the organizational order around the telegraph was still undecided can provide insights into the imaginations that technologies spur when they are still new. Also, it can make us aware of the potential blind spots that can arise from theorizing about technologies while they develop. In the case of the telegraph, there is no better source to such thoughts and imaginations than the writings of the so-called Chicago School.

The Chicago School was not really a school, but a name that William James in 1904 gave to the work of John Dewey and other sociologists who shared his interest in understanding how human and social consciousness is shaped by the way organisms experience the world. Writing in the aftermath of both the invention of the telegraph and Darwin's evolutionary theory, the Chicago theorists believed that the quality of experiences—and thereby consciousness—were conditioned by the way communication technology transmitted signals from the social environment. As put by Charles Horton Cooley:

> A man's social environment embraces all persons with whom he has intelligence or sympathy, all influences that reach him...the existing system of communication determines the reach of the environment.   (Cooley, 1897: 73–4)

The Chicago theorists saw intelligence as both organic and material and they thought of electrically wired communication as a form of selection mechanism that enabled new forms of social organization. Their central point was that selection was not necessarily located in an individual consciousness but rather in a 'larger mind' (Cooley, 1909) that gets more and more intelligent when it is capable of making useful selections of symbols and thoughts:

> [By] fixing certain thoughts at the expense of others, [the] system of communication is a tool, a progressive invention, whose improvements react upon mankind and alter the life of every individual and institution.   (Cooley, 1909: 64)

Dewey and Cooley were fascinated by the speed with which signalling and selection could occur when mediated through electricity. Whereas selection of information had previously been rooted in the local community, the pace of the telegraph—they thought—would enable people to have instant awareness of the externalities of their actions. This would make it possible for people to adapt to each other's actions in more intelligent ways. The Chicago theorists saw such adaption as an unprecedented freedom of the mind:

> Communication must be full and quick in order to give that promptness in the give-and-take of suggestions upon which moral unity depends...the recent marvellous improvements in communicative machinery makes a free mind on a great scale even possible.   (Cooley, 1909: 54)

> [I believe that] the intellectual forces which have been gathering since the Renascence and the Reformation, shall demand complete free movement and, by getting the physical leverage in the telegraph and the printing press, [they] shall, through

free inquiry in a centralized way, demand the authority of all other so-called authorities.   (Dewey quoted in Czitrom, 1982: 106)

In these quotes, we see a tendency to deduce social and moral order from the material features of communication technologies such as the telegraph. The telegraph was theorized as an instrument for a new form of democratically organized intelligence that would push democracy beyond expert rule, i.e., as a force of social progress. This way of linking technology, organization, and democratic progress was not foreign to the times. In fact, organic metaphors about communication as well as analogies between electricity and intelligent nervous systems were widely used to describe the potential effects of electricity on social order (Marvin, 1990). An example comes from Morse himself who in 1838 anticipated later talks about the 'global village':

> ...the whole surface of this country would be channelled for those nerves which are to diffuse, with the speed of thought, a knowledge of all that is occurring throughout the land; making, in fact, one neighborhood of the whole country
> (Morse quoted in Czitrom, 1982: 11–12)

Reading Dewey and Cooley's thoughts on electric communication illustrates how connections between technology and social order are inevitably interpreted from within specific cultural traditions. Seeing communication technology as a force of progress was not far-fetched in nations built around the postal system. Embracing the possibilities of quick transmission of information over vast distances and seeing it as a source of intelligence for a population that was growing too large to organize itself through physical meetings made a lot of sense considering the organizational challenges facing a young nation (John, 2010).

In a sense, the work of the Chicago School can be read as a response to a challenge that was already formulated by Benjamin Franklin when he signed the constitution. When asked what the constitution gave the American people, his answer was 'A Republic—if you can keep it'. Formulating a philosophy of an electrically wired nation with new infrastructures for symbolic interaction was a priority for the Chicago School. However, rather than accepting their progressive philosophy of technology, we can productively use the blind spots in their thinking to sharpen our contemporary engagements with media, technology, and organization.

The Telegraph and its Present-Day Relevance

As Carey (2009) reminds us, the telegraph did many things—much more than what was written about in Chicago. It led to a major patent struggle, it marked the beginning of the electrical goods industry, it spurred the development of a modern, 'objective' news industry, it brought about a new language—the Morse code—and it separated communication from transportation, which was 'a watershed in communication' (Carey, 2009: 156). The telegraph also predates contemporary issues in telecommunications, such as the development of—and struggles over—standards and interoperability (Gasser and Palfrey, 2012) and the intersections between military, financial, and technological forces.

Most importantly, the material and ideational history of the telegraph reminds us of the tendency to conflate technological and social transformations, that is, to see technologies as drivers of progressive social change. This argument was often voiced without much material qualification by writers like Cooley. Extending these ideas, we suggest that the telegraph offers a number of insights about the relationship between technologies and social ordering, which have present-day relevance. More specifically, the telegraph allows us to highlight three issues of relevance to contemporary concerns about media, technology, and social transformation.

Media Technologies as Reconfiguring Time and Space

As the first technology to separate communication from physical transportation—and thereby transform its temporal and spatial properties—the telegraph in many ways signals the beginning of globalization and global media (Scholte, 2005; Thussu, 2000). Before the telegraph, communication had to travel by roads, sea, or train tracks (see De Cock, this volume). As distribution and interaction became able to travel without the limitations and constraints that physical objects usually have, it became possible to share thoughts, send orders, and construct collective realities across distance. These developments not only fascinated the Chicago theorists, but also predate what we think of as the virtual organization today, and created the foundations for distributed forms of work and outsourcing.

However, it is crucial to remember that such reconfigurations of time and space always are shaped by existing spatial and temporal organization. The social order is always already configured and changes must be discussed on this background. For instance, in the United States—until the advent of the telegraph—the federal Post Office Department controlled the speed of information transfer. This ensured a more or less level playing field when it came to the transmission of important information such as market prices on crops and news about political affairs. Even though 'private expresses' that carried news faster than the Post Office were in operation, the nation was organized in such a way that individual merchants, journalists, and politicians all relied on the same infrastructure to transmit the intelligence they based their decisions upon. This changed with the privatization of the telegraph—it opened for the possibility that wealthy individuals could obtain 'intelligence' faster than others. Such questions are also on our mind today when we consider whether 'net neutrality' is under threat, giving service providers the option to let some digital information travel faster than others.

When it comes to the relations between the telegraph and the organization of the market, it was a concern that the provision of intelligence through private wires increased the risk of insider trading in the sense that buyers and sellers might not have synchronous access to the same information about the goods they were trading. For instance, a seller might get knowledge about prices on agricultural staples in overseas markets faster than buyers (John, 2010: 21). The competent merchant was therefore a person who knew *when* to buy or sell. This was radically different from the previous competence of knowing *where* to buy and sell. In the so-called arbitrage-trading system, you could make profit by moving goods around—buying them on one city and selling them in another where price was higher because of local conditions. As information through the telegraph evened out such local market differences, it moved trading from a focus on space to a focus on time (Carey, 2009).

In the context of news, the increased speed of information spurred an interest in 'live reporting' that shifted journalistic priorities from editorial opinion pieces to fact-based corresponding from the field. New professional roles—such as the 'telegraph reporter'—emerged (Czitrom, 1982: 15–19) with the result that the function of news-gathering was no longer performed by the local party press. This meant that the criteria for relevant news became less local and more standardized. Genres such as situated hoaxes and ironic commentaries gave way to language that could work across the wire system. Also, the costs of sending words through the wire made reporting an exercise in moderation and in many ways separated observation of facts from the actual writing of prose (Carey, 2009: 162). The way new media technologies enabled novel genres of news to arise is also of huge importance to current discussions about the way platforms like Facebook and YouTube enable producers of 'fake news' to do mash-ups of invented content and graphical layout of well-known newspapers (e.g., using the logo of the *Financial Times* to circulate fake stories).

In the private sphere, the reconfiguration of time and space opened for new types of impersonal interactions such as 'telegraph-weddings' where marriage between a bride and a groom could be simultaneously witnessed by people located elsewhere. This made it possible to be more selective about the circle of guests attending a ceremony. It even made it possible for a priest to carry out a ceremony despite the fact that neither bride, groom, nor priest were located in the same place. In fact, such a wedding was annulled in court in 1883 after the wife learned that her new husband was 'a colored man' (Marvin, 1990). The importance of physical co-presence was challenged in a community that had previously organized itself around physical meeting-points such as the church and the school.

These examples from the contexts of markets, news, and private relations illustrate that the telegraph reconfigured time and space rather than 'annihilated' these dimensions. The configuration of time and space was still of huge importance—and to a large extent ambivalent—even after the technology was accepted as standard. This ambivalence was especially clear in the suggestion to coordinate telegraphic scheduling of trains through the introduction of standardized time. Before 1883 the railroads in North America were coordinated though fifty-eight local time zones keyed to the largest cities. The introduction of standard 'railroad time' in 1883 reduced these to four zones in order to optimize the flow of transportation (see Gregg and Kneese, this volume). However, this also meant that religious practices established with reference to local movements of the sun were challenged as organizing principles for everyday life. Time was not annihilated—it was reorganized in an ambivalent way. Accordingly, Cooley's focus on the promises of reach and speed above seems in hindsight one-sided. However, this one-sidedness is a good reminder for present analyses of media, technology, and organization to prioritize inquiries into opportunities and challenges arising from the way contemporary media technologies reconfigure time and space. The telegraph not only allowed for the transmission of information, but was central to wide-reaching transformations having to do with time, space, perception, and more structural aspects of social life. Through such processes of institutionalization, the telegraph—and present-day technologies—come to shape processes of social ordering.

Media Technologies as Metaphors for Society

In both the writings of the Chicago School and statements from Morse himself, we have seen that the telegraph raised hopes about connections and integration, and about technologically enabled unity. Humans may always have hoped and expected that new tools and technological innovation would solve problems and reconfigure society for the better. If we create a direct link between technological and societal transformation, we start to look for moments where technological features are cast as societal or organizational features. These are situations where technological possibilities are translated into social imaginations and metaphors for the good and just society. In the context of the telegraph such translations were made in several different ways.

First, Morse himself made explicit connections between the material design of the electrical telegraph and a distinctive American mode of societal organization. Whereas he saw the optical telegraph as a monarchical technology, he believed the electrical telegraph to be more republican. One reason was that it could leave traces of signals, which enabled a bureaucratic organization that ensured checks and balances on decisions. Another reason was the flexibility in the code language, which made it possible to formulate news and other messages in the words of the transmitter rather than restricting it to a predefined book of phrases. Both of these reasons frame the electric telegraph as more democratic because it affords more transparency and free speech then the optical variant. As the late nineteenth century was also marked by a belief in the genius to invent technologies for moral progress, such proposed links between technology and social order by an 'inventor' were important. (John, 2010: 44–5).

Second, imaginaries about the features of electricity as a natural force were also translated into beliefs about this medium could—while tamed in technologies like the telegraph—be a social force. Such imaginaries were deeply shaped by religious metaphors in both positive and negatives ways. Taming electricity would—some argued—enable a new form of society where, for example, processes of work were not limited by the setting of the sun. There was a quasi-religious faith in the possibilities of electricity to overcome natural constrains (Carey, 2009) and ultimately provide possibilities for telepathy and electro-therapy (Simon, 2005). On the other hand, the advent of electricity also generated anxieties and there was a public scepticism towards introducing this strange force into their private homes. In 1909 when Cooley wrote about social organization, only 10 per cent of American homes were wired and had simple technologies such as light bulbs (Simon, 2005).

More recent accounts of the consequences of digital transformations also reproduce such direct linkages between technological and social forms. One example is how the network form inherent in digital technologies such as the Internet is interpreted as the driver of networked organizations and network societies (Castells, 2000). Such technology-centric accounts of social ordering remain widespread today, such as when people and organizations dream about fast computers and data-sharing as the drivers of innovation, collaboration, and progress.

Media Technologies and Monopolization

The history of the telegraph also reminds us of the importance of *the emergence of monopolies*. We may have become used to the idea that technologies require regulation and coordination, but at the time of the telegraph, such links were not really in place. In terms of monopolization, the telegraph is notable because it shifted from a federal project to a private monopoly within few decades. This was especially the case in the United States where Congress funded the first demonstration project whereas the later partnership of Western Union and the Associated Press led to a *de facto* monopoly on news dissemination. Investigations by the US Senate documented that censorship of critical accounts of the telegraph and aggressive price setting made it difficult for new publishers to participate in the dissemination of news (Czitrom, 1982: 26–7).

One explanation of this development is that anti-monopoly movements were slow to grasp the telegraph as an important issue. For instance, advocates behind the crucial 'cheap postage legislations' of 1845 and 1851 never saw the telegraph as important as mail when it came to ensuring free public communication (John, 2010). There were virtually no petitions on cheap telegraphy until 1875 and the speed of signals was not a major public concern. Only around the time when Guild acquired the telegraph in 1881 did anti-monopolization movements focus on the telegraph (Czitrom, 1982: 21). When the public arrived with its concerns, it was too late. The wires were under private control. It seems that once infrastructures are in place, they become difficult to question and alter—an observation that ought to guide us in present-day discussions about the dominance of Internet companies and the sorts of social infrastructures they strive to become (see Ridgway, this volume).

Comparing this with the writings of the Chicago theorists, it is hard to recognize, for example, Cooley's dreams about a free mind and organic selection mechanisms in organizational settings. Rather than a means of public communication, the telegraph was most of all a flow of signals controlled by a few powerful organizations.

This was something that was not discussed much in Chicago. On the other hand, both Dewey and Cooley formulated theories about the democratic potentials in high-speed information—something that the existing public did not care much about until quite late.

Looking at today's media environment, we see monopolization developments with the emergence and consolidation of Internet behemoths such as Google and Facebook. In many cases these companies are able to lock down and dominate entire domains of society, such as searching for information, staying in touch with friends and family, or selling advertisements. Like the telegraph the infrastructures spawned by these companies in many ways represent a move towards becoming the primary platforms of communication. The story of the telegraph thus pre-empts present-day developments such the monopoly and anti-trust cases against Internet giants that buy up and otherwise seek to dominate bigger and bigger chunks of the digital domain.

In terms of institutionalized regulatory responses to developments in technology, the telegraph may also point to emergent reactions to the dominance of Internet giants.

Furthermore, the telegraph is notable because it led to early experiments with new forms of institutionalized governance and regulatory regimes. The telegraph became the driver of some of the first intergovernmental agreements in the telecommunications area. The International Telegraph Union, formed in 1865 to oversee, develop, and coordinate the development of the telegraph, was one of the first intergovernmental organizations to emerge. In 1932, it became the International Telecommunication Union (ITU), a UN organization that regulates telephony and radio and other telecommunications services (Braithwaite and Drahos, 2000; Siochrú and Girard, 2002). Like the International Telegraph Union, these international agreements were set up to preserve and strengthen, not replace national monopolies and control, and it is only recently that this 'national, regulatory containment of telecommunications is in the process of breaking down, often in dramatic ways' (Braithwaite and Drahos, 2000: 322).

While similar developments are embryonic in the Internet domain, we do see a growing focus on the responsibilities and societal demands that we expect Internet companies to accept, such as in the areas of taxation, data protection, and the shaping of public spheres (Flyverbom et al., 2019).

REFERENCES

Balbi, Gabriele and Richard R. John. 2015. 'Point-to-point: telecommunications networks from the optical telegraph to the mobile telephone'. In Lorenzo Cantoni and James A. Danowski (eds), *Communication and Technology*. Berlin: Walter de Gruyter.

Braithwaite, John and Peter Drahos. 2000. *Global Business Regulation*. Cambridge: Cambridge University Press.

Carey, James W. 2009. 'Technology and Ideology: The Case of the Telegraph'. In James W. Carey, *Communication as Culture*. New York: Routledge, 155–177.

Carpenter, F. G. 1892. 'Henry Clay on Nationalizing the Telegraph'. *The North American Review*, 154(424), 380–2.

Castells, Manuel. 2000. *The Rise of the Network Society: The Information Age: Economy, Society, and Culture*. New York: John Wiley.

Cooley, Charles H. 1897. 'The Process of Social Change'. *Political Science Quarterly*, 12(1), 63–81.

Cooley, Charles H. 1909. *Social Organization: A Study of the Larger Mind*. New York: Charles Scribner's Sons.

Czitrom, Daniel J. 1982. *Media and the American Mind: From Morse to McLuhan*. Durham, NC: University of North Carolina Press.

Flyverbom, Mikkel, Ronald Deibert, and Dirk Matten. 2019. 'The Governance of Digital Technology, Big Data, and the Internet: New Roles and Responsibilities for Business'. *Business & Society*, 58(1), 3–19.

Gasser, Urs and John Palfrey. 2012. *Interop: The Promise and Perils of Highly Interconnected Systems*. New York: Basic Books.

Hackett, Edward J., Olga Amsterdamska, Michael E. Lynch, and Judy Wajcman. 2007. *The Handbook of Science and Technology Studies*. Cambridge, MA: MIT Press.

Jasanoff, Sheila and Sang-Hyan Kim (eds). 2015. *Dreamscapes of Modernity: Sociotechnical Imaginaries and the Fabrication of Power*. Chicago, IL: University of Chicago Press.

John, Richard R. 2010. *Network Nation*. Cambridge, MA: Harvard University Press.

Latour, Bruno. 2005. *Reassembling the Social: An Introduction to Actor-Network-Theory*. New York: Oxford University Press.

Marvin, Carolyn. 1990. *When Old Technologies Were New: Thinking about Electric Communication in the Late Nineteenth Century*. New York: Oxford University Press.

Scholte, Jan Aart. 2005. *Globalization—A Critical Introduction*. New York: PalgraveMacmillan.

Simon, Linda. 2005. *Dark Light: Electricity and Anxiety from the Telegraph to the X-ray*. Boston, MA: Houghton Mifflin Harcourt.

Siochrú, Seán Ó and Bruce Girard. 2002. *Global Media Governance: A Beginner's Guide*. New York: Rowman & Littlefield.

Thussu, Daya Kishan. 2000. *International Communication: Continuity and change*. London: Arnold Publishers.

CHAPTER 41

....

TYPEFACE

....

ROBIN HOLT

THIS is typeface. Like human beings, typeface is essentially homeless; it lives to be something else, and can become almost anything. Its omnipresent prodigality is so commonplace as to be barely recognized. It works silently and perfectly time after time, without decay, and carrying within itself all the meaning, menace, and love a world might muster.

ROME

....

Perhaps in the past, before it was being photoset and digitized through computers, typeface was more 'thingly'. At the opening of the twentieth century say, when it was bound to the movable type used in letterpress printing and laid out in a compositor's board, one inked character standing in and stepped against another, or set next to metal blanks (white space had also to be made present), all of which arrangement was of a significantly material order. Indeed the term font—a particular set of a typeface, say Arial 12 point in bold—comes from the French *fonder*: to melt and mould the metal for the type, hard enough to last, soft enough to take ink and press but not cut the paper.[1] The typeface was part of a wider distribution of often heavy, intricate, and cumbersome machinery allowing type to move and text to fragment into repeatable units (Innis, 1950; McLuhan, 2013: 8). When broken up and held in solitary isolation the pieces of type look out of their element, as might a catch of herring find themselves on land, wide-eyed and gulping at dry air. Every piece is replaceable, the blackening silver outline of one as precise and coolly schematic as the next, each having been measured against the kind of point scale outlined by Pierre-Simon Fournier in his *Tables of Proportions*. Fournier's scale (there were competitor scales) declared absolutely the dimensions by

---

[1] Type in wood/metal was first carved/moulded/cast in Korea, China, and Japan around the thirteenth century (see Innis, 1950). Acknowledging the limits of its author, this entry lingers with Western typeface.

which typeface compositors and printers should be bound, along with the punch cutters, from whose punches the type was cast and for whom he devised a measuring tool—'the prototype' (Rider, 1998).

Defining typeface geometrically, not materially, according to principles of substitutability, predictability, and replication, was a long way from Guttenberg's initial hopes for typeface as a decorative rival to the curlicues of monkish scribes. It was the clarifying innovations of Venetian (Nicolaus Jensen) and Veronese Renaissance printers, rather than the elaborations being curried by Guttenberg, which advanced typeface above the handwritten. Their innovations, like many, came from historical imitation; the future was the past. Felice Feliciano (b. 1433), for example, having been seduced by the crisp elegance of Roman inscriptions on marble ruins, set out on a peregrination from his native Verona to measure and copy the letters of an old empire. He divined the carvers' preference for placing letters within a square-framed circle and balancing the width and height of strokes in a divinely human ratio of 1:10 (Rider, 1998), a timeless verity to which typography, like any form of human design, ought to comport itself: out of typeface came truth, and truthful living. Fournier's grid system distils this understanding, it does not abstract from it: the 'point measures' were still based on the carved and written letters that Feliciano found so beguiling.

TYPOCIDE

Centuries later, and compelled by this association of typeface, and essential, truthful form, the book binder Thomas Cobden-Sanderson established the Doves Press, named after a nearby pub in Hammersmith, London. He had an intensity of manner, much like his friend William Morris, whose Kelmscott Press was just down the road. The Doves Press had the singular objective of creating the most beautiful books in the world. For Cobden-Sanderson, and his business partner Emery Walker, this required a small, dedicated workforce of craftspeople attending to every aspect of book production equally: the stitching, paper, and inks, the robustness and feel of the cover. Yet as they worked, they found the typeface becoming first amongst these equals, for here was the moment where the book, its maker, author, and reader became part of a greater whole of potentially life-long, meaningful communion. Much of this he had learned from Morris, but it was in the form a typeface should take that he differed. He felt Morris' efforts with his 'Golden Type' were sensually excessive. There was something wanton about the sheer quantity of ink and mixed exuberance of line, as though the text was conscious of its being materially present and anxious to show itself, forcing meaning into the shadows. Cobden-Sanderson argued a typeface that gives unto life—all of life—was necessarily simple, uncluttered, one whose beauty lay in the very invisibility of strokes and spaces (see Vinken, this volume). Only then would the text, voice, eye, and meaning become an organized whole: the typeface lives and dies by its mediating power working as a relational whole of weight, stresses, accents, crossing, and counter-spacing.

The whole duty of Typography, as of Calligraphy, is to communicate to the imagination, without loss by the way, the thought or image intended to be communicated by the Author. And the whole duty of beautiful typography is not to substitute for the beauty or interest of the thing thought and intended to be conveyed by the symbol, a beauty of interest of its own, but, on the one hand, to win access for that communication by the clearness & beauty of the vehicle, and, on the other hand, to take advantage of every pause or stage in that communication to interpose some characteristic & restful beauty in its own art.

(Cobden-Sanderson, 1900: 6)

The Doves Press typeface would be Roman. Like Feliciano, Cobden-Sanderson obsessed over earlier forms of typeface (though from Venetian books, not Roman 'originals') trying to discern their originating beauty amid the over inking and imperfect press work, and eventually coming up with drawings of a refined version of Jensen's type that was handed over Robert Prince who carved the punches. Machines were available to do this work, but Cobden-Sanderson wanted them hand-cut, without conceptual guidance, and with Prince's punches the individual drawings materialized into something unified and strikingly beautiful (Green, 2015). The punch was to be impressed into a softer metal casting mould or matrix, from which a type mould was then made into which molten metal was poured, forming typecast, the pieces that were then arranged into lines of text using a composing stick, being slid from the stick onto a composing table, building up a page which was then placed on the press bed, and inked to receive the paper.

The result was a typeface that fell into the page as elegantly and naturally as light into water, and which allowed Cobden-Sanderson free rein when it came to typographically composing pages of unsurpassed restraint in which text and white space set forth in animated flow (Figure 41.1).

Relief to the monotony of black on white text was given by the calligrapher Edward Johnston's designs for monumental headings and large initial letters that were then cut into woodblocks by Eric Gill and C. E. Keats and printed in red ink. Into some titles Johnston also inked directly in very occasional flourishes of green or gold ink, vestiges of a human signature whose distinction arose from the surrounding uniformity.

Cobden-Sanderson's obsessions meant the liminal venture called the Doves Press was commercially precarious. In 1909 the now acrimonious partnership between Cobden-Sanderson and Emery Walker was dissolved; the business had been haemorrhaging money for years. The form of the legal winding-up resulted in Cobden-Sanderson being given free use of the company's type for the rest of his life (the type deemed the only thing of any value), and on his dying its possession would pass to Emery Walker. Cobden-Sanderson agreed but remained aghast at the prospect that the type—his type—could be used by his erstwhile partner to make books that were less than beautiful, for example by using a mechanical press to print books of dubious literary quality. Unable to bear this likelihood, and fully aware that he was breaking a commercial agreement and so exposing his family to being sued on his death, Cobden-Sanderson decided to furtively destroy the type. If he could not use it, no one would. He started

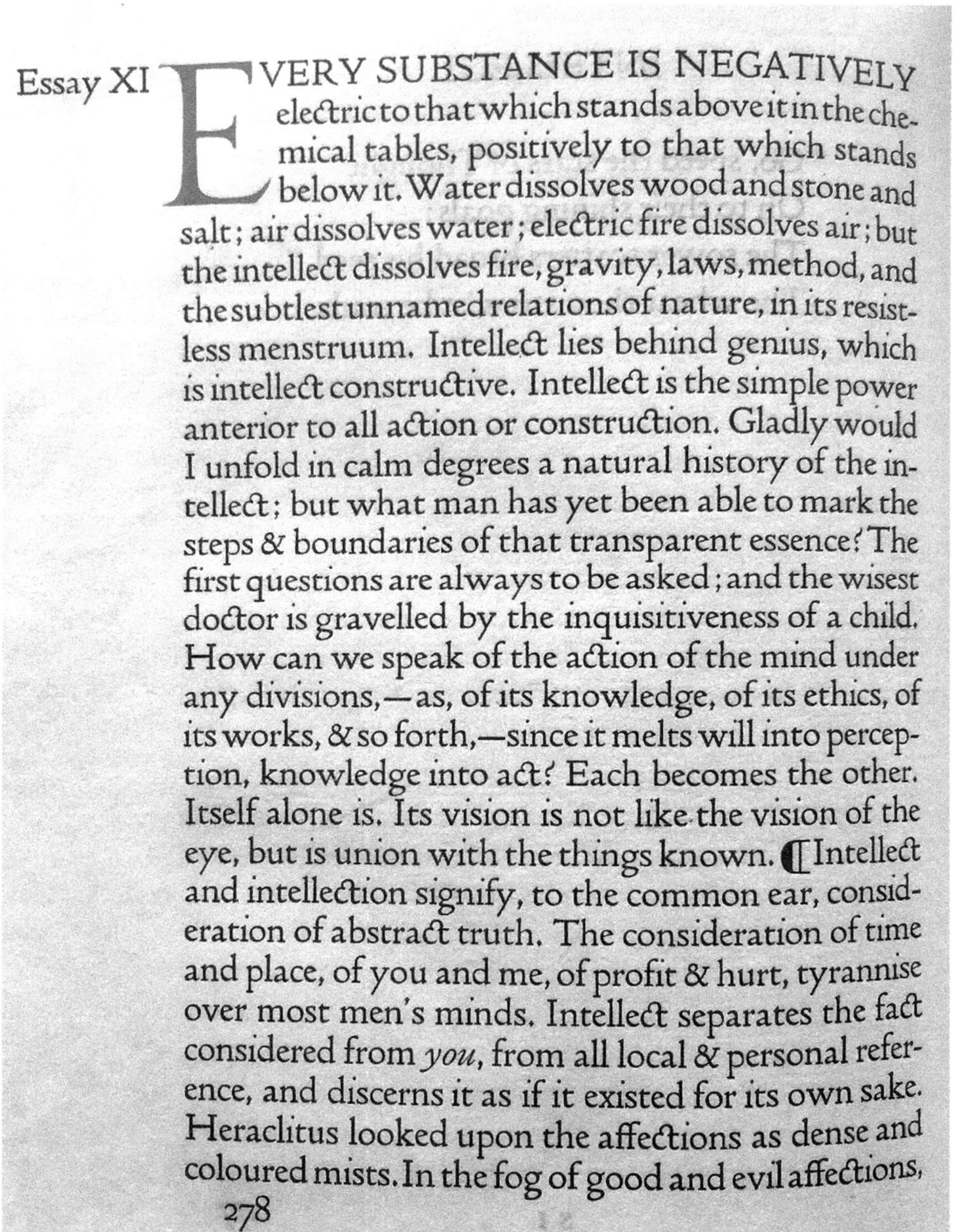

Essay XI EVERY SUBSTANCE IS NEGATIVELY electric to that which stands above it in the chemical tables, positively to that which stands below it. Water dissolves wood and stone and salt; air dissolves water; electric fire dissolves air; but the intellect dissolves fire, gravity, laws, method, and the subtlest unnamed relations of nature, in its resistless menstruum. Intellect lies behind genius, which is intellect constructive. Intellect is the simple power anterior to all action or construction. Gladly would I unfold in calm degrees a natural history of the intellect; but what man has yet been able to mark the steps & boundaries of that transparent essence? The first questions are always to be asked; and the wisest doctor is gravelled by the inquisitiveness of a child. How can we speak of the action of the mind under any divisions,—as, of its knowledge, of its ethics, of its works, & so forth,—since it melts will into perception, knowledge into act? Each becomes the other. Itself alone is. Its vision is not like the vision of the eye, but is union with the things known. ❡ Intellect and intellection signify, to the common ear, consideration of abstract truth. The consideration of time and place, of you and me, of profit & hurt, tyrannise over most men's minds. Intellect separates the fact considered from *you*, from all local & personal reference, and discerns it as if it existed for its own sake. Heraclitus looked upon the affections as dense and coloured mists. In the fog of good and evil affections, 278

FIGURE 41.1 Emerson's *Essays*, Doves Press (1906)

(Photo by the author)

in 1913, taking the punches and matrices to nearby Hammersmith Bridge and throwing them into the Thames. The metal type itself was to follow. This extended 'typocide' took over 170 trips. The bags were heavy, and Cobden-Sanderson, aged 76, got tired quickly. The typocide had to be committed with no one noticing until the very end, so the presses might keep working, but using an increasingly diminishing supply of type founts that could never be recast—the numerous vowels, the infrequent 'j' and the really rather rare ligature 'qu', they all went into the water. He finished in 1917 and died soon afterwards, poor, but vindicated: his type would never be sullied by the organizing impress of industry to which Emery Walker, like all business figures, was subject. Almost uniquely amongst typeface designers, Cobden-Sanderson had stipulated how his would be used, by killing it.

TRUTH

Doves type is beautiful handwork, straining to fulfil Cobden-Sanderson's ambition that it gather a communion of reader, author, material, and meaning. Machine work upsets this mutual occasioning. Cobden-Sanderson understood that typeface mediated thought (McLuhan, 2013: 11, 272). To consider such a claim, the documentary maker Errol Morris (2015) set up an experiment in the *New York Times*. He had been reading Saul Kripke's *Naming and Necessity* that argued there was a determining intimacy between meaning and the rule-bound structures of its expression. Morris wondered if typeface counted as a structure. At his request the newspaper printed a quote from a mathematician who argued that even if an asteroid were to come close to earth, humanity was in a position to defend itself because of science. Readers were asked whether they believed the statement. Some 45,000 responded. Unbeknownst to them the questionnaire was printed in five different typefaces. One typeface—Baskerville—elicited a higher belief rating than the others. In Baskerville we trust. It was in this typeface, designed by the Birmingham printer John Baskerville in 1750s, that Morris printed his results.

Baskerville exudes a grounding appeal. Take the word:

Quisling

How does it appeal? There is little appeal in the referent: the Norwegian Nazi stooge Vidkun Quisling, and his duplicitous ilk. But as type it has appeal. It may be the slightly incomplete curve on the 'g', arresting, the zig-zagged tail on the upper case 'Q', or the solid, angled stroke of the serif feet and ascender heads which nod confidently towards the touch of the pen. Throughout comes an open, rounded curvature displaying a mannered awareness of proportional balance.

Baskerville, like all typeface designers, obsessed over approachability. Height is broken down into x-height (in 'Quisling' the height of the 'u' and 'n', or the top part of the 'g'), ascenders—in those characters whose lower case extends to the height of an upper case, as in the 'l'—and descenders, where characters go beneath the base line of both upper and lower case, as in the 'g' and 'Q'. The distance from the top of the highest ascender to the bottom of the deepest descender defines the size of the typeface (measured in points). Width is a function of line weight and slant, of the flowing nudge offered to the reader by serifs that give onto one another, of letter spacing, as well as letter form, so a 'u' is wider than an 'i' (save in monotypes such as the Courier typeface used on mechanical typewriters). Width is also formed by kerning: the space into which characters might spillover (or be offset) allowing a more proportional feel, where 'u' intrudes on 'i' for example, with some cases warranting the creation of ligatures or parings, like 'Qu'.

Morris' experiment hits on something basic about typeface: as a medium for information/fact/truth/vision, it helps constitutes them: it is mediation as transformation not just transmission. The oft-used *Times New Roman* (a variant on Baskerville) explicitly makes play with this quality, having been cut for *The Times* newspaper in London in 1931 to render an authoritative and legible feeling in the reader (Kinross, 1992: 79), and subsequently used in academic publishing, official papers, and to script vast colonial histories in which the victor, apparently, is decency. Such a typeface exudes confidence that what is being stated carries the gravity of its being said properly, gathering around it a gravity that is lacking in something as fleeting as spoken utterance (Innis, 1950: 168–9).

A stark example of this complicity between typeface, authorial forms, and political authority was played out in Germany when the Third Reich decreed the Germanic typeface *Faktura* (an official, long-dominating form of *Schwabacher*) to be too ponderous, aged, and laboured for a thrusting, dynamic race of people blood-bound for world domination. They wanted a break from the past, notably from Bismarck, who was forever whingeing from beneath his mastodon moustache about having to read German text in Latin script. He wanted his *Schwabacher*, a typeface that had emerged from goose quills, its sharp edges suited to hard, vertical ducts (strokes), which weakened and deflected considerably when asked to deviate into curves. For Bismarck the heavy, direct, and fulsome *Faktura* was the quintessence of German character: indomitable, unabashed, its block form laden with the hierarchy and gravity of those for whom seriousness is a serious advance on the directionless curves of play. In contrast Latin typefaces like *Antiqua* (again a variant on Baskerville and Times New Roman) were too curvaceous, superficial, and airy. By 1941, however, these associations were upended, with Martin Bormann declaring the classic, open lines of this Roman typeface to be the new standard letter for all official communication (see Kinross, 1992: 101–2), putting truth amid the shadows.

FUNCTIONALISM

The Roman typeface designs from the likes of Baskerville and Fournier had set in train a host of derivative and deviating forms of which *Times* and *Antiqua* were but two. By the start of the twentieth century continuous casting machinery allowed type to be set and automatically cast in lines (slugs), confining human involvement to type design and operating the keyboard of the typesetting systems. With a keyboard and hot metal casting creating monotonously spaced courier typeface and assertive full stops these machines enforced a certain kind of meaning-making on users: speculation and digression gave way to clear exposition (Seigert, 2013). With typeface meaning was becoming hierarchically structured, storable, movable in tightly figured blocks. Type arrested the human eye, then imprisoned it in linear, punctuated movements of left-right, top-down orthogonal compliance.

Whilst some lamented this technologically-led encroachment upon humanist forms, others embraced it; indeed they called for fonts that befitted the seamless, indifferent, tireless chirr of standardized production. Foremost amongst such designs was Paul Renner's *Futura*, branded as 'the typeface of our time' by the Bauer type foundry who were making and distributing the type matrices to feed the growing number of continuous casting machines. Renner was a child of the first machine age, a devotee of Peter Behren's *Werkbund* movement in whose neatly canalized flow a typeface, like all products of thoughtful design, was to embody the predictable productiveness of machine work, the individuality and attentiveness of craft, and the atmosphere of calm accomplishment emanating from the financial capital of upstanding industrialists. For Renner the form type took was of profound consequence for it was in the printed word, and especially in books, that a culture revealed itself to itself: the letters, punctuation, and numerals revealed the spirit of a culture, and during the 1920s the German *Geist* was enduring a battering. The First World War had impoverished what was an outmoded imperial order, and with the 1920s came the economic pain, but also the promise of a new order in which the productive capital of machinery and industrial finance, and not aristocrats, organized human activity. Renner wanted a typeface that encountered, admitted, and even embraced this emerging technological order in which function, raw material, and mechanized processes determined the type form, the designer being a shaping conduit of such, nothing more. A good typeface was beholden to the job of efficient communication. No calligraphic flourishes, no superfluous details, no historical vestiges. *Futura*'s form (the strict shape of the letters) was denuded of archaic style (the weight given to strokes as they might thicken and narrow their slant as they press into papers of different thickness, or a fluidity of movement as one letter might cursively suggest the next with a generous serif). Renner wanted the form to be without style, splitting from the touch of a hand and stripped of history.

Though Renner espoused such a split from history, he was not able to deliver, at least according to Jan Tschichold (1998 [1928]: 74–5) in whose precision-obsessed eyes *Futura* still carried humanist eruptions, imperfections, and digressions. The main problem was Paul Renner himself, an author, and any typeface worthy of a machine age must subsume individuality to the functional demands of legibility, readability, and universality, and be designed anonymously, by engineers:

> The engineer shapes our age. Distinguishing marks of his work: economy, precision, use of pure constructional forms that correspond to the functions of the

object.... They have created a new—our own—attitude to our surroundings.... The collective whole already largely determines the material existence of every individual. The individual's identical needs are met by standardized products[.]

These products configure 'us' as individuals with self-similar needs. Rationalization, standardization, mechanization: this holy trinity defined an age where the machine needed no rhetorical validation, it was already within everything. Typeface, inevitably, would comply. Designed by anonymous collectives facelessly economizing on the means of expression, on the ease of use and replication, on the comprehensibility of message, and the democracy of appearance, all of this gathered to realize the single goal to which all human endeavour must be put:

> **Unity of Life!**
> So the arbitrary isolation of a part is no longer possible for us—every part belongs to and harmonizes with the whole. Where slackness is still the rule, we must make it our work to fight against laziness, envy, and narrow mindedness.
>   (Tschichold, 1998 [1928]: 13)

True creativity comes with the geometry imposed by logical, impersonal means, entirely free of the individualistic line of artistic thinking (Tschichold, 1998 [1928]: 28). It is constructed through a dependency on purpose, levels of demand, manufacturing constraints, and raw materials:

> It is essential to give pure and direct expression to the contents of whatever is printed: just as in the works of technology and nature, "form" must be created out of function. Only then can we achieve a typography which expresses the spirit of modern man. The function of printed text is communication, emphasis (word value), and the logical sequence of the contents.   (Tschichold, 1998 [1928]: 67)

If not Renner then maybe the Bauhaus' Hebert Bayer got closer to such an ideal typeface with his 1925 *Universalschrift* (Figure 41.2). Where Renner's typeface hinted at the prospect of eternal time, Bayer's suggested infinite space. As an engineer Bayer's intent was to strip the typeface bare and hose it down—no serifs, no flourishes, no stylistic embellishment, no *groszer* letters—making it simple to use for compositors, typographers, and readers. Being universal, *Universalschrift* was to communicate how society, nature, and history were being organized in processes of substitutability: the perfectly ordered constitution of needs and means (products), each adjusting silently and utterly to the

quisling

FIGURE 41.2 *Universalschrift*

other. It was a typeface without prejudice or even purpose. It was purely purposive—an ordering for its own sake and which, sprung from an abstract space, could be used anywhere.

Except, of course, it could not. Its form of black and white shapes, of strokes and spaces, as with Tschichold's and Renner's, was a thoroughbred of industrialized reasoning that had a specific place: all meaning was reducible to the demands placed on things by what we might call the representations of function (relating with things in order to do something) and economy (relating with things with a maximal degree of clarity, or bracketing) (Bayer, 1981). Beauty was abandoned (quite clearly). With Renner's *Futura*, the blank, black flank of the lines of functionalist type emerged from a constructivist logic calling for a correct and proper assembly of parts, and with Tschichold and Bayer's *Universalschrift* they became condensed into a supremacist logic in which a mechanized purity of form was to seer through and steer the seamless organization of all life. The typeface was, like all typefaces, a form of writing, the mark of the hand was always 'there', it has to be, found in how the white spaces govern the placing of the black strokes, and in stroke contrasts (Noordzij, 2005 [1991]: 15–19). Even *Universalschrift* has a specific contrast and rhythm (one letter follows the next, one word follows the next, into a line) and so is constituted into a system of writing of which handwriting is, too, a part, and which in turn constitutes the system of reading.

The universal aspiration of the typeface would only be met if it ceased to be a form of writing, somehow admitting the possibility that writing is to become redundant once more, more a pulse amongst pulses: machines do the memory and writing work now, and we can consult and converse within their scripted confines. Here *Universalschrift* would succeed with its own demise upon admission to a singular, seamless universe in which, ideally, it and its users would cease to be a distinct part, and instead be replaced by a second-order system of signs consisting of 1 and 0 (representing 'ink' or 'no ink' in the pixels, presence or absence), the one being the condition of creation, the zero the vanishing point at which being ceases, and the second order being organized entirely mechanically with a shutting on and off, or more lately a computer mediated on/off (Kittler, 2001). There would then be no stroke, and no white space, and no writing, and no history and no means of infusing the real with the functional.

Expression

Whilst functionalism was blithely riffing off idealizations of perfect forms of information transmission, other, more ironic responses to the intimacy between meaning, accident, and typographic expression were also in play. These experiments had been catalysed by Stéphane Mallarmé's *A Throw of the Dice Will Never Abandon Chance* in which text of different size and form comes spread across both pages in ways that elicit

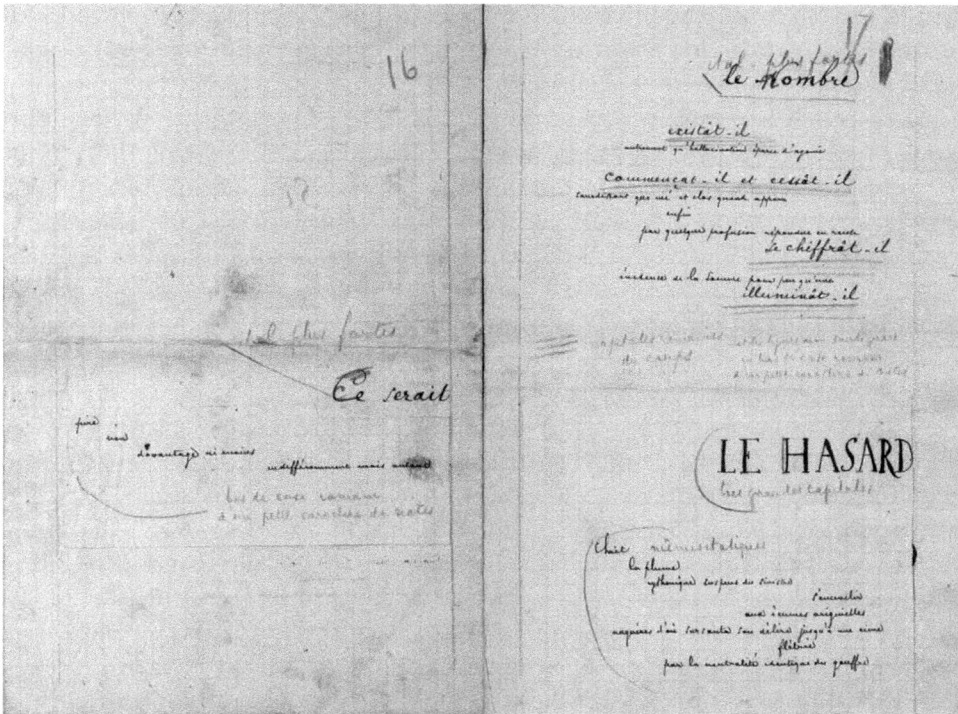

FIGURE 41.3 Autograph layout of Mallarmé's *Un Coup de Dés Jamais N'Abolira Le Hasard* (1896)
(*Source*: https://commons.wikimedia.org)

pause and rapidity in readering, forcing moments of concentration or hurrying the words along until they trip into a jumble (Figure 41.3).

Revealing the affects of typeface by upsetting their habitual structures, Mallarmé is toying with how the writer, typographer, and reader conform with habituated rules of composition, transmission, and reception. Their actions—including the functional idealists—never entirely or evenly theirs because they belong to a performance of which they are often an unwitting part, are pulled into ordered shapes from which wider social norms can be inferred: designing and using a typeface is a cultural technique not only configuring a prevailing aesthetic style but an entire system of organization: using list-like layouts, for example, or composing on standard sheet sizes that can be filed, stored, and distributed (Vismann, 2013).

That Mallarmé is alive to this, and plays with it, is to counter the determinations of machine-based writing system: chance gets woven into the text machine. There is something of the Romantic in Mallarmé's showing the disruptive potential of typeface's embodying force. His is a text machine gone haywire, signifiers floating across and over, determining their own flow irrespective of any narrative intent. It is this kind of rich experiment in script-images (*Schriftbild*) that entranced Walter Benjamin, to the point he envisaged relinquishing the elusive and suggestive plasticity of his fountain pen were a machine made that could deliver the kind of graphically sophisticated picture-writing (*Bildlichkeit*) and moving script (*Wandelschrift*) that could be delivered by handwork.[2]

More compelling still, it excited Tschichold's sometime business partner Kurt Schwitters to find a far less functional history for typeface. Wandering through the wastelands of Hanover in the aftermath of the First World War Schwitters had picked up a battered advert for Kommerz und Privatbank. Though shabby, the letters 'merz' remained intact, and he sandwiched them between daubs of colour, creating the montage style of art which became *MERZbild*, or just Merz (Schwitters, 2000 [1920]). It occurred to him that typeface was an artistic material as valid as any other, a democracy of things he would assemble in twisting, spoiling arrangements using gluing, cutting, and over-painting. The result was an image in which ruined, left over, and partial things (lengths of cotton, spent ration cards, buttons, adverts) became lines, planes, surfaces, rents, and unities. Yet the elements retained their previous life; no matter how insistent the deformation, the paintings could always be read socially, historically, as well as artistically, giving the collages a peculiar liveliness as they work through what always remains a hesitant, disturbed, playful presentation of continually disturbed meaning. Bus tickets become blocks of raw colour but the municipal typeface still indicates the route, on which date and with what omnibus company, and food labels became curves but remained aspirant emblems of a commercial product. New aspects dawn without the old ones disappearing and the whole has the appearance of barely clinging on to its unity, a fragility struggling to hold together the different objects and meanings vying for attention, beyond any control the artist might have intended to have. 'Make connexions' urged Schwitters, 'preferably between all things'. McLuhan would have concurred.

There are hints, allusions, crannies, titles that startle but just as readily cloud over, convinced as Schwitters was that 'every form is the frozen instantaneous picture of a process'. In making these works Schwitters expressed without consciously expressing himself; the found materials and his paired down, formal idiom of collage allowed him to avoid becoming himself too much the centre of things.

Apparently random, and certainly lacking a unifying coherence, Merz works as a singular orientation to the everyday world using the very thing—typeface—that mediates this world, foreclosing what is textually being made present. The found type has no content, or more accurately it has content once lost now recovered and endlessly transformed, giving form to Jose Saramago's (2012 [1976]: 153) calligraphic observation: 'the expression of incoherence demands a great deal of organization'.

---

[2] Machinic is also chaotic. Paul de Man (1996: 181) describes text as behaving sometimes like a machine, but by which he meant an ordering of repetitious detachment from authorial direction, a machine of derangement '...you are writing a fine compliment for somebody and without your knowledge, just because words have a way of doing things, it's sheer insult and obscenity. There is a machine there, a text machine, an implacable determination and a total arbitrariness, *unbedingter Willkür*, he [Schlegel] says, which inhabits words on the level of the play of the signifier, which undoes any narrative consistency of lines.' A deranged arbitrariness that perhaps also attracted Umberto Eco, who, in typical ludic style, often named his novels' characters after fonts.

FIGURE 41.4 Kurt Schwitters (German, 1887–1948). 'Difficult', 1942/43

(Albright-Knox Art Gallery, Buffalo, New York; Seymour H. Knox Foundation (1965: 14).
© Estate of Kurt Schwitters / Artists Rights Society (ARS), New York/VISDA Denmark)

One wonders what Schwitters would have made of Doves type had he found it during low-tide, walking the banks of the Thames. He was, after all, in the area for a while. After being designated a degenerate artist he left Germany first for Norway and then, in 1941, for England, to be interred as a POW on the Isle of Mann before then spending the rest of the war in West London. The Merz picture 'Difficult' (Figure 41.4) was made from the bombed out, scanty, and torn materials found round-and-about the streets of Hammersmith. Was Cobden-Sanderson right in opposing the functionalist, business machinery by killing what was made with love? Surely he was, for wasn't this where functionalist industrialization had got us—blowing one another to bits. But in the bits... Schwitters finds something.

REFERENCES

Bayer, H. 1981. 'Oral history interview with Herbert Bayer', 3 October. Archives of American Art, Smithsonian Institution.
Cobden-Sanderson, Thomas James. 1900. *The Ideal Book*. London: Doves Press.

De Man, Paul. 1996. *Aesthetic Ideology.* Minneapolis: University of Minnesota Press.

Green, Robert. 2015. 'History of the Doves Type', 25 July. https://typespec.co.uk/doves-type-history [Accessed 28 August 2017].

Innis, Harold. 1950. *Empire & Signs.* Oxford: Oxford University Press.

Kinross, Robin. 1992. *Modern Typography: An Essay in Critical History.* London: Hyphen Press.

Kittler, Friedrich. 2001. 'Perspective and the Book' (trans. Sara Ogger). *Grey Room*, 5, 38–53.

McLuhan, Marshall. 2013. *Understanding Media: The Extensions of Man.* Berkeley, CA: Ginko Press.

Morris, Errol. 2015. *Pentagram Papers 44 Hear All Ye People: Hearken, O Earth.* https://www.pentagram.com/work/pentagram-papers-44-hear-all-ye-people-hearken-o-earth.

Noordzij, Gerrit. 2005 [1991]. *The Stroke: Theory of Writing*, trans. Peter Enneson. London: Hyphen Press.

Rider, Robin. 1998. 'Shaping Information: Mathematics, Computing and Typography'. In Timothy Lenoir (ed.), *Inscribing Science: Scientific Texts and the Materiality of Communication.* Stanford, CA: Stanford University Press, 39–54.

Saramago, José. 2012 [1976]. *Manual of Painting and Calligraphy*, trans. Giovanni Pontiero. Boston: First Mariner Books.

Schwitters, Kurt. 2000 [1920]. 'Merz'. In Robert Herbert (ed.), *Modern Artists on Art*, 2nd edn. New York: Dover Publications, 66–75.

Seigert, Bernard. 2013. 'Cultural Techniques: Or the End of the Intellectual Postwar Era in German Media Theory' (trans. Geoffrey Winthrop-Young). *Theory, Culture & Society*, 30(3), 48–65.

Tschichold, Jan. 1998 [1928]. *The New Typography*, trans. Ruari McLean. London: Faber & Faber.

Vismann, Cornelia. 2013. 'Cultural Techniques and Sovereignty' (trans. Ilinca Irascu). *Theory, Culture & Society*, 30(3), 83–93.

....

CHAPTER 42

WHITEBOARD, FLIPCHART

....

JÖRG METELMANN
TRANSLATED BY ERIK BORN

In my office, a whiteboard (Figure 42.1) and a flipchart stand side-by-side in silent correspondence. They are separated by a bookcase, a door, and a window onto the hallway, which according to those responsible for the administration of the architectural infrastructure was intended to express our university's 'institutional transparency'. Bookcase, door, window—these are an organization's spatial realities. Like the whiteboard and the flipchart, they are not easy to transport and cannot be turned into data without an additional step (e.g., a smartphone photo). Whenever guests come to my office, they immediately acquire a disconcerted look upon catching sight of these seemingly obsolete fixtures, the massive media of storage and presentation flanking the entrance to my office. 'Honestly, who still writes on paper anymore,' their eyes seem to say, 'or on any surface that isn't a smartboard?'

USE PRACTICES (DISPOSITIVE 1): SCENES

As someone trained in literary studies, I am familiar with this concerned look, though for different reasons. For a long time, I was unable to make any headway with flipcharts or whiteboards, if only for the simple fact that they were not yet common in university classrooms during my studies. In the early 1990s, the dominant object in every classroom was the green chalkboard, which would only ever be used for jotting down some obscure name or a room change. To my knowledge, there is no 'organizational scene' in literary and cultural studies involving the flipchart, whiteboard, or even the chalkboard; even among interdisciplinary scholars and those working on media practice, these organizational media do not play an essential role in the process of negotiation and decision making. In recent years, the concept of the scene has been

FIGURE 42.1 Whiteboard and flipchart

(Photo: Jörg Metelmann)

taken up again by Jacques Rancière (2013) and can be described for the present context as 'singular occurrences, set in different times and places, around which a network of perceptions, ideas, speculations and connections can be woven' (Beyes, 2017: 146). A scene can be understood as both an indicator and an instantiation, as an expression of a large, collective organizational structure whose interpretive system does not determine it completely. The concept of the scene makes apparent what, where, and by whom something is done.

I experienced my first organizational scene with the flipchart as a project assistant for the German branch of the global management consulting firm McKinsey & Company in the mid-2000s. At a brainstorming session for a future workshop, one consultant suddenly stood up and sketched out a scenario on a flipchart. It was a precisely-timed, creative exchange of potential approaches, applying the method made fashionable by San Francisco-based IDEO Labs (essentially, allowing every idea to stand as such, and postponing criticism of cost, feasibility, and other considerations). Still, my colleague's scribbles stood in stark contrast to our polished client presentation. If the flipchart meant improvisation, the PowerPoint slide show represented a kind of perfection, which cost plenty of money even with an in-house division for graphic design.

The organizational scene changed in my experience working at a start-up company involved in labour market mediation and advanced training. In these fields, the flipchart's emphasis on innovation and improvisation took on connotations of self-actualization. To the unemployed workers whose cases our company was managing, the flipchart in our workshop room was often a complete unknown, or at the very least something unfamiliar enough to make anything written on it seem more important. There was something scholarly about the flipchart; it could be authoritative without evoking the school chalkboard's dustiness and pedantry. For many participants, the challenge of making a plan on a flipchart, whether for a course of action in continuing education or the steps back to gainful employment, amounted to overcoming a twofold obstacle. They had to collect their dreams and desires from the realm of wishful thinking, and they were required to stand up and present them to the group in concrete form. In this social setting, the flipchart can be read in a fairly straightforward manner as a projection space for one's own future. At the same time, the organizational scene also encapsulated the European labour market policy of promoting training for getting promotions, which was developed in the UK under New Labour and in Germany under Gerhard Schröder's 'New Middle'.

The third main scene featuring these media of organization came with my experience working at a business school. In every classroom, the university had pre-installed, in standard series, not only a chalkboard but also a flipchart and a whiteboard (along with projectors for the unavoidable PowerPoint presentations.) In this particular setting, the flipchart and the whiteboard took on even stronger persuasive qualities derived from their typical usage. For the most part, the ageing visual aids were only ever used when speakers departed from the text of their lectures (a rare occurrence) or deviated from their PowerPoint presentation (almost always the case) to explain some point in further detail (often at the request of the audience). This explanation would inevitably play out at the flipchart because, as we would say, 'it's just better for illustrating things', which more precisely meant 'for showing things more freely, on a more human scale, with a more human touch'.

While all three of these scenes required a certain ability to adhere to standardized phrases, the university demanded something that would have been impossible in consulting (since it would have been perceived as a lack of preparation)—namely, the ability to persuade an audience, in freestyle, through the careful handling of a flipchart or a whiteboard (of course, not only through those visual media, since there was frequently also verbal debate).

FRAMING 1: NOTE-TAKING, ACCORDING TO FREUD

In each of these scenes, the flipchart and the whiteboard exhibit their capacity, as technologies for producing knowledge, to put things in order. They inform the kind of writing (large or small; handwriting!), the spatial relationship between the writer and the audience (you step to the front of the room, in front of everyone else; at which distance are things no longer legible?), and word–image structures (many YouTube videos show the best practices for persuasive chart design). In this respect, the flipchart and the whiteboard have a medial impact on the way information can be stored, processed, and organized.

One early attempt to think through the medial functions of note-taking systems can be found in Sigmund Freud's classic essay on the Wunderblock, a child's toy comprised of a wax tablet and a sheet of wax paper. Freud's 'Note upon the "Mystic Writing Pad"' (1925) opens with a statement about the potential for improving one's memory through two different note-taking procedures, which the Wunderblock combines:

> On the one hand, I can choose a writing-surface which will preserve intact any note made upon it for an indefinite length of time—for instance, a sheet of paper which I can write upon in ink. I am then in possession of a "permanent memory-trace." The disadvantage of this procedure is that the receptive capacity of the writing-surface is soon exhausted. The sheet is filled with writing, there is no room on it for any more notes, and I find myself obliged to bring another sheet into use, that has not been written on. Moreover, the advantage of this procedure, the fact that it provides a "permanent trace," may lose its value for me if after a time the note ceases to interest me and I no longer want to "retain it in my memory." The alternative procedure avoids both of these disadvantages. If, for instance, I write with a piece of chalk upon a slate, I have a receptive surface which retains its receptive capacity for an unlimited time and the notes upon which can be destroyed as soon as they cease to interest me, without any need for throwing away the writing-surface itself. Here the disadvantage is that I cannot preserve a permanent trace. If I want to put some fresh notes upon the slate, I must first wipe out the ones which cover it. Thus an unlimited receptive capacity and a retention of permanent traces seem to be mutually exclusive properties in the apparatus which we use as substitutes for our memory: either the receptive surface must be renewed or the note must be destroyed.   (Freud, 1940 [1925]: 207–8)

In Freud's interpretation, the mutually exclusive properties of the technical apparatus reveal the exceptional performance of the mental apparatus: 'it has an unlimited receptive capacity for new perceptions and nevertheless lays down permanent—even though not unalterable—memory-traces of them' (Freud, 1940 [1925]: 208). As a new medium around 1900, the Wunderblock confirmed Freud's conception of the mental apparatus, since it explained the double function of the mind, for both memory and perception, in an easy-to-understand material arrangement (Pichler, 1989).

For our context, Freud's essay is relevant in two ways. On the one hand, the flipchart and the whiteboard can easily be read as further examples of the media technologies mentioned near the start of his essay. If the flipchart stands for the permanent trace and quick exhaustion of the material (i.e., a lack of paper), then the whiteboard represents an unlimited recording capacity with the loss of any permanent trace. Under certain circumstances, some of which are perhaps still familiar today, Freud's analogy extended even further: should one happen to grab hold of the wrong pen (e.g., a permanent marker instead of a dry-erase marker) and not have the right cleaning agent at hand, then the loss of the permanent trace would not have been total. Instead, the inscription would have remained preserved, in the best sense of the Wunderblock, on a permanent, more or less visible level. In other words, miniature palimpsests would emerge showing traces of both the text and other marks made on the surface over time, which subsequent scholars in cultural studies repeatedly interpreted as signs of the human capacity for memory and creativity (e.g., Genette, 1997).

USE PRACTICES (DISPOSITIVE 2): POSITIONS

The flipchart is a medium of the twentieth century. While the first US patent for a flipchart was filed in 1913, the pictorial record dates back a year earlier to when the legendary management innovator John Henry Patterson, the owner of the National Cash Register Company and creator of the 'American Selling Force', gave a presentation to the 100 Point Club using a proto-flipchart. With the growing popularity of the whiteboard in schools and businesses in the 1990s, as well as the general spread of laptops and projectors, the flipchart's career came to an end around the year 2000.

The most well-known form of flipchart, with a pad of paper on a small whiteboard, was invented by Peter Kent in the 1970s. As a variation on the classic lecture, the flipchart changed the presentation setting, which was no longer a matter of only speaking and listening, but also of seeing and writing. As a result, the presenter's handwriting and drawing skills became either a resource or a natural boundary, either too small, too messy, verging on chicken scratch, or attractive, attention-grabbing, and perhaps even persuasive (today, there are countless how-to films on YouTube about writing on the board well and above all legibly). In doing so, the flipchart establishes a crucial bond with the speaker's body and assigns the position of power to the object rather than the subject. After establishing a relationship with the board, the speaker is expected to keep on using it for the remainder of the presentation; it takes up too much space and attention for something to be written on it once and then dismissed. The device has to be played with, and thus itself plays along in the task of persuading an audience. The writing duo of human and machine can release great power, or it can be dismissed as uninspired scribbling and nonsensical babbling.

A similar positional dynamic can develop at meetings without any explicit presentation. In my experience, whenever there was a flipchart in the room, someone would inevitably stand up, go to the flipchart, and say, 'I'll just go ahead and take notes.' The flipchart served not only to document the meeting, which would have been equally possible with a private record of the minutes; creating a public note on an oversized pad of paper also made the state of knowledge collective through a widely accessible visualization. Whatever was put on display in this manner acquired a certain authority: beyond providing a simple record of the ephemeral, notes made on a flipchart made knowledge objective. After the fact, nothing was deemed sayable that was not previously brought up for discussion and noted down on the board. Hence, standing up and taking notes provided an integral structure in the process of exchanging ideas.

Even after deciding to step forward to the flipchart and note something down, one would immediately be faced with further questions about where and how to position oneself: keep the pen in hand or leave it resting in the little slat below the chart? Stand next to the flipchart or behind it, so as not to obstruct anyone's view? Or, on second thought, maybe go back to one's seat and listen some more before heading off again (which would have been ideal, though it would have undermined the chart's very purpose)? As a movable material object, the flipchart organizes the space of a presentation differently than the (unmovable) whiteboard or the (often immobile) projector. Merely repositioning a flipchart (carrying or rolling it) can bring about a change of focus (e.g., putting the flipchart in front of those who have yet to contribute anything to the discussion). The blank page is to the scene of writing as the blank notepad is to the scene of presentation: whether a source of anxiety or motivation, the blank page extends a call to participation, a call to action.

Whenever a meeting or seminar was at an end, the question would inevitably arise of how to create a collective record of the results that had been negotiated collectively and often written down collectively on the flipchart in the first place. More often than not, the responsibility for making the record would be determined either hierarchically (i.e., the secretary would do it) or democratically (i.e., whoever's turn it was). Next, the oversized pages would have to be gathered together, spread out on the office floor, written up in a Word document or on PowerPoint slides, and finally sent around in an email. In January 2007, the smartphone (see Whyte, this volume) broke into this transitional world of mixed-methods organization with the striking force of a hammer blow: each page could be quickly photographed and sent out as an email attachment. Now, every participant would have everything available before their eyes and would be able to select whatever they deemed significant. It was no longer necessary to copy everything down a second time only to make it available electronically—an enormous gain in efficiency. Nor was it necessary to retain untranscribed pages over an extended period (e.g., in a multi-day compact seminar). With cell phone photos and a (cloud) storage folder, any used flipchart paper could be dropped in the recycling bin on the way out the door.

USE PRACTICES (DISPOSITIVE 3): PENS

Digital photography, especially with a smartphone, represented a decisive turning point for dealing with the whiteboard, too, since the possibility of a photographic record would prevent the potential loss of current data through erasure. This had been the whiteboard's clear disadvantage in comparison to the flipchart, just as its much less dusty and dirty operation was a clear advantage over any chalk-based medial support. The rapid transition from the blackboard to the whiteboard, which was based on the invention of a dry-erase marker in 1975, may have also severed an affective bond to chalk captured in common memories of the smell of chalk, the way it sounded on the board, and in some cases even elementary school chalk fights. Today, the chalkboard is still covered with a certain nostalgic charm, which calls to mind the good old days—even to the extent that a media personality like Glenn Beck can write out his abstruse conspiracy theories with chalk so as to evoke some traditional world in which men were the only people deemed capable of explaining things. The legitimacy of chalk, as it were.

The whiteboard exhibits its own affects, and what fascinate people are once again the writing implements. The permanent marker is the topic of surprisingly many online videos—or, more precisely, how to remove it from a whiteboard. Putting aside the potential variables in whiteboard materials (which are also connected to the question of its receptivity), the following solution applies to most off-the-shelf melamine products. If one happened, by accident or out of disregard, to grab the wrong pen, which cannot be washed off easily, then various chemicals can restore the white of the board (see Hjorth, this volume). These differ from each other in terms of their degree of residue: apparently, acetone and even plain old mouthwash do not leave any residue, whereas isopropyl alcohol and disinfectant wipes leave streaks that occasionally even turn grey (which is a little unbelievable in the case of isopropyl alcohol, since acetone is derived from it).

However, there is another method for removing permanent markers, which brings us back to Freud. Many users report, in YouTube videos of themselves beaming with joy, that a perfectly ordinary whiteboard marker can also be made to function as an eraser. One needs only to apply it in a very thick layer to the spot in question so that the whiteboard marker completely covers the permanent marker. Then, both layers can easily be wiped off. The procedure recalls the magical effect of the writing pad in Freud's text: 'If one wishes to destroy what has been written, all that is necessary is to raise the double covering-sheet from the wax slab by a light pull, starting from the free lower end' (Freud, 1940 [1925]: 210). Just as the wax paper loses the letters that were engraved through it onto the wax tablet, the permanent marker disappears on account of a chemical reaction caused by painting multiple layers. However, the trace does not disappear completely after following this method; little vestiges, palimpsests, remain in place. The whiteboard becomes the Wunderblock, even if only unintentionally. These unsightly blemishes (since of course, you happen to grab the wrong pen over and over again) are also the most common reason for wanting and needing to change the whiteboard. Unless one keeps it for nostalgic reasons, in which case the whiteboard creates an effect, once hung on the wall, almost like an office accessory in a museum. It becomes a kind of placeholder, comparable to the obligatory family photo (see, again, the image of my office), and now displays a semi-permanent slogan or a couple of keywords.

FRAMING 2: MEDIA ARCHAEOLOGY

Looking at Freud's Wunderblock essay made the flipchart and whiteboard legible as media technologies and highlighted their particular affordances. If the flipchart provides a permanent trace at the expense of a limited receptive capacity, the whiteboard provides unlimited receptivity at the expense of the permanent trace. For Freud, the Wunderblock seemed to provide both affordances. Today, smartboards (i.e., interactive whiteboards) are also thought capable of doing more than their predecessors. They can be written on with pens (and, in the case of touchscreens, even fingers); they supply any desired number of pages, which can be refreshed at the push of a button; and these numerous pages can be saved, so that unlimited receptivity does not come at the expense of the loss of the permanent trace.

By connecting the whiteboard to the computer and the projector, the smartboard appears to opens up a wide range of possibilities for education. Teachers are able not only to write, paint, and draw, as on the classic board, but even to do so in colour (which already applied to the ordinary whiteboard). They are also able to work with materials (e.g., filling in forms, correcting text, writing over things) that they already prepared at home (which did not yet apply to the ordinary whiteboard; keyword: 'loss of the permanent trace'). Consequently, the smartboard seems to be replacing the classic handout, which may potentially lead to the loss of writing competency; students no longer have to fill out materials themselves, but rather only look up to the smartboard at the front of the classroom. Another adverse effect of the smartboard is that concentration on a screen prolongs the same lecture-style, teacher-centred, or frontal instruction that had been taken to be obsolete. At the same time, more interactive teaching formats based on individual technologies (e.g., tablets or smartphones, which everyone has anyway) have the disadvantage of quickly distracting students, since they make available a great variety of options (for work and entertainment). On the whole, however, media competency, additional features, and basic curiosity may ultimately tip the scales in favour of the new tool. Why, then, should one still write on an ordinary whiteboard or even on plain old paper? They are unforgivably outdated, one might claim.

From the perspective of media praxis, we should not forget the effect Canadian journalist David Sax recently termed 'the revenge of analog', citing Joanne McNeish's studies of elementary students in Canada and Israel who prefer printed textbooks to digital textbooks for reasons other than nostalgia (Sax, 2016: 187–9). The desire for the tangible as a sign of the real can also be found in organizations with measurably older personnel than the students involved in these studies (the whiteboard is a different case entirely). As in the case of Glenn Beck's spiritual exercises with a chalkboard, the material praxis of writing on the board involves a sensorimotor affect, which is part of the lived context of the organizational scenes discussed above.

From the perspective of media theory, the affective sense of the flipchart may consist primarily in the fact that it is sturdy without being firmly-mounted (like the whiteboard). In this respect, the flipchart makes the space of knowledge dynamic, since any arbitrary place can now become a space for writing in large letters, visible to all, or performing calculations for information, training, or sales. If, around 1800, the pairing of the student's board with the teacher's board served the end of the students' autonomy (Bosse, 1985), then the emergence of the flipchart, around 1900, freed large-format presentations literally from the wall and figuratively from the whiff of the classroom.

The whiteboard, in turn, appears to be a missing link in media theory that only reveals its true functional meaning in retrospect, seen from the perspective of the smartboard. Naturally, the chalkboard had many inconveniences (moisture, a stinking sponge), some of which were alleviated by the invention of the dry-erase marker in 1975. However, the whiteboard did not have any functional advantage over the chalkboard in that people remained tied to its place of installation. Only around 2000 did the smartboard change this situation: it made space dynamic once again in the form of a *virtual* space, connecting writing on a surface to the presentation of pre-prepared slides and the possibility of access to the World Wide Web.

As a media archaeologist, Freud would have probably viewed this development with some bemusement, though he generally considered human beings to be deficient in dealing with too many impressions and realities (Elsaesser, 2011: 113). Still, the idea that the medial support is more than a projection (and here the smartboard, in principle, merges with that of Google Glass), that the medial support becomes a surface containing the external world with all its images, words, and other contents, and that this world can be made to disappear at the push of a button—all of this seems to fit perfectly into Freud's analogy. Near the end of his Wunderblock essay, Freud speaks of certain 'cathectic innervations', concentrating internal energy on some external object:

> It is as though the unconscious stretches out feelers, through the medium of the system Pcpt.-Cs., towards the external world and hastily withdraws them as soon as they have sampled the excitations coming from it. Thus the interruptions, which in the case of the Mystic Pad have an external origin, were attributed by my hypothesis to the discontinuity in the current of innervation; and the actual breaking of contact which occurs in the Mystic Pad was replaced in my theory by the periodic non-excitability of the perceptual system.   (Freud, 1940 [1925]: 212)

Updating Freud's analogy, the World Wide Web would be the global imaginary, which is currently humanity's unconscious—doubtless that of digital natives, who grew up with nothing else. In this model, there would always be competition between memory and perception, provided that the unconscious (here the World Wide Web) can be 'innervated' (i.e., made accessible and excitable). At the same time, consciousness would not be able to switch off and would have to crash sooner or later: when the Viennese artist Stefanie Sargnagel describes herself as an '*Internet Waachhirn*' (Internet mushhead), she describes this precise lack of focus—and timing. In Freud's model, the conception of time arises through the 'discontinuous method of functioning of the system Pcpt.-Cs.', which is not discontinuous per se, but rather because cathectic energy gets sent and received 'in rapid period impulses'. If these come continuously, one's inner world becomes blurry—to put it polemically, in a 24/7 pulp. Against this background, studying uses of the flipchart and the whiteboard as medial practices capable of producing and transferring knowledge can provide a useful theoretical building block for a future media anthropology of the twenty-first century. In my office, the flipchart and the whiteboard will retain their place, as illustrations of themselves, at least as long as I still have any say in the matter (or as long as there are still paper and markers).

References

Beyes, Timon. 2017. 'Color and Organization Studies'. *Organization Studies*, 38(10), 1467–82.

Bosse, Heinrich. 1985. ' "Die Schüler müßen selbst schreiben lernen" oder Die Einrichtung der Schiefertafel'. In D. Boueke and N. Hopster (eds), *Schreiben—Schreiben lernen*, edited by. Tübingen: Narr, 164–99.

Elsaesser, Thomas. 2011. 'Freud and the Technical Media: The Enduring Magic of the Wunderblock'. In Erkki Huhtamo and Jussi Parikka (eds), *Media Archeology: Approaches, Applications, and Implications*. Berkeley: University of California Press, 95–115.

Freud, Sigmund. 1940 [1925]. 'A Note upon the "The Mystic Writing Pad" ' (trans. J. Strachey). *International Journal of Psycho-Analysis*, 21, 469–74.

Genette, Gérard. 1997. *Palimpsests: Literature in the Second Degree*, trans. C. Newman and C. Doubinsky. Lincoln, NE: University of Nebraska Press.

Pichler, Cathrin (ed.). 1989. *Wunderblock. Eine Geschichte der modernen Seele*. Vienna: Löcker.

Rancière, Jacques. 2013. *Aisthesis: Scenes from the Aesthetic Regime of Art*, trans. Z. Paul. London: Verso.

Sax, David. 2016. *The Revenge of Analog: Real Things and Why They Matter*. New York: Perseus.

CHAPTER 43

....

WIKI

....

OLGA RODAK, TOMASZ RABURSKI,
AND DARIUSZ JEMIELNIAK

HISTORY OF WIKI

THE first wiki was created in 1994 and launched in 1995 by Ward Cunningham, Oregon-based software programmer, the owner of the software consultancy Cunningham & Cunningham. It was an addition to Portland Pattern Repository, a website documenting knowledge about a certain approach to programming, of which Cunningham was a pioneer. The purpose of creating a wiki was to support collaboration among programmers developing the Design Patterns approach ('Wiki History'). When Cunningham noted ineffectiveness of the listsserv communication, he set himself a goal of creating a tool that would enable users to easily modify the content of pages and to follow other contributors' changes (Cunningham, 2014). Cunningham named the service 'WikiWikiWeb' after 'Wiki Wiki Shuttle', a bus serving passengers he had noticed at the international airport in Honolulu: 'Wiki' is Hawaiian for 'quick' ('Correspondence...').

On the website supporting distribution of wiki and discussion of Bo Leuf's and Cunningham's book *The Wiki Way: Quick Collaboration on the Web*, published six years after the first wiki's launch, one may read a precise and comprehensive definition of wiki, reflecting original intentions of its creator:

Wiki is in Ward's original description: *The simplest online database that could possibly work.*

Wiki is a piece of server software that allows users to freely create and edit Web page content using any Web browser. Wiki supports hyperlinks and has a simple text syntax for creating new pages and crosslinks between internal pages on the fly.

Wiki is unusual among group communication mechanisms in that it allows the organization of contributions to be edited in addition to the content itself.

Like many simple concepts, »open editing« has some profound and subtle effects on Wiki usage. Allowing everyday users to create and edit any page in a Web site is exciting in that it encourages democratic use of the Web and promotes content composition by nontechnical users   ('What is Wiki?').

There are two striking things about this definition. The first one is that it encapsulates software technology as a bundle of features which enable users of particular actions, defined by technology designers. The specificity of this technology can be recognized against already existing software: wiki's predecessors were pre-Web hypertext systems, developed to enhance communication between software developers. One of them, HyperCard, directly inspired Ward, who in fact enriched it for himself with the three-field stack and possibility of creating links to non-existent cards; features to become a landmark of wiki ('History of wikis'; 'Interview...'). This definition of wiki reveals the inevitably social character of technology: the social context of its creation—experience of using particular other technologies and the values underpinning the hacker culture—is inscribed into the heart of the wiki design.

Secondly, the definition touches upon what the technology creator considered the desirable broader 'sociological' consequences of wiki's adaptation, that is, democratization of participation in online culture and fostering the emergence of social movements (Konieczny, 2009). This echoes the early Internet years' dream of the 'free' and 'collaborative Web' that would transform the prevailing model of cultural and economic production, centred around copyright, market, and private ownership of goods (Benkler, 2006). But to what extent does wiki fulfil the hopes placed in it? Is technology itself able to reorganize society in such a radical way?

At the beginning, wiki was applied and developed by software programmers' communities—egalitarian, tech savvy, and treating knowledge and software as commons. Cunningham's WikiWikiWeb was growing dynamically, slightly modified according to users' needs, but controversies around open editing and curation of content forced the community to work out regulatory norms and led to the emergence of sister projects ('Wiki History'). In the meantime, Cunningham published a public version of wiki software for free use and modification—Wiki Base. As a result, the original version was enriched with modifications, but also budded into numerous 'clones', that is, alternative wiki software applications, written in various programming languages and involving different features to support content management, such as syntax or database types. It is difficult to estimate the number of existing wiki applications: Wikipedia gives a list of around forty of them ('Comparison...').

One of these applications from 2001 was a basis of what would be later on named 'the community and collaboration on a scale never seen before' (Grossman, 2006)—Wikipedia, the free online encyclopaedia created by 'you', a casual Internet user. Wikipedia became a synonym of open collaboration and seemingly the fulfilment of Cunningham's prophecy. However, researchers scrutinizing actual organizational practices around Wikipedia revealed that technical features which were supposed to 'encourage' open editing, may paradoxically stymie it. For example, the 'rollback' function became a source of power of experienced users over novices attempting to publish their first contribution, and the edit counter enabled accumulation of social capital translating into higher position in the community (Jemielniak, 2014).

The success of Wikipedia significantly influenced the direction of development of wiki technology. Wikimedia Foundation, a non-profit institution governing Wikipedia and its spin-offs, in cooperation with the Wikipedia community sets the direction of development of Media Wiki, the only currently important open wiki application. Wikimedia leaders created also a private company providing hosting for wikis based on Media Wiki, but not being an official Wikimedia project. Simultaneously, Wikipedia popularized wiki technology and inspired its use beyond the circle of software developers. New open projects emerged, aiming at gathering knowledge in various areas: from Javapedia and Wikitravel to Memory Alpha, devoted to *Star Trek*. Competing encyclopaedias were created, or even Wikipedia parodies.

However, self-organized, loose, meritocratic networks based on 'free' and generous knowledge exchange conquered also the imagination of management theorists and practitioners. Organizational implementation of a wiki was perceived to be one way to trickle down some of the 'wealth of networks' (Benkler, 2006) into organizational realms. The interest in wiki was coupled with the hype around knowledge management systems, designed in order to extract employees' knowledge. Awazu and Desouza (2004) proclaimed the necessity to learn from open source communities and called for the rise of 'open knowledge management'—for doing it 'the wiki way'. What wiki brought to the table was the shift from using proprietary, inflexible, and hierarchical intranets to adopting social media tools, known to employees from outside of the workplace. 'Conversational knowledge management's' advantage was making tacit knowledge explicit in the course of social interactions, visualizing and ordering organizational knowledge, and identifying gaps to be fulfilled (Wagner, 2004). However, corporate applications of wiki were by no means like Wikipedia, as they were proprietary, closed, and customized in a way that reflects organizational hierarchy and constrains the seemingly 'democratizing' properties of technology.

FEATURES AND AFFORDANCES
OF WIKI TECHNOLOGY

The diversity of software solutions and organizational implementations of wikis is so large that it is very hard to enumerate basic features of this technology. Basically, wiki is a system of interconnected web pages, where information can be easily added and modified by multiple users. However, even this basic description does not fit all implementations of wiki technology. There are many desktop wikis, which only imitate the working of web pages, but do not use the browser (e.g., Wiki Mode for Emacs, WikidPad, ConnectedText, ZIM). Tiddlywiki, a wiki-style notebook, is a single one, very long webpage, which connects with links to different parts of itself. There is also a whole branch of personal wikis, software, and pages dedicated for the single-user. Wikis do not even need specialized software. Every application that allows the creation of internal hyperlinks can be turned into wiki-like mode (e.g., Emacs, Evernote, or Microsoft

Onenote). Many features do not appear in small-scale wiki projects. In consequence, on a purely technical level wiki can be described as a technology of organizing primarily textual information, in a web-like hyperlinked environment, which parts can be easily edited by one or multiple users.

What is more, even though the popular image of wiki technology is dominated by Wikipedia, there are many proprietary, hierarchical, and control-based wiki applications. Therefore, open and closed wikis may differ dramatically in terms of what actions are enabled and what kind of systems emerge in the interaction between technology and users. The way wiki-based systems work is a result of a combination of technological infrastructure, a few basic constitutional rules for editors, and the superstructure of community-negotiated secondary rules, standards, and customs. Every wiki system will develop its own normative order, based on its starting premises, features of the community, and the overall goal of the wiki. For example, Wikipedia is an encyclopaedia, written from a neutral point of view, free and open-source, respecting other people, with no firm rules and standards.

Since basic features of wiki cannot be easily identified and the same technological solution may display different social outcomes depending on the context, it is more fruitful to talk about affordances of wiki rather than its features. Rather than concentrating on features of technology and what they allegedly enable or constrain, or on social actors' agency in giving a meaning to technology, it is what happens between a bundle of material elements of technology and an actor that should be the focus of analysis:

> ... [A]ffordances are both functional (artifact has a material presence) in the sense of enabling and constraining action with the technology, and relational (differs from one person to another and based on context of use). Thus, we can conclude that, within organizations, the affordance of artifacts is not simply based on their materiality but also on the relational properties that arise because of the symbolic and social nature of the setup.... Because these mutuality relations are situated and emergent in practice, there is a potential for the existence of multiple affordances of the same artifact depending on the focal context.   (Faraj and Azad, 2012: 253–4)

In the following paragraphs, we suggest some affordances of wiki, bearing in mind that they can differ from one application to another.[1] We follow the line of action of a user who is attempting to edit content (see Figures 43.1 and 43.2).

1. Social roles of users

These affordances are specific to open collaboration wikis. Most of the company wikis assign rigid roles to their users and do not endorse changing them.

[1] The research that informed this chapter was funded by the Polish National Science Centre (project number DEC-2012/05/E/HS4/01498).

1.1 Open collaboration wikis pull their users into more active roles. Passive users are encouraged to make minor changes, to become regular contributors, and finally, to embrace more responsibility for the project.

1.2 The attention of active editors is drawn into administrative tasks, starting from small administrative tasks (organization of content, standardization, control of recent changes) to more responsible ones, and finally into becoming an overseeing administrator. Many administrative tasks are almost indistinguishable from regular editing. This distinguishes wiki from other types of software, in which administration tools are hidden from casual users.

1.3 Editing activities of all users are automatically saved and visible to everybody. The user profile is a bundle of his/her privileges and history of edits. His/her contributions are automatically copyrighted (or free-licensed). Although the user is informed about the copyright issues, in most of the wikis, he/she cannot choose the type of intellectual property protection he/she wishes.

2. Content creation

2.1 One of the most distinguishing features of wiki technology is the easiness of creating and modifying the content. Minor edits, moving the paragraphs, and major redrafting are all easy to make.

2.2 Whereas content creation is very easy, the structuring of the content is hard, as is the search feature. Search results are unstructured and formal queries are available only in few wikis. Thus, structural issues pose an important problem for the users.

The most common form of order offered by wiki is a complex network of interlinked pages. This network is often overly complex and inefficient. Poorly-linked pages or orphans (pages without links to other pages) may be virtually invisible to users. Only some of the wikis (e.g., recently Wikipedia) have a completion proposal feature for internal links, reducing the problem of different naming conventions.

In addition to a network structure, many wikis organize their content in a system of hierarchical categories (very few wikis use the tagging system). However, whilst the categorization of content is easy, it is virtually impossible to get a consistent categorization system in open collaboration wikis. Different concepts of order compete with each other. Different parts of wiki become organized according to different principles. Category trees become recursive and overly intertwined. Therefore, in large, open wikis dynamic forces of disorganization and flow of new content are very strong, and need to be constantly controlled. This can only be done fully in closed wikis. Their structure is often pre-planned and fixed, and wiki technology is used only as a convenient method of adding the new content.

2.3 Wikis support a particular way of content development. It is much easier to develop content horizontally (content at the same semantic level, e.g., another biography) and vertically descending (giving more details about existing entries), than vertically ascending (general overviews and summaries).

If pages grow too large, they can be easily divided into smaller units. Page merging is harder and often requires rewriting. The topics that are fundamental or most general are in consequence often underdeveloped.

2.4 With the exception of some company wikis, the development of content is not strictly planned and is a result of the shifting attention of the users or an answer to current needs. Thus, the content's quality varies across the project.

2.5 Although wiki's version control system allows full control of the changes, it doesn't guarantee that the quality of the content will be improving, and not declining. The progressive development has to be supported by the non-technological means (community). If community is weak, wiki becomes prone to decline.

3. Control and communication

3.1 Wikis (both open and closed ones) offer a wide variety of features and instruments allowing extensive, universal and dispersed control. Every change in content creates a distinct version of a page, with its own stamp of authorship and copyright, and it can be easily viewed or reverted. Open wikis give the reverting power to nearly every user, making it the most convenient instrument of control.

3.2 Reverting is much easier and faster than modifying or developing content. If a new entry has flaws, it tends to be reverted by other users. In consequence, the content is resilient to changes.

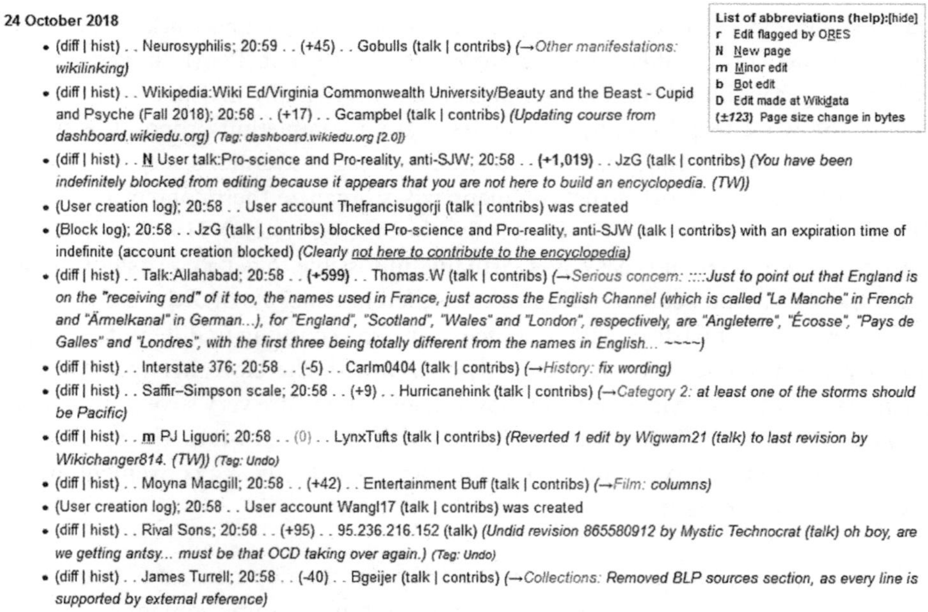

---

24 October 2018

List of abbreviations (help):[hide]
r   Edit flagged by ORES
N New page
m Minor edit
b Bot edit
D Edit made at Wikidata
(±123) Page size change in bytes

- (diff | hist) . . Neurosyphilis; 20:59 . . (+45) . . Gobulls (talk | contribs) (→Other manifestations: wikilinking)
- (diff | hist) . . Wikipedia:Wiki Ed/Virginia Commonwealth University/Beauty and the Beast - Cupid and Psyche (Fall 2018); 20:58 . . (+17) . . Gcampbel (talk | contribs) (Updating course from dashboard.wikiedu.org) (Tag: dashboard.wikiedu.org [2.0])
- (diff | hist) . . N User talk:Pro-science and Pro-reality, anti-SJW; 20:58 . . (+1,019) . . JzG (talk | contribs) (You have been indefinitely blocked from editing because it appears that you are not here to build an encyclopedia. (TW))
- (User creation log); 20:58 . . User account Thefrancisugorji (talk | contribs) was created
- (Block log); 20:58 . . JzG (talk | contribs) blocked Pro-science and Pro-reality, anti-SJW (talk | contribs) with an expiration time of indefinite (account creation blocked) (Clearly not here to contribute to the encyclopedia)
- (diff | hist) . . Talk:Allahabad; 20:58 . . (+599) . . Thomas.W (talk | contribs) (→Serious concern: :::::Just to point out that England is on the "receiving end" of it too, the names used in France, just across the English Channel (which is called "La Manche" in French and "Ärmelkanal" in German...), for "England", "Scotland", "Wales" and "London", respectively, are "Angleterre", "Écosse", "Pays de Galles" and "Londres", with the first three being totally different from the names in English... ~~~~)
- (diff | hist) . . Interstate 376; 20:58 . . (-5) . . Carlm0404 (talk | contribs) (→History: fix wording)
- (diff | hist) . . Saffir–Simpson scale; 20:58 . . (+9) . . Hurricanehink (talk | contribs) (→Category 2: at least one of the storms should be Pacific)
- (diff | hist) . . m PJ Liguori; 20:58 . . (0) . . LynxTufts (talk | contribs) (Reverted 1 edit by Wigwam21 (talk) to last revision by Wikichanger814. (TW)) (Tag: Undo)
- (diff | hist) . . Moyna Macgill; 20:58 . . (+42) . . Entertainment Buff (talk | contribs) (→Film: columns)
- (User creation log); 20:58 . . User account Wangl17 (talk | contribs) was created
- (diff | hist) . . Rival Sons; 20:58 . . (+95) . . 95.236.216.152 (talk) (Undid revision 865580912 by Mystic Technocrat (talk) oh boy, are we getting antsy... must be that OCD taking over again.) (Tag: Undo)
- (diff | hist) . . James Turrell; 20:58 . . (-40) . . Bgeijer (talk | contribs) (→Collections: Removed BLP sources section, as every line is supported by external reference)

FIGURE 43.1 Page showing recent changes on Wikipedia. Every change is a distinct entry, attributed to a user and very easy to track or revert

3.3 Unlike in later developed social networking services, social relations and communication do not occupy a central position in wikis. All wikis are content-focused (Klobas, 2006). Social aspects of a wiki are important, but secondary and subsidiary to content-oriented activities. Communities are built around information systems, which they develop, look after, and defend against vandalization.

3.4 Direct communication between users is inconvenient and ineffective. Wikis did not develop distinctive forms of direct communication, and the same tools used for editing content are used for discussing. Editors leave their messages on 'discussion pages'. Each message is a new version of a page. Communication is indirect, stretched out across time and (sometimes) different places. One can leave a message on one's talk page and the recipient will be notified, when he or she logs into the wiki. However, a reply may be left in the same place, and as long as the person replying does not make additional effort to notify the interlocutor, it is his or her responsibility to watch the page. Long discussions with many participants become hard to follow and inconclusive.

3.5 As the result, many company wikis implemented communication features borrowed from other types of software (e.g., social media). In open wiki communities, alternative channels of communications arise spontaneously (e.g., chats, Facebook groups).

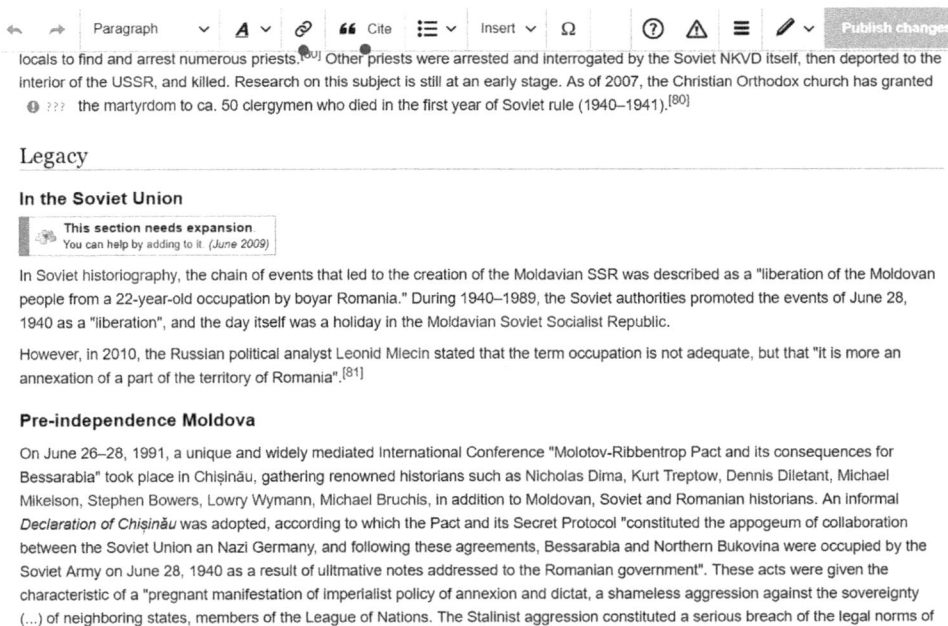

locals to find and arrest numerous priests.[80] Other priests were arrested and interrogated by the Soviet NKVD itself, then deported to the interior of the USSR, and killed. Research on this subject is still at an early stage. As of 2007, the Christian Orthodox church has granted the martyrdom to ca. 50 clergymen who died in the first year of Soviet rule (1940–1941).[80]

Legacy

In the Soviet Union

This section needs expansion. You can help by adding to it. (June 2009)

In Soviet historiography, the chain of events that led to the creation of the Moldavian SSR was described as a "liberation of the Moldovan people from a 22-year-old occupation by boyar Romania." During 1940–1989, the Soviet authorities promoted the events of June 28, 1940 as a "liberation", and the day itself was a holiday in the Moldavian Soviet Socialist Republic.

However, in 2010, the Russian political analyst Leonid Mlecin stated that the term occupation is not adequate, but that "it is more an annexation of a part of the territory of Romania".[81]

Pre-independence Moldova

On June 26–28, 1991, a unique and widely mediated International Conference "Molotov-Ribbentrop Pact and its consequences for Bessarabia" took place in Chișinău, gathering renowned historians such as Nicholas Dima, Kurt Treptow, Dennis Diletant, Michael Mikelson, Stephen Bowers, Lowry Wymann, Michael Bruchis, in addition to Moldovan, Soviet and Romanian historians. An informal Declaration of Chișinău was adopted, according to which the Pact and its Secret Protocol "constituted the appogeum of collaboration between the Soviet Union an Nazi Germany, and following these agreements, Bessarabia and Northern Bukovina were occupied by the Soviet Army on June 28, 1940 as a result of ulitmative notes addressed to the Romanian government". These acts were given the characteristic of a "pregnant manifestation of imperialist policy of annexion and dictat, a shameless aggression against the sovereignty (...) of neighboring states, members of the League of Nations. The Stalinist aggression constituted a serious breach of the legal norms of

FIGURE 43.2 Traditionally, wikis have separate modes for editing and displaying content. Editing was relatively easy, but required basic knowledge of wikicode. Modern wikis often implement WYSIWYG editors, lowering the entry barriers for new users

(*Source*: Reproduced from Wikipedia, 'Soviet occupation of Bessarabia and Northern Bukovina'. Distributed under the terms of the Creative Commons Attribution-ShareAlike 3.0 Unported License (CC BY-SA 3.0). https://creativecommons.org/licenses/by-sa/3.0/)

3.6 Archiving of all conversations is a consequence of assuming a straight legal approach to copyrights: after all, it allows precise tracking of each edit. However, since it also applies to discussions, it has an additional social effect. As a result, even one-on-one conversations are conducted with a public audience in mind, and sometimes take a semi-official tone.

3.7 In open collaboration, the community is often sustained not through the strong ties among its members, but rather through establishing ties between the members and the community itself (Wasko et al., 2009). The very fact that these procedures are usually created and modified by the community itself definitely adds to the social glue that keeps the community together.

Organizing Properties of Wiki

Organizational scholars are interested in wiki systems for two reasons. Firstly, they have wanted to understand how successful open wikis are governed. The combination of technical features and ideological premises led to the emergence of a specific organizational model, known as open collaboration (OC) of which Wikipedia is a quintessential example. Computer scientists Andrea Forte and Cliff Lampe defined an open collaboration system as an 'an online environment that (a) supports the collective production of an artifact (b) through a technologically mediated collaboration platform (c) that presents a low barrier to entry and exit and (d) supports the emergence of persistent but malleable social structures' (Forte and Lampe, 2013: 536). Management in OC systems resembles the one in common pool resources (Ostrom, 1990), but proliferation of global projects based on this logic was enabled by the dissemination of information and communication technologies.

Secondly, they aim at inquiring whether innovative features of new organizational forms can be transferred to conventional organizations to make them more efficient. Argyris and Ransbotham (2016) noticed that management literature on corporate applications of wikis can be divided into two strands. The first one consists of conceptual papers 'advocat[ing] the benefits of wiki' (Argyris and Ransbotham, 2016: 228). Their authors describe technical features of wikis and predict how they potentially contribute to organizational effectiveness. These papers are deterministic in the sense that they usually assume smooth link between implementation and specified consequences. The latter group consists of papers based on on-site empirical studies of actual technology adoption in various social contexts. This part of literature is still more interested in the 'social-engineering' side and indicating conditions of successful implementation of technology, but it gives some interesting insights into organizational effects of technology use as well.

Based on the latter literature, Argyris and Ransbotham identified the pattern that successful wiki implementation relied on 'leaders who exploited the wiki affordances for wiki-based knowledge' processes (2016: 228). On the other hand, what constrained the process was lack of clear incentives to share knowledge and a tension between seemingly 'democratizing' features of technology and hierarchical structure of organizations. For this reason, many organizational wikis are failures: editing senior employees' contributions is interpreted as judgemental and avoided, and middle-level managers were observed to consciously sabotage wikis in the fear of flattening organizational structure (Argyris and Ransbotham, 2016). Additionally, there was a certain balance between formalization and the much needed bureaucratic rules of conduct, and spontaneous, a-hierarchical, and 'leaderless' disruptive governance (Jemielniak, 2016).

Argyris and Ransbotham described a classical success story of wiki implementation at NBC. Based on the concept of an institutional entrepreneur, the authors indicated that it was 'knowledge entrepreneurship' of one employee that led to the successful adoption of technology. Victoria was exhausted by newcomers asking her the same questions again and again, so she transferred a ready-made, free solution from a partnering General Electrics. Firstly, she fulfilled a wiki with a 'critical mass' of valuable content. Consequently, she encouraged key members of other departments to take ownership over the project by mapping tasks and assigning leaders to particular areas of expertise. Once they started contributing, they could see how easy, educative, and enjoyable it was. At NBC 'it's in the wiki', became 'the culture we want to create. It's the type of thing that people come in and embrace and that's the kind of people we want here', as noted by one of the company's directors (Argyris and Ransbotham, 2016: 234). As noted elsewhere in reference to moving the majority of a company's projects to a wiki, 'employing wiki pages as a collaboration tool does more than just integrate the KM [knowledge management] system into an employee's work process; it essentially becomes the actual work process' (Gonzalez-Reinhart, 2005: 6).

Another prominent example is a wiki implementation at NASA, for example for flight controllers and instructors from NASA's EVA (Extravehicular Activity) Operations group. It was a classical organizational change with management involved and making sure that everyone has the feeling of ownership over the new system. EVA's wiki is being constantly customized depending on the users' (management?) needs, including setting of diversified access permissions to satisfy organizational hierarchy. Management encourages software extensions, among them WatchAnalytics—a tool that increases participation and users' accountability by tracking changes on the page and suggesting desirable actions to users (Johnston et al., 2016). Technology's panopticism helps keep control over the project, but also makes it much harder to procrastinate.

The NASA case is a great illustration of how one important advantage of wiki systems is their cheapness and potential for evolution, resulting from their open source status. Numerous companies, such as NBC mentioned above, only start experimenting with KM thanks to minimal front-end costs of applying wiki in comparison with extreme costs of proprietary solutions. And there is a big chance a proprietary system fails, given its lack of flexibility in confrontation with organizational structures. Still, open source software has also its disadvantages, as noted by surveyed IT executives who hesitated to apply wikis due to lack of vendor support. Consequently, since at least

1998 open source is openly a business dedicated solution; providers of further generations of wiki software attempt to monetize 'the common'—not by 'enclosure', however, but by assuring various services around it (Gonzalez-Reinhart, 2005).

Paradoxically though, the same technology may be at the same time a basis for anarchistic, purely anti-systemic projects, such as Wikipedia, and support greater effectiveness of 'business as usual'.

REFERENCES

Argyris, Young A. and Sam Ransbotham. 2016. 'Knowledge Entrepreneurship: Institutionalising Wiki-Based Knowledge-Management Processes in Competitive and Hierarchical Organisations'. *Journal of Information Technology*, 31(2), 226–39.

Awazu, Yukika and Kevin C. Desouza. 2004. 'Open Knowledge Management: Lessons from the Open Source Revolution'. *Journal of the American Society for Information Science and Technology*, 55(11), 1016–19.

Benkler, Yochai. 2006. *The Wealth of Networks: How Social Production Transforms Markets and Freedom*. New Haven and London: Yale University Press.

'Comparison of Wiki Software'. Wikipedia. https://en.wikipedia.org/wiki/Comparison_of_wiki_software [Accessed 22 May 2019].

'Correspondence on the Etymology of Wiki'. http://c2.com/doc/etymology.html [Accessed 24 May 2019).

Cunningham, Ward. 2014. 'Wiki Design Principles'. http://wiki.c2.com/?WikiDesignPrinciples [Accessed 23 February 2018].

Faraj, Samer and Bijan Azad. 2012. 'The Materiality of Technology: An Affordance Perspective'. In Paul M. Leonardi, Bonnie A. Nardi, and Jannis Kallinikos (eds), *Materiality and Organizing: Social Interaction in a Technological World*. Oxford: Oxford University Press, 237–58.

Forte, Andrea and Cliff Lampe. 2013. 'Defining, Understanding and Supporting Open Collaboration: Lessons from the Literature'. *American Behavioral Scientist*, 57(5), 535–47.

Gonzalez-Reinhart, Jennifer. 2005. 'Wiki and the Wiki Way: Beyond a Knowledge Management Solution'. Information Systems Research Center, Houston, Texas.

Grossman, Lev. 2006. 'You—Yes, You—Are TIME's Person of the Year'. *Time*, 25 December. http://content.time.com/time/magazine/article/0,9171,1,570,810,00.html.

'History of wikis'. Wikipedia. https://en.wikipedia.org/wiki/History_of_wikis [Accessed 23 February 2018].

'Interview: Wikinewsie Kim Bruning discusses Wikimania'. https://en.wikinews.org/wiki/Interview:_Wikinewsie_Kim_Bruning_discusses_Wikimania [Accessed 23 February 2018].

Jemielniak, Dariusz. 2014. *Common Knowledge? An Ethnography of Wikipedia*. Stanford, CA: Stanford University Press.

Jemielniak, Dariusz. 2016. 'Wikimedia Movement Governance: The Limits of A-hierarchical Organization'. *Journal of Organizational Change Management*, 29(3), 361–78.

Johnston, Stephanie S., Brian K. Alpert, Edwin James Montalvo, Lawrence Daren Welsh, Scott Wray, and Costa Mavridis. 2016. 'EVA Wiki – Transforming Knowledge Management for EVA Flight Controllers and Instructors. Presentation at the 46th International Conference on Environmental Systems ICES-2016-405 10-14 July 2016, Vienna, Austria, 1–12.

Klobas, Jane. 2006. *Wikis: Tools for Information Work and Collaboration*. Amsterdam: Elsevier.

Konieczny, Piotr. 2009. 'Wikipedia: Community or Social Movement?' *Interface: A Journal for and about Social Movements*, 1(2), 212–32.

Ostrom, Elinor. 1990. *Governing the Commons: The Evolution of Institutions for Collective Action*. Cambridge: Cambridge University Press.

Wagner, C. 2004. 'Wiki: A Technology for Conversational Knowledge Management and Group Collaboration'. *Communications of the Association for Information Systems*, 13, 265–89.

Wasko, Molly McLure, Robin Teigland, and Samer Faraj. 2009. 'The Provision of Online Public Goods: Examining Social Structure in an Electronic Network of Practice'. *Decision Support Systems*, 47(3), 254–65.

'What is Wiki?' http://wiki.org/wiki.cgi?WhatIsWiki [Accessed 23 February 2018].

'Wiki History'. http://wiki.c2.com/?WikiHistory [Accessed 23 February 2018].

CHAPTER 44

BY MEANS OF WHICH: MEDIA, TECHNOLOGY, ORGANIZATION

TIMON BEYES, ROBIN HOLT, AND CLAUS PIAS

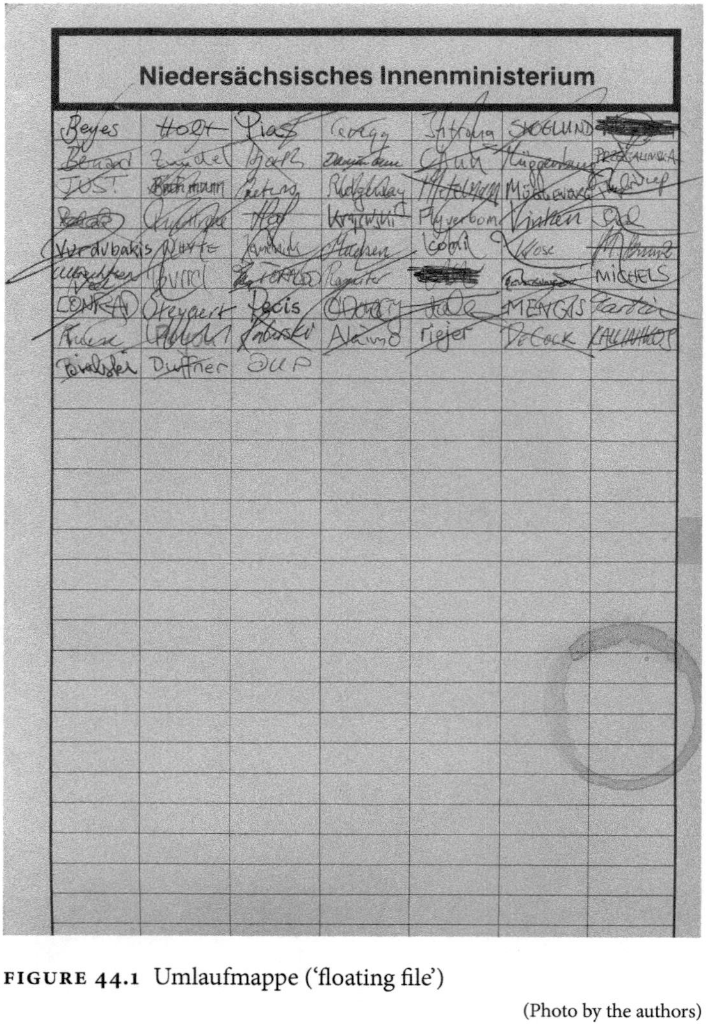

FIGURE 44.1 Umlaufmappe ('floating file')

(Photo by the authors)

CONSIDER the medium in front of your eyes. Preferably, perhaps, a proper book, nicely bound, maybe a bit too bulky and heavy, but a thing with tactile qualities that lends weight to your reading. Or a text on the screen, then, so much easier to access and circulate, and to search for key terms, and follow links, and copy and paste, or skip altogether. Let's assume for a moment that it is a proper material book. You're reading the last entry of a *Handbook of Media, Technology, and Organization Studies*.

What is being held is a structurally, materially, and symbolically mediated object. Its paper, type, imagery, and ink represent information gleaned from other books, images, and files, or observation, and recorded, stored, searched, analysed, and written about through various media. The information is factual, but the facts adorn and give life to the narrative truth claims made in the entries, they do not constitute them. The facts are, suggests Cornelia Vismann (2008: 5–8), more like leaves that rustle in the wind and absorb light, transforming experience into understanding of shared and verifiable expression sustained in a branchwork of categories and concepts. A certain type of reader (student, or academic) follows these narratives, pursuing work patterns to which the contents might be enlisted. There are other users of the Handbook employed in different work patterns: copyeditors, indexers, typesetters, publishers, librarians, accountants, and as time passes, the nature of these work patterns morph, extending to include conservation or disposal, or to making scholarly historical comparisons of what was once known and how. Without even reading their content, the nature of the entries can be ascertained by a panoply of mediating forms: the publisher's insignia evoking long-standing scholarly tradition, the authors' academic titles symbolizing scholarly training, the distribution networks organized through university libraries, card-based 'paper machines' (Krajewski, 2011), and search 'engines'. Even its sheer weight, in physical form, projects an impression of dry details being given carefully considered, speculative life.

Both in paper and digitized form the Handbook is designed to endure. As an example of what Katherine Hayles (2002: 22–4) calls an inscription technology, the Handbook is readable and searchable. The page turns of the paper form are organized by the folding and cutting conventions of book binding, by the way pulped wood can absorb ink without leaching and remain sufficiently opaque to be printed on both sides.[1] Conforming with the Western standard, left-right, top-down linearity of printed text, the Handbook carries within it a set of narratives which, as a reference work, more obviously encourage interrupted use than one moving from beginning to end; like files, this book 'will move in roundabout ways' (Vismann, 2008: 13)—in this sense like an *Umlaufmappe*, a 'floating file' (Figure 44.1). The entries are organized by an index and table of contents, the kind of list that 'sorts and engenders circulation' and which functions administratively, embodying and enacting a command as both imperative and information (look further, look here) (Vismann, 2008: 5–8). Each entry can be taken alone, or cross referenced by editorial nudges (see Krajewski, this volume), and the references can throw readers beyond the pages of the Handbook and into different sources.

---

[1] The 'mulchy, messy pulp' which is the material middle between the two sides of the page of physical paper, writes the artist and novelist Tom McCarthy, would then be the space of literature, 'which is neither one nor the other; it's this messy, unresolved *between*' (McCarthy, 2016: 50; orig. emphasis).

In its digital form this semiotic and semantic searching, reading, and jumping becomes more in the way of a wandering and meandering series of clicks governed by access rights, connection speeds, crafty software permeated by invisible recommender systems, page aesthetics, the physical limits to cloud storage in 'personal accounts', or structural filtering of search results by academic research listing sites.

It is increasingly through a digitized medium that such a Handbook will be used, displayed on a screen, perhaps sitting on a cheaply veneered office desk, mediated materially and meaningfully as a readable, searchable inscription technology. Into this performative scene we might imagine surrounding objects coming to play roles: an ergonomically compliant office chair and, in the near (physical) distance, a coffee machine could be percolating, through whose caffeine promise the desk can sustain itself as a pinion space, a blank space of stultifying ennui for the lightly-drugged, academic 'office worker' who, sitting there a while, is in the process of conducting a systematic review of scholarly literature published on the topic of technology, media, and organization studies. Despite having a background in literary studies, the academic has found herself working in a Business School (with its elevators, office plants, and shoddy acoustic tiles designed to dampen the noise and hide the infrastructures of lighting, ventilation, and power supply) but finds it hard to concentrate on this algorithmically ordered task of systematic review, one which amounts to little more than reading, summarizing, and ordering the abstracts of scholarly work filtered by bibliographic search engines. Being thoroughly mediated by tiles, chair, coffee, desk, and computer, her scholarly activity has become a process of self-governed filing, little more:

> At the interface of computer and user, material files turn into icons, which a mouse, replacing the hand, "opens" and "closes" with a click. The very terminology of computer surfaces is designed to remind users seated before screens of the familiar world of files. The menu tab offering options like "list," "format," "thesaurus," "table" and the instructions copy, delete, save turn users into virtual chanceries or chancellors. By condensing an entire administrative office, the computer implements the basic law of bureaucracy according to which administrative techniques are transferred from the state to the individual: from the specialized governmental practices of early modern chanceries to the "common style," from absolutist administrative centers to individual work desks, from the first mainframe computers in defense ministries to the desktop at home.   (Vismann, 2008: 163)

To compensate for the boredom setting in, she begins thinking of other desks, in particular the small, twelve-sided, walnut occasional table used by Jane Austen to hold even smaller sheets of paper that could be covered easily lest someone approaching catch her in the unwomanly act of writing novels of comic genius. Austen's was hardly a desk, more something used as such, and it was placed near the front door of a rented cottage, holding itself invisibly as a liminal space of profound creative autonomy that had escaped the patriarchal ministrations of a calculating order (Tomalin, 2008). In contrast, the academic office worker has a desk in full view of her co-workers, each set to similar tasks at the behest of senior co-authors whose names will appear in the review papers, simply by virtue of administrative process. She is both indignant and resigned, a kind of digital cynical realism has set in. The academic finds herself being organized, utterly.

But what of resistance? Surely, given that our equipmental relationships with objects are never fixed, we can always find them becoming otherwise and so open to new expressions of subjectivity? The academic could, for example, dedicate her screen time to becoming a social media 'influencer' upon whose comment others hang their own opinions, momentarily; or turn her analytic skills into commercially lucrative pattern matching that would enable social media companies to better apprehend and manage the desires of their users, or allow financial traders to more rapidly spread micro investments within the interstices of global trading exchanges. She could help making handbooks public and free by uploading them into databases housed in servers in Iceland, say, or into the Russian library system. Or, in trying to upload, she might simply become transfixed by the spinning wheel icon indicating bouts of buffering that resemble provisional interruptions, suspensions or recesses, in which, for a moment, 'the entire time of the world... is put on hold' (McCarthy, 2016: 45).

Yet these are hardly transformative options, and the mediating technology by which such creative expression becomes possible is becoming so crowded, varied, and ephemeral that there is little space left for distinction-making intervention. Indeed even gaining a sense of individuality is becoming problematic. As Giorgio Agamben (2009: 21) notices, in the recent past processes of subjectification meant people were able to sense a kind of life-narrative of becoming something, of attainment, of acquiring an identity emerging from personal and social development. Nowadays, however, technologies carry their own logic in which the subject is being continually de-subjectified. Digital media technologies have configured us as units of on/off presence: access code; social media rankings; re-booted avatars; bibliographic identifier numbers; productivity rates; biometric rhythm. Identity becomes synonymous with being recorded as such by objects that are very much distinct from human beings,[2] and being on the record simply means, in the words of Vismann (2008: xi), being processed 'on file':

> When files or records are referred to in the plural, their content rarely seems to matter; it is buried underneath their materiality. And even when files are opened to reveal their contents, they are not simply read. Files are processed, just like stones and other such matter.

Processed by other objects such as computers, yet files still retain their own identity, their deletion also requiring a process, and leaving a recorded trace. Humans too, but like files, as objects affected by other objects. And when the files are digitized these affects are everywhere. We are always inside, living *by means of which*: technology

---

[2] 'For the administrations of the Western world, a life without files, without any recording, a life *off the record*, is simply unthinkable' (Vismann, 2008: xii).

has, it seems, come to 'indicate the evolution of living by other means than life' (Stiegler, 1994: 135).[3]

In words closer to our academic's semi-fictional workspace, and perhaps your's, these 'other means than life' are forces of organization, if by organization we mean the creation of centres, gists, and hubs towards which the best, the noble, and the effective gravitate, and by which they are defined. Which means that as scholars of organization and as scholars of media, we need to give these means attention, and, of course, a weighty Handbook. We need to interrogate media, technology, and organization in their foundational relations, their forms, and their constraining and loosening effects and affects. Emphasis must be given to the very technological materials and objects that enable and shape organization (and that are enabled and shaped by organizational processes in return).

As practical means of going on, then, what to make of 'technology', 'media', and 'organization'?

TECHNOLOGY

Humans are woven and wired with technology. Since their mythical inception under the faltering protection of the Prometheus, tools—useful objects found ready to hand—have been what marks them. We were given tools in compensation for having been overlooked by Prometheus' brother, Epimetheus, the Titan god in charge of assigning all living things their distinguishing feature before being let loose upon the world. By the time he remembered us, Epimetheus' basket of traits was empty, leaving us 'naked, unshod, unarmed and un-bedded'. Devoid of inner qualities Prometheus stepped in, stole fire from Hephaestus and wisdom in artifice from Athena and gifted them to humans. Though without any qualities of our own, we now had immortal qualities of intelligence and practical manipulation through which we might acquire and share the qualities of all other things (Plato, *Protagoras*, 320d–322a).[4] Humanity became distinct by becoming the only creature whose individuation emerged through connectivity to others, to the immortals on one side, and the world on the other:

> And now that man was partaker of a divine portion, he, in the first place, by his
> nearness of kin to deity, was the only creature that worshipped gods, and set him-

---

[3] This carries echoes of Michel Foucault's (1977: 188–205) discussion of sovereignty and disciplinary abeyance to its centrality, specifically how in making itself invisible sovereignty inscribes visibility, measurement, and objectivity into subjects in such a way as to render them as lifeless as an a regimental troop of soldiers drilled to the point of utter replicability and replaceability: ' "[V]ery good," Grand Duke Mikhail once remarked of a regiment, after having kept it for one hour presenting arms, "only *they breathe*" '.

[4] Thought follows from productiveness of the body, *techne* grounds *logos*, tool-use forms a background of practical wisdom from which justice and reverence eventually emerge (though only with further divine intervention).

self to establish altars and holy images; and secondly, he soon was enabled by his skill to articulate speech and words, and to invent dwellings, clothes, sandals, beds, and the foods that are of the earth. Thus far provided, men dwelt separately in the beginning.   (Plato, *Protagoras*, 322a)

In this was established the middle space that humans have believed themselves creating and embodying ever since, the in-between space touched by angels and spirits from one side and sublime and sentient nature on the other, but still always alone in their own lifeworld: timelessness and immortals over one horizon, unthinking stones, plants, gases, and animals over the other. Though humans gained their gift through a god's sympathy and guile, their fate was to always find the immortal world an inscrutable horizon, a realm of beguiling hints or terrible retribution. Nature, however, was more approachable for our tools, and we dug in.

Bernard Stiegler (1994) uses this story to emphasize how our most immediate empirical experience has become that of being amid objects-for-us and discovering how they might align, more or less easily, with us, also 'as' objects. In this sense, we relate to and understand the world as an extension of ourselves: we are reckoners, measurers, calculators, and designers and the world is what appears within our sensory grasp. We either find ourselves moving smoothly and rhythmically as we cohere with this appearing world around us, or we are stumbling as objects resist us and we try and correct or compensate by treating the object as an obstacle, in the process using other objects with which we are already familiar. The basic tools that dig, protect, or magnify are apprehended prosthetically as extensions of a physiological and sensory apparatus presumed to be commanded by the profaning force of human will; making things with tools became a subject of study and the application of learning: it became technology. Through technology, it appears, the world becomes an extension of ourselves; it is smoothed out and mapped and occupied across measured coordinates of time and space. What does not fit with this coordinated organization of human activity is declared random chance, repressed, ignored, or we simply do not recognize it.

So much so that method—the learned and applied activities of measuring appearances—has become the world itself, though through refinement its ubiquity is often concealed. Something is not something until it appears on a radar, is caught in a net, is classified as a member, is staked or buried in a plot, is stored and recalled in a search. Through technology human and human/non-human patterns of interaction become blurred, and modified in relation to immediate experience and acquired habit: what starts with a swing and a grunt, can, for a brief while, become a flourish and nuanced counter thrust, before ending up in predictable input/output patterns. With tools and their use comes the immersion of human beings into practices which, in turn, can encourage the use of human beings, also as tools, as users seek to bring the world and information about it into managed and expressive coherence. From the simple use of simple tools comes, then, the whole panoply of human activity and thought from whose conditioning patterns emerge processes of organization as these patterns become subjects of time (they gather histories of appreciated past, an experienced present, and envisaged future) and space (they are located, carrying with them an atmospheric resonance of belonging and occupation).

Yet this is only part of the story. In Ernst Kapp's *Elements of a Philosophy of Technology* which dates from 1877 (contrary to what is often claimed these days, much of this kind of thinking is far from new), tools are never just an extension of the human body; our understanding of this very body, as well as its practices, are predicated on these tools (as Prometheus configured it). So the folding of humans and technology works both ways (Kapp called this the theory of 'organ projection'): human bodies, we might say, can be apprehended prosthetically as extensions of technologies (Kapp, 2018 [1877]). Attention thus shifts from *social* organization (which implies, still, a human primacy) to the technical means of organizing the (techno-)social. As we suggested in the case of our academic office worker, far from making her more present and distinct, the increasingly adept forms of organizing have found her becoming little more than the consciously structured residue of technological operations: *logos* has become *techne*, and in turn *techne* has become machinic and coded. She comes to understand herself through objects (including her colleagues and even friends) as tools, and the way, in return, these relate to—or *mediate*—her action and thought. There is neither a distinctly human middle ground, nor gods, only mediation.

MEDIA

So to consider technology is to also to consider its mediating and mediated presence extended through human practices, a presence that is, being mediated, invisible. In the words of John Durham Peters (2015: 22) media organize civilization: 'Wherever data and world are managed, we find media', and media, being the condition by which material reality and knowledge of that reality are brought together 'are our infrastructures of being, the habitats and materials through which we act and are' (Peters, 2015: 15). 'Media' are, thus, not reducible to their current colloquial isolation as 'social' or 'mass' media or IT systems; indeed to think of them as some*thing* at all (as in *the media*) is a case of grammatical confusion. For Friedrich Kittler (1999), the term 'media' does not apply to a certain predefined group of objects, but to any object (technology, artefact, or apparatus) that conditions the structure of a certain situation and the specific possibilities of perceiving, acting, and thinking in it. Media 'affect conditions of possibility in general' (Galloway et al., 2014: 1), they allow objects to have edges by making them sensible, approachable, and manageable, without really revealing themselves other than through the working of the objects they make apparent. We might say that something becomes media by being epistemologically productive as an order of materiality and technological or technologically influenced structures of communication, interaction and affect through which material, energy, and information are brought into continual commerce at a scale whose organization is beyond the scope of measurement and hence recognition. There is no end to such mediation. There are only settlements of agreed use from which materiality and meaning emerge, and if we begin with this understanding of technology and media as fundamental, conditional, and infrastructural, then how organization is made or how organization takes place becomes a question that calls for an engagement with such apparatuses. Media configure organizational relations that are in-built into the devices and apparatuses of organizational life.

Now digitized, this media-technological condition has created a new organizational complex perhaps best called 'surveillance capitalism' (Zuboff, 2015), but not just surveillance 'of' objects, but also, and increasingly, surveillance as its own object. The technologies (recommender systems, auction devices, bots) are becoming ever more invisible. Indeed, as Geert Lovink and Ned Rossiter (2011: 280) have pointed out, in the age of ubiquitous digital technology and smart machines, 'the media a priori is so obvious that it seems to have drifted into the realm of the collective unconscious'.

Ontologically speaking, then, 'there are no media' (Horn, 2007: 7); there is no subject area, no ontologically identifiable domain that could be called 'media' (Siegert, 2015: 3; Vogl, 2008). But there are objects and object-bound processes of technical mediation that constitute the somewhat arbitrary conditions of their own continuing mediation. Thinking about technical media thus implies *not* focusing on what is represented ('in the media'), or why something is represented and not something else. Rather, the focus falls on the very material conditions of representation (Vogl, 2008). But if media do not constitute ontological objects, how are they to be approached? By reconstructing how such mediation organizes, and how organizing takes place around it; by revealing the material specificities of organization and tracing how mediation takes place. This is what the entries in this Handbook do. Hence the notion of a 'medial *a priori*' or a 'technical *a priori*' of organizing: we need to focus on the often unprepossessing, partly invisible objects and the processes generated around them that render organization possible. As this Handbook's list of objects indicates, this can lead to an 'aesthetic of unlikely topics' (Pias, 2011: 24). The topics are aesthetic because what we are studying are affects of forming in which objects always and only become themselves in relation to other objects. In this forming (*aesthesis*) the idea of separate entities set in spatial relation to one another gives way to a more disorienting sense of continually interacting objects whose affective power is apprehended as a force of propensity and performative probability. The topics are unlikely because typically the objects function as equipment (they mediate), and in doing so they function invisibly: they are the means not the object of inquiry. The Handbook makes these means into the object of inquiry, making them present-at, rather than ready-to, hand. So the entries can lend the objects a sense of uncanniness, for no matter how exhaustive and diligently written, they can only ever approach them from within other objects: talk about media is itself always mediated, and as one object approaches another, the other always pulls away (Harman, 2018).

Broadly put, then (and shortcutting a long and rich history of medial thought), the collective singular of 'media' points to a fundamental technical form, a generalized mediality that has conditioned human making and thinking from its very origin (Mitchell and Hansen, 2010). The plural 'media' points to objects as technological apparatuses of mediation that form the infrastructural conditions and contexts of perception, experience, and agency. So to consider media is to break the comforting narrative in which technology, organization, and civilization progress. In considering media comes awareness that just as humans reach into the world, the world reaches into them ('organ projection'), prosthesis becomes *aphairésis*, a loosening of individuality and anthropomorphic elevation as objects find themselves re-instated, whether in the most meagre human gesture, or the grandest strategic scheme.

An elevator, for example (and we could isolate many other objects to make the same point), lives as part of a communication and transportation network. It is designed to move vertically, predictably, and to be used thoughtlessly and invisibly. In being designed and used like this it has the functional sincerity of what Gilbert Simondon calls a network object. And its operation is not limited to its performance within a single network, for it converges with other networks (most obviously those in energy and information/surveillance, but perhaps also those of leisure and consumption, or of productive enterprise) constituting a vast spread or bloom of technologically mediated organization in which the part of human agency is spatially and temporally dissipated: its own space and time are fed into the myriad other expressions and experiences of space and time, all of them modifying continually. This networked condition alters the sense of human agency quite fundamentally: in pushing the button we do not decide, act, and acknowledge the consequences, rather we are insinuated into an endless array of other objects, each sliding into one another as noiselessly as the elevator doors.

ORGANIZATION

Such, at least, is the ideal functional picture of perfectly mediated organization, a picture where the collective and individual, the particular and general, merge seamlessly: living beings and machines converge, in a flattened-out ontology constituted organizationally, through endless mediation. There are, however, inefficiencies: not just glitches, mechanical breakdowns, and of course (themselves mediated) obfuscations (Brunton and Nissenbaum, 2016), but forces that Simondon (2012: 3, 10) associates with the lingering presence of social, historical, and psychological structures that allude to unities that cannot be easily unfolded and endlessly extended across networks. These are encounters in which an object remains obdurate, refusing to partake of the mutual transforming, distributing, extracting, and storing that typically characterizes object relations.

For example, the office plants decorating a lobby area onto which an elevator spills might, just might, be encountered as living plants. Not the plant encountered as comforting decoration, not as a symbol of nature designed to elicit a sense of corporate environmental and social responsibility, not as part of an interior designer's symbolic capital, and not as a natural filter to better purify the air and enhance the productivity of office workers as they continually busy themselves. None of these. Instead it is the plant encountered as a plant, an object suggesting its unshakeable, inscrutable unity as a living thing and which refuses to be parsed into wider and wider networks of mediated influence. The plant as plant whose substance is primary, eliding from predicates, refusing to be cloned, duplicated, or switched.

It is just that this kind of encounter with a plant is increasingly unlikely, as is, more broadly, any encounter with an object-as-such, even fellow human beings, because, it seems, the world has become utterly mediated. Even when objects are treated as singular units—for example the accused person in a criminal trial or the original piece of art that cannot broken into parts, or replaced with like units, and remain the same (Simondon, 2012: 4)—they are units whose essential quality it is to be qualified in some way by others' judgements in right or wrong, or taste. We are becoming habituated into processes of structuration in which objects are only and forever circulating as active and functional parts of wider networks of objects. And there is a quickening of this circulation; objects are being refused time off. The office plants, in a choice of different green hues, come and go with increased, unseasonal rapidity, their care and management are no longer the responsibility of themselves as 'non-sentient life'. They are no longer the responsibility of an assigned person even, one whose dilatory or forgetful nature might find the plant gathering dust under a pall of indifference, or whose caring nature might find the plants being given undue (costly and diverting) attention. Rather the plants are becoming properly managed through outsourced systems of atmospheric governance in which they, the plants, become invisible as objects in themselves, and instead become much more a performative form, an active and ongoing planting, which in turn is part of wider performative networks associated with climate management that extend, say, to the shipping containers in which plants are distributed across the world, and to the climate itself, which is warming up unpredictably from all these attempts at climatic control.

And as it is with elevators, office plants, and push buttons, so it is with all the objects gathered in this Handbook: the objects that litter our practices, a motley of ordinary equipment and instrumentation by which human life folds into other objects and is, in turn, folded: the swell and surge of *prosthesis* and *aphairésis*. The Handbook entries consider objects as having use value, as being tools, potentially: a perfectible bundle of resources and possibilities held in relations working both ways, projecting out from the body and back in. And with incision and incursion arrives a great host of structural and social forces that loosen any anthropocentric conceit that it is the human agent who is in control, or at the centre of things. As mediated objects they can only be considered in their possible use, and in these operations comes an ever expanding range of possible settlements whose edges abut other edges, inwards and onwards.

What, again, of organization under these conditions? The implications are obvious, or so we think. First, that there is a media-technological *a priori* of organizing is so glaring, to return to Lovink and Rossiter's point, that we simply need to think organization from media technologies (Beyes et al., 2019). That this kind of rationale sometimes touches upon a rather infantile technological determinism should, of course, be taken into account, yet being infantile it is beside the point: the point is to identify and stick with objects that organize, and that mediate and remediate organization, and to move on from here.[5] The objects are real; they cannot be negotiated or spirited away.

Second, there is a history of thinking organization as entangled with, or even predicated upon, technology. And it is not a minor or additional one, as is sometimes conveniently claimed, perhaps to push the agenda of the so-called (socio-)material turn (e.g., Orlikowski, 2010). Advocates of this turn point to organization studies suffering *Dingvergessenheit* (an obliviousness to 'things'), yet their diagnosis overlooks the fact that the relation of technology and organization is one of the main strands of the history of organizational thought (Bonazzi, 2008). Max Weber, for one, directly related the forces of rationalization and bureaucratization (and thus of disenchantment) to 'technical and economic conditions of machine production which today determine the lives of all individuals who are born into this mechanism' (Weber 1958: 181). And as Vismann points out, it was Weber who saw that modern, industrial office management is predicated on written documents and files (Vismann, 2008: 91). The study of organization had never, in this sense, lost track of objects. Yet these objects were seen as predicated upon a given notion and framework of what an organization is: technologies of production, for instance (e.g., Woodward 1958), or information technologies (e.g., Simon 1965), or accounts of how technology is socially constructed in organizational life (e.g., Pettigrew 1973). By and large, there was no need to bother with the everyday objects and ubiquitous processes of mediation and their organizational force. A media-technological *a priori* opens up the study of organization once more, but perhaps this time more purposefully, to a simultaneously much broader and more nuanced approach to how organization takes place with, and through, technologies.[6]

Third, it follows that we need to refrain from a scholarly bias of assuming that organizational entities and their agents are there to use objects and tools, an assumption that finds its way even into so-called socio-material approaches, for even here an organization is posited first, from which to then explore human/non-human constellations of interaction. We are thus not even sure, that is, that 'speaking of organizations in their own language' that come into being '*by dint of association* with the beings of technology [and fiction, and reference]' (Latour, 2013: 372; orig. emphasis) is a sensible way of proceeding, of, for example, tracing Stiegler's 'life by other means'. With such socio-material approaches comes the assumption of symmetrical or 'flat' chains of influence that verges on the metaphysical, while of course opening up the study of organization to its 'quasi-objects'. Likewise, in the face of technologies and constant

---

[5] The most infantile media determinism, it seems to us, is not to be found in recent theories of media and technology, but in the either glorified or dystopian popular discourse around Silicon Valley and the wonders of digital connectivity. It resides for instance in the mottoes and, apparently, belief systems of the corporate behemoths that have taken control of searching and networking (Barbrook and Cameron, 1996).

[6] We find a precursor of such medial thinking of organization (yet without any material specificity) in the work of Robert Cooper, whose notion of 'cyborganization' saw organization as made up of heterogeneous and partly incompatible circuits of information processing (Cooper and Law, 2016; Parker and Cooper, 2016).

mediation, we hesitate to assume that there is little or 'almost nothing solid or durable' in organizational life (Latour, 2013: 388)—those files we encountered with Vismann at the beginning of this essay are of course in motion, yet they are certainly solid and they certainly endure. It is not that they are simply motion, nor that they are entities in motion, but that as objects they are forever exposed to the edges of what they are not, other objects, setting in train forces of modification, transformation, and withdrawal. It is perhaps no surprise that advocates of socio-material approaches such as Latour turn quite happily (and with some sense of relief) to the measuring technique of 'organizational scripts', and thus a narrative paradigm, in order to make the question of organization conceptually manageable. But what if we take 'technology, media, and organization' (even) more seriously? What if we adopt a medial *a priori* in order to precisely get closer to the experience of being-organized, and of organizing? In other words, what if we indeed 'ballast…scripts technologically', since the narratives of quasi-objects and quasi-subjects that we comfort ourselves with are too cushy, too comforting (Latour, 2013: 413)? Consider our exemplary academic, or actor: what symmetrical relation can we assume between her humanoid agency (which is already unthinkable without technological mediation) and the sheer power and noise of the concatenation of objects and technological scripts that govern her perceptions and conduct? At the very least, it behoves us to pose the question of how technological mediation organizes us. And we would do well, we think, to begin our narrative storytelling and script-writing with technologies and mediation.

Fourth, then, we should not be too quick in assuming what the raw material of organization-mediation is. It is easy to claim communication as the last ground of media, technology, and organization studies. But what kind of communication, or communicative atmosphere even, could precede mediation? Should we not at least investigate? The point being, we should not exclusively posit communication as grounding organizational technologies without tracing how mediation takes place. As Walter Benjamin (2002) pointed out almost a century ago, technological apparatuses foremost work on the mediation of the human sensorium and everyday sense experience.

Fifth, the Handbook amply demonstrates the promise of turning to both historical media-technological developments and 'old' objects in order to trace how they organize and keep on organizing.

LIFE ON THE GRID

The technological innovations of digital mediation do raise important concerns with regard to the relation of media, technology, and organization. One thing to be learned from the Handbook entries is that at the digital level 'objects' are often not evident as 'objects' in the traditional sense—compare the cloud with the office plant, or the elevator with the filtering system. In the words of Peters (2015: 7) 'digital media traffic less in content, programs, and opinions than in organization, power, and calculation'. He goes on:

> Digital media serve more as logistical devices of tracking and orientation than in providing unifying stories to the society at large. Digital media revive ancient navigational functions: they point us in time and space, index our data, and keep us on the grid.

They also define the grid, and what lies off-grid is unfiled, and hence unlived. They organize for its own sake, so rapid, varied, and hasty that they are now bypassing any concern for the 'content' of subjectification, the gridded pointing and moving and indexing is everything.

Yet we can still find room to move into the world of accident and material surprise. Apart from the perpetual breakdowns of digitally mediated operations, their dirty and corrupt data, as well as already-now arcane operations laid down in millions of lines of codes and thus beyond comprehension (Beyes and Pias, 2019), apart from willed resistance, sabotage, and withdrawal (obfuscation, anonymity, disconnection); Kapp's organ projection invariably works both ways. Maybe even with the Handbook, which is organized 'as' something in relation to other things, held together in what Fisher (1991: 243) calls materialized acts of participation. Given objects like Handbooks are arriving and pulling away in processes of transformation, and given their material entanglements and their meaning are never fixed they are also constituted by what they are *not*, by their mediated motions of postponement; they can always become something else and are never just what they are (Bryson, 1988: 99; Fisher, 1991: 243–4; Cooper, 2014).

To return again to our exemplary academic, such a postponement is experienced in a hesitation (which may be prolonged) from within the circulation of networked objects. Yet (as with all workers, indeed all mediated roles and offices and scripted identifications) this hesitation, or tarrying, is often a bewildering and exhausting experience (Vogl, 2011). What forms of hesitation are open to a scholar given the task of conducting systematic reviews (the upshot of a complex array of pushed buttons) and for whom academic work has become nothing more than something to be counted as an influencing ouput, over and over? So often in such mediated conditions distinction only comes to those who come to count by finding new ways to make things countable. Being utterly mediated, the academic lives in a quantitative condition that Stuart Elden (2006: 147) has likened, ironically, to the quality by which the *quale* or essence of things is no longer something distinct (despite Epimetheus' best efforts) but what they have in common—their being subjects in continuously renewing orders of countability. Hesitation has become impossible outside of this technological mediation, and within it the pause easily becomes another potentially calculated action. Through such technological mediation life has become an extended flow of aggregates, features, amalgams, types, with neither soul to warm nor head to think about how one might intervene in the processes of our own actual and potential subjugation. And with the digitalization of this flow these aggregates, features, amalgams, types are being constantly transferred, transformed under the watchful vision of the machinic eye, an endless exhaustless surveillance. Under its impress every aspect of life is being offered for update, creating a restlessness whose only rationale is its own replication (Chun, 2016). Under its impress we, like the academic, worry we are not being enterprising enough, that we lack the productivity against which the measured must be counted. Under its impress we resign ourselves to the constant suspicion that we are carriers of contagion, whether bacterial or dogmatic. Under its impress politics and civility have given way to a calculated order that has assumed the role of governing the world and which has, in fact, become the coldest of cold parodies of government and civilization (Agamben, 2009).

And yet, to recover life—by which we might mean nothing more metaphysical or profound than finding the edges to, and spaces in, the aimless bloom of technological ordering (Agamben, 2009: 22)—is to find ways to work with rather than within technology. This is not a case of finding better (more effective and efficient) uses of desks, smartphones, presentation software, or high heels, for all of these have modes of subjectification (increasingly accelerated and obsolescent) scripted into them by which any proper use is governed: use is an extension of mediating structuring. Rather, it is remaining alive to the hesitations already provided: the glitches, accidents, misuses, and alternative projections, and to wander and wonder with them. We do so because we are already thoroughly mediated and hence used to how objects are supposed to be used, and being well practised we can become aware of when and where to experiment, one object complicit with other objects, all swinging into the future of collaborative functioning which is more or less tense, more or less coherent, more or less dangerous. In this swing the object always becomes subject to what Massumi (2012: 30) calls 'the material event of taking form', and the taking of form is a process of endless mediation.

References

Agamben, Giorgio. 2009. *What is an Apparatus?*, trans. David Kishik and Stefan Pedatella. Stanford, CA: Stanford University Press.

Barbrook, Richard and Andy Cameron. 1996. 'The Californian Ideology'. *Science as Culture*, 6(1), 44–72.

Benjamin, Walter. 2002. 'The Work of Art in the Age of Its Technological Reproducibility: Second Version'. In *Walter Benjamin: Selected Writings, Volume 3, 1935–1938*, ed. Howard Eiland and Michael W. Jennings. Cambridge, MA: Belknap Press, 101–33.

Beyes, Timon, Lisa Conrad, and Reinhold Martin. 2019. *Organize*. Minneapolis/Lüneburg: University of Minnesota Press/Meson Press.

Beyes, Timon and Claus Pias. 2019. 'The Media Arcane'. *Grey Room*, 75, 84-107.

Bonazzi, Guiseppe. 2008. *Geschichte des organisatorischen Denkens*, ed. Veronika Tacke. Wiesbaden: VS.

Brunton, Finn and Helen Nissenbaum. 2016. *Obfuscation: A User's Guide for Privacy and Protest*. Cambridge, MA: MIT Press.

Bryson, Norman. 1988. 'The Gaze in the Expanded Field'. In Hal Foster (ed.), *Vision and Visuality*. Seattle, WA: Bay Press, 87–108.

Chun, Wendy H. K. 2016. *Updating to Remain the Same: Habitual New Media*. Cambridge, MA: MIT Press.

Cooper, Robert. 2014. 'Process and Reality'. In Jenny Helin, Tor Hernes, Daniel Hjorth, and Robin Holt (eds), *The Oxford Handbook of Process Philosophy and Organization Studies*. Oxford: Oxford University Press, 585–604.

Cooper, Robert and John Law. 2016. 'Organization: Distal and Proximal Views'. In Gibson Burrell and Martin Parker (eds), *For Robert Cooper: Collected Work*. London: Routledge, 199–235.

Elden, Stuart. 2006. *Speaking Against Number: Heidegger, Language and the Politics of Calculation*. Edinburgh: Edinburgh University Press.

Fisher, Philip. 1991. *Making and Effacing Art: Modern American Art in a Culture of Museums*. New York: Oxford University Press.

Foucault, Michel. 1977. *Discipline and Punish: The Birth of the Prison*, trans. Alan Sheridan. Reprinted in *The Foucault Reader*, ed. Paul Rabinow. New York: Pantheon, 1984.

Galloway, Alexander, Eugene Thacker, and McKenzie Wark. 2014. *Excommunication: Three Inquiries in Media and Mediation*. Chicago, IL: University of Chicago Press.

Harman, Graham. 2018. *Object Oriented Ontology: A New Theory of Everything*. London: Pelican.

Hayles, N. Katherine. 2002. *Writing Machines*. Cambridge, MA: MIT Press.

Horn, Eva. 2007. Editor's Introduction: 'There Are No Media'. *Grey Room*, 29, 6–13.

Kapp, Ernst. 2018 [1877]. *Elements of a Philosophy of Technology: On the Evolutionary History of Culture*, trans. Lauren K. Wolfe. Minneapolis: University of Minnesota Press.

Kittler, Friedrich. 1999. *Gramophone, Film, Typewriter*, trans. Geoffrey Winthrop-Young and Michael Wutz. Stanford, CA: Stanford University Press.

Krajewski, Markus. 2011. *Paper Machines: About Cards & Catalogues, 1548–1929*, trans. Peter Krapp. Cambridge, MA: MIT Press.

Latour, Bruno. 2013. *An Inquiry into Modes of Existence: An Anthropology of the Moderns*, trans. Catherine Porter. Cambridge, MA: Harvard University Press.

Lovink, Geert and Ned Rossiter. 2011. 'Urgent Aphorisms: Notes on Organized Networks for the Connected Multitudes'. In Mark Deuze (ed.), *Managing Media Work*. London: Sage, 279–90.

McCarthy, Tom. 2016. *Recessional—Or, the Time of the Hammer*, ed. Elisabeth Bronfen. Zurich: Diaphanes.

Massumi, Brian. 2012. 'Technical Mentality Revisited'. In Arne de Boerver, Alex Murray, Jon Roffe, and Ashley Woodward (eds), *Being and Technology*. Edinburgh: Edinburgh University Press, 19–36.

Mitchell, Warren J. and Mark B. N. Hansen. 2010. 'Introduction'. In W. J. Mitchell and M. B. N. Hansen (eds), *Critical Terms for Media Studies*. Chicago: University of Chicago Press, vii–xxii.

Orlikowski, Wanda. 2010. 'The Sociomateriality of Organisational Life: Considering Technology in Management Research'. *Cambridge Journal of Economics*, 34(1), 125–41.

Parker, Martin and Robert Cooper. 2016. 'Cyborganization: Cinema as Nervous System'. In Gibson Burrell and Martin Parker (eds), *For Robert Cooper: Collected Work*. London: Routledge, 236–52.

Peters, John Durham. 2015. *The Marvellous Clouds: Toward a Philosophy of Elemental Media*. Chicago, IL: University of Chicago Press.

Pettigrew, Andrew M. 1973. *The Politics of Organizational Decision-Making*. London: Tavistock.

Pias, Claus. 2011. 'Was waren Medien-Wissenschaften? Stichworte zu einer Standortbestimmung'. In Claus Pias (ed.), *Was waren Medien?* Zurich: Diaphanes, 7–30.

Siegert, Bernhard. 2015. *Cultural Techniques: Grids, Filters, Doors, and Other Articulations of the Real*, trans. Geoffrey Winthrop-Young. New York: Fordham University Press.

Simon, Herbert A. 1965. *The Shape of Automation for Men and Management*. New York: Harper & Row.

Simondon, Gilbert. 2012. 'Technical Mentality', trans. Arne de Boever. In Arne de Boerver, Alex Murray, Jon Roffe, and Ashley Woodward (eds), *Being and Technology*. Edinburgh: Edinburgh University Press, 1–18.

Stiegler, Bernard. 1994. *Technics and Time: The Fault of Epimetheus*. Stanford, CA: Stanford University Press.

Tomalin, Claire. 2008. 'Writers' Rooms: Jane Austen'. *The Guardian*, 12 July.

Vismann, Cornelia. 2008. *Files*, trans. Geoffrey Winthrop-Young. Stanford, CA: Stanford University Press.

Vogl, Joseph. 2008. 'Becoming-Media: Galileo's Telescope'. *Grey Room*, 29, 14–25.

Vogl, Joseph. 2011. *On Tarrying*, trans. Helmut Müller-Sievers. London: Seagull Books.

Weber, Max. 1958. *The Protestant Ethic and the Spirit of Capitalism*, trans. Talcott Parsons. New York: Charles Scribner.

Woodward, Joan. 1958. *Management and Technology*. London: H. M. Stationery Office.

Zuboff, Shoshana. 2015. 'Big Other: Surveillance Capitalism and the Prospects of an Information Civilization'. *Journal of Information Technology*, 30(1), 75–89.

INDEX

Note: Figures are indicated by an italic 'f' following the page number.